DMV Seminar
Band 31

Infinite Dimensional Kähler Manifolds

Alan Huckleberry
Tilman Wurzbacher
Editors

Springer Basel AG

Editors:

Alan Huckleberry
Ruhr-Universität Bochum
Fakultät für Mathematik
Gebäude NA 4
44780 Bochum
Germany

e-mail: ahuck@cplx.ruhr-uni-bochum.de

Tilmann Wurzbacher
Institut de Recherche Mathématique Avancée
Université Louis Pasteur et CNRS
7, rue René Descartes
67084 Strasbourg
France

e-mail: wurzbach@math.u-strasbg.fr

2000 Mathematical Subject Classification 17B68, 22E65, 22E67, 32Mxx, 53Cxx, 53Dxx, 58Bxx, 81R10

A CIP catalogue record for this book is available from the
Library of Congress, Washington D.C., USA

Deutsche Bibliothek Cataloging-in-Publication Data
Infinite dimensional Kähler manifolds / Alan Huckleberry ; Tilman Wurzbacher, ed..
Basel ; Boston ; Berlin : Birkhäuser, 2001
   (DMV-Seminar ; Bd. 31)
   ISBN 978-3-7643-6602-5   ISBN 978-3-0348-8227-9 (eBook)
   DOI 10.1007/978-3-0348-8227-9

Contents

Infinite-dimensional Groups and their Representations

Karl-Hermann Neeb

Borel-Weil Theory for Loop Groups

Karl-Hermann Neeb

Coadjoint Representation of Virasoro-type Lie Algebras and Differential Operators on Tensor-densities

Valentin Yu. Ovsienko

From Group Actions to Determinant Bundles Using (Heat-kernel) Renormalization Techniques

Sylvie Paycha

Fermionic Second Quantization and the Geometry of the Restricted Grassmannian

Tilmann Wurzbacher

Preface

Infinite dimensional manifolds, Lie groups and algebras arise naturally in many areas of mathematics and physics. Having been used mainly as a tool for the study of finite dimensional objects, as for example in four-dimensional gauge theory or in the study of closed geodesics or trajectories of Hamiltonian flows, the emphasis has changed and they are now frequently studied for their own independent interest.

Examples include the representation theory of loop groups and current groups and their Lie algebras, diffeomorphism groups and Lie algebras of vector fields, e.g., the Virasoro algebra, the geometric approach to bosonic and fermionic second quantization, and the analysis and geometry of spaces of paths and loops in finite dimensional Riemannian manifolds. From this point of view, finite dimensional geometry and representation theory becomes more of a helpful guideline rather than the goal of mathematical study of infinite dimensional objects.

Being interested in communicating parts of this highly active subject to advanced students and young researchers, the first named editor and Alexander Kirillov organized a DMV-Seminar on "Infinite dimensional Kähler manifolds" which took place November 19–25, 1995, at the Mathematisches Forschungsinstitut Oberwolfach. Unfortunately, administrative problems prevented Kirillov from attending the seminar. In response, the second named editor joined in the preparations of the DMV-Seminar which finally consisted of a longer introductory course given by Alan Huckleberry and several shorter courses on selected specialized subjects. Let us take the opportunity to mention that, besides those lectures which led to contributions in this volume, there were others which were given by Askar Dzhumadil'daev, Peter Heinzner, Patrick Iglesias and the late Giorgio Valli.

We now briefly describe the contents of this volume. At the outset it should be emphasized that, while the basic themes were at least touched upon in the lectures, the contributions here go significantly further. These range from being foundational in nature to expositions which describe recent results which were proved after the time of the seminar. The "Introduction to group actions in symplectic and complex geometry" by Alan Huckleberry is of the former type. Here the theory of differentiable manifolds and group actions is developed almost from scratch. The availability of many good text books on Riemannian geometry and personal taste led the author to emphasize the complex and the symplectic aspects of Kähler geometry, resulting in a concise, though accelerated presentation of the basics of complex and symplectic manifolds as well as Lie group actions on them. As an application a rather detailed account of the geometric realization of the irreducible respresentations of compact Lie groups, i.e., Borel-Weil theory in finite dimensions, is given. Though principally written as a "crash course", notably the aforementioned last part can equally well be read as a reminder and a preparation for the Borel-Weil theory for loop groups.

In "Infinite dimensional groups and their representations" Karl-Hermann Neeb extends the basic differential calculus on Fréchet manifolds and Lie groups to the more general setting of manifolds modelled on sequentially complete locally convex topological vector spaces, in the spirit of, e.g., John Milnor. After recalling some refined material from functional analysis, foundational results on the topology of continuous, smooth and holomorphic functions on infinite dimensional manifolds are proved. This provides both a rigorous and very general framework for the theory of actions and representations of infinite dimensional Lie groups and allows the study of "generalized coherent state representations" beyond the Hilbert space situation.

Neeb applies this again in his contribution "Borel-Weil theory for loop groups", where the theory of irreducible unitary positive energy representations is developed in a geometric manner, i.e., in terms of holomorphic sections of line bundles over Kähler manifolds acted upon transitively by loop groups. Although this approach has been rather well-known since the appearance of the book *Loop groups* by Andrew Pressley and Graeme Segal, the thorough exposition should make this contribution a useful reference. Notably, the unitary structure is obtained quite naturally by use of the theory of positive definite functions.

In "Coadjoint respresentations of Virasoro-type Lie algebras and differential operators on tensor-densities" Valentin Ovsienko explains a fundamental example of an action in infinite dimensions, namely the coadjoint action of the (Bott-)Virasoro group corresponding to the Virasoro Lie algebra. A complete but accessible account on the relations between this action and numerous important geometric and algebraic concepts, such as the Schwarzian derivative, projective structures on the circle, periodic Sturm-Liouville operators and their monodromy, and Lie super-algebras is given. Along with simple proofs of the central theorems of the subject, in a concluding chapter some natural generalizations of the Virasoro algebra are described.

Although the contribution of Ovsienko is not focussed on it, we would like to recall here the beautiful result that the coadjoint action of the Virasoro group yields invariant Kählerian structures on the Fréchet manifold $\mathrm{Diff}^+(S^1)/\mathrm{Rot}(S^1)$, where $\mathrm{Rot}(S^1)$ denotes the rotation subgroup in the group $\mathrm{Diff}^+(S^1)$ of all orientation-preserving diffeomorphisms of S^1.

Sylvie Paycha addresses in "From group actions to determinant bundles using (heat-kernel) renormalization techniques" an important issue of infinite dimensional geometry, namely how to give a sense to quantities that in finite dimensions are defined in terms of traces and determinants. After reviewing and relating different methods of "regularization" or "renormalization" they are applied to two general situations: the problem of characterizing "minimal orbits" in infinite dimensions and the geometry of determinant bundles.

As an example, relevant notably to string theory, the action of the diffeomorphism group of a closed oriented two-dimensional manifold Z on the infinite dimensional (weakly) Hermitian manifold of metrics of constant curvature equal to -1 on Z is considered. It is now classical that the L^2-metric on the space of metrics, or equivalently on the space of almost-complex structures on Z, induces the (Kählerian) Weil-Petersson metric on Teichmüller space upon going to the quotient space. Here, in Sylvie Paycha's contribution, the regularized quantities arising in the orbit picture, respectively, in the determinant bundle picture, are compared.

In "Fermionic second quantization and the geometry of the restricted Grassmannian" Tilmann Wurzbacher gives a detailed account of another important infinite dimensional Kähler manifold: G_{res}, the so-called restricted Grassmannian of a polarized Hilbert space. This contribution starts with the Klein-Gordon equation and traces the way to G_{res} via the Dirac equation, the negative energy problem and fermionic second quantization, i.e., the respresentation theory of the CAR-algebra on certain Fock spaces. The homogeneous Kähler geometry of G_{res} and the numerous closely related central extensions of infinite dimensional Lie groups and Lie algebras are then studied in detail.

A discussion of symplectic manifolds and group actions in the framework described by Neeb in "Infinite dimensional groups and their representation" allows an explanation of the (non-equivariant) moment map associated to the action of the restricted unitary group on G_{res}. Several other examples of infinite dimensional moment maps are sketched.

In the last chapter a C^*-algebraic geometric approach to the Grassmannian and its determinant bundle is developed. We expect these methods to be very useful for the study of other infinite dimensional Kähler manifolds.

Let us take this opportunity to thank those involved in the organization of the DMV-Seminar, in particular the *Deutsche Mathematiker-Vereinigung* and the *Mathematisches Forschungsinstitut Oberwolfach*. Notably, we would like to thank the director, Matthias Kreck, and the entire staff of the Oberwolfach institute for providing us with optimal conditions during this intense week in November, 1995. Finally, we would like to thank the lecturers whose talks did not lead to a contribution in this volume and all participants for their curiosity and enthusiasm.

Alan Huckleberry and Tilmann Wurzbacher   Bochum and Strasbourg
                                                   November 1, 2000

Introduction to Group Actions in Symplectic and Complex Geometry

Alan Huckleberry

I. Finite-dimensional manifolds

In this preparatory chapter certain basic results on differentiable manifolds are outlined. Standard references should include [Sp] and [War].

1. Vector space structures

1.1 Alternating forms

Let V denote a finite-dimensional $\mathbb{R}$-vector space and $\omega : V \times V \to \mathbb{R}$ an *alternating bilinear form*, i.e., linear in both factors and $\omega(v, w) = -\omega(w, v)$ for all $v, w \in V$.

Example. Let $V := \mathbb{R}^2$ and define $\omega(v, w) = det(v, w)$. More explicitly, if $v = (a, b)$ and $w = (c, d)$ are regarded as column vectors, then $\omega(v, w) = ad - bc$. Thus ω is the Euclidean area of the parallelogram which is spanned by the two vectors.

Denote the dual space of all *linear functionals* $F : E \to \mathbb{R}$ on E by E^*. Then the direct sum $V := E \oplus E^*$ possesses a canonical form

$$\omega : V \times V \to \mathbb{R}, \quad ((v, v^*), (w, w^*)) \mapsto w^*(v) - v^*(w).$$

Note that the determinant in the above example can be regarded as such a form.

If ω is alternating, then $\omega(v, v) = 0$ for all $v \in V$. Thus ω can not be regarded as the square of a norm. However, the notion of *perpendicular* makes sense. One must only become accustomed to the fact that a vector is always orthogonal to itself. If W is a subspace of V, then

$$W^{\perp_\omega} := \{v \in V : \omega(v, w) = 0 \text{ for all } w \in W\}.$$

The subspace $V^{\perp_\omega}$ is referred to as the *degeneracy* of ω on V. If $V^{\perp_\omega} = \{0\}$, then ω is said to be *non-degenerate*. In this case V, equipped with ω, i.e., the pair (V, ω), is called a *symplectic* vector space.

Exercise. Show that the canonical form on $E \oplus E^*$ is non-degenerate.

Remark. All of the above makes good sense for infinite-dimensional spaces as well. For example, let H be the vector space of complex-valued continuous functions on the unit interval $I = [0,1]$. For $f, g \in H$ define

$$\omega(f,g) := Im\Big(\int_0^1 f\bar{g} \Big).$$

It follows that (H, ω) is a symplectic vector space.

Exercise. Show that ω in the preceding remark is non-degenerate on H.

A bilinear form ω induces a natural linear map

$$\varphi_\omega : V \to V^*, \quad w \longmapsto F_w, \text{ where } F_w(v) := \omega(v, w).$$

In general φ_ω is neither injective nor surjective. The *kernel* $Ker(\varphi_\omega)$ is the degeneracy $V^{\perp\omega}$ and a functional in the image is said to *representable* by ω. In the finite-dimensional setting the non-degeneracy of ω is equivalent to φ_ω being an isomorphism, i.e., every functional is uniquely represented.

A form ω on V may be pulled back to any subspace V_1 : For $v_1, w_1 \in V_1$, define $\omega_1(v_1, w_1) := \omega(v_1, w_1)$. If V_1 is a subspace which is *complementary* to the degeneracy $V^{\perp\omega}$, i.e., $V = V_1 \oplus V^{\perp\omega}$, then (V_1, ω_1) is a symplectic vector space.

Exercise. Show that ω_1 is non-degenerate on V_1.

A subspace V_1 of a symplectic vector space (V, ω) is called a *symplectic subspace* if (V_1, ω_1) is symplectic. If V_1 is symplectic and $dim_{\mathbb{R}} V_1 = 2$, then one often refers to V_1 as a *hyperbolic plane*. Given a symplectic space (V, ω) and $v \in V \setminus \{0\}$, there exists $w \in V$ with $\omega(v, w) \neq 0$. It follows that $V_1 := ((v, w))$ is a hyperbolic plane. If $W := V_1^\perp$, then $V = V_1 \oplus W$ and $\omega|W$ is non-degenerate. The following fact can be proved by iterating this procedure.

Proposition. *A symplectic vector space is a direct sum of hyperbolic planes.*

Corollary. *The dimension of a symplectic vector space is even.*

If $b : V \times V \to \mathbb{R}$ is a bilinear form and $(e_1, \ldots, e_m)$ is a basis, then the matrix $B = (b_{ij})$ of b in this basis is defined by

$$b_{ij} := b(e_i, e_j).$$

Thus, if $v, w \in V$ have coordinates $x, y \in \mathbb{R}^n$, then

$$b(v, w) = {}^t x B y.$$

Note that B is symmetric, i.e., $B = {}^t B$, whenever b is symmetric and is *skew-symmetric*, i.e., $B = -{}^t B$, if b is an alternating form.

Example. Let (V, ω) be a hyperbolic plane and choose a basis $((e_1, e_2))$ such that $\omega(e_1, e_2) = 1$. In a basis of V the matrix M_ω of ω has entries $\omega(e_i, e_j)$. In this case it is given by

$$M_\omega := \begin{pmatrix} 0 & 1 \\ -1 & 0 \end{pmatrix}.$$

If $V = V_1 \oplus \cdots \oplus V_n$ is displayed as a sum of hyperbolic planes, then such a basis can be chosen in each plane and

$$M_\omega = M_1 \oplus \cdots \oplus M_n,$$

where each (2×2)-Matrix M_i is as above.

Let $V = \mathbb{R}^n$ be equipped with the standard basis $\{e_1, \ldots, e_n\}$ and V^* with the associated dual basis. Then the matrix of the canonical symplectic form on $V \oplus V^* \cong \mathbb{R}^n \oplus \mathbb{R}^n = \mathbb{R}^{2n}$ is given by

$$M_\omega = \begin{pmatrix} 0 & Id \\ -Id & 0 \end{pmatrix},$$

where Id is the $(n \times n)$-identity matrix.

The induced structure on $\mathbb{R}^{2n}$ which is defined by $(x, y) \mapsto {}^t x M y$ is called the standard symplectic structure and is denoted by ω_{std}. A normal form for a symplectic vector space is an isomorphism

$$\varphi : (V, \omega) \to (\mathbb{R}^{2n}, \omega_{std}), \text{ i.e. } \varphi^*(\omega_{std}) = \omega.$$

Proposition. *Every symplectic vector space has a normal form.*

Proof. Decompose $V = V_1 \oplus \cdots \oplus V_n$ into a sum of hyperbolic planes and choose a basis $((v_i, w_i))$ of V_i with $\omega(v_i, w_i) = 1$. Then $((v_1, v_2, \ldots, v_n, w_1, \ldots, w_n))$ is a basis of V and φ can be defined by $\varphi(v_i) = e_i$ and $\varphi(w_i) = e_{n+i}, 1 \leq i \leq n$. $\square$

1.2 Riemannian and Hermitian structures

A complex vector space is a pair (V, J), where V is an $\mathbb{R}$-vector space and $J : V \to V$ is an endomorphism with $J^2 = -Id$. The map J is called a complex structure on V. For such a vector space there is a canonical decomposition of its *complexification* $V_\mathbb{C} := V \otimes_\mathbb{R} \mathbb{C}$ which is defined by the $+i$, resp. $-i$, eigenspace $V^{1,0}$, resp. $V^{0,1}$, of the $\mathbb{C}$-linear extension $J : V_\mathbb{C} \to V_\mathbb{C}$:

$$V_\mathbb{C} = V^{1,0} \oplus V^{0,1}.$$

The induced complex structure on $V^{1,0}$ is given by multiplication by i. The map $V \to V^{1,0}, v \mapsto v_\mathbb{C} := \frac{1}{2}(v - iJv)$, is $\mathbb{C}$-linear, i.e. $(Jv)_\mathbb{C} = iv_\mathbb{C}$.

A symmetric bilinear form $g : V \times V \to \mathbb{R}$ on an $\mathbb{R}$-vector space V is called a Riemannian structure if and only if it is *positive-definite*, i.e., $\|v\|_g^2 := g(v, v) > 0$ for all $v \in V \backslash \{0\}$. Consider a pairing $h : V \times V \to \mathbb{C}$ on a complex vector space (V, J) which is $\mathbb{C}$-linear in the first argument and $\mathbb{C}$-antilinear[1] in the second argument. Note that if V^* is equipped with the dual complex structure $J^* \in End(V^*), f \mapsto f \circ J^{-1}$, then the induced map

$$\varphi_h : V \to V^* \quad , w \mapsto F_w, \text{ where } F_w(v) = h(v, w),$$

---

[1] A map $F : V \to W$ between $\mathbb{C}$-vector spaces is said to be $\mathbb{C}$-antilinear if it is $\mathbb{R}$-linear and $F \circ J_V = -J_W \circ F$.

is $\mathbb{C}$-linear. If in addition h is positive-definite in the sense that $\|v\|_h^2 := h(v,v) > 0$ for all $v \in V\backslash\{0\}$ and $h(v,w) = \overline{h(w,v)}$, then it is called a Hermitian structure on V. Note that $h(Jw, Jv) = h(w, v)$, i.e. the $\mathbb{C}$-structure is an isometry of the Hermitian structure.

Exercise. Let $J_{std} : \mathbb{R}^{2n} \to \mathbb{R}^{2n}$ be defined by $J_{std}(e_i) = e_{n+i}$ and $J_{std}(e_{n+i}) = -e_i$, $1 \le i \le n$, and let g_{std} be the standard Riemannnian structure, i.e.

$$g_{std}(e_i, e_j) = \delta_{ij} := \begin{cases} 1, & i = j \\ 0, & \text{otherwise.} \end{cases}$$

Show that there exists a unique Hermitian structure h_{std} on $(\mathbb{R}^{2n}, J_{std})$ with $g_{std} = Re(h_{std})$. Show that $\omega_{std} = -Im(h_{std})$.

The content of the above exercise can be formulated as follows.

Proposition. *Let (V, J) be a complex vector space. A Riemannian structure g on V is the real part $Re(h)$ of a Hermitian structure if and only if J is a g-isometry, i.e., $g(Jv, Jw) = g(v, w)$ for all $v, w \in V$. In this case*

$$h = g - i\omega, \text{ where } \omega(v, w) := g(v, Jw)$$

is a symplectic structure on V. A given symplectic structure ω is the imaginary part $Im(h)$ of a Hermitian structure if and only if (i) $\omega(Jv, Jw) = \omega(v, w)$ for all $v, w \in V$ and (ii) $\omega(v, Jv) > 0$ for all $v \in V\backslash\{0\}$.

If the above conditions (i) and (ii) are satisfied, then J is said to be ω-compatible. If only (ii) is satisfied, then J is said to be ω-tame.

Exercise. Let (V, ω) be a symplectic vector space. Then there exists an ω-compatible complex structure $J : V \to V$ on V. If $W \subset V$ is a complex subspace, i.e. $J(W) = W$, and J is ω-tame, then W is a symplectic subspace.

2. Local theory

Let U be an open subset of $\mathbb{R}^n$ and denote by $\mathcal{E}(U)$ the $\mathbb{R}$-algebra of $\mathbb{R}$-valued C^∞-functions on U. The corresponding algebra of $\mathbb{C}$-valued functions is denoted by $\mathcal{A}(U)$. It is often convenient to consider a *cube*

$$W := \{x = (x_1, \ldots, x_n) \in \mathbb{R}^n : -1 < x_i < 1\}.$$

Fundamental Lemma. *For $f \in \mathcal{E}(W)$ there exist functions $f_1, \ldots, f_n \in \mathcal{E}(W)$ such that*

$$f = f(0) + x_1 f_1 + \cdots + x_n f_n.$$

For the proof it is necessary to recall the fundamental theorem of calculus, i.e., Stokes' Theorem in the 1-dimensional case: If $\gamma : [0, 1] \to \mathbb{R}^n$ is a smooth curve, and f is a smooth function defined in a neighborhood of the image of γ, then

$$f(1) - f(0) = \int_0^1 \langle \nabla f(\gamma(t)), \dot{\gamma}(t) \rangle dt,$$

where $\nabla f = (\frac{\partial f}{\partial x_1}, \dots, \frac{\partial f}{\partial x_n})$, $\dot{\gamma}(t) = (\dot{\gamma}_1(t), \dots, \dot{\gamma}_n(t))$, and $\langle, \rangle$ is the standard scalar product in $\mathbb{R}^n$. As a consequence, if γ is a closed curve, then

$$\int_0^1 \langle \nabla f(\gamma(t)), \dot{\gamma}(t) \rangle dt = 0.$$

This implies for example that if $f \in \mathcal{E}(W), x \in W$ and γ is a curve beginning at 0 and ending at x, then

$$f(x) = f(0) + \int_0^1 \langle \nabla f(\gamma(t)), \dot{\gamma}(t) \rangle.$$

Proof of the Fundamental Lemma. Choose $\gamma = \gamma_1 + \cdots + \gamma_n$, where $\gamma_i(t) = (x_1, \dots, x_{i-1}, t, 0, \dots, 0)$. $\square$

Remark. It follows immediately that $f_i(0) = \frac{\partial f}{\partial x_i}(0), 1 \le i \le n$. If the cube $W = W(x^0)$ is centered about a general point $x^0 = (x_1^0, \dots, x_n^0)$, then the Fundamental Lemma takes the form

$$f = f(x^0) + (x_1 - x_1^0)f_1 + \cdots + (x_n - x_n^0)f_n$$

and $f_i(x^0) = \frac{\partial f}{\partial x_i}|_{x=x^0}$.

If $I = (i_1, \dots, i_n)$ is a multi-index, let $|I| := i, + \cdots + i_n, I! = i_1! \cdots i_n!, x^I := x_1^{i_i} \cdots x_n^{i_n}$ and $\frac{\partial^{|I|}}{\partial x^I} = \frac{\partial^{|I|}}{\partial x_1^{i_1} \dots \partial_n^{i_n}}$.

Exercise. In this notation formulate and derive the basic result on Taylor series.

3. Global differentiable objects

In the present chapter all manifolds are finite-dimensional. A (countable) exhaustion of a topological space S is an increasing family $\{U_k\}$ of open subsets such that $U_k \subset\subset U_{k+1}$ and $\lim_{k \to \infty} U_k = S$. If S has such an exhaustion, then it is said to be paracompact. A space S is Hausdorff if any two points in S possess disjoint open neighborhoods. Manifolds are always assumed to be Hausdorff and we always assume that a finite-dimensional manifold is paracompact.

If S is a topological space and U is an open subset, then a homeomorphism $\psi : U \to V \subset \mathbb{R}^n$ onto an open subset V is said to give *coordinates* on U. If $x = (x_1, \dots, x_n)$ is the standard coordinate map on $\mathbb{R}^n$, then, by abuse of notation, x is regarded as being defined via ψ on U. The pair (U, x) is called a coordinate chart.

Suppose (U_α, x_α) and (U_β, x_β) are charts. Then there is a uniquely defined homeomorphism $\psi_{\alpha\beta}$ so that $x_\beta \circ \psi_{\alpha\beta} = x_\alpha$.

Exercise. Define $\psi_{\beta\alpha}$ in precise terms.

The coordinates x_α and x_β are said to be C^∞- compatible if $\psi_{\alpha\beta}$ is a diffeomorphism, i.e., it and its inverse are C^∞-maps.

Definition. A finite-dimensional manifold is a paracompact Hausdorff space with a covering $\mathcal{U} = \{U_\alpha, x_\alpha\}$ of coordinate charts any two of which are C^∞- compatible.

Remarks. (1) We will always replace $\mathcal{U}$ by a maximal covering by compatible coordinate charts. For example, without further ado we may refine coverings.
(2) It can be proved that each connected component of M has a well-defined dimension, i.e., for connected manifolds the charts are in a fixed vector space $\mathbb{R}^n$. In this case, $dim_\mathbb{R} M =: n$.

Since the coordinates are compatible, if U is an open subset of a manifold M, the algebra $\mathcal{E}(U)$, resp. $\mathcal{A}(U)$, of C^∞-functions is well-defined : f is a C^∞- function if and only if on any coordinate chart U_α there exists $f_\alpha \in \mathcal{E}(U_\alpha \cap U)$ such that $f = f_\alpha \circ x_\alpha$.

One of the primary goals of mathematics is to understand the set of solutions of a given equation. This is indeed one of the origins of the study of manifolds. For example, consider C^∞-functions $f_1, \ldots, f_k \in \mathcal{E}(\mathbb{R}^n)$ and

$$M := \{x \in \mathbb{R}^n : f_i(x) = 0, \ i = 1, \ldots, n\}.$$

Even though C^∞-functions are *smooth*, any closed set M can arise in this way. One way of distinguishing *reasonable* solution sets from the others is to introduce a linearization principle. The next sections serve as preliminaries for a discussion of this matter.

3.1 Vector fields and tangent vectors
Definition. A *vector field* X on M is a linear map

$$X : \mathcal{E}(M) \to \mathcal{E}(M)$$

which satisfies Leibnitz's rule: $X(fg) = fX(g) + gX(f)$. The $\mathcal{E}(M)$-module of vector fields on M is denoted by $Vect(M)$.

Exercise. Let U be an open subset of M and observe that, given a compact subset $K \subset U$, there exists a function $\chi \in \mathcal{E}(M)$ with $0 \le \chi \le 1$, $\chi|K \equiv 1$ and compact support $supp\,\chi$ contained in U. Using such cut-off functions, show that it makes sense to restrict vector fields, i.e., the restriction map $r_U : Vect(M) \to Vect(U)$ is well-defined with the obvious desired properties.

Exercise. Let W be a coordinate chart which is modelled on a cube. Then, for all $X \in Vect(W)$ there exist C^∞-functions $a_i \in \mathcal{E}(W), 1 \le i \le n$, such that

$$X = \sum a_i \frac{\partial}{\partial x_i}.$$

(Hint: Use the Fundamental Lemma.)

In particular the above exercise shows that a vector field is locally nothing more than a smoothly varying prescription of directional derivatives. To make this precise, for $p \in M$ define the *evaluation map* $\epsilon_p : \mathcal{E}(M) \to \mathbb{R}, \quad f \longmapsto f(p)$. Thus, for $X \in Vect(M)$, $X(p) := \epsilon_p \circ X : \mathcal{E}(M) \to \mathbb{R}$ is a linear functional which satisfies the Leibnitz-rule

$$X(p)(fg) = f(p)X(p)(g) + g(p)X(p)(f).$$

In other words $X(p)$ *is a directional derivative at* p.

Exercise. Show that if $f, g \in \mathcal{E}(M)$ and there exists a neighborhood $U = U(p)$ such that if $f|U = g|U$, then $X(p)(f) = X(p)(g)$.

For $p \in M$ define the equivalence relation $\sim_p$ on $\mathcal{E}(M)$ by $f \sim_p g$ if and only if there exists U as above. Denote by f_p the equivalence class of the function f and refer to it as the *germ of f at p* and let $\mathcal{E}_p$ denote the algebra of such germs.

Remark. Using a cut-off function it is possible to extend a function which is only defined in a neighborhood of $p \in M$ to a function defined on all of M which agrees with the original function in some smaller neighborhood. Thus, for the definiton of *germs* it is only necessary to consider functions which are defined in some neighborhood of the given point.

Definition. A tangent vector t at $p \in M$ is a linear map $t : \mathcal{E}_p \to \mathbb{R}$ which satisfies the Leibnitz rule $t(fg) = f(p)t(g) + g(p)t(f)$.

It follows from the Fundamental Lemma that the *tangent space at p*, i.e., the $\mathbb{R}$-vector space T_p of all such tangent vectors, has $\{\frac{\partial}{\partial x_1}|_p, \ldots, \frac{\partial}{\partial x_n}|_p\}$ as a basis, where $(x_1, \ldots, x_n) = x$ is a local coordinate mapping.

Exercise. If $\gamma : [-\epsilon, \epsilon] \to M$ is a smooth curve with $\gamma(0) = p$, then

$$t_p(f) := \frac{d}{dt}\Big|_{t=0} f \circ \gamma(t)$$

defines a tangent vector $t_\gamma \in T_p$. Show that in this way a tangent vector can be regarded as an equivalence class of curves.

The *curve definition* of a tangent vector is for certain purposes more convenient and more intuitive. For example, if $F : M \to N$ is a smooth map of manifolds, then

$$F_* : T_p(M) \to T_{F(p)}(N)$$

can be defined by the induced map of curves: $F_*(\gamma) = F \circ \gamma$.

Remark. In local coordinates, i.e., in the coordinate induced bases of T_p, resp. $T_{F(p)}$, the matrix of the linear map F_* is either the Jacobian or something closely related to it, depending on notational tradition.

Exercise. Show that the *push-forward* mapping F_* is not necessarily defined at the level of vector fields.

3.2 Differential forms

Differential forms are multi-linear functionals defined on vector fields. For a precise definition recall that if $V_1, \ldots, V_k$ are $\mathbb{R}$-vector spaces, then an element $F \in V_1^* \otimes \cdots \otimes V_k^*$ is a functional

$$F : V_1 \times \cdots \times V_k \to \mathbb{R}$$

which is linear in each of its entries. The space $V_1^* \otimes \cdots \otimes V_k^*$ contains *pure tensors* $f_1 \otimes \cdots \otimes f_k$ which are defined by

$$(f_1 \otimes \cdots \otimes f_k)(v_1, \ldots, v_k) = f_1(v_1) \cdot \ldots \cdot f_k(v_k).$$

If the spaces V_i^* come equipped with bases, then the pure tensors built from the basis elements of the respective summands form a basis of the tensor product.

In the situation at hand we consider $\otimes^k V^* = V^* \otimes \cdots \otimes V^*$, i.e the k-fold tensor product of V. If $\{e_1, \ldots e_n\}$ is a basis, then

$$\{e_{i_1} \otimes \cdots \otimes e_{i_k} : 1 \leq i_1, \ldots, i_k \leq n\}$$

is a basis of $\otimes^k V$.

Now consider the role of *symmetry*. Let the *symmetric group* $\mathfrak{S}_k$ of all permutations act on the k-fold product $\times^k V$ in the natural way, i.e., for $\tau \in \mathfrak{S}_k$,

$$\tau(v_1, \ldots, v_k) := (v_{\tau(1)}, \ldots, v_{\tau(k)}),$$

and consider the induced representation on $\otimes^k V^*$: $\tau(F) := F \circ \tau^{-1}$.

The *symmetric tensors* in $\otimes^k V^*$ are just those which are invariant,

$$S^k(V^*) := \{F \in \otimes^k V^* : \tau(F) = F \ \forall \tau \in \mathfrak{S}_k\}.$$

The tensors which are appropriate for the purposes of measuring area are the *alternating tensors*:

$$\Lambda^k V^* := \{F \in \otimes^k V^* : \tau(F) = sgn(\tau)(F) \ \forall \ \tau \in \mathfrak{S}_k\}.$$

Recall that a permutation $\tau \in \mathfrak{S}_k$ can be written as a composition of permutations which leave all elements but two fixed and those two will be exchanged. It follows that $sgn(\tau) = (-1)^n$, where n is the number of the simple permutation in the composition.

Another possible definition, where it is clear that the notion is well-defined, is

$$\prod_{i \neq j}(x_{\tau(i)} - x_{\tau(j)}) := sgn(\tau) \prod_{i \neq j}(x_i - x_j)$$

where $x = (x_1, \ldots, x_k)$ is the standard coordinate chart in $\mathbb{R}^k$.

Exercise. Show that $sgn : \mathfrak{S}_k \to \mathbb{Z}_2$ is a homomorphism.

The homomorphism sgn can also be interpreted in terms of orientations of a vector space: Define two ordered bases b_0 and b_1 of a vector space V to be equivalent if and only if there is a *continuous curve of bases* which connects the one to the other, i.e., if $b_0 = ((e_0^1, \ldots, e_0^n))$ and $b_1 = ((e_1^1, \ldots, e_1^n))$, then there exist continuous curves e_t^i so that $b_t := ((e_t^1, \ldots, e_t^n))$ is a basis for all $t \in [0, 1]$. An *orientation* is an equivalence class of such bases.

Observe that the group $Gl_n(\mathbb{R})$ acts in a natural way on ordered bases: If $A \in Gl_n(\mathbb{R})$ is matrix and $b = ((e_1, \ldots, e_n))$ is a basis, then $A(b) = ((f_1, \ldots, f_n))$, where $f_i = \sum A_{ij} e_j$.

Proposition. *In its topology as a subset of $Mat(n \times n, \mathbb{R}) \subset \mathbb{R}^{n^2}$ the group $Gl_n(\mathbb{R})$ has two connected components. These are distinguished by the sign of the determinant.*

Note that $det : Gl_n(\mathbb{R}) \to \mathbb{R}^* = \mathbb{R}\backslash\{0\}$ is a continuous homomorphism, that $\mathbb{R}^{>0}$ is a normal subgroup of $\mathbb{R}^*$ and that $q : \mathbb{R}^* \to \mathbb{R}^*/\mathbb{R}^{>0} \cong \mathbb{Z}_2$ is also a homomorphism.

Exercise. Let b_0 and b_1 be an ordered basis. Show that there is a matrix $A \in Gl_n(\mathbb{R})$ so that $b_1 = A(b_0)$. Define $sgn(A) = q \circ det(A) \in \mathbb{Z}_2$ and show that b_0 and b_1 have the same orientation if and only if $sgn(A) = +1$. Finally, show that $sgn(A_\tau) = sgn(\tau)$ for $\tau \in \mathfrak{S}_n$, where $\mathfrak{S}_n$ is embedded in $Gl_n(\mathbb{R})$ via its action on the standard basis.

Returning to the matter of *differential forms*, note that the natural structure on $Vect(M)$ is that of an $\mathcal{E}(M)$-module, i.e., if $X, Y \in Vect(M)$ and $f, g \in \mathcal{E}(M)$, then $fX + gY$ is in $Vect(M)$.

A *differential 1-form* on M is an $\mathcal{E}(M)$-linear functional $\alpha : Vect(M) \to \mathcal{E}(M)$. One can regard a vector field as a prescription of tangent vectors which varies smoothly and a 1-form as a prescription of functionals $\alpha(p) \in T_p^*$ which is likewise varying smoothly. More precisely, if $t \in T_p$, then there exists a vector field $X \in Vect(M)$ which $X(p) = t$ and

$$\alpha(p)(t) := \alpha(X)(p)$$

is well-defined independent of the choice of the extension X.

Notation. The dual space T_p^* is referred to as the *cotangent space* at the point $p \in M$.

Although vector fields and 1-forms are dual to each other, there is a certain lack of symmetry. Recall that the *push-forward* F_* is only definable at the level of tangent spaces. However, differential forms can be *pulled back at the global level*: Let $F : M \to N$ be a smooth map and define $F_p^* : T_{F(p)}^* \to T_p^*$ by $F_p^*(\alpha)(t_p) = \alpha(F_*(t_p))$. This definition can be globalized as follows.

Let $\mathcal{E}^1(M)$ denote the $\mathcal{E}(M)$-module of 1-Forms and define $F^* : \mathcal{E}^1(N) \to \mathcal{E}^1(M)$ by

$$F^*(\alpha)(X)(p) = \alpha(F(p))(F_*((X)(p))).$$

A simple calculation, e.g., in a basis given by coordinate charts, shows that F^* maps smooth forms to smooth forms. As in the case of F_*, the matrix of F^* is closely related to the Jacobian.

Basic Example. The natural pairing

$$\mathcal{E}(M) \times Vect(M) \to \mathcal{E}(M), \quad (f, X) \mapsto X(f),$$

yields an $\mathbb{R}$-linear map

$$d : \mathcal{E}(M) \to \mathcal{E}^1(M) \text{ defined by } df(X) = X(f).$$

Since vector fields are derivations, $d(fg) = f\,dg + g\,df$.

Exercise. Show that in local coordinates

$$df = \sum \frac{\partial f}{\partial x_i} dx_i, \text{ where } \{dx_1, \ldots, dx_n\}$$

is the $(\mathcal{E}(U)$-module)basis dual to $\{\frac{\partial}{\partial x_1}, \ldots, \frac{\partial}{\partial x_n}\}$ for the vector fields. Show that the differentials of coordinate functions satisfy $d(x_i) = dx_i$.

Notation. Let $A, B,$ and C be algebraic objects of the same type, e.g. vector spaces, groups, $\ldots$, and let $\alpha : A \to B$ and $\beta : B \to C$ be homomorphisms. The *sequence* $A \overset{\alpha}{\to} B \overset{\beta}{\to} C$ is said to be *exact* if image (α) = kernel (β). There are two special cases: $0 \to A \to B$ and $B \to C \to 0$. Exactness of the first means that α is injective and exactness of the seond means that β is surjective.

Exercise. Let M be connected and $i : \mathbb{R} \to \mathcal{E}(M)$ be the canonical map defined by $i(a) =: f_a$, where $f_a \equiv a$. Show that $0 \to \mathbb{R} \overset{i}{\to} \mathcal{E}(M) \overset{d}{\to} \mathcal{E}^1(M)$ is exact.

The p-fold exterior product $\wedge^p V^*$ was discussed above at the level of vector spaces. The exact same discussion can be carried over to the $\mathcal{E}(M)$-module $\mathcal{E}^1(M)$, i.e., we have $\mathcal{E}^p(M) := \bigwedge^p \mathcal{E}^1(M)$ and an element $\alpha \in \mathcal{E}^p(M)$ is referred to as a differential p-form.

It is possible to think of a p-form α as follows: Let $X_1, \ldots, X_p \in Vect(M)$ and regard $(X_1, \ldots, X_p)$ as an oriented p-dimensional parallelogram which is moving around on M. At each $x \in M$ the number

$$\alpha(x)(X_1(x), \ldots, X_p(x))$$

is something like a *generalized volume* of the p-fold parallelogram which depends in a smooth way on $x \in M$.

There are special *pure wedge products* which should be mentioned: If V^* is module, then the pure tensors $f_1 \otimes \cdots \otimes f_p \in \otimes^p V^*$ were defined above. It is important to note that there exists a natural projection

$$\text{alt}: \otimes^p V^* \to \wedge^p V^*, \quad T \mapsto \frac{1}{p!} \sum_{\tau \in \mathfrak{S}_p} sgn(\tau)\tau(T).$$

Exercise. Determine $ker(\text{alt})$ in the case $p = 2$. Show that the case of $p > 2$ is essentially different.

It follows that $\{\text{alt}(f_{i_1} \otimes \cdots \otimes f_{i_p}) : 1 \le i_1 < \cdots < i_p \le n\}$ is a basis for $\wedge^p V$.

Notation. Define $f_1 \wedge \cdots \wedge f_p := c_p \cdot \text{alt}(f_1 \otimes \cdots \otimes f_p)$, where c_p is a constant.

Exercise. Based on mathematical and/or physical common sense, suggest a value for c_p.

If (U, x) is a coordinate chart, then

$$\{dx_{i_1} \wedge \cdots \wedge dx_{i_p} : 1 \le i_1 < \cdots < i_p \le n\}$$

is a basis for $\mathcal{E}^p(U)$. In multi-index notation, a differential p-form α is locally described by

$$\alpha = \sum_I \alpha_I dx_I, \text{ where } dx_I = dx_{i_1} \wedge \cdots \wedge dx_{i_p}$$

for $I = (i_1, \ldots, i_p)$ and the coefficients α_I are smooth functions.

3.3 The de Rham complex

The map $d : \mathcal{E}(M) \to \mathcal{E}^1(M), df(X) := X(f)$, is defined for any open set $U \subset M$ and, since the necessary compatibility conditions are fulfilled, is well-defined at the level of germs. Move precisely, if $f_m \in \mathcal{E}_m$ is a germ which has a representive f in some neighborhood of m, then the germ of the 1-form df in $\mathcal{E}_m^1$ is independent of the choice of the representative.

There is a unique extension of d to the full *exterior algebra* $\oplus_{p=0}^n \bigwedge^p \mathcal{E}^1(M), n :=$ $dim_{\mathbb{R}} M$, by requiring the validity of Leibnitz's rule for $d : \mathcal{E}^p(M) \to \mathcal{E}^{p+1}(M)$, i.e. if $f \in \mathcal{E}(M)$ and $\alpha \in \mathcal{E}^p(M)$, then

$$d(f\alpha) = df \wedge \alpha + f d\alpha,$$

and $d \circ d = 0$. Just as at the level of 1-forms, this yields a natural definition $d : \mathcal{E}^p(U) \to \mathcal{E}^{p+1}(U)$ for any open set $U \subset M$. For $\alpha = \sum \alpha_I dx_I$ in local coordinates, it follows that $d\alpha = \sum d\alpha_I \wedge dx_I$.

Notation. A sequence of morphisms

$$A_1 \xrightarrow{\alpha_1} A_2 \xrightarrow{\alpha_2} A_3 \xrightarrow{\alpha_3} \dots$$

is called a *complex* whenever $Im(\alpha_i) \subset Ker(\alpha_{i+1})$, i.e. $\alpha_{i+1} \circ \alpha_i = 0$ for all i.

Thus, for any $U \subset M$, or at the level of germs, we have the *de Rham complex*

$$0 \to \mathbb{R}(U) \to \mathcal{E}(U) \to \mathcal{E}^1(U) \to \cdots \to \mathcal{E}^n(U) \to 0.$$

Here $\mathbb{R}(U)$ represents the vector space of functions which are locally constant on U.

The de Rham complex for germs is in fact exact, i.e., for every $m \in M$

$$\frac{Ker(d : \mathcal{E}_m^p \to \mathcal{E}_m^{p+1})}{Im(d : \mathcal{E}_m^{p-1} \to \mathcal{E}_m^p)} = \{0\} \quad \text{for } 1 \leq p \leq dim_{\mathbb{R}} M.$$

This means that, given a p-form α with $d\alpha = 0$ in some neighborhood of $m \in M$, there exists a $(p-1)$-form η with $d\eta = \alpha$ in some (possibly smaller) neighborhood of m.

To formalize these considerations it is convenient to introduce the $p - th$ *de Rham cohomology* space

$$H_{deR}^p(M) := \frac{Ker(d : \mathcal{E}^p(M) \to \mathcal{E}^{p+1}(M))}{Im(d : \mathcal{E}^{p-1}(M) \to \mathcal{E}^p(M)).}$$

The exactness of the de Rham complex of germs follows from the

Poincaré Lemma. *For W a cube,*

$$H_{deR}^p(W) = \{0\}$$

for $1 \leq p \leq dim_{\mathbb{R}} W$.

Before giving the proof it is convenient to introduce a *homotopy principle* for differential forms.

Let $I = [0, 1]$ be the unit interval, M be an n-dimensional manifold and $H : I \times M \to M$ a smooth map. For $t \in I$, let $i_t^* : \mathcal{E}^p(I \times M) \to \mathcal{E}^p(M)$ be defined by the embedding $i_t : M \hookrightarrow I \times M$, $m \mapsto (t, m)$.

Define

$$K : \mathcal{E}^{p+1}(I \times M) \to \mathcal{E}^p(M) \quad , \quad 1 \leq p \leq n,$$

by

$$K(\omega) := \int_I \omega,$$

where the integral is carried out with respect to $t \in I$, i.e.,

$$K(\omega)(X_1, \ldots, X_p) := \int_I \omega\left((i_t)_*(X_1), \ldots, (i_t)_*(X_p), \frac{d}{dt}\right) dt.$$

If $\omega = \omega_1 \wedge dt + \omega_2$, where $\omega_j(\frac{\partial}{\partial t}, *) = 0$, $j = 1, 2$, then $d\omega = d_2\omega_1 \wedge dt + d_2\omega_2 + \frac{\partial \omega_2}{\partial t} \wedge dt$, where $d_2\omega_1$, and $d_2\omega_2$ are differentials of the corresponding forms on $I \times M$ which are defined by regarding them as forms on M depending on a parameter $t \in I$ and $\frac{\partial \omega_2}{\partial t}$ is just the derivative along such a curve of forms.

Since $K(\omega_2) = K(d_2\omega_2) = 0$, it immediately follows that

$$K(d\omega) - dK(\omega) = \int_0^1 \frac{\partial \omega_2}{\partial t} \wedge dt = i_1^*(\omega) - i_0^*(\omega).$$

Now apply this homotopy of forms to a geometric homotopy.

Homotopy Lemma. *Let $H : I \times M \to M$ be a smooth map and $\omega \in \mathcal{E}^p(M)$ a closed form, i.e. $d\omega = 0$. For $t \in I$ define $H_t : M \to M$ by $m \mapsto H(t, m)$. Then*

$$H_1^*(\omega) = H_0^*(\omega) - dK(H^*(\omega)).$$

The proof of this follows immediately from the above property of K and the fact that $H_t = H \circ i_t$. $\qquad \square$

Corollary. *A geometric homotopy induces the identity on de Rham cohomology.*

A space M is *homotopic to a point* $m_0 \in M$ if there is a smooth map

$$H : I \times M \to M$$

so that H_0 is the identity and $H_1(m) = m_0$ for all $m \in M$.

Corollary. *If M is homotopic to a point, then*

$$H_{deR}^p(M) = \{0\} \text{ for all } p \geq 1.$$

Since cubes are homotopic to their centers, Poincaré's Lemma follows immediately. $\qquad \square$

The de Rham cohomology $H_{deR}^*(M)$ reflects a certain part of the geometry of M. In order to most efficiently discuss examples of this, it is useful to have the tool of integration.

4. A sketch of integration theory

For the purposes of integration theory it is convenient to express the notion of an *orientation* in terms of diffential forms. For this let M be a connected n-dimensional manifold. An *orientation form* is a non-vanishing (*top-dimensional*) form $\lambda \in \mathcal{E}^n(M)$.

A basis $\{v_1, \ldots, v_n\}$ of a tangent space $T_m, m \in M$, is said to be *positively oriented* with respect to λ if $\lambda(m)(v_1, \ldots, v_n) > 0$.

Exercise. If b_1 and b_2 are bases of T_m, then we have already introduced the notion that they have the same (or different) orientation. Show that this is the same as the following notion given by *any* n-form λ which does not vanish at $m : b_0$ and b_1 *have the same orientation if* $\lambda(m)(b_0)$ *and* $\lambda(m)(b_1)$ *have the same sign.*

If $U \subset M$ is an open subset, e.g., a coordinate chart, then it is possible that there are vector fields $X_1, \ldots, X_n \in Vect(U)$, $n := dim_\mathbb{R} M$, so that $\{X_1(m), \ldots X_n(m)\}$ is a basis of $T_m M$ for all $m \in U$. In this case $(X_1, \ldots, X_n)$ is called a *frame* on U. Such a frame is said to be *postively*, resp. *negatively*, oriented if $\lambda(X_1, \ldots, x_n)$ is a positive, resp. negative, function on U.

We equip the manifold $\mathbb{R}^n$ with its *standard orientation* $\lambda_{std} = dx_1 \wedge \ldots \wedge dx_n$, where $x = (x_1, \ldots, x_n)$ is the standard coordinate mapping.

Finally, if (M_1, λ_1) and (M_2, λ) are oriented manifolds and $\varphi : M_1 \to M_2$ is a smooth map, then φ is said to be *orientation preserving* if $\varphi^*(\lambda_2) = f_\varphi \lambda_1$, where f_φ is a positive function.

If $\alpha \in \mathcal{E}^p(M)$, then $supp(\alpha)$ is defined to be the topological closure of $\{m \in M : \alpha(m) \neq 0\}$. Let $\mathcal{E}_c^p(M)$ denote the space of forms with *compact support*. An integral is an $\mathbb{R}$-linear functional

$$\int : \mathcal{E}_c^n(M) \to \mathbb{R}$$

which is defined on an oriented n-dimensional manifold M.

Remark. If λ is a fixed orientation form, then it is possible to define the integral at the level of functions:

$$\int f := \int f\lambda.$$

The essential property of the integral is the *transformation rule*: Let $\varphi : M_1 \to M_2$ be an orientation preserving diffeomorphism. Then

$$\int_{M_2} \eta = \int_{M_1} \varphi^*(\eta) \quad \text{for all } \eta \in \mathcal{E}_c^n(M_2).$$

This property is of course reflected in the construction of the integral.

Let W be the standard cube in $\mathbb{R}^n$ equipped with the standard orientation and let (M, λ) be an oriented manifold.

An orientation preserving diffeomorphism

$$C : W \to M$$

onto an open subset of M is called an *oriented n-cell* in M.

Remark. Since all constructions allow *shrinking*, it is possible to assume that C is a diffeomorphism of some neighborhood of $\overline{W}$.

A covering $\mathcal{U} = \{U_\alpha\}$ by oriented n-cells is given by orientation preserving diffeomorphisms $C_\alpha : W \to U_\alpha$.

For an n-cell C we have the formal definition

$$\int_C \eta := \int_W C^*(\eta).$$

Using the compatibility which is implicit in a covering by oriented n-cells, it is possible to *glue together* the integrals which are defined on the individual cells to obtain an integral on the full manifold M.

The main tool for *gluing together smooth objects* is a *partition of unity*. For this we consider a locally finite covering $U = \{U_\alpha\}$, i.e., for every compact set $K \subset M$ there are only finitely many U_α's such that $K \cap U_\alpha \neq \emptyset$. Since M is paracompact, every covering has a locally finite refinement. Thus the restriction to locally finite covers is of no essential consequence.

Using cut-off functions on coordinate charts, the existence of which is an exercise with concrete functions, it is possible to construct a

Partition of Unity. Let $U = \{U_\alpha\}$ be a locally finite covering of manifold. Then there exists a collection of functions $\{\chi_\alpha\}$ which satisfy the following conditions:
  (i) $\chi_\alpha \in \mathcal{E}_c(U_\alpha) \subset \mathcal{E}(M)$
  (ii) $0 \leq \chi_\alpha \leq 1$
  (iii) $\sum_\alpha \chi_\alpha \equiv 1$.

Remark. Partitions of unity are far from being unique.

If $U = \{U_\alpha\}$ is a covering of oriented n-cells and $\{\chi_\alpha\}$ is a partition of unity, then

$$\int_M \eta := \sum_\alpha \int_{C_\alpha} (\chi_\alpha \cdot \eta) = \sum_\alpha \int_W C_\alpha^*(\chi_\alpha \cdot \eta) .$$

Remarks. (1) Since the form $\alpha \in \mathcal{E}_c^n(M)$ has compact support, the summation has only finitely many non-zero terms and consequently convergence need not be discussed.

(2) Of course the integral is defined in terms of a covering and a partition of unity. The *main theorem* in the subject is that it depends on neither of these. This result is in effect a sort of transformation rule which follows, for example, from the transformation rule for the Riemann integral for domains in $\mathbb{R}^n$.

Without further discussion, we assume that the integral has been shown to be well-defined. It is obviously an $\mathbb{R}$-linear functional

$$\int : \mathcal{E}_c^n(M) \to \mathbb{R}.$$

The two basic properties of the integral, i.e. the *transformation rule* and *Stokes' Theorem*, immediately follow from these results for cubes in $\mathbb{R}^n$. It should be underlined that in the proof of the well-definedness one must for example prove at least a local version of version of the transformation rule.

To prove the global transformation rule for an orientation preserving diffeomorphism $\varphi : M_1 \to M_2$, assuming the analogous result for cubes, just note that for η in $\mathcal{E}_c^n(M_2)$ and $\{\chi_\alpha\}$ a partition of unity subordinate to the open covering by n-cells of M_2 given by $\{C_\alpha(W)\}$ we have:

$$\int_{M_1} \varphi^*(\eta) = \sum_\alpha \int_{\varphi^{-1} \circ C_\alpha} (\chi_\alpha \circ \varphi) \cdot \varphi^*(\eta) =$$

$$= \sum_\alpha \int_{\varphi^{-1} \circ C_\alpha} \varphi^*(\chi_\alpha \eta) = \sum_\alpha \int_{C_\alpha} \chi_\alpha \eta = \int_{M_2} \eta.$$

For a statement of Stokes' Theorem it is necessary to discuss the notion of a (piecewise) *smooth boundary* $\partial\Omega$ of an open set $\Omega \subset M$ and its induced orientation.

5. Smooth submanifolds

Let $f_1, \ldots, f_k \in \mathcal{E}(M)$ and consider

$$V(f_1, \ldots, f_k) := \{m \in M : f_1(m) = \cdots = f_k(m) = 0\}$$

For $m_0 \in V(f_1, \ldots, f_k)$ assume that $df_1 \wedge \ldots \wedge df_k(m_0) \neq 0$. In particular this implies that $k \leq n = dim_\mathbb{R} M$.

Using local coordinates at m_0 we choose additional functions $f_{k+1}, \ldots, f_n \in \mathcal{E}(M)^{2)}$ so that

$$(*) \qquad\qquad\qquad df_1 \wedge \ldots \wedge df_n(m_0) \neq 0$$

If $x = (x_1, \ldots, x_n)$ is a coordinate map, it follows that

$$df_1 \wedge \ldots \wedge df_n = det\left(\frac{\partial f_i}{\partial x_j}\right) dx_1 \wedge \ldots \wedge dx_n.$$

Thus $(*)$ is equivalent to the non-vanishing of the Jacobian determinant $J := det(\frac{\partial f_i}{\partial x_i})$ in some neighborhood of m_0.
Recall the *inverse mapping theorem*:

Let $f_1, \ldots, f_n$ be smooth functions which are defined in some neighborhood of 0 in $\mathbb{R}^n$. Assume that $f_i(0) = 0$, $1 \leq i \leq n$. Let $F := (f_1, \ldots, f_n)$ be the associated map. Then there are open neighborhoods $U(0)$ and $V(0)$ such that

$$F : U \to V$$

---

2) In reality it is only necessary to choose these functions in a neighborhood of m_0.

is a diffeomorphism if and only if

$$J(F)(0) := det(\frac{\partial f_i}{\partial x_j})(0) \neq 0.$$

Corollary. (Implicit function theorem). *Let $m_0 \in V(f_1, \ldots, f_k)$ be such that $df_1 \wedge \ldots \wedge df_k(m_0) \neq 0$. Then there are coordinates $x = (x_1, \ldots, x_n)$ on a neighborhood $U = U(m_0)$ so that*

$$N := V(f_1, \ldots, f_k) \cap U = \{p : x_1(p) = \ldots = x_k(p) = 0\}.$$

Remark. The coordinates $x' = (x_{k+1}, \ldots, x_n)$ define a manifold structure on N.

Exercise. Show that this structure is independent of the defining functions $f_1, \ldots, f_k$.

A closed subset N of a manifold M which is locally defined as above, i.e., for $m \in N$ there exists an open neighborhood $U = U(m)$ and $f_1, \ldots, f_k \in \mathcal{E}(U)$ with $df_1 \wedge \ldots \wedge df_k(m) \neq 0$ and $N \cap U = V(f_1, \ldots, f_k)$, is called a (closed) *submanifold* of M. The above exercise shows that the manifold structure of N which comes from the application of the implicit function theorem is well-defined.

Remark. The number $codim_m N := k$ may vary with m, but is locally constant and therefore is constant on the connected components of N.

If N is a connected submanifold of M with $codim\, N = 1$, then it is a candidate to be a *boundary* $\partial\Omega$ of a domain $\Omega \subset M$.

Exercise. Show that there are curves in certain 2-dimensional manifolds which are not boundaries. For example, consider the unit circle in the annulus

$$M := \{z \in \mathbb{C} : \frac{1}{2} < |z| < 2\}.$$

Let Ω be an open set in a manifold M. The boundary $\partial\Omega$ is said to be smooth if at every $m_0 \in \partial\Omega$ there exists an open neighborhood $U = U(m_0)$ in M and a function $\rho \in \mathcal{E}(U)$ such that
(i) $d\rho(m) \neq 0$ for all $m \in U$
(ii) $U \cap \Omega = \{m : \rho(m) < 0\}$.

In particular a smooth boundary is a 1-codimensional submanifold of M. Therefore, locally in the appropriate coordinates, $\partial\Omega = \{x_1 = 0\}$.

For any submanifold $i : N \hookrightarrow M$, the push-forward map $i_* : T_m N \to T_m M$ realizes the tangent spaces of N as subspaces of the tangent spaces of M.

Let $X \in Vect(M)$ and $f \in \mathcal{E}(N)$. For some extension $\tilde{f}$ of f to M consider $X(\tilde{f})$. It is desirable to determine those vector fields for which the map $f \mapsto X(\tilde{f})$ is independent of the choice of the extension.

Exercise. Discuss this matter for $M = \mathbb{R}^2, N = \{x = 0\}$ and $X = \frac{\partial}{\partial y}$.

Proposition. *Let $N = V(f_1, \ldots, f_k)$ in some neighborhood U, where $df_1 \wedge \ldots \wedge df_k$ is nowhere vanishing and let $X \in Vect(U)$. Then $f \mapsto X(\tilde{f})$ is well-defined independent of the extension $\tilde{f}$ if and only if $df_i(X)|_N = 0$, $1 \leq i \leq k$.*

Proof. Let g and h be two extensions of f. Then, for every $m \in N$, an application of the Fundamental Lemma in a coordinate neighborhood $W = W(m) \subset U$ yields the existence of functions $l_1, \ldots, l_k \in \mathcal{E}(W)$ such that

$$g - h = f_1 l_1 + \ldots + f_k \cdot l_k.$$

Consequently

$$X(g)|_N = X(h)|_N + l_1 df_1(X) + \ldots + l_k df_k(X) = X(h)|_N$$

whenever $X(f_i) = 0$ for all i.

The converse direction is proved by noting that any extension $\tilde{f}$ can be modified by adding a function f_i and, if $X(f_i)|_N \neq 0$, then this yields a change in $X(\tilde{f})$. $\square$

Remark. We say that the forms $df_1, \ldots, df_k$ define the tangent spaces of N in U : $T_m N = \bigcap_j Ker(df_j)(m)$ for all $m \in N \cap U = V(f_1, \ldots f_k)$.

If $x = (x_1, \ldots x_n)$ is a local coordinate mapping on U, then $df_j = \sum \frac{\partial f_j}{\partial x_i} dx_i$ and one may think of the df_j as a *gradient vector* $\nabla f_j = (\frac{\partial f_j}{\partial x_1}, \ldots, \frac{\partial f_j}{\partial x_n})$ with respect to this system of coordinates. In this way $T_m N = ((\nabla f_1(m), \ldots, \nabla f_k(m)))^\perp$ with respect to the standard scalar product in $\mathbb{R}^n$.

6. Induced orientation and Stokes' theorem

Let (M, λ) be an oriented manifold and $\Omega \subset M$ open subset. If $\partial\Omega$ is smooth and $\rho \in \mathcal{E}(U)$ is a local defining function for $\partial\Omega$ near $m \in \partial\Omega$, then $\nabla\rho(m)$ can be regarded as an *outward pointing* vector which, in the Euclidean geometry induced by the coordinate space, is orthogonal to $T_m N$. We use the normal fields $\nabla\rho$ to construct an induced orientation form on $\partial\Omega$.

Cover $\partial\Omega$ by open sets O_α which are the intersection of coordinate charts U_α with $\partial\Omega$. The boundary $\partial\Omega \cap U_\alpha$ is defined by $\rho_\alpha \in \mathcal{E}(U_\alpha)$. We may assume that the coverings $\{O_\alpha\}$ and $\mathcal{U} = \{U_\alpha\}$ are locally finite. For simplicity of notation we set $U = U(\partial\Omega) = \bigcup_\alpha U_\alpha$.

Now let $\{\chi_\alpha\}$ be a partition of unity for $\mathcal{U}$ and define $\rho = \sum_\alpha \chi_\alpha \rho_\alpha$. Then, after shrinking U if necessary, it follows that $d\rho$ is nowhere zero on U and that ρ defines $\partial\Omega$ i.e. ρ is a *global* defining function.

Finally, since $d\rho$ is nowhere zero, $\lambda = d\rho \wedge \mu$ for some $\mu \in \mathcal{E}^{n-1}(U)$. If $i : \partial\Omega \hookrightarrow M$ is natural embedding, then we let

$$\lambda_{\partial\Omega} := i^*(\mu).$$

Exercise. Show that $\lambda_{\partial\Omega}$ defines an orientation which depends only on the orientation of the ambient space and the convention that defining functions should have outward pointing gradients.

The orientation $\lambda_{\partial\Omega}$ is referred to as the *induced orientation* of the boundary $\partial\Omega$.

Notation. If $i : \partial\Omega \hookrightarrow M$ is the canonical embedding and $\mu \in \mathcal{E}_c^{n-1}(M)$, then

$$\int_{\partial\Omega} \mu := \int_{\partial\Omega} i^*(\mu),$$

where the integral is defined by the induced orientation.

Stokes' Theorem. *Let Ω be a domain with smooth boundary $\partial\Omega$ in an oriented manifold (M, λ). For $\mu \in \mathcal{E}_c^{n-1}(M)$ it follows that*

$$\int_{\partial\Omega} \mu = \int_{\Omega} d\mu.$$

Remark. Since α has compact support, it is enough to prove this for relatively compact domains Ω in M.

For the proof of Stokes' Theorem it is convenient to cover $\partial\Omega$ with *partial* boundaries of n-cells which are contained in Ω. More precisely, such a cell

$$C_\alpha : W = \{(x_1, \ldots, x_n) \in \mathbb{R}^n : |x_j| < 1\} \longrightarrow U_\alpha = C_\alpha(W) \subset M$$

should be an orientation preserving diffeomorphism so that

$$C_\alpha | \{x_1 = 0\} : W \cap \{x_1 = 0\} \longrightarrow U_\alpha \cap \partial\Omega$$

is orientation preserving for the induced orientation. The boundary can be covered by such neighborhoods and, by choosing additional n-cells U_γ which are relatively compact in Ω, we extend this to a covering $\mathcal{U}$ of $\overline{\Omega}$.

Proof of Stokes' Theorem. Let $\mathcal{U}$ be a locally finite covering of the above type and let $\mathcal{V} = \{V_\alpha\}$ the cover given by $V_\alpha = U_\alpha$ for $U_\alpha \subset\subset \Omega$ and $V_\alpha = C_\alpha(W \cap \{x_1 > 0\})$ for $U_\alpha \cap \partial\Omega \neq \emptyset$.

Let $\{\gamma_\alpha\}$ be a partition of unity for $\mathcal{V}$. Then

$$\int_\Omega d\mu = \int_\Omega d(\sum_\alpha \gamma_\alpha\mu) = \sum_\alpha \int_\Omega d(\gamma_\alpha\mu) = \sum_\alpha \int_{V_\alpha} d(\gamma_\alpha\mu).$$

We may now apply Stokes' Theorem for cubes in $\mathbb{R}^n$. For the relatively compact U_α's, since $\gamma_\alpha\mu$ has compact support,

$$\int_{V_\alpha} d(\gamma_\alpha\mu) = \int_W C_\alpha^*(d(\gamma_\alpha\mu)) = \int_W dC_\alpha^*(\gamma_\alpha\mu) = \int_{\partial W} C_\alpha^*(\gamma_\alpha\mu) = 0.$$

For those V_α's with $\partial U_\alpha \cap \partial\Omega \neq \phi$, the same calculation shows that

$$\int_{V_\alpha} d(\gamma_\alpha\mu) = \int_{\{x_1=0\}} C_\alpha^*(\gamma_\alpha\mu).$$

Let $\partial\mathcal{U}$ be the induced covering of $\partial\Omega$ by the partial boundaries $bV_\alpha := C_\alpha(\{x_1 = 0\})$. It follows that

$$\int_\Omega d\mu = \sum_{\alpha : U_\alpha \cap \partial\Omega \neq \emptyset} \int_{bV_\alpha} \gamma_\alpha\mu = \int_{\partial\Omega} \sum \gamma_\alpha\mu = \int_{\partial\Omega} \mu. \qquad \square$$

Remark. It is clear that, assuming the existence of a partition of unity, the general Stokes' Theorem and in fact the entire integration theory on manifolds is a formal consequence of the analogous results for cubes in $\mathbb{R}^n$. However, one should not underestimate the discovery of the notion of a partition of unity.

7. Functionals on de Rham cohomology

Let (N, λ) be a compact oriented p-dimensional manifold and let $C^\infty(N, M)$ be the set of smooth maps $F : N \to M$. Integration on N defines a pairing

$$(*) \qquad C^\infty(N, M) \times \mathcal{E}^p(M) \to \mathbb{R}, \quad (F, \alpha) \mapsto \int_N F^*(\alpha).$$

Since $F^*(d\beta) = dF^*(\beta)$ and N is compact, Stokes' Theorem implies that $(*)$ is defined at the level of deRham cohomology:

$$C^\infty(N, M) \times H^p_{deR}(M) \to \mathbb{R}, \quad (F, [\alpha]) \mapsto \int_N F^*(\alpha).$$

Exercise. Two maps $F_0, F_1 : N \to M$ are said to be homotopic if there is a map

$$H : I \times N \to M \text{ with } H_0 = F_0 \text{ and } H_1 = F_1.$$

The map H_t is defined by $x \mapsto H(t, x)$. Such a map is called a *homotopy* from F_0 to F_1.

Exercise. Show that *homotopy equivalence* is an equivalence relation and that $(*)$ is defined at the level of *homotopy classes* of maps.

Remark. Homotopy equivalence is a very strong notion and in fact $(*)$ can be defined for much weaker notions of geometric equivalence.

The pairing $(*)$ gives us a method for showing that a given cohomology class, resp. homotopy class of maps, is non-trivial. The most obvious example of this is the top-dimensional class generated by an orientation form.

Let (M, λ) be a compact oriented manifold and, in the formalism of the pairing $(*)$, let $N = M, \alpha = \lambda$ and $F = Id$. Since $\int_M \lambda > 0$, it follows that $[\lambda] \neq 0 \in H^n_{deR}(M)$ and $[Id]$ is a non-trivial homotopy class of maps.

Remark. It can in fact be shown that

$$\int_M : H^n_{deR}(M) \to \mathbb{R}$$

is an *isomorphism* for a compact oriented n-dimensional manifold M. From certain points of view, an isomorphism $H^n_{deR}(M) \simeq \mathbb{R}$ *is the defintion of an integral.*

Exercise. Let $\mathcal{E}^{n-1}_c(M)$ denote the compactly supported forms of degree one less than the dimension of M. If α is such a form, then Stokes' Theorem shows that $\int d\alpha = 0$. Define $H^n_c(M)$ to be the quotient of $\mathcal{E}^n_c(M)$ by $d\mathcal{E}^{n-1}_c(M)$. Show that, up to a multiplicative constant, $\int$ is the unique linear functional on $H^n_c(M)$.

Example. Let $M = \mathbb{R}^2\backslash\{(0,0)\}$ and $N := \{z : |z| = 1\}$ be the unit circle in $\mathbb{C} \cong \mathbb{R}^2$ equipped with its induced orientation. The 1-form

$$\mu = \frac{1}{x^2 + y^2}(-ydx + xdy)$$

defines a cohomology class $\xi = [\mu] \in H^1_{deR}(M)$. Let $F : N \to M$ be the natural embedding. Then $F^*(\mu)$ is calculated as $d\varphi$ in polar coordinates: Note $d\varphi$ is not d of a 1-form, because

$$\int_N F^*(\mu) = 2\pi.$$

Consequently $[F]$ is not the trivial homotopy class of maps and $\xi \neq 0 \in H^1_{deR}(M)$.

II. Elements of Lie groups and their actions

1. Introduction to actions and quotients

The purpose here is to set the notation.

1.1 Basic definitions

A paracompact differentiable manifold G is called a Lie group if it is a group and the group structure $G \times G \to G$, $(g,h) \mapsto gh^{-1}$, is a C^∞ – map. Note that, although a Lie group might not be connected, it has at most countably many components, all of which have the same dimension.

Notation. If H is a subgroup of G, we write $H < G$.

One of the foundational results in the subject is that a Lie group has a unique C^∞-compatible $\mathbb{R}$-analytic structure. In other words, there is a covering by C^∞-coordinate charts so that change of coordinates is $\mathbb{R}$-analytic, i.e., the functions defining these diffeomorphisms can be convergently developed in power series expansions. This is the unique $\mathbb{R}$-analytic structure on the Lie group G such that the above multiplication map is $\mathbb{R}$-analytic.

If $H < G$ is a closed subgroup, then in fact H is a Lie subgroup, in particular it is a closed $\mathbb{R}$-analytic submanifold of G.

An action of a group G on a set S is a map

$$G \times S \to S, \quad (g,s) \mapsto g(s),$$

such that $g(h(s)) = (gh)(s)$ and $e(s) = s$ for all $s \in S$ and $g, h \in G$. Here $e \in G$ denotes the identity. If M is a manifold equipped with the action $G \times M \to M$ of Lie group and the action map is smooth, then we refer to it as a smooth G-action.

Given a G-action on S and $s \in S$, then $G.s := \{g(s) : g \in G\}$ denotes the G-orbit of the point s. Let $G_s := \{g \in G : g(s) = s\}$ denote the *isotropy subgroup* at s.

Occasionally there will be elements of G which do not act at all: Let $I := \{g : g(s) = s$ for all $s \in S\}$ denote the *ineffectivity* of the G-action on S. If $g \in G$ and $s \in S$, then it follows that G_s and $G_{g(s)}$ are conjugate: $G_{g(s)} = gG_sg^{-1}$. While the isotropy may vary along a G-orbit, the ineffectivity is constant; in fact it is a normal subgroup.

1.2 Set-theoretic quotients

Let G act on a set S and define an equivalence relation on S by $s \sim t$ if and only if there exists $g \in G$ with $g(s) = t$. Thus s and t are equivalent if and only if $G.s = G.t$, i.e., the equivalence classes are the G-orbits.

Notation. The natural quotient $X \rightarrow X/\sim$ is denoted by $X \rightarrow X/G$ and is referred to as the (set-theoretic) quotient of X by G. The quotient space X/G is often called the orbit space.

If G is a group and $H < G$, then H acts on G by both left- and right-multiplication. For the sake of definiteness, choose right-multiplication: $H \times G \rightarrow G, (h, x) \mapsto xh^{-1}$. For $x \in G$ the H-orbit $H.x$ is often denoted by xH in order to underline which action is being discussed. Thus the orbit space G/H is the set of right-cosets.

Since left- and right-multiplication commute, the quotient $S = G/H$ is equipped with the induced action $G \times S \rightarrow S, xH \mapsto gxH$.

An action $G \times S \rightarrow S$ is called *transitive* if S/G consists of a single point, i.e., given two points, $s, t \in S$ there exists $g \in G$ so that $g(s) = t$. Of course the actions of G on itself by left- and right-multiplication are transitive. Consequently, the induced left-action of G on $S = G/H$ is likewise transitive.

Notation. If G acts transitively on a set S, then we refer to S as a G-*homogeneous space*.

If G acts on X and Y, then a map $\varphi : X \rightarrow Y$ is said to be G-*equivariant* if $\varphi \circ g = g \circ \varphi$ for all $g \in G$.

Note that if we equip S/G with the trivial G-action, then the quotient map $S \rightarrow S/G$ is equivariant.

For $s \in S$ the orbit map $O_s : G \rightarrow S, g \mapsto g(s)$, is equivariant with respect to left-multiplication of G on itself. Furthermore, it is invariant under the action of G_s on G by right-multiplication, i.e. $O_s(gh^{-1}) = O_s(g)$ for all $h \in G_s$. Thus O_s factors through the homogeneous space $G/G_s : G \times G/G_s \rightarrow G.s \hookrightarrow S$. We summarize this in the following

Proposition. *The orbit map O_s yields a canonical G-equivariant bijection $G/G_s \rightarrow G.s$.*

2. Examples of Lie groups

Although a Lie group is a differentiable manifold, it should not be forgotten that a finite or countably infinite group is a perfectly good example. These are the 0-dimensional Lie groups. Symmetry groups of regular figures occur in crystallography, chemistry, physics, ... F. Klein wrote a whole book on the icosahedral group – it seems to come up everywhere one looks.

A key point is that groups are not just abstract sets satisfying a certain minimal set of axioms. Rather, groups *act*, and understanding their actions is essential for many considerations in modern science.

Many groups occur as the stabilizer of some structure. For example, let G act on itself by conjugation, i.e., $x \mapsto gxg^{-1}$. Then the isotropy group G_x is the centralizer of the element x. If $\mathcal{S}(G)$ is the set of subgroups of G, then conjugation induces a G-action on $\mathcal{S}(G)$. In this case the isotropy at a subgroup H is its normalizer. The G-fixed points under this action are the normal subgroups.

For example, if V is an $\mathbb{R}$-vector space, then the group of structure preserving transformations, i.e., the linear isomorphisms, is denoted by $Gl_\mathbb{R}(V)$. For the special vector space $\mathbb{R}^n$ this is abbreviated by $Gl_n(\mathbb{R})$. The latter group, equipped with the manifold structure defined by the open embedding $Gl_n(\mathbb{R}) \hookrightarrow Mat(n \times n, \mathbb{R})$ is clearly a Lie group.

A basis of V yields a group theoretic isomorphism $Gl_\mathbb{R}(V) \cong Gl_n(\mathbb{R})$. In this way $Gl_\mathbb{R}(V)$ is a Lie group and its structure is independent of the basis.

Remark. If $G \times X \to X$ is an action by continuous mappings on a topological space X, then G can be equipped with the *compact-open* topology: For K compact and U open in X, sets of the form $V(K,U) := \{g \in G : g(K) \subset U\}$ form a subbasis.

For example, if V is equipped with its canonical topology as a finite-dimensional vector space and $Gl_\mathbb{R}(V)$ is equipped with the compact-open topology, then this yields a Lie group structure which is the same as that defined by the identification with $Gl_n(\mathbb{R})$.

Let (V, λ) be an oriented vector space. Then the group of orientation preserving transformations in $Gl_\mathbb{R}(V)$ is the connected component $Gl_\mathbb{R}(V)^\circ$ which contains the identity.

Recall that $det : Gl_\mathbb{R}(V) \to \mathbb{R}^*$ is naturally defined independent of an identification with $Gl_n(\mathbb{R})$ and that $Gl_\mathbb{R}(V)^\circ = \{T \in Gl_\mathbb{R}(V) : det(T) > 0\}$. Here we have an example of a group which is defined by an algebraic *inequality*.

If V is equipped with an orientation form λ, then the group of volume preserving transformations in $Gl_\mathbb{R}(V)$ does not depend on the choice of λ. It is the group

$$Sl_\mathbb{R}(V) := \{T \in Gl_\mathbb{R}(V) : det(T) = 1\}.$$

This is naturally a Lie subgroup of $Gl_\mathbb{R}(V)$.

Notation. The group $Sl_\mathbb{R}(V)$ is called the *special* linear group. In general, if G is a group of linear transformations, then SG denotes the elements $T \in G$ with $det(T) = 1$.

2.1 Orthogonal groups

If $g : V \times V \to \mathbb{R}$ is a Riemannian structure, then

$$O_\mathbb{R}(V, g) := \{T \in Gl_\mathbb{R}(V) : g(Tv, Tw) = g(v, w) \text{ for all } v, w \in V\}.$$

This is referred to as the *orthogonal group* of the Riemannian structure g. The elements $T \in O_\mathbb{R}(V, g)$ are called g-isometries. Note that in order for $T \in Gl_\mathbb{R}(V)$ to be an isometry it is sufficient that $\|Tv\|_g^2 = \|v\|_g^2$ for all $v \in V$.

The *special orthogonal group* consisting of the volume preserving elements in $O_\mathbb{R}(V, g)$ is denoted by $SO_\mathbb{R}(V, g)$. This is perhaps a slightly confusing point, because it might seem impossible for an isometry not to be volume preserving. However, reflections are isometries which do not preserve volume, because they are not orientation preserving. In fact $SO_\mathbb{R}(V, g)$ is the connected component $O_\mathbb{R}(V, g)°$.

Let us consider the vector space $\mathbb{R}^n$ equipped with the standard Riemannian structure given by the scalar product, i.e., if v has coordinates $x = (x_1, \ldots, x_n)$ and w has coordinates $y = (y_1, \ldots, y_n)$, then

$$g(v, w) := {}^t x \cdot y = \sum x_j y_j$$

The orthogonal group is given by

$$O_n(\mathbb{R}) = \{A \in Mat(n \times n, \mathbb{R}) : {}^t A \cdot A = Id\}.$$

Consider the vector space $M = Mat(n \times n, \mathbb{R})$ equipped with the norm $\|A\|^2 = \sum a_{ij}^2$ and observe that

$$\|A\|^2 = tr({}^t A \cdot A).$$

Thus $O_n(\mathbb{R})$ is a closed submanifold of the sphere of radius $\sqrt{n}$ in M. In particular, we have the following

Proposition. *The orthogonal group $O_\mathbb{R}(V, g)$ and the special orthogonal group $SO_\mathbb{R}(V, g)$ are compact Lie groups.*

2.2 Complex linear transformations

If (V, J) is a complex vector space, then

$$Gl(V, J) := \{T \in Gl_\mathbb{R}(V) : T \circ J = J \circ T\}.$$

If V is identified with $V^{1,0}$ by $v \mapsto v^\mathbb{C} = \frac{1}{2}(v - iJv)$, then $Gl(V, J)$ is just set of $\mathbb{R}$-linear isomorphisms T which satisfy $T(iv) = iT(v)$, i.e. T is complex linear. Since the condition $T \circ J = J \circ T$ is even linear, it is clear that $Gl(V, J)$ is a Lie group.

Remark. In fact $Gl(V, J)$ is a basic example of a complex Lie group, because it has compatible *complex analytic* structure which can be defined, e.g., by its realization as $Gl_n(\mathbb{C})$, which in turn is an open subset of $Mat(n \times n, \mathbb{C})$. We shall return to this point after introducing the basic notions for *complex manifolds*.

2.3 Unitary groups

Let $h : V \times V \to \mathbb{C}$ be a Hermitian structure on a complex vector space (V, J) and define the *associated unitary group* to be the isometries of this structure which are also compatible with J.

$$U(V, h) := \{T \in Gl(V, J) : h(Tv, Tw) = h(v, w) \text{ for all } v, w \in V\}.$$

The *special unitary group* $SU(V, h)$ is denoted in the usual way.

Note that $S^1 = \{z \in \mathbb{C} : |z|^2 = 1\}$ is naturally contained in $U(V, h)$, because, for $z \in S^1$,

$$g(zv, zw) = |z|^2 g(v, w) = g(v, w).$$

In fact, this group of transformations which acts by scalar multiplication is exactly the *center* of $U(V, h)$.

Just as in the case of orthogonal groups, we consider a unitary basis, i.e., a basis $\{e_1, \ldots, e_n, Je_1, \ldots, Je_n\}$ for V with $h(e_i, e_j) = \delta_{ij}$. Via the identification with $V^{1,0}$ this realizes

$$U(V, h) := U_n = \{A \in Gl_n(\mathbb{C}) : {}^t A . \bar{A} = Id\}.$$

The standard Hermitian norm on $M := Mat(n \times n, \mathbb{C})$ is given by

$$\|A\|^2 = tr({}^t A \bar{A}) = \sum_{i,j} |a_{ij}|^2.$$

Thus we realize $U(V, h)$ as a compact submanifold of a sphere in M.

Warning. Although it naturally lives in a the complex vector space $End(V, J)$, the unitary group $U(V, h)$ is not a complex submanifold.

Exercise. Show that both U_n and SU_n are connected.

2.4 Symplectic groups

Let (V, ω) be a symplectic vector space and define

$$Sp(V, \omega) := \{T \in Gl_{\mathbb{R}}(V) : \omega(Tv, Tw) = \omega(v, w) \; \forall v, w \in V\}.$$

For $v, w \in V$ we regard $\omega(Tv, Tw) - \omega(v, w) = 0$ as a (quadratic) defining equation. Thus the *symplectic group* is a closed Lie subgroup of $Gl_{\mathbb{R}}(V)$.

As in Chapter I., let $V = V_1 \oplus \cdots \oplus V_n$ be a decomposition of V into a sum of hyperbolic planes and let $\{e_i, f_i\}$ be a basis of V_i with $\omega(e_i, f_i) = 1$. This gives us an ordered basis $\{e_1, \ldots, e_n, e_{n+1}, \ldots, e_{2n}\}$, where $e_{n+i} := f_i$. A basis of this type, i.e., $\omega(e_i, e_{n+i}) = 1$ and $\omega(e_i, e_j) = 0$ for $i < j \neq n + i$, will be called a *symplectic basis* for (V, ω).

If $\{e_i, \ldots, e_n, e_{n+1}, \ldots e_{2n}\}$ is a symplectic basis and $J \in Gl_{\mathbb{R}}(V)$ is defined by $J(e_i) := e_{n+i}$ and $J(e_{n+i}) := -e_i$, $i = 1, \ldots, n$, then J is an ω-compatible complex structure. Let $h = g - i\omega$ be the associated Hermetian structure.

Proposition. *If J is an ω-compatible complex structure with associated Hermitian structure h, then the associated unitary group $U(V, J, h) =: K$ is a maximal compact subgroup of $Sp(V, J)$.*

Before proceeding with the proof, it is necessary to make some remarks on *maximal compact* subgroups of a Lie group.

As the terminology indicates, a maximal compact subgroup of G is a compact subgroup K which is not contained in a larger compact subgroup. The following is a basic classical result.

Theorem. *If G has finitely many connected components, then it has a maximal compact group K and any two maximal compact groups are conjugate. Furthermore, there exists a connected, closed, solvable subgroup $S < G$ so that the map $K \times S \to G, (k,s) \mapsto k \cdot s$, is a diffeomorphism. The subgroup S is diffeomorphic to some $\mathbb{R}^n$.*

Remark. In particular this shows that G is homotopically equivalent to a maximal compact subgroup K.

In the next section we shall show that an n-dimensional compact group possesses a unique non-vanishing form $\lambda \in \mathcal{E}^n(K)$ which is left- and right-invariant[3] and such that the associated measure dk is a probability measure, i.e., $\int_K dk = 1$.

Proof of the Proposition. Let $h = g - i\omega$ be the $U = U(V, J, h)$-invariant Hermitian structure. Since $Sp(V, \omega)$ is defined by algebraic equations, it has only finitely many components and therefore has a maximal compact subgroup $K \supset U$.

Of course g is U-invariant, but a priori is not K-invariant. However, it is possible to apply the *averaging principle*: Define

$$\tilde{g}(v, w) := \int_{k \in K} g(k(v), k(w))dk.$$

It follows that $\tilde{g} : V \times V \to \mathbb{R}$ is a K-invariant Riemannian structure on V.

By choosing a symplectic basis, we may identify V with $\mathbb{C}^n$ and U with the unitary group U_n. Note that the orbits of U_n in $\mathbb{C}^n \backslash \{0\}$ are the spheres $\{z \in \mathbb{C}^n : \|z\|_{std}^2 = r^2\}$. In particular these orbits are 1-codimensional.

If $K \times M \to M$ is a smooth action of a compact group, then the K-orbits in M are compact submanifolds. Note that the connected component of the identity K^0 contains U_n. Since the U_n orbits in $V \backslash \{0\}$ are connected real hypersurfaces and the K^0-orbits cannot be open, it follows that the K^0-orbits and the U_n-orbits in $V \backslash \{0\}$ are the same. In particular the level sets of the norm functions $\| \cdot \|_{\tilde{g}}^2$ and $\| \cdot \|_{g_{std}}$ are the same and consequently there exists $c > 0$ so that $\tilde{g} = c g_{std}$.

Therefore g_{std} is K-invariant and as a result

$$K \subset Sp(\mathbb{R}^{2n}, \omega_{std}) \cap O_{\mathbb{R}}(\mathbb{R}^{2n}, g_{std}) = U_n. \qquad \square$$

Corollary. *The symplectic group $Sp(V, \omega)$ is connected.*

---

[3] For $g \in G$ let $\ell(g)$, resp. $r(g)$, be the maps defined by left-, resp. right-, multiplication. A form λ is left-, resp. right-, invariant if $\ell(g)^*(\lambda) = \lambda$, resp. $r(g)^*(\lambda) = \lambda$, for all $g \in G$.

Proof. By the above Theorem $Sp(V, \omega)$ is a product of a cell and a maximal compact subgroup K. But $K \cong U_n$ is connected. $\qquad\qquad\qquad\qquad\qquad\qquad\qquad\qquad\square$

3. Smooth actions of Lie groups

An important example of a Lie group is the additive group of real numbers, $G = (\mathbb{R}, +)$. If $\mathbb{R} \times M \to M$ is a smooth action, $(t, x) \mapsto t(x)$, and $f \in \mathcal{E}(M)$, then we denote

$$X(f)(x) := \frac{d}{dt}\Big|_{t=0} f(t(x)).$$

It follows that $X \in Vect(M)$.

Conversely, if $X \in Vect(M)$ is a vector field on a manifold M, then there is a *local* $\mathbb{R}$-action on M which defines X in the above way. A local action is a map $\alpha : U \to M$ from an open neighborhood U of $\{0\} \times M$ in $\mathbb{R} \times M$ such that $\alpha(0, x) = x$ for all $x \in M$ and

$$(*) \qquad\qquad\qquad \alpha(t, \alpha(s, x)) = \alpha(t + s, x)$$

wherever it makes sense. Note that if $\alpha : U \to M$ is a local action such that $U \cap (\mathbb{R} \times \{x\})$ contains a fixed interval $I = (-\epsilon$ ` for all x, then α can be extended to a global action $\alpha : \mathbb{R} \times M \to M$.

Let G be a Lie group acting on itself by left-multiplication, i.e., $g(x) = \ell(g)(x) = g \cdot x$. In a group there is one distinguished point, namely the identity $e \in G$, and we consider its tangent space T_e. The *left-action* translates this to other tangent spaces: $\ell(g)_* : T_e \to T_g$. For $v \in T_e$ define $X_v(g) := \ell(g)_*(v)$. Thus $X_v \in Vect(G)$ is a left-invariant vector field: $\ell(g)_*(X_v(h)) = X_v(gh)$. The analogous construction can be made for the right action.

Remark. If X is a left-invariant vector field and $g \in G$, then $r(g)_*(X)$ is also left-invariant.

Let $\mathfrak{g}$ (or $Lie(G)$) denote the vector space of left-invariant vector fields G. Since such a vector field is determined by its value at a point, evaluation at the identity establishes an isomorphism $\mathfrak{g} \cong T_e G$. If $X, Y \in \mathfrak{g}$, then the Lie bracket $[X, Y] = XY - YX$ is likewise left-invariant. Thus $(\mathfrak{g}, [,])$ is a Lie subalgebra of $Vect(G)$. It is referred to as the Lie algebra of the group G.

A vector field $X \in \mathfrak{g}$ is *globally* integrable on G, because the neighborhood $U \subset \mathbb{R} \times G$ for the associated action can be chosen to be G-invariant. Thus, if $I \times \{g_0\} \subset U$ for some $g_0 \in G$, then $I \times \{g\} \subset U$ for every g. For $X \in \mathfrak{g}$ let $\alpha_X : \mathbb{R} \times G \to G$ be the associated $\mathbb{R}$-action.

Exercise. Show that the orbit through the identity $\alpha_X(t, e), t \in \mathbb{R}$, is a subgroup, and show that there is $1-1$ correspondence between connected 1-dimensional subgroups of G and the elements of $\mathfrak{g}$.

For brevity we write $\alpha_X(t) := \alpha_X(t, e)$ and observe that the vector field X is defined by the right-action $g \mapsto g \cdot \alpha_X(t)$, i.e.,

$$X(f)(g) = \frac{d}{dt}\Big|_{t=0} f(g \cdot \alpha_X(t)).$$

Examples. (1) The map $\alpha : \mathbb{R} \to Sl_2(\mathbb{R}), t \mapsto \begin{pmatrix} \cos t & \sin t \\ -\sin t & \cos t \end{pmatrix}$, defines a compact subgroup isomorphic to S^1. In fact K is a maximal compact subgroup of $Sl_2(\mathbb{R})$.

(2) The map $\alpha : \mathbb{R} \to Sl_2(\mathbb{R}), t \mapsto \begin{pmatrix} 1 & t \\ 0 & 1 \end{pmatrix}$, defines a 1-parameter subgroup. Here α is an isomorphism onto its image.

(3) Let $G = S^1 \times S^1 = \{(z, w) \in \mathbb{C}^* \times \mathbb{C}^* : |z| = |w| = 1\}$. Here we regard G with multiplication group structure

$$(z_1, w_1) \cdot (z_2, w_2) = (z_1 z_2, w_1 w_2).$$

The map $\alpha_\lambda : \mathbb{R} \to S^1 \times S^1, t \mapsto (e^{it}, e^{i\lambda t})$ defines a 1-parameter subgroup for all $\lambda \in \mathbb{R}$. If $\lambda \in \mathbb{R}\backslash\mathbb{Q}$, then α_λ is an isomorphism onto its image. However, unlike the situation in (2) or the case where $\lambda \in \mathbb{Q}$, this image is dense in G.

Remark. If G_1 and G_2 are Lie groups and their connected components of the identity are isomorphic [4], $\varphi : G_1^0 \to G_2^0$, then $\varphi_* : \mathfrak{g}_1 \to \mathfrak{g}_2$ is an isomorpism of Lie algebras. Furthermore, if G_1 is a Lie group with a discrete normal subgroup $\Gamma < G_1$ and $G_2 = G_1/\Gamma$ is equipped with the natural quotient structure, then $\mathfrak{g}_1$ and $\mathfrak{g}_2$ are canonically isomorphic.

One point above is that the Lie algebra $\mathfrak{g}$ does not determine G uniquely. However, the two types of phenomena indicated in the above remark are the only things that can go wrong.

Proposition. *Let G_j, $j = 1, 2$, be simply-connected, connected Lie groups. Then an isomorphism $\varphi_* : \mathfrak{g}_1 \to \mathfrak{g}_2$ of their Lie algebras induces a unique isomorphism $\varphi : G_1 \to G_2$.*

There is in fact an existence result.

Theorem. *Given a Lie algebra $\mathfrak{g}$ there exists a unique connected, simply-connected Lie group G with $Lie(G) \cong \mathfrak{g}$.*

These foundational results can be found in any basic text on Lie groups.

Remark. Let us recall that in these preparations our manifolds and Lie groups are always finite dimensional.

A manifold is said to be simply-connected if its fundamental group $\pi_1(M)$ vanishes. Since the universal cover of a Lie group G is a Lie group $\tilde{G}$ with universal covering map $\pi : \tilde{G} \to G$ being a Lie group homomorphism, the notion of *simply-connected* for Lie groups can be formulated in another way: A connected Lie group G is

---

[4] An isomorphism of Lie groups is a diffeomorphism $\varphi : G_1 \to G_2$ which is a group theoretic isomorphism. It is automatically $\mathbb{R}$-analytic.

simply-connected if and only if for any connected Lie group $\tilde{G}$ with discrete normal subgroup $\Gamma < \tilde{G}$ such that $\tilde{G}/\Gamma \cong G$ it follows that $\Gamma = \{e\}$.

Now let $G \times M \to M$ be an abitrary smooth action. For $\xi \in \mathfrak{g}$ with associated 1-parameter group $g_\xi(t)$ we define the vector field $X_\xi \in Vect(M)$ by $X_\xi(f)(x) := \frac{d}{dt}|_{t=0} f(g_\xi(t)(x))$. The map $\mathfrak{g} \to Vect(M), \xi \to X_\xi$, is a Lie algebra homomorphism up to a sign.

Example. Let $G = \mathbb{R}$ act on $\mathbb{R}^2$ by $(x, y) \mapsto (e^t x, e^{-t} y)$. Then the vector field $\xi = \frac{\partial}{\partial t} \in \mathfrak{g}$ corresponds to X_ξ which is calculated as follows:

$$X_\xi(f)(x, y) = \frac{d}{dt}\Big|_{t=0} f(x + tx + o(t), y - ty + o(t)) = x\frac{\partial f}{\partial x} - y\frac{\partial f}{\partial y}.$$

Thus $X_\xi = x\frac{\partial}{\partial x} - y\frac{\partial}{\partial y}$.

3.1 Representations

A particularly important way for a Lie group to act is via a *representation*. A real, resp. complex, representation of a group G is a homomorphism

$$\rho : G \to Gl_{\mathbb{R}}(V), \quad \text{resp.} \ Gl_{\mathbb{C}}(V).$$

Here we shall at first be most interested in representations on finite-dimensional vector spaces where $Gl_{\mathbb{R}}(V)$ is itself a Lie group and we require that ρ be smooth, i.e., a Lie group homomorphism. Thus a representation is equivalent to a smooth action

$$G \times V \to V, \quad v \mapsto \rho(g)(v),$$

by linear transformations.

Given an action of a group on, e.g., a manifold M we have the induced representations on the linear spaces which are associated to M, e.g., spaces of functions, vector fields, differential forms, cohomology, etc. These spaces are quite often infinite dimensional, but due to the special nature of the action, certain finite dimensional subspaces may be stabilized. Here we consider several important examples of this.

Let $G = Gl_{\mathbb{C}}(V)$ act in the natural way on V by complex linear transformations. We have the induced *dual action* $G \times V^* \to V^*, (T, f) \mapsto f \circ T^{-1}$. This is a particular case of a finite-dimensional space being stabilized due to special properties of an action: If $G \times S \to S$ is a group action on any set, then one has the induced complex linear representation on the vector space of complex valued functions on S.

If $G \times M \to M$ is a smooth action, then we have the representation on the C^∞-functions $\mathcal{E}(M)$. For example, for the standard action of $Gl_{\mathbb{C}}(V)$ on V we have the induced action on $\mathcal{E}(V)$.

Now V has algebraic structure, e.g., we have the complex polynomials $\mathbb{C}[V]$. Concretely, if $V = \mathbb{C}^n$ with standard coordinates $z = (z_1, \ldots, z_n)$, then a polynomial $P \in \mathbb{C}[V]$ is of the form $P = \sum_{|I|=0}^{k} a_I z^I, a_I \in \mathbb{C}$. In this case $deg\, P = k$.

A *homogeneous* polynomial P of degee k is of the form $P = \sum_{|I|=k} a_I z^I$. The space
of homogeneous polynomials of degree k is $\mathbb{C}[V]_k := S^k(V^*)$, where the latter
denotes the subspace of symmetric tensors in $\otimes^k V^*$. Thus the space of polynomials
naturally decomposes,

$$\mathbb{C}[V] = \bigoplus_{k \geq 0} S^k(V^*),$$

into a direct sum of finite-dimensional subspaces each of which is stabilized by
the induced G-action on $\mathbb{C}[V]$. Hence we obtain an infinite sequence of associated
finite-dimensional representations.

Even the simplest of situations may have surprising importance. For example, take
the standard representation of $Gl_\mathbb{C}(V)$ and consider the induced representation on
$End(V) = V \otimes V^*$, given by $(g, x) \mapsto gxg^{-1}$.

Thus $W := End(V)$ is itself a vector space equipped with the induced $Gl_\mathbb{C}(V)$-
representation. This in turn yields a representation on $\mathbb{C}[W]$.

Notation. If a group G acts on a set S, then S^G denotes the set of fixed points,
i.e., $S^G = \{s : g(s) = s \ \forall g \in G\}$.

The set of fixed points V^G of a linear representation on V is a subspace of V. This
is certainly the simplest possible piece of a representation.

In the above case, the vector space $\mathbb{C}[W]^G$ of fixed points is a subalgebra of $\mathbb{C}[W]$.
The functions in $\mathbb{C}[W]^G$ are just the invariants of conjugation: $P(x) = P(gxg^{-1})$.
The facts that $\mathbb{C}[W]^G$ is a finitely generated algebra and that generators can be
chosen in concrete terms, e.g., if $dim_\mathbb{C} V = n$, then

$$\mathbb{C}[W]^G = \mathbb{C}[P_1, \ldots, P_n], \text{ where } P_k(x) = tr(\Lambda^k x),$$

are fundamental for many areas of mathematics.

If G is a Lie group, then we have the natural induced representations $G \times Vect(G) \to$
$Vect(G), (g, X) \to \ell(g)_*(X)$ and $(g, X) \to r(g^{-1})_*(X)$, given by left- and right-
multiplication.

For example, the vector space $Vect(G)^G$ of invariants of left-multiplication is the
Lie algebra $\mathfrak{g}$ itself. Now the action on $Vect(G)$ given by right-multiplication stabilizes $\mathfrak{g}$ and yields the fundamental adjoint representation

$$Ad : G \to Gl_\mathbb{R}(\mathfrak{g}), \ g \mapsto r(g^{-1})_*.$$

Exercise. Let G act on itself by conjugation, i.e., if $M := G$, define $G \times M \to M$
by $(gx) \mapsto gxg^{-1} =: int(g)(x)$. Note that $e \in M^G$ and consider the induced
representation

$$\rho : G \to Gl_\mathbb{R}(T_e M), \quad g \mapsto (int(g)_*)(e).$$

Show that $Ad(g) = (int(g)_*)(e)$.

Let $(\mathfrak{g}, [,])$ be a finite-dimensional Lie algebra. An endomorphism $T \in End(\mathfrak{g})$ is said to be a derivation if it satisfies Leibnitz' rule: $T([f, g]) = [Tf, g] + [f, Tg]$. The vector space of all derivations is denoted by $Der(\mathfrak{g})$.

For example, if $X \in \mathfrak{g}$, then the *Jacobi identity* is exactly the statement that $ad(x) := [X, \cdot]$ is a derivation.

Now if $\varphi : G_1 \rightarrow G_2$ is any Lie group homomorphism, then the linear map $\varphi_* : \mathfrak{g}_1 \rightarrow \mathfrak{g}_2$ is a Lie algebra homorphism, i.e., $\varphi_*([v, w]) = [\varphi_*(v), \varphi_*(w)]$.

Exercise. Show that for the adjoint representation $Ad : \mathfrak{g} \rightarrow Gl_{\mathbb{R}}(\mathfrak{g})$, the derivative $\varphi_* : \mathfrak{g} \rightarrow Lie(Gl_{\mathbb{R}}(\mathfrak{g}) = End(\mathfrak{g})$ is given by $\varphi_* = ad$, i.e., $Ad_* = ad$.

Remark. The identification $Lie(Gl_{\mathbb{R}}(V)) = End(V)$ is derived by noticing that $Gl_{\mathbb{R}}(V)$ is naturally realized as an open subset of $End(V)$:

$$Lie(Gl_{\mathbb{R}}(V)) = T_e(Gl_{\mathbb{R}}(V)) = T_e(End(V)) = End(V).$$

If $G \times M \rightarrow M$ is a smooth action, then we have the associated representations on the spaces $\mathcal{E}^k(M)$ of differential forms. Again, certain distinguished finite-dimensional subspaces may be stabilized.

For example, if $M := G$ and $dim_{\mathbb{R}} G = n$, then we may consider the space $\mathfrak{g}^*$ of left-invariant 1-forms in $\mathcal{E}^1(M)$ as well as the higher order invariant forms $\Lambda^k \mathfrak{g}^* \subset \mathcal{E}^k(M)$. Thus, inside the full tensor algebra $\oplus_k \otimes^k \mathfrak{g}^*$ we have a G-stable finite-dimenstional vector space $\oplus_{k=0}^n \Lambda^k \mathfrak{g}^*$.

The dual representation

$$Ad^* : G \rightarrow Gl_{\mathbb{R}}(\mathfrak{g}^*), \quad g(f) := f \circ Ad(g^{-1}),$$

which is called the *coadjoint* representation, defines a representation on the exterior algebra $\Lambda \mathfrak{g}^*$.

If $\rho : G \rightarrow Gl(V)$ is a representation with an *invariant* non-degenerate bilinear pairing $(,)$, i.e., $(g(v), g(w)) = (v, w)$, then ρ and its dual representaion ρ^* are equivalent.[5]

Proposition. *If K is a compact group, then Ad and Ad^* are isomorphic representations.*

Proof. By averaging an arbitary Riemannian structure on $\mathfrak{g}$ we obtain an invariant Riemannian structure.   ☐

In general Ad and Ad^* are not isomorphic. For example, this is the case for the 3-dimensional Heisenberg group $G = \left\{ \begin{pmatrix} 1 & a & b \\ 0 & 1 & c \\ 0 & 0 & 1 \end{pmatrix} : a, b, c \in \mathbb{R} \right\}$.

Exercise. Show that in the case of the Heisenberg group $\mathfrak{g}^G$ is 1-dimensional, whereas $(\mathfrak{g}^*)^G$ is 2-dimensional.

---

[5] Representations of G on vector spaces V and W are said to be equivalent whenever there is an equivariant isomorphism $V \rightarrow W$.

3.2 Proper actions, principal bundles and homogeneous spaces

If $G \times X \to G$ is a continuous action of a topological group, then the set-theorietic quotient X/G can be equipped with the *quotient topology*. Thus, if $\pi : X \to X/G$ is the canonical projection, a set $U \subset X/G$ is open if and only if its pre-image $\pi^{-1}(U)$ is open. The map π is continuous by definition.

Note that the *closed points* in X/G correspond to the closed G-orbits in X.

Exercise. Construct an example of a smooth G-action $G \times M \to M$ with all orbits closed, but where the quotient M/G is not Hausdorff. (Hint: Consider the $\mathbb{R}$-action on $\mathbb{R}^2 \backslash \{(0,0)\}$ defined by $(x,y) \mapsto (e^t x, e^{-t} y)$.)

Exercise. Discuss the invariant continuous functions on X in the context of the quotient $\pi : X \to X/G$.

A continuous action $G \times X \to X$ is called *proper* if the map $G \times X \to X \times X, (g, x) \mapsto (g(x), x)$, is proper. [6] If we are in a situation where it is sufficient to only consider sequences, this means the following: Let $\{x_n\} \subset X$ and $\{g_n\} \subset G$ be sequences such that $x_n \to x$ and $g_n(x_n) \to y$. Then there exists a convergent subsequence $g_{n_k} \to g$ with $g(x) = y$.

Suppose $G \times X \to X$ is a proper action on a Hausdorff space X such that each point in X has a countable local neighborhood basis. Let $x, y \in X$ have neighborhoods $V_n = V_n(x)$ and $W_n = W_n(y)$ with $\bigcap V_n = \{x\}$ and $\bigcap W_n = \{y\}$. Take $\tilde{V}_n := G.V_n$ and $\tilde{W}_n := G.W_n$ to be the corresponding neighborhoods in the orbit space. If $\tilde{V}_n \cap \tilde{W}_n \neq \emptyset$ for all n, then ther exists $x_n \to x$ and $g_n \in G$ with $g_n(x_n) \to y$. The properness of the G-action then implies that $y \in Gx$. This proves the following

Proposition. *If $G \times X \to X$ is a proper action on a Hausdorff space, then the orbit space X/G is Hausdorff.*

Exercise. The above proposition was proved under the additional condition that the topology is locally countable. Is this necessary?

Now consider a proper smooth action $G \times M \to M$. For $x \in M$ observe that the isotropy group $K := G_x$ is compact. Let $\rho : K \to Gl_{\mathbb{R}}(T_x M)$ be the natural representation, $\rho(k) := k_*$.

Notation. A representation $\rho : G \to GL(V)$ is called *faithful* if $Ker(\rho) = \{e\}$.

Proposition. (Linearization of compact isotropy). *Let $K \times M \to M$ be a smooth action of a compact Lie group on a connected manifold M and let $x \in M^K$ be a fixed point. Let $\rho : K \to GL(T_x M)$ be the natural representation. Then*

$$Ker(\rho) = \{k \in K : k(m) = m \quad \forall m \in M\}.$$

Proof. Let g be a Riemannian metric on M [7]. Using the averaging procedure we may assume that g is K-invariant. Suppose $k \in K$ acts trivially on $T_x M$. Then

---

[6] A mapping is proper if and only if the preimage of every compact set is compact.
[7] See the next section for the precise definition.

k stabilizes the local geodesics eminating from x and acts trivially on each such geodesic. Therefore an open neighborhood of x is pointwise fixed.

Let U be a maximal open neighborhood of x which is pointwise fixed by K and let $y \in \partial U$. Of course $y \in M^K$ and by continuity it follows that k acts trivially on $T_y M$. Thus, as above, it acts trivially in a neighborhood of y and consequently $\partial U = \emptyset$. $\qquad\square$

Information from the linearized isotropy action can be transported to M by the exponential map of an invariant metric: Suppose that K fixes x and g is an invariant Riemannian metric as above. Choose a K-invariant neighborhood $U = U(0) \subset T_x M$ so that

$$exp : U \to V \subset M$$

is a diffeomorphism onto its image $V = V(x)$. Clearly $k(exp(tv)) = exp(k_*(tv))$, i.e., exp. is K-equivariant.

Now, let $G \times M \to M$ be a proper smooth G-action and, for $x \in M, K := G_x$. Let $T_x M = T_x(G.x) \oplus N$ be an orthogonal splitting with respect to a K-invariant metric and define the *slice* $S := exp(N \cap U)$. Note, that S is a K-stable smooth submanifold of V which is *transversal* to the orbit $G.x$ at x.

Consider the mapping $E : G \times S \to M$, $(g, s) \mapsto g(s)$. Of course at the point (e, x) it has maximal rank and thus, by shrinking U if necessary, we may assume that E is an open map onto a G-invariant open neighborhood W of $G.x$. Observe that E is *invariant* with respect to the *free* K-action on $G \times S$ which is defined by $(g, s) \mapsto (gk^{-1}, k(s))$. Furthermore, E is G-equivariant with respect to its action on the first factor.

Proposition. *If U is chosen small enough, $E : G \times S \to W \subset M$ is an open mapping of maximal rank whose fibers are exactly the K-orbits in $G \times S$.*

Proof. Let Σ be a submanifold of a small neighborhood of $e \in G$ which contains e, which is transversal to K at e and such that $\Sigma \times K \to G$, $(\sigma, k) \mapsto \sigma \cdot k$, is a diffeomorphism onto an open neighborhood of K in G. Note that it follows that $\Sigma \to G.x$, $\sigma \mapsto \sigma(x)$, is a diffeomorphism onto an open neighborhood of x in the orbit $G.x$. Therefore, if S and Σ are chosen small enough, we may assume that $\Sigma \times S \to M$, $(\sigma, s) \mapsto \sigma(s)$, is a diffeomorphism onto an open neighborhood of x in M.

To prove the desired result it is enough to show that if S is chosen small enough, then $g_1(s_1) = g_2(s_2)$ implies that $g_1^{-1} g_2 \in K$. Suppose that this is not the case. Then there exists a sequence $s_n \in S$ with $s_n \to x$ and $g_n(s_n) = t_n \to x$ with $g_n \notin \Sigma.K \subset G$. But, by the properness of the action, we may assume that $g_n \to g$. This is a contradiction, because $g(x) = x$ whereas $g_n \notin \Sigma.K$ for all n. $\qquad\square$

Since K acts diagonally on $G \times S$, the quotient mapping $G \times S \to (G \times S)/K$ is denoted by $G \times S \to G \times_K S$. The above proposition states that the map E realizes $G \times_K S$ G-equivariantly as a G-invariant neighborhood of the orbit $G.x$.

As is the case with the diagonal K-action above, let $G \times M \to M$ be a *free* proper G-action and consider the quotient $\pi : M \to M/G$. For $x \in M$ let S be a slice at x for the orbit $G.x$ as in the proposition. Since $K = \{e\}$, in this case the map $E : G \times S \to M$, $(g, s) \mapsto g(s)$, is a diffeomorphism onto a G-invariant open neighborhood W_S. In particular $U_S := \pi(S)$ is an open neighborhood of $\pi(s)$ and $\pi : S \to U_S$ is a homeomorphism.

Now, no matter how it was constructed in the first place, S can be chosen small enough, so that it is an open set in some coordinate space $\mathbb{R}^k$, where $k := codim_\mathbb{R} G \cdot x$, and via π we may equip U_S with these coordinates.

Suppose that $\tilde{S}$ is a k-dimensional submanifold of $G \times S$ which is transveral to the G-orbits and such that the quotient map $G \times S \to \tilde{S}$ induces a homeomorphism. Then this homeomorphism is a diffeomorphism. This shows that if we cover the quotient M/G with coordinate charts of the type U_S, then the change of coordinate homeomorphisms are in fact diffeomorphisms.

Proposition. *If $G \times M \to M$ is a smooth G-action which is proper and free, then the quotient M/G has a unique structure of a differentiable manifold such that $\pi : M \to M/G$ is a differentiable map which has rank $codim_\mathbb{R} G.x = dim_\mathbb{R} M/G$ at every $x \in M$.*

In the situation of the above Proposition the quotient $\pi : M \to M/G$ is called a smooth *principal bundle*. The following is an equivalent defintion:

Let G be a Lie group, $\pi : E \to B$ an everywhere maximal rank surjective map between differentiable manifolds. Suppose that for every $b \in B$ there exists an open neighborhood $V := V(b)$ such that $\pi^{-1}(V) =: U$ is G-equivariantly diffeomorphic to the product $V \times G$ so that the projection on the first factor yields the restriction $\pi|U$.

If V_α and V_β are two such open sets in B and $V_{\alpha\beta} := V_\alpha \cap V_\beta \neq \emptyset$, then the induced coordinate change $V_{\alpha\beta} \times G \to V_{\alpha\beta} \times G$ is given by

$$(b, g) \mapsto (b, g_{\alpha\beta}(b) \cdot g)$$

where $b \mapsto g_{\alpha\beta}(b)$ defines a differentiable map $V_{\alpha\beta} \to G$. The bundle $\pi : E \to B$ is said to be *trivial over the open sets* V_α and the diffeomorphisms $\pi^{-1}(V_\alpha) \cong V_\alpha \times G$ are said to be *trivializations*.

Exercise. Let $\pi : E \to B$ be a principal G-bundle in the second sense above. Show that right-multiplication $g(b, h) := (b, hg^{-1})$ on the trivalizations defines a free, proper smooth action on E. Show that the two definitions of a principal bundle are the same.

We have already seen a basic important example of a principal bundle: If $G \times M \to M$ is a proper smooth action, then every orbit $G.x$ has a slice neighborhood

$W := G \times_K S$ which is the base space of a K-principal bundle

$$E : G \times S \to G \times_K S.$$

The G-action on W is induced from the G-action on the first factor in $G \times S$.

A second basic example is given by a closed Lie subgroup H of a Lie group G. In this case consider the H-action on G which is given by right-multiplication, i.e., $h(g) := gh^{-1}$. This is a free, proper smooth action. The properness of the action is equivalent to the assumption that H is a *closed* subgroup.

Thus the quotient $\pi : G \to G/H$ realizes G as an H-principal bundle space over G/H. Since the action of G on itself by left-multiplication is smooth and commutes with the H-action, the induced G-action on G/H is smooth.

Exercise. Let $\pi : M \to N$ be a surjective, everywhere maximal rank smooth mapping between differentiable manifolds. Suppose $G \times M \to M$ is a smooth G-action such that the transformations $g \in G$ map π-fibers to π-fibers. Does there exist a smooth G-action $G \times N \to N$ such that $\pi : M \to N$ is equivariant?

The *local sections* in the principal bundle $\pi : G \to G/H$ can be defined in a particularly useful way. For this recall that for $\xi \in \mathfrak{g}$ we have 1-parameter group homomorphism $t \mapsto g_\xi(t) \in G$. This is usually denoted by $t \mapsto exp(\xi t)$. In this way the homomorphism property looks familiar:

$$exp(\xi(t + s)) = exp(\xi t)exp(\xi s),$$

where the multiplication is group multiplication in G.

By fixing $t = 1$ one defines $exp : \mathfrak{g} \to G$, $\xi \mapsto exp(\xi)$. The derivative $exp_*(0) :$ $\mathfrak{g} \to T_eG$ is simply the identity and consequently there exists a neighborhood $U = U(0) \subset \mathfrak{g}$ so that $exp|U$ is a diffeomorphism.

Exercise. Recall that $Lie(Gl_n(\mathbb{R})) = Mat(n \times n, \mathbb{R})$. Show that in this case the exponential map $exp : Lie(Gl_n(\mathbb{R})) \to Gl_n(\mathbb{R})$ is given by $A \mapsto e^A = Id + A + \frac{A^2}{2!} + \ldots + \frac{A^n}{n!} + \ldots$

Using the same method of proof as that which was used for the basic slice theorem above, we have the following

Proposition. *Let H be a closed subgroup of a Lie group G and let $\mathfrak{m}$ be a subspace of $\mathfrak{g}$ such that $\mathfrak{g} = \mathfrak{h} \oplus \mathfrak{m}$. Then, if $V = V(0) \subset \mathfrak{m}$ is open and small enough, it follows that $S := exp(V)$ is a slice for the action of H on G by right-multiplication, i.e. $E : H \times S \to G$, $(h, s) \mapsto sh^{-1}$ is a diffeomorphism onto an H-invariant open neighborhood of $e \in G$.*

4. Fiber bundles

Let E and B be differentiable manifolds and $\pi : E \to B$ an everywhere maximal rank surjective differentiable map. Let F be a differentiable manifold with $\pi^{-1}(b) \cong F$ for every $b \in B$. We say that $\pi : E \to B$ is a *locally trivial bundle* with *base* B, *fiber* F and *total space* E if and only if there exists a covering $V = \{V_\alpha\}$ of the base such that for $U_\alpha := \pi^{-1}(V_\alpha)$ the restriction $\pi|U_\alpha : U_\alpha \to V_\alpha$ is a trivial product with fiber F. This means that there is a diffeomorphism $U_\alpha \cong V_\alpha \times F$ so that $\pi|U_\alpha$ is given by projection on the first factor.

As in the case of principal bundles, trivializations over two different open sets V_α and V_β are connected by a change of trivializations $(x, v) \mapsto (x, g_{\alpha\beta}(x)(v))$, where $g_{\alpha\beta} \to Diff(F)$ is a smooth map. Furthermore, the cocycle conditions

$$g_{\alpha\beta} \cdot g_{\beta\gamma} \cdot g_{\gamma\alpha} = Id_{V_{\alpha\beta\gamma}} \text{ and } g_{\alpha\beta} = g_{\beta\alpha}^{-1}$$

are satisfied.

Remark. We will not discuss here the differentiable structure on the infinite-dimensional group $Diff(F)$. We simply define $g_{\alpha\beta}$ to be smooth if

$$(x, v) \mapsto g_{\alpha\beta}(x)(v)$$

is a smooth map of finite dimensional manifolds.

Exercise. Let $\{g_{\alpha\beta}\}$ satisfy the cocycle conditions above. Construct a bundle $\pi : E \to B$ with $\{g_{\alpha\beta}\}$ as transition diffeomorphisms.

Let G be a Lie group and $G \times F \to F$ a smooth action. If the transition diffeomorphisms can be chosen in G, i.e., $g_{\alpha\beta} : V_{\alpha\beta} \to G$ is smooth and the cocycle conditions are satisfied, then we say that $\pi : E \to B$ has structure group G.

The properties of the fiber F which are invariant under the structure group are then preserved in the *family* E of fibers over B. For example, if $F = \mathbb{R}^n$ and $G = Gl_n(\mathbb{R})$, then each fiber of $\pi : E \to B$ has a well-defined vector space structure, i.e., addition of points in the fiber and scalar multiplication.

Exercise. Show that operations of addition and scalar multiplication are given by smooth maps $E \times E \to E$ and $\mathbb{R} \times E \to E$ over B.

If $\pi : E \to B$ is a fiber bundle, then a section τ in E is a differentiable map $\tau : B \to E$ such that $\pi \circ \tau = Id_B$. Note that a principal bundle has a section if and only if it is equivariantly diffeomorphic to the trivial principal bundle.

Notation. Let $\Gamma(B, E) = \Gamma_{C^\infty}(B, E)$ denote the set of smooth sections of a fiber bundle $\pi : E \to B$.

Note that $\Gamma(B, E)$ is the subset of $C^\infty(B, E)$ which is defined by the *equation* $\pi \circ \tau = Id_B$. Thus it can be equipped with an infinite-dimensional differentiable structure. If $E \to M$ is a vector bundle, then $\Gamma(M, E)$ is itself a vector space.

Example. (Tangent bundle) Let M be a manifold and consider the *set* $TM = \bigcup_{x \in M} T_x M$ and let $\pi : TM \to M$ denote the natural projection. Of course this is a vector bundle in the category of sets and vector fields are sections. It is desirable to equip TM with a smooth vector bundle structure so that $Vect(M) = \Gamma(M, TM)$. For this recall that if V_α is a coordinate chart in M with coordinates $x_\alpha := (x_\alpha^1, \ldots, x_\alpha^n)$, then

$$Vect(V_\alpha) = ((\frac{\partial}{\partial x_\alpha^1}, \ldots, \frac{\partial}{\partial x_\alpha^n}))_{\mathcal{E}(V_\alpha)}$$

So $\pi^{-1}(V_\alpha)$ is then realized as the trivial product by

$$t \mapsto a_\alpha^1 \frac{\partial}{\partial x_\alpha^1}\Big|_x + \ldots + a_\alpha^n \frac{\partial}{\partial x_\alpha^n}\Big|_x,$$

where the coefficients $(a_\alpha^1, \ldots, a_\alpha^n)$ are determined uniquely for every $t \in T_x M$. If $a_\alpha(t) = (a_\alpha^1, \ldots, a_\alpha^n) \in \mathbb{R}^n$, then we have the bijective correspondence

$$\varphi_\alpha : \pi^{-1}(V_\alpha) \to V_\alpha \times \mathbb{R}^n, \quad t \mapsto (\pi(t), a_\alpha(t)).$$

Since V_α is a coordinate chart for M, we may regard $(\varphi_\alpha, \pi^{-1}(V_\alpha))$ to be a coordinate chart for TM. It remains to check the smoothness and linearity of the change of coordinates

$$(x, v) \mapsto (x, g_{\alpha\beta}(x)(v)),$$

where $g_{\alpha\beta}(x)$ is just the basis change from

$$\left(\frac{\partial}{\partial x_\alpha^1}\Big|_x, \ldots, \frac{\partial}{\partial x_\alpha^n}\Big|_x\right) \quad \text{to} \quad \left(\frac{\partial}{\partial x_\beta^1}\Big|_x, \ldots, \frac{\partial}{\partial x_\beta^n}\Big|_x\right).$$

But this is clearly a smooth map, $g_{\alpha\beta} : V_{\alpha\beta} \to GL_n(\mathbb{R})$, and consequently the *tangent bundle* $\pi : TM \to M$ possesses smooth vector bundle structure.

Exercise. Show that $Vect(M) = \Gamma(M, TM)$. Think about this question in the case where M is infinite dimensional.

Let $\pi : E \to B$ be a fiber bundle with fiber F and structure group G, a finite dimensional Lie group. Since the transition diffeomorphisms $\{g_{\alpha\beta}\}$ satisfy the co-cycle conditions, we have the associated principal bundle $P \to B$ with structure group G, i.e., with trivializations *glued together* by $(x, g) \mapsto (x, g_{\alpha\beta}(x).g)$.

Quite often the associated principal bundle has itself an interesting interpretation.

Example. (Frame bundle) Let $E \to M$ be an $\mathbb{R}$-vector bundle with structure group $G = Gl_n(\mathbb{R})$. A *frame* e_α on an open set $V_\alpha \subset M$ is a smooth map $e_\alpha : V_\alpha \to E \oplus \ldots \oplus E$ such that $e_\alpha(x) = (e_\alpha^1(x), \ldots e_\alpha^n(x))$ is a basis for each $x \in V_\alpha$.

If e_α and e_β are frames given by local coordinates, then $e_\alpha = g_{\alpha\beta} e_\beta$, where $g_{\alpha\beta} \cdot V_{\alpha\beta} \to Gl_n(\mathbb{R})$ is the transition matrix for the tangent bundle.

Now define the frame bundle $\mathcal{F} \to M$ by gluing frames. Note that $G = Gl_n(\mathbb{R})$ acts on the right, i.e., $e_\alpha g^{-1}$, and this action realizes the frame bundle as the principal bundle associated to the tangent bundle.

Above, given a fiber bundle $\pi : E \to B$ with structure group G acting on its fiber, $G \times F \to F$, we constructed and associated principal bundle $P \to B$. This procedure can be reversed.

Let $P \to B$ be a G-principal bundle and $G \times F \to F$ a smooth action. This yields a free proper (diagonal) G-action on the product $P \times F$ and we have the quotient $(P \times F)/G$ which we denote by $P \times_G F$. Let us analyze the resulting situation:

$$
\begin{array}{ccc}
P \times F & \longrightarrow & P \times_G F \\
\pi \downarrow & & \downarrow \\
B & \overset{Id}{\longrightarrow} & B
\end{array}
$$

Suppose $P \to B$ is trivialized on a cover $\mathcal{U} = \{U_\alpha\}$. Thus $\pi^{-1}(U_\alpha) \cong U_\alpha \times (G \times F)$ and the G-action is defined by $(x, g, v) \to (x, g.h^{-1}, h(v))$. Consequently the embedding $U_\alpha \times F \to \pi(U_\alpha)$, $(x, v) \mapsto (x, e, v)$, realizes $U_\alpha \times F$ as a section of the quotient $\pi^{-1}(U_\alpha) \to \pi^{-1}(U_\alpha)/G$. Thus $P \times_G F \to B$ is a locally trivial fibration with fiber F and it remains to consider its structure group.

Let $\{g_{\alpha\beta}\}$ denote the transition diffeomorphisms for $P \to B$ which are defined by the trivialization over $U = \{U_\alpha\}$. It follows that the transition diffeomorphisms for $P \times F \to B$ are given by $(x, g, v) \mapsto (x, g_{\alpha\beta}(x) \cdot g, v)$. As a result, transition for $P \times_G F$ is given by

$$U_\alpha \times F \ni (x, v) \mapsto (x, e, v) \mapsto (x, g_{\alpha\beta}(x), v) \sim (x, e, g_{\alpha\beta}(x)(v))$$

$$\mapsto (x, g_{\alpha\beta}(x)(v)) \in U_\beta \times F.$$

This can be summarized as follows.

Proposition. *If $P \to B$ is a principal G-bundle with transition diffeomorphisms $\{g_{\alpha\beta}\}$ on a trivialization $\mathcal{U} = \{U_\alpha\}$ and $G \times F \to F$ is a smooth action, then $P \times_G F \to B$ is a smooth fiber bundle with fiber F and structure group G. The trivialization of $P \to B$ yields a trivialization of $P \times_G F \to B$ in a canonical way with transition diffeomorphisms $\{\rho(g_{\alpha\beta})\}$, where ρ is given by the smooth action. If $E \to B$ is given with transition maps $\{\rho(g_{\alpha\beta})\}$, then $P \times_G F \cong E$.*

Remark. We have purposely avoided a seemingly small point which, in certain situations, plays an important role: It is possible that the diffeomorphisms $\{\rho(g_{\alpha\beta})\}$ satisfy the cocycle condition and $\{g_{\alpha\beta}\}$ do not! This difficulty does not come up in the sequel, but the reader should be warned.

Example. (Associated Tensor-Bundles). Let $TM \to M$ be the tangent bundle of a differentiable manifold M. Thus we have the associated $Gl_n(\mathbb{R})$-principal bundle $P \to M$. For brevity let $G := Gl_n(\mathbb{R})$ and $V = \mathbb{R}^n$. For $p, q \in \mathbb{N}^+$ we have the associated representation on the full tensor algebra $\bigoplus_{p,q} (\bigotimes^p V \otimes \bigotimes^q V^*) =: T(V)$. This is of course an infinite-dimensional representation space, but the are many G-invariant finite dimentional subspaces, e.g., $\bigotimes^p V \otimes \bigotimes^q V^*, S^q(V), \Lambda^p V^*, \ldots$.

If $W \subset T(V)$ is such a space, then we have the associated vector bundle $P \times_G W \to M$, e.g., $\bigotimes^q TM \otimes \bigotimes^p T^*M, S^k(TM), \Lambda^p T^*M$, etc.

Vector bundles may come equipped with additional structures which have the effect of *reducing the structure group* to a smaller group. We close this chapter with several examples of this.

A Riemannian metric g on a vector bundle $E \to M$ is a smooth map $g : E \oplus E \to \mathbb{R}$ over M to the trivial $\mathbb{R}$-bundle $\mathbb{R} \to M$ such that $g_x : E_x \oplus E_x \to \mathbb{R}_x = \mathbb{R}$ is a Riemannian metric on the fiber E_x for all $x \in M$. One often says g *is a metric on the fibers of E which varies smoothly from fiber to fiber.* A symmetric bilinear form on a vector space V is an element of $S^2(V^*)$. Thus

$$g \in \Gamma(M, S^2(T^*M))^+ := \{g \in \Gamma(M, S^2(T^*M)) : g(v, v) > 0 \quad \forall v \in TM_x \setminus \{0\}\}.$$

Note that $\Gamma(M, S^2(T^*M))^+$ is an open cone in $\Gamma(M, S^2(T^*M))$.

If $E \to M$ is a vector bundle, then $\Gamma(M, E)$ is a module over the C^∞-functions on the base, i.e., if $f, g \in \mathcal{E}(M)$, then $f\sigma + g\tau$ is again a section for any two sections $\sigma, \tau \in \Gamma(M, E)$. For example, if $\mathcal{U} = \{U_\alpha\}$ is a covering of M and $\{\sigma_\alpha\}$ is a set of sections, then a partition of unity $\{\chi_\alpha\}$ yields a globally defined section

$$\sigma := \sum_\alpha \chi_\alpha \sigma_\alpha.$$

In this way it is possible to construct a Riemannian metric: Let $\{U_\alpha\}$ be a covering of M over which the tangent bundle $\pi : E \to M$ is trivial. Over each U_α let $e_\alpha = (e_\alpha^1, \dots, e_\alpha^n)$ be a frame and define g_α by $g_\alpha(e_\alpha^i, e_\alpha^j) = \delta_\alpha^{ij}$. [8]

Remark. Conversely, given a Riemannian metric on $E \to M$, it is possible to construct frames $\{e_\alpha\}$ on a covering $\mathcal{U} = \{U_\alpha\}$ so that $g(e_\alpha^i, e_\alpha^j) = \delta^{ij}$. Such frames are called *orthonormal*.

Notation. A metric g on TM is called a Riemannian metric on M and (M, g) is called a Riemannian manifold.

Let $E \to M$ be a vector bundle which is equipped with a metric and let $\{e_\alpha\}$ be a set of orthonormal frames on a covering $\mathcal{U} = \{U_\alpha\}$. The transition maps $g_{\alpha\beta} \to Gl_n(\mathbb{R})$ are defined by $e_\alpha = g_{\alpha\beta} e_\beta$. But a linear transformation which maps an orthonormal basis to an orthonormal basis is an orthonormal matrix, i.e. $g_{\alpha\beta} : U_{\alpha\beta} \to O_n(\mathbb{R})$. In this way we observe that metric reduces the structure group to $O_n(\mathbb{R})$.

A *complex vector bundle* $E \to M$ comes equipped with a smoothly varying complex structure in its fibers. Precisely speaking, a complex structure J on E is a section $J \in \Gamma(M, End(E))$ such that $J^2 = -Id$. Here $Id \in \Gamma(M, End(E))$ is the identity section. Of course $End(E) \cong E^* \otimes E$ is a vector bundle and therefore $\Gamma(M, End(E))$ is a very big space. However, $J^2 = -Id$ is a non-linear equation which may have no solutions. For example, if $M = S^n$ is the n-dimensional sphere and $E := TM$, then E possesses a complex structure only in two cases, $n = 2, 6$. In

---

[8] The symbol δ^{ij} denotes a *function* which is either identically 1 or 0, depending on $i = j$ or $i \neq j$.

general, a complex structure on the tangent bundle TM is referred to as a *complex structure* on M.

Exercise. Show that a complex structure on E yields a reduction of the structure group to $Gl_n(\mathbb{C}) \hookrightarrow Gl_{2n}(\mathbb{R})$. Here $Gl_n(\mathbb{C})$ is embedded in $Gl_{2n}(\mathbb{R})$ as the set of matrices which commute with the standard structure

$$J_{std} = \begin{pmatrix} 0 & -Id \\ Id & 0 \end{pmatrix}.$$

If $E \to M$ is a complex vector bundle, then there is a covering $\mathcal{U} = \{U_\alpha\}$ and trivializations $\pi^{-1}(U_\alpha) =: V_\alpha \cong U_\alpha \times \mathbb{C}^n$, i.e., there exist complex valued frames $\{e_\alpha\}$. Define a Hermitian metric h_α over U_α by

$$h_\alpha(e_\alpha^i, e_\alpha^j) = \delta^{ij} \text{ and let } h := \sum_\alpha \chi_\alpha h_\alpha$$

as in the case of a Riemannian metric. Thus (E, h) is a Hermitian bundle, i.e., $h : E \oplus E \to \mathbb{C}$ is a smooth map over M which is Hermitian in the fibers.

Exercise. Show that a Hermitan metric on a complex vector bundle yields a reduction of the structure group to U_n.

If (E, h) is a Hermitian bundle, then $h = g - i\omega$ and $g := Re(h)$ is a Riemannian metric and $\omega = -Im(h)$ a non-degenerate alternating form on the fibers of E. In this case the complex structure is compatible with ω, i.e., $\omega(Ju, Jv) = \omega(u, v)$ and $\omega(w, Jw) > 0$ for all u, v and $w \neq 0$. Even if there is no complex structure at hand, it is still possible to discuss the notion of a *symplectic bundle*.

A non-degenerate[9] alternating form $\omega \in \Gamma(M, \Lambda^2 E^*)$ is called a symplectic structure on the bundle E. We refer to (E, ω) as a symplectic bundle.

Remark. In the next chapter we shall show that if (E, ω) is a symplectic bundle, then E possesses an infinite-dimensional space of complex structures which are ω-compatible.

If (E, ω) is a symplectic bundle, then on a trivializing cover $\mathcal{U} = \{U_\alpha\}$ there exist *standard symplectic frames* $\{e_\alpha\}$, i.e., $\omega(e_\alpha^i, e_\alpha^{i+n}) = -\omega(e_\alpha^{i+n}, e_\alpha^i) = 1$, $1 \leq i \leq n$, and $\omega(e_\alpha^i, e_\alpha^j) = 0$ otherwise. To produce such frames one may, for example, carry out the procedure of decomposing a symplectic vector space into a direct sum of hyperbolic planes in a smooth way depending on $x \in U_\alpha$.

If (E, ω) is a symplectic bundle with standard symplectic frames $\{e_\alpha\}$ and $\{g_{\alpha\beta}\}$ is the associated set of transition maps, i.e., $e_\alpha = g_{\alpha\beta}e_\beta$, then $g_{\alpha\beta} : U_{\alpha\beta} \to Sp_{2n}(\mathbb{R}) \subset Gl_{2n}(\mathbb{R})$, where

$$Sp_{2n}(\mathbb{R}) := \{A \in Gl_{2n}(\mathbb{R}) : {}^tA \cdot J_{std} \cdot A = J_{std}\}$$

is the (real) symplectic group. Thus we have the following

---

[9] This means that $E_x^{\perp \omega_x} = \{0\}$ for all $x \in M$.

Proposition. *A symplectic structure on a vector bundle $E \to M$ yields a reduction of the structure group to $Sp_{2n}(\mathbb{R})$.*

Remark. Above it has been shown that an additional structure on a vector bundle yields a reduction of the structure group. This has been indicated in four examples, i.e., Riemannian metrics, complex structures, Hermitian metrics and symplectic structures. These structures yield reductions of the structure groups to $O_n(\mathbb{R}), Gl_n(\mathbb{C}), U_n$ and $Sp_{2n}(\mathbb{R})$ respectively. In each case we have the associated principal bundle.

Exercise. Show that a reduction of a structure group yields an additional structure on E, e.g., a reduction to $O_n(\mathbb{R})$ yields a Riemannian metric. Define the notion of an oriented bundle. What does this mean in terms of structure group? Does such a structure always exist?

III. Manifolds with additional structure

In applications manifolds usually come equipped with additional structure. Here we are oriented toward those manifolds which are relevant for, e.g., symplectic geometry, Hamiltonian systems and quantization problems.

1. Geometric structures on vector spaces

As a first step a basic parameter space is described

1.1 Grassmannians

Let V be a finite-dimensional $\mathbb{R}$-vector space and $k \in \mathbb{N}$ with $k \le dim_\mathbb{R} V$. Define

$$Gr_k(V) := \{W : W \text{ is a subspace of } V \text{ with } dim_\mathbb{R} W = k\}.$$

This is a priori just a set, but, since our intuition tells us that subspaces can *move about in a smooth way*, we wish to equip it with a natural manifold structure. As is typical for many mathematical considerations, we first impose more structure in order to simplify matters and then we drop the additional structure.

A *parameterization* of a k-plane $W \in Gr_k(V)$ is an isomorphism $T : \mathbb{R}^k \to W$. Thus the set of parameterized k-planes is

$$P_k(V) := \{T \in Hom(\mathbb{R}^k, V) : rank(T) = k\}.$$

Now $Hom(\mathbb{R}^k, V)$ carries the differentiable vector space structure and $P_k(V)$ is an open subset. In fact $P_k(V)$ is the complement of the closed set $S_k(V)$ which is defined by the *equation* $rank(T) < k$, i.e., $Te_1 \wedge \ldots \wedge Te_k = 0$, where $\{e_1, \ldots, e_k\}$ is the standard basis in $\mathbb{R}^k$.

Of course the same k-plane can be parameterized in many different ways. The choices correspond to the action of $L := Gl_k(\mathbb{R})$ on $P_k(V)$ which is defined by $T \mapsto T \circ g^{-1}$. Thus $Gr_k(V)$ can be identified with the quotient $P_k(V)/L$.

Exercise. Show that L acts freely and properly on $P_k(V)$.

Using this result, we equip $Gr_k(V)$ with the canonical quotient structure as a differentiable manifold.

Note that $G := Gl_{\mathbb{R}}(V)$ also acts on $P_k(V)$, $T \mapsto g \circ T$, and it is clear that this G-action is smooth and commutes with the L-action. Thus we have the induced smooth G-action

$$G \times Gr_k(V) \to Gr_k(V)$$

on the Grassmannian.

Exercise. Show that this action is transitive.

Thus the Grassmann manifold $Gr_k(V)$ can be differentiably and G-equivariantly identified with the homogeneous space G/H, where H is the stabilizer of some k-plane $W_0 \subset V$.

Consider the L-principal bundle $\pi : P_k(V) \to Gr_k(V)$ and the product $P_k(V) \times \mathbb{R}^k$ equipped with the diagonal L-action. Define the *evaluation* map

$$\epsilon : P_k(V) \times \mathbb{R}^k \longrightarrow V, \qquad (T, v) \mapsto T(v).$$

We sort out these images by defining an induced map to the trivial bundle.

$$\rho : P_k(V) \times \mathbb{R}^k \longrightarrow Gr_k(V) \times V, \quad p \mapsto (\pi(p), \epsilon(p)).$$

Exercise. Show that ρ realizes the quotient $P_k(V) \times \mathbb{R}^k \to P_k(V) \times_L \mathbb{R}^k$ by the diagonal L-action $(T, v) \mapsto (T \circ g^{-1}, g(v))$.

Thus $E_k := P_k(V) \times_L \mathbb{R}^k \cong Im(\rho) \hookrightarrow Gr_k(V) \times V \to Gr_k(V)$. Viewing the vector bundle $E_k \subset Gr_k(V) \times V$, it is clear that the fiber of $E_k \to Gr_k(V)$ over a *point* $W \in Gr_k(V)$ is the vector space $W \subset V$. Thus

$$E_k = \{(W, w) \in Gr_k(V) \times V : w \in W\}.$$

It is referred to as the *tautological* vector bundle on $Gr_k(V)$.

The above can be formalized as follows:

Proposition. *The tautological bundle $E_k \to Gr_k(V)$ is the bundle $P_k(V) \times_L \mathbb{R}^k$ associated to the L-principal bundle $P_k(V) \to Gr_k(V)$ of parameterized k-planes by the standard action of $L := Gl_k(\mathbb{R})$ on $\mathbb{R}^k$.*

If the vector space V is equipped with additional structure, then there may be special types of k-planes.

Example. Let (V, J) be a complex vector space and choose k to be an even number. An element $W \in Gr_k(V)$ is said to be complex if $J(W) = W$, i.e., the k-plane W inherits the complex struture J. Now $Gl_{\mathbb{C}}(V) := \{g \in G : g \circ J = J \circ g\}$ is naturally embedded in $G = Gl_{\mathbb{R}}(V)$ and therefore also acts smoothly on $Gr_k(V)$.

Exercise. Show that if W_0 is a complex plane, then the orbit $Gl_{\mathbb{C}}(V).W_0$ consists of the set of *all* complex planes. Prove that this orbit is a *closed submanifold* of $Gr_k(V)$.

A plane $W \in Gr_k(V)$ in a complex vector space may be *partially complex*, i.e., the dimension $d_J(W)$ of $W \cap J(W)$ may be positive, and of course this structure is also preserved by the $Gl_{\mathbb{C}}(V)$-action on $Gr_k(V)$.

If $2k \leq dim_{\mathbb{R}} V$, then there exist k-planes W with $W \cap J(W) = \{0\}$ which are called *totally real*. If $W = ((w_1, \dots, w_k))_{\mathbb{R}}$ and $U = ((u_1, \dots, u_k))_{\mathbb{R}}$ are totally real planes, then we may consider their complexifications $W \oplus JW$, resp. $U \oplus JU$, as complex subspaces of V. The linear map which sends $w_i \mapsto u_i$ and $Jw_i \mapsto Ju_i$ is a complex linear isomorphism which can be extended to an element $g \in Gl_{\mathbb{C}}(V)$. Thus $Gl_{\mathbb{C}}(V)$ *acts transitively on the set of totally real subplanes.*

Exercise. If $2k > dim_{\mathbb{R}} V$, then no k-plane is totally real. What takes the place of the totally real planes in this case?

Similar considerations to those above show that the $Gl_{\mathbb{C}}(V)$-orbits are exactly the sets $O_\ell := \{W : d_J(W) = \ell\}$, i.e., the only invariant is the dimension $d(W)$ of the partial complex structure. In particular, $Gl_{\mathbb{C}}(V)$ has only finitely many orbits in $Gr_k(V)$ and only one of them is open, i.e., that corresponding to the notion discovered in the above exercise. Since the other orbits are locally closed lower-dimensional submanifolds, this orbit is dense.

To complete the picture, the following should also be noted.

Proposition. *The Grassmannian $Gr_k(V)$ is compact.*

Proof. Equip V with a Riemannian metric g which defines an orthogonal group $K := O_{\mathbb{R}}(V, g) \subset Gl_{\mathbb{R}}(V)$. Since K is compact and acts transitively on $Gr_k(V)$, it follows that $Gr_k(V)$ is compact. $\square$

Corollary. *The set of complex planes in $Gr_k(V)$ is compact.*

As a second example of a vector space with additional structure we consider a *symplectic vector space* (V, ω). Here we have the subgroup

$$Sp(V, \omega) := \{g \in Gl_{\mathbb{R}}(V) : g^*(\omega) = \omega\}$$

which preserves the invariants which are of a symplectic nature.

A subspace $i : W \hookrightarrow V$ is said to be *symplectic* if the pull-back $i^*(\omega)$ is non-degenerate. Of course if $k := dim_{\mathbb{R}} W$ is an odd number, then there are *no* symplectic subspaces in $Gr_k(V)$. If k is even, then there are symplectic subspaces.

Exercise. Show that the set of symplectic subspaces in $Gr_k(V)$ is the unique open $Sp(V, \omega)$-orbit. – Of course k is assumed to be even.

If $W \in Gr_k(V, \omega)$, then, analogous to the case of a complex vector space, we define $d_\omega(W) := dim(W \cap W^{\perp \omega})$. For example, $d_\omega(W) = 0$ means that W is a symplectic subspace. This is an *open dense condition*: Provided k is even, in every neighborhood of $W_0 \in Gr_k(V)$ there exist elements W with $d_\omega(W) = 0$ and small perturbations of symplectic subspaces are symplectic.

Just as in the case of complex vector spaces, the invariant $d_\omega(W)$ parameterizes the $Sp(V, \omega)$-orbits in $Gr_k(V)$. We suggest proving this as an exercise.

Proposition. *For $W_0 \in Gr_k(V, \omega)$, it follows that $Sp(V, \omega) \cdot W_0 = \{W \in Gr_k(V) : d_\omega(W) = d_\omega(W_0)\}$.*

As was indicated above, group actions such as the $Gl_\mathbb{C}(V)$-action on $Gr_k(V, J)$ and the $Sp(V, \omega)$-action on $Gr_k(V, \omega)$ are extremely well-behaved. This is due to the fact that these actions are *algebraic*. We will discuss algebraic actions in some detail later on, but for now the reader should begin to think as follows: If G is acting algebraically, then the topological closure of a G-orbit is formed by adding a (possibly mildly singular) set of smaller dimension which consists of certain smaller-dimensional orbits. If there are only finitely many orbits, as in cases above, then the closure is formed by adding finitely many orbits of smaller dimension.

For applications in symplectic geometry it is useful to have notation for certain special kinds of subspaces of (V, ω):

- W symplectic $:\Leftrightarrow i^*(\omega)$ non-degenerate
- W isotropic $:\Leftrightarrow i^*(\omega) = 0$
- W Lagrangian $:\Leftrightarrow W$ isotropic and maximal dimensional with this property, i.e., $2 \cdot dimW = dimV$.
- W coisotropic $:\Leftrightarrow W^{\perp\omega} \subset W$.

Note that W is Lagrangian if and only if $W^{\perp\omega} = W$. If $V = W \oplus \tilde{W}$ is a decomposition of V into Lagrangian subspaces, then $\tilde{w} \mapsto \omega(\cdot, \tilde{w})$ identifies $\tilde{W}$ with the dual space W^* and (V, ω) has the form of the canonical symplectic vector space $(W \oplus W^*, \omega_{std})$.

1.2 Homogeneous spaces of geometric structures

Let V be a $\mathbb{R}$-vector space and consider the set $\mathcal{R}(V)$ of symmetric positive definite bilinear forms on V. Of course $\mathcal{R}(V)$ is an open positive cone in the vector space $S^2(V^*)$. Although this is a useful description, it is also important to describe $\mathcal{R}(V)$ and various other geometric structures as homogeneous spaces.

Let $g_0 \in \mathcal{R}(V)$ be a fixed Riemannian structure, i.e., a *base point* for our considerations, and $\{e_1^0, \ldots, e_n^0\}$ a g_0-orthonormal basis. If $\{e_1, \ldots, e_n\}$ is any other basis, then there exists a unique transformation $T \in Gl_\mathbb{R}(V)$ such that $T(e_i^0) = e_i$, $1 \leq i \leq n$. If $g \in \mathcal{R}(V)$ and this basis is g-orthonormal, then

$$T(g_0)(e_i, e_j) := g_0(T^{-1}e_i, T^{-1}e_j) = g_0(e_i^0, e_j^0) = \delta_{ij} = g_0(e_i, e_j).$$

Thus the action of $Gl_\mathbb{R}(V)$ on $\mathcal{R}(V)$, $g \mapsto g \circ T^{-1}$ is transitive. If we identify $\mathcal{R}(V)$ with the orbit of g_0, then $\mathcal{R}(V) \cong Gl_\mathbb{R}(V)/O_\mathbb{R}(V, g)$, and if a basis is chosen as above, $\mathcal{R}(V) \cong Gl_n(\mathbb{R})/O_n(\mathbb{R})$.

We now proceed with similar arguments for other geometric structures.

In order to compute the homogeneous space $\mathcal{J}(V)$ of all *complex structures* on an $\mathbb{R}$-vector space V, we let J_0 be a base point and choose a standard basis

$\{e_1^0, \ldots, e_n^0, J_0 e_1^0, \ldots, J_0 e_n^0\}$. If $J \in \mathcal{J}(V)$ is given, then, after choosing a J-standard basis $\{e_1, \ldots, e_n, Je_1, \ldots, Je_n\}$, we define $T \in Gl_{\mathbb{R}}(V)$ by $e_i^0 \mapsto e_i$ and $J_0 e_i^0 \mapsto Je_i$, $1 \leq i \leq n$. It follows that $T(J_0) := T \circ J_0 \circ T^{-1} = J$.

Consequently, the natural action of $Gl_{\mathbb{R}}(V)$ by conjugation is transitive and $\mathcal{J}(V) = Gl_{\mathbb{R}}(V)/Gl_{\mathbb{C}}(V)$, where $Gl_{\mathbb{C}}(V)$ is identified with the isotropy group at the base point, i.e., $Gl_{\mathbb{C}}(V) \cong \{T \in Gl_{\mathbb{R}}(V) := T \circ J_0 = J_0 \circ T\}$

Let (V, J_0) be a fixed complex vector space and $\mathcal{H}(V, J_0)$ the set of *Hermitian structures*:

$$h \in \mathcal{H}(V, J_0) :\Longleftrightarrow \left(h(v, w) = \overline{h(w, v)} \text{ and } h(J_0 v, w) = ih(v, w) \right).$$

In this case we choose a J_0-standard basis $\{e_1^0, \ldots, e_n^0, J_0 e_1, \ldots, J_0 e_n^0\}$ which is h_0-unitary, i.e., $h_0(e_i, e_j) = h_0(J_0 e_i, J_0 e_j) = \delta_{ij}$ and $h_0(e_i, J_0 e_j) = 0$ for $1 \leq i, j \leq n$. If $h \in \mathcal{H}(V, J_0)$ is given, then the transformation $T \in Gl_{\mathbb{R}}(V)$ which maps this basis to a J_0-standard basis for h is in $Gl_{\mathbb{C}}(V)$, i.e., $T \circ J_0 = J_0 \circ T$. Thus

$$\mathcal{H}(V, J_0) = Gl_{\mathbb{C}}(V, J_0)/U(V, J_0),$$

where $U(V, J_0) = \{T \in Gl_{\mathbb{C}}(V, J_0) : T(h_0) = h_0\}$ is the *unitary group* at the base point.

Let us now turn to symplectic structures. If V is an even-dimensional $\mathbb{R}$-vector space, then we let $\Omega(V) := \{\omega : (V, \omega) \text{ symplectic}\}$ denote the set of *symplectic structures*. Of course $Gl_{\mathbb{R}}(V)$ acts on this space by $\omega \mapsto T(\omega) = \omega \circ T^{-1}$. As usual we let $\omega_0 \in \Omega(V)$ be a base point and choose an ω_0-standard basis $\{e_1^0, \ldots, e_n^0, e_{n+1}^0, \ldots, e_{2n}^0\}$, i.e., $\omega_0(e_i^0, e_{n+j}^0) = \delta_{ij}$ and $\omega_0(e_i^0, e_j^0) = \omega(e_{n+i}^0, e_{n+j}^0) = 0$ for $1 \leq i, j \leq n$.

The same argument as used in the various cases above shows that $Gl_{\mathbb{R}}(V)$ acts transitively on $\Omega(V)$ and thus $\Omega(V) = Gl_{\mathbb{R}}(V)/Sp(V, \omega_0)$, where $Sp(V, \omega_0) := \{T \in Gl_{\mathbb{R}}(V) : T(\omega_0) = \omega_0\}$ is the isotropy group of ω_o-symplectic isomorphisms.

For $\omega \in \Omega(V)$ we consider the space $\mathcal{J}_\omega(V) := \{J \in \mathcal{J}(V) : J \text{ is } \omega - \text{compatible}\}$. Recall that J is ω-compatible if $\omega = \omega \circ J$ and $\omega(v, Jv) > 0$ for all $v \in V \backslash \{0\}$.

Beginning as usual, we take a base point J_0 and an basis $\{e_1^0, \ldots, e_n^0, J_0 e_1^0, \ldots, J_0 e_n^0\}$ which is ω_0-standard. If J is ω_0-compatible, then there exists also a basis $\{e_1, \ldots, e_n, Je_1, \ldots, Je_n\}$ which is J-standard and ω_0-standard. Consider the transformation $T \in GL_{\mathbb{R}}(V)$ which is defined by $T(e_i^0) = e_i$ and $T(J_0 e_i^0) = Je_i$. It follows that $T(J_0) = TJ_0 T^{-1} = J$ and $T(\omega_0) = \omega_0$. Thus the symplectic group $Sp(V, \omega_0)$ acts transitively on $\mathcal{J}_{\omega_0}(V)$ and

$$\mathcal{J}_{\omega_0}(V) := Sp(V, \omega_0)/U(V, J_0).$$

Here the isotropy group is the unitary group $Sp(V, \omega_0) \cap Gl_{\mathbb{C}}(V, J_0) = U(V, J_0)$; in particular, it is a maximal compact subgroup of $Sp(V, \omega_0)$.

Remark. (Topology of homogeneous spaces). Let $M = G/H$ be a homogeneous space which is a quotient of a Lie group G, which has at most finitely many components, by a subgroup H which likewise has at most finitely many components. Let

K be a maximal compact subgroup of G which contains a given maximal compact subgroup L of H. In this situation one regards the compact homogeneous space $N = K/L$ as being the essential topological part of M. This can be made precise by the following result of Mostow: The space M can be realized as a K-equivariant vector bundle over N.

In particular, a minimal K-orbit $N \cong K/L$ (which is a section), is a strong deformation retract of M.

We apply this to the above homogeneous spaces. Note that if the isotropy subgroup H contains a maximal compact subgroup of G, then it follows that the homogeneous space G/H is a cell. This can be applied to the following cases:

$$\mathcal{R}(V) = Gl_{\mathbb{R}}(V)/O_{\mathbb{R}}(V, g),$$
$$H(V, J_0) = Gl_{\mathbb{C}}(V, J_0)/U(V, J_0)$$

and

$$\mathcal{J}_{\omega_0}(V) = Sp(V, \omega_0)/U(V, J_0).$$

In the cases of $\mathcal{J}(V) = Gl_{\mathbb{R}}(V)/Gl_{\mathbb{C}}(V)$ and $\Omega(V) = Gl_{\mathbb{R}}(V)/Sp(V, \omega_0)$ the manifold $N = K/L$ is the quotient $O_{\mathbb{R}}(V, g)/U(V, J_0)$ which is a compact manifold with non-trivial topology.

We close this section by considering natural bundles whose fibers are homogeneous spaces of the above types.

Let M be a differentiable manifold with a smooth vector bundle $E \to M$ and let $P \to M$ be the associated $Gl_n(\mathbb{R})$-principal bundle. Recall that this can be identified with the bundle of n-frames on the n-dimensinal manifold M. Note that a section $\sigma \in \Gamma(M, P)$ would correspond to a global trivialization of the bundle E.

Exercise. Experience shows that, e.g., most compact manifolds have non-trivial tangent bundles. Pick your favorite compact manifold which would seem to have a non-trivial tangent bundle and prove it.

In each case above we defined a notion of a standard basis which we now extend to the frame-level. This is motivated by the following

Example. Let $e = (e_1, \dots, e_n)$ be a frame over an open set U for the bundle E. Pointwise, all other frames are abtained by applying the free right-action of $Gl_n(\mathbb{R})$ to this or any particular basis. Of course such a frame determines a metric g in the fibers of E over U by $g(e_i, e_j) = \delta_{ij}$. The orbit $O_n(\mathbb{R}).e$ by the right-action consists of those frames which give the same metric. Thus we consider the fiber bundle $\mathcal{R}(E) \to M$ which is defined by $P/O_n(\mathbb{R}) =: \mathcal{R}(E)$. The sections $\Gamma(M, \mathcal{R}(E))$ are the Riemannian structures on E.

In exactly the same way as in case of metrics we consider the bundle $\mathcal{J}(E) \to M$ of complex structures and the bundle $\Omega(E) \to M$ of symplectic structures. The fibers of $\mathcal{J}(E)$ are isomorphic to the homogeneous space $Gl_{2n}(\mathbb{R})/Gl_n(\mathbb{C})$ which is homotopically equivalent to the compact homogeneous manifold $O_{2n}(\mathbb{R})/U_n$ and a section $J \in \Gamma(M, \mathcal{J}(E))$ is a bundle endomorphism $J : E \to E$ which satisfies $J^2 = -Id$.

If such a section is prescribed, we refer to (E, J) as a complex vector bundle.

Remark. As we noted earlier, certain even-dimensional manifolds do not possess almost complex structures, i.e., there is no complex structure on the tangent bundle. Thus the existence of a section $J \in \Gamma(M, \mathcal{J}(E))$ is already something.

The fiber of $\Omega(E) \to M$ is the homogeneous space $Gl_{2n}(\mathbb{R})/Sp_{2n}(\mathbb{R})$ which is again homotopically equivalent to $O_{2n}(\mathbb{R})/U_n$. If $\omega \in \Gamma(M, \Omega(E))$, then (E, ω) is referred to as a symplectic bundle. The section ω can be thought of as an alternating form in the fibers of E which varies smoothly from fiber to fiber.

Let (E, ω) be a symplectic bundle. Instead of considering the principal bundle of all frames of E, we may consider those frames which are ω-standard, i.e., $\{e_1, \ldots, e_n, e_{n+1}, \ldots, e_{2n}\}$ with $\omega(e_i, e_{n+j}) = \delta_{ij}$ and $\omega(e_i, e_j) = \omega(e_{n+i}, e_{n+j}) = 0$ for $1 \le i, j \le n$. Any two differ by a transformation $g \in Sp_{2n}(\mathbb{R})$. Thus the associated principal bundle is an $Sp_{2n}(\mathbb{R})$-bundle.

Now if $(e_1, \ldots, e_n, e_{n+1}, \ldots, e_{2n})$ is an ω-standard frame, then we may define an ω-compatible complex structure by $J(e_i) = e_{n+i}$ and $J(e_{n+i}) = -e_i$. Two such complex structures are the same if and only if they differ by a unitary transformation $g \in U_n$. Thus the bundle of ω-compatible complex structures on (E, ω) is the quotient $\mathcal{J}_\omega(E) := P_\omega/U_n$ of the associated $Sp_{2n}(\mathbb{R})$-principal bundle by the unitary group U_n. It is important to notice that the fiber of this bundle is a cell. If $J \in \Gamma(M, \mathcal{J}_\omega(E))$, then one says that J is an ω-compatible complex structure on E.

Exercise. Carry through the analogous discussion for Hermitian structures on a complex bundle.

In the cases of $\mathcal{R}(E)$, $\mathcal{H}(E, J)$ and $\mathcal{J}_\omega(E)$ the fibers are naturally identifiable with cells. Thus the following is of particular relevance.

Proposition. *If $E \to B$ is a fiber bundle with F a cell, then $\Gamma(B, E) \neq \{0\}$.*

Remark. This is a statement in the smooth category, i.e., $\Gamma_{C^\infty}(B, E) \neq \{0\}$. In fact $\Gamma(B, E)$ carries the structure of an infinite-dimensional manifold. Depending on degrees of regularity, its charts are in Hilbert, Banach or Fréchet spaces. These types of manifolds turn up quite naturally in various contexts, e.g. non-linear differential equations.

2. The elements of function theory

The fact that complex manifold structure is important is already transparent in the 1-dimensional case.

Let $D \subset \mathbb{C}$ be a domain in the complex plane. The space of complex valued smooth forms on D is denoted by $\mathcal{A}^p(D)$, $p = 0, 1, 2$. The complex version of the Fundamental Lemma shows that every 1-form $\alpha \in \mathcal{A}^1(D)$ can be written

$\alpha = fdz + gd\bar{z}$, where $z : D \to \mathbb{C}$ is the standard coordinate function. The dual basis at the level of complex vector fields is given by $(\frac{\partial}{\partial z}, \frac{\partial}{\partial \bar{z}})$, where

$$\frac{\partial}{\partial z} = \frac{1}{2}(\frac{\partial}{\partial x} - i\frac{\partial}{\partial y}) \text{ and } \frac{\partial}{\partial \bar{z}} = \frac{1}{2}(\frac{\partial}{\partial x} + i\frac{\partial}{\partial y}).$$

These coordinates yield a splitting $d = \partial + \bar{\partial}$, where

$$\partial f = \frac{\partial f}{\partial z}dz \text{ and } \bar{\partial}f = \frac{\partial f}{\partial \bar{z}}d\bar{z}.$$

Thus the differential $df(p)$ is $\mathbb{C}$-linear on a tangent space T_pD if and only if $\bar{\partial}f(p) = 0$.

Definition. A function $f \in \mathcal{A}(D)$ is said to be *holomorphic* on $D, f \in \mathcal{O}(D)$, whenever $\bar{\partial}f = 0$.

Let $TD \to D$ be the tangent bundle of D and define the almost complex structure J on D by using the global frame $(\frac{\partial}{\partial x}, \frac{\partial}{\partial y})$, where $z = x + iy : J(\frac{\partial}{\partial x}) = \frac{\partial}{\partial y}$ and $J(\frac{\partial}{\partial y}) = -\frac{\partial}{\partial x}$. In this way $T_p^{1,0}D = ((\frac{\partial}{\partial z}\big|_p))_{\mathbb{C}}$ and $T_p^{0,1}D = ((\frac{\partial}{\partial \bar{z}}\big|_p))_{\mathbb{C}}$ are the $+i$- and $-i$-eigenspaces of the complex linear version of J on the complexified tangent bundle $T^{\mathbb{C}}(M) := TM \otimes_{\mathbb{R}} \mathbb{C}$.

Exercise. Show that $f : D \to \mathbb{C}$ is holomorphic if and only if $f_* : T_pD \to T_{f(p)}\mathbb{C}$ is complex linear in the sense that $f_*(Jv) = Jf_*(v)$ for $v \in T_pD$ and all $p \in D$.

Forms of the type $\alpha = fdz$, resp. $\alpha = fd\bar{z}$, are called $(1,0)$-forms, resp. $(0,1)$-forms. The spaces of such forms are denoted by $\mathcal{A}^{1,0}(D)$ and $\mathcal{A}^{0,1}(D)$ respectively.

Note that a 2-form $\omega \in \mathcal{A}^2(D)$ is automatically of type $(1,1)$, i.e., $\omega = fdz \wedge d\bar{z}$ where $f \in \mathcal{A}(D)$.

Of course we have the de Rham complex

$$0 \to \mathbb{C} \to \mathcal{A}(D) \xrightarrow{d} \mathcal{A}^1(D) \xrightarrow{d} \mathcal{A}^2(D) \to 0$$

for $\mathbb{C}$-valued forms and the associated cohomology spaces $H^k_{deR}(D, \mathbb{C})$, $k = 0, 1, 2$.

Exercise. Show that the class $\xi = [\omega] \in H^1_{deR}(D)$, $\omega = \frac{1}{2\pi i}(\frac{dz}{z})$ is a particularly good generator of the cohomology of $D := \{z \in \mathbb{C} : 0 < |z| < 1\}$.

Remark. For M an connected, orientable, non-compact n-dimensional manifold one has $H^n_{deR}(M) = \{0\}$. Thus, for domains in the plane, the only interesting de Rham cohomology is $H^1_{deR}(D)$.

The $\partial-$ and $\bar{\partial}$-operators define various complexes that are of importance for complex analysis. For simplicity we only discuss those which are defined by $\bar{\partial}$. First, we consider the Dolbeault complex

$$0 \to \mathcal{O}(D) \to \mathcal{A}(D) \xrightarrow{\bar{\partial}} \mathcal{A}^{0,1}(D) \to 0$$

defined by the natural injection $\mathcal{O}(D) \hookrightarrow \mathcal{A}(D)$ and the $\bar{\partial}$-operator, $\bar{\partial} : \mathcal{A}(D) \to \mathcal{A}^{0,1}(D)$. The last map is in fact $\bar{\partial} : \mathcal{A}^{0,1}(D) \to \mathcal{A}^{0,2}(D) = \{0\}$. To remain in harmony with the higher-dimensional situation, one defines

$$H^{0,1}_{Dolb}(D) = \frac{Ker(\bar{\partial} : \mathcal{A}^{0,1}(D) \to \mathcal{A}^{0,2}(D))}{Im(\bar{\partial} : \mathcal{A}(D) \to \mathcal{A}^{0,1}(D))}.$$

Remark. In fact $H^{0,1}_{Dolb}(D) = \{0\}$ for D a domain in the complex plane. This non-trivial result leads to the solution of numerous important problems for such domains. One of the main goals of the higher-dimensional theory is to determine when such *vanishing theorems* hold.

The map $\bar{\partial} : \mathcal{A}^{1,0}(D) \to \mathcal{A}^{1,1}(D)$ is non-trivial and one defines $\Omega^1(D) := Ker(\bar{\partial} : \mathcal{A}^{1,0}(D) \to \mathcal{A}^{1,1}(D))$ to be the space of holomorphic 1-forms. This yields the complex

$$0 \to \Omega^1(D) \to \mathcal{A}^{1,0}(D) \to \mathcal{A}^{1,1}(D) \to 0$$

and the analogous cohomology

$$H^{1,1}_{Dolb}(D) := \frac{Ker(\bar{\partial} : \mathcal{A}^{1,1}(D) \to \mathcal{A}^{1,2}(D))}{Im(\bar{\partial} : \mathcal{A}^{1,0}(D) \to \mathcal{A}^{1,1}(D))}.$$

The understanding of integrals of, e.g., holomorphic 1-forms was one of the strong motivating factors in the initial stage of development of complex analysis. Of course Stokes' Theorem is an essential tool.

2.1 Cauchy Integral Theorem and immediate consequences

Theorem. *Let $D \subset \mathbb{C}$ be a relatively compact domain with smooth boundary ∂D and let $\omega \in \Omega(\overline{D})$ be a holomorphic 1-form on some neighborhood of the closure $\overline{D}$. Then*

$$\int_{\partial D} \omega = 0.$$

Proof. Since $d\omega = \bar{\partial}\omega = 0$, this is an immediate consequence of Stokes' Theorem.
$\square$

Exercise. The splitting $d = \partial + \bar{\partial}$ and the various spaces of forms, e.g., $\Omega(D)$, $\mathcal{A}^{0,1}(D)$, etc. are important for considerations in complex analysis. Show that these operators, spaces, etc. are well-defined independent of the choice of *holomorphic* coordinates. More precisely, a diffeomorphism $\varphi : U \to V$ between domains is said to be holomorphic if

$$\varphi_* : (T_p, J_p) \to (T_{\varphi(p)}, J_{\varphi(p)})$$

is complex linear for all $p \in U$, i.e., $\varphi_* \circ J_p = J_{\varphi(p)} \circ \varphi_*$. Show that the above concepts are well-defined independently of coordinate change by a holomorphic diffeomorphism.

Forms of the type $\omega := \frac{1}{2\pi} \frac{dz}{z}$ are certainly not holomorphic on a neighborhod of $0 \in \mathbb{C}$, but their integrals can usually be computed by a limiting process.

For example, let $\Delta := \{z \in \mathbb{C} : |z| < 1\}$ and $\Delta_\epsilon := \{z : \epsilon < |z| < 1\}$. Then $\partial \Delta = \partial \Delta_\epsilon + \partial \Delta(\epsilon)$, where $\Delta(\epsilon) := \{z : |z| < \epsilon\}$. It follows that

$$\frac{1}{2\pi i} \int_{\partial \Delta} \frac{1}{z} dz = \frac{1}{2\pi i} \int_{\partial \Delta_\epsilon} \frac{1}{z} dz + \lim_{\epsilon \to 0} \int_{\partial \Delta(\epsilon)} \frac{1}{2\pi i} \frac{dz}{z} = 0 + 1 = 1.$$

More generally, we have the

Cauchy Integral Formula. *Let D be a relatively compact domain in the complex plane with smooth boundary ∂D. For $f \in \mathcal{O}(\overline{D})$ and $\zeta \in D$ it follows that*

$$f(\zeta) = \frac{1}{2\pi i} \int_{\partial D} \frac{f dz}{(z - \zeta)}.$$

Proof. Let $\Delta(\epsilon) := \{z : |z - \zeta| < \epsilon\}$ be a small disk in D about ζ. As above

$$\frac{1}{2\pi i} \int_{\partial D} \frac{f dz}{(z - \zeta)} = \lim_{\epsilon \to 0} \frac{1}{2\pi i} \int_{\partial \Delta(\epsilon)} \frac{f dz}{z - \zeta} = \lim_{\epsilon \to 0} \int_0^{2\pi} f(\zeta + \epsilon e^{i\varphi}) \frac{d\varphi}{2\pi} = f(\zeta). \quad \square$$

Suppose for example that $f \in \mathcal{O}(\overline{\Delta})$, where Δ is the unit disc. Then

$$f(\zeta) = \frac{1}{2\pi i} \int_{\partial \Delta} \frac{f(z) dz}{z - \zeta} = \frac{1}{2\pi i} \int_{\partial \Delta} \frac{f(z)}{z} \cdot \frac{1}{1 - \zeta/z} dz =$$

$$= \frac{1}{2\pi i} \int_{\partial \Delta} f(z) \sum_{n=0}^\infty \frac{\zeta^n}{z^{n+1}} dz = \sum_{n=0}^\infty a_n \zeta^n, \text{ where } a_n := \frac{1}{2\pi i} \int_{\partial \Delta} \frac{f(z)}{z^{n+1}} dz.$$

Exercise. The power series expansion

$$f(\zeta) = \sum a_n \zeta^n$$

is uniformly convergent on compact sets in Δ. This and the fact that summation and integration may be interchanged are consequences of the fact that $\zeta \in \Delta$ and $z \in \partial \Delta$ and therefore $|\zeta \cdot z^{-1}| < 1$. Fill out the details of this argument.

We summarize this in the following

Corollary. *If $f \in \mathcal{O}(\Delta)$, then the Taylor series*

$$\sum_{n=0}^\infty \frac{f^{(n)}(0)}{n!} \zeta^n$$

converges uniformly to f on all compact subsets of Δ.

Exercise. Relate the coefficients a_n and the derivatives $f^{(n)}(0)$, $0 \le n < \infty$.

As a result of the above corollary we have the following

Proposition. (Power Series Characterization Holomorphic Functions) *A complex valued function f is holomorphic on a domain D in the complex plane if and only if it can be locally represented by a uniformly convergent Taylor series in the variable z alone.*

The correct topology on the function space $\mathcal{O}(D)$ of holomorphic functions is that of convergence on compact subsets. To make this precise, let $| \cdot |_K$ denote the sup-norm on a compact subset $K \subset D$. Define the topology on $\mathcal{O}(D)$ by these seminorms. Hence a sequence $\{f_n\}$ is Cauchy if and only if it is $|\cdot|_K$-Cauchy for all K compact in D.

Proposition. *The space $\mathcal{O}(D)$ is complete.*

Proof. Take $K = \overline{\Delta}$ to be a closed disk in D. Since the space $(C(K), |\cdot|_K)$ of continuous functions on K is complete, it follows that a Cauchy sequence $\{f_n\}$ of holomorphic functions converges to a function f which is at least continuous on K. But

$$(*) \qquad f(\zeta) = \lim_{n \to \infty} f_n(\zeta) = \lim_{n \to \infty} \frac{1}{2\pi i} \int_{\partial \Delta} \frac{f_n(z)}{z - \zeta} dz = \frac{1}{2\pi i} \int_{\partial \Delta} \frac{f(z)}{z - \zeta} dz.$$

Since the integral $\frac{1}{2\pi i} \int_{\partial \Delta} \frac{f(z)}{z-\zeta} dz$ is a holomorphic function of ζ (one may exchange $\bar\partial$ and $\int$), it follows that $f \in \mathcal{O}(\Delta)$. Finally, the notion *holomorphic* is a local property and therefore it is enough to verify it on all such disks. $\qquad\Box$

Exercise. Justify the last step in $(*)$ above.

2.2 Local Theory
We close our review of complex analysis in one variable by turning to the basic local result.

Normal Form Theorem. *Let f be a holomorphic function defined in some neighborhood of $0 \in \mathbb{C}$. Suppose that $f(0) = 0$, but that $f \neq 0$. Then there exists a holomorphic coordinate $w = w(z)$ and a number $n \in \mathbb{N}^{>0}$ such that*

$$f(w) = w^n.$$

The number n is defined to be the order of vanishing $o(f)(0)$ at zero and is well-defined independent of the choice of holomorphic coordinates.

Before proving this we would like to underline the holomorphic version of the inverse mapping theorem.

Recall that a necessary and sufficient condition for a smooth map $\varphi : U \to V$ to be a diffeomorphism in a neighborhood of a point $p \in U$ is that its derivative

$$\varphi_*(p) : T_p U \to T_{\varphi(p)} V$$

is a linear isomorphism.

If U and V are domains in the complex plane and $\varphi(z) = u(z) + iv(z)$, i.e., in the standard $\mathbb{R}$-coordinates φ is given by $(x, y) \mapsto (u(x,y), v(x,y))$, then the matrix of φ_* in the basis $\{\frac{\partial}{\partial x}, \frac{\partial}{\partial y}\}$ is $\begin{pmatrix} u_x & v_x \\ u_y & v_y \end{pmatrix} =: Jac_{\mathbb{R}}(\varphi)$.

Suppose now that φ is holomorphic. This means that as a matrix φ_* commutes with $J = \begin{pmatrix} 0 & -1 \\ 1 & 0 \end{pmatrix}$ or as a $\mathbb{C}$-valued function $\bar\partial\varphi = 0$. In the standard $\mathbb{R}$-coordinates this is equivalent to the validity of the *Cauchy-Riemann-equations*

$$u_x = v_y$$
$$u_y = -v_x.$$

Hence, if $\varphi : U \to V$ is a holomorphic map between domains in the complex plane, it follows that

$$det(Jac_{\mathbb{R}}(\varphi)) = \left|\frac{\partial\varphi}{\partial z}\right|^2 = u_x^2 + u_y^2 = v_x^2 + v_y^2.$$

Proposition. *Let $\varphi : U \to V$ be a holomorphic map between domains in the complex plane. Then φ is a holomorphic diffeomorphism[10) in a neighborhood of $p \in U$ if and only if $\frac{\partial\varphi}{\partial z}(p) \neq 0$.*

Proof. Since $det(Jac_{\mathbb{R}}(\varphi)) = |\frac{\partial\varphi}{\partial z}|^2$, the non-vanishing of $\frac{\partial\varphi}{\partial z}(p)$ is certainly a necessary condition. Conversely, if $\varphi'(p) := \frac{\partial\varphi}{\partial z}(p) \neq 0$, then φ is a diffeomorphism in a neighborhood of p with smooth inverse ψ. Differentiating the the identity $\varphi \circ \psi(z) = z$ with respect to $\frac{\partial}{\partial \bar{z}}$, one obtains

$$0 = \frac{\partial}{\partial \bar{z}}(\varphi \circ \psi(z)) = \varphi'(\psi(z)) \cdot \frac{\partial\psi}{\partial \bar{z}}(z).$$

The holomorphicity of ψ in a neighborhood of $\varphi(p)$ follows from the non-vanishing of from φ'.   □

A second fact which is needed in the proof of the *Normal Form Theorem*, and which is central for many other considerations in complex analysis, is the *existence of a logarithm*.

Consider the power series

$$E(z) = \sum_{n=0}^{\infty} \frac{z^n}{n!}$$

which converges uniformly on compact subsets of the complex plane. Elementary calculations show that $E : \mathbb{C} \to \mathbb{C}$ is the unique holomorphic solution to the differential equation $E' = E$ with $E(0) = 1$ and therefore E is a *homomorphism*, i.e., $E(z + w) = E(z) \cdot E(w)$.

It follows immediately that $E : \mathbb{C} \to \mathbb{C}^*$ and that the derivative E' is nowhere zero, i.e., E is locally biholomorphic. Its image $E(\mathbb{C})$ is an open subgroup of the connected Lie group $\mathbb{C}^*$ and consequently $E : \mathbb{C} \to \mathbb{C}^*$ is surjective.[11)

Proposition. *The exponential map $E : \mathbb{C} \to \mathbb{C}^*$ is a locally biholomorphic surjective homomorphism with*

$$Ker(E) = 2\pi i \cdot \mathbb{Z}.$$

[10) This means that φ^{-1} exists and that φ and φ^{-1} are holomorphic. Such maps are called *biholomorphic*.

[11) This map is sometimes denoted by $z \mapsto e^z$, where e is the real number $e := E(1)$.

A *logarithm* $L(z) = log\, z$ is a local inverse of E, i.e., a locally defined holomorphic function such that $E \circ L(z) = z$. The existence of such is guaranteed by the above holomorphic inverse mapping theorem. If it exists, then its differential satisfies

$$dL = \frac{dz}{z}$$

and consequently the existence question can be formulated in terms of de Rham cohomology: given $f \in \mathcal{O}(D)$ with $\{f = 0\} = \emptyset$, let

$$\omega_f := \frac{df}{f} \text{ and } \xi_f := [\omega_f] \in H^1_{deR}(D).$$

Proposition. *Suppose that* $H^1_{deR}(D) = \{0\}$, *e.g., that* D *is homotopic to a point. Then, for every* $f \in \mathcal{O}(D)$ *with* $\{f = 0\} = \emptyset$, *there exists a holomorphic function* $L_f \in \mathcal{O}(D)$ *with* $E \circ L_f = f$.

Proof. Since $H^1_{deR}(D) = \{0\}$ and $\{f = 0\} = \emptyset$, it follows that ω_f is well-defined and $\omega_f = dL_f$ for some smooth function L_f. Note that $\omega_f = \frac{df}{f} = \frac{f'}{f}dz$ is a $(1,0)$-form. Thus $\bar{\partial}L_f = 0$, i.e., $L_f \in \mathcal{O}(D)$. Furthermore, if $h := \frac{E \circ L_f}{f}$, then $dh \equiv 0$. Therefore, by adding an appropriate constant to L_f, it follows that $h \equiv 1$ and $E \circ L_f = f$ as desired. $\square$

Exercise. Discuss the unicity of L_f.

Proof of the Normal form Theorem. Let f be holomorphic in a neighborood of $0 \in \mathbb{C}$ and assume that $f(0) = 0$ and that $f \not\equiv 0$. Then there exists $n \in \mathbb{N}^{>0}$ and $a_n \neq 0$ such that

$$f = a_n z^n + a_{n+1} z^{a+1} + \ldots \quad = z^n \cdot u(z),$$

where u is also holomorphic near 0 and has the additional property $u(0) \neq 0$.

Since u is continuous, there exists a disk $\Delta = \Delta(0)$ with $\{u|_\Delta = 0\} = \emptyset$. Replace u by $u|_\Delta$ and consider the holomorphic function $v = v(z)$ defined by

$$v(z) = e^{h(z)}, \text{ where } h(z) = \frac{1}{n}log\, u(z).$$

It follows that v is an *nth root* of u, i.e. $v^n = u$, and

$$f = (zv(z))^n.$$

Of course $v(0) \neq 0$ and, for $\varphi := z \cdot v(z)$, it follows that $\varphi'(0) \neq 0$. Thus φ is locally biholomorphic in a neighborhood of $0 \in D$ and, using the coordinate $w = \varphi(z)$, we have $f = w^n$. The integer $n \in \mathbb{N}^{>0}$ is well-defined, because it is the smallest such with $\frac{d^n}{dz^n}(f)(0) \neq 0$. $\square$

The fact that holomorphic functions have such a simple normal form has a number of important consequences. We now derive three of these.

Open Mapping Theorem. *Suppose* $f \in \mathcal{O}(D)$ *is nowhere locally constant. Then* $f : D \to \mathbb{C}$ *is an open mapping.*

Proof. Since f is nowhere locally constant, at every $p \in D$ there is a choice of a local coordinate w on D such that

$$f(w) = f(p) + w^n \text{ for some } n \in \mathbb{N}^{>0}.$$

The result then follows from the fact that $w \mapsto w^n$ is an open map.   □

Maximum Principle. *Suppose that $f \in \mathcal{O}(D)$ is nowhere locally constant. Then for all $p \in D$ and $U = U(p) \subset\subset D$ it follows that*

$$|f(p)| < |f|_U := \sup_{q \in U} |f(q)|.$$

Proof. This follows immediately from the fact that, under the assumptions, $f(U)$ is an open neighborhood of $f(p)$.   □

Identity Principle. *Let D be a connected domain in $\mathbb{C}$ and $f \in \mathcal{O}(D)$. Then either f is identically constant or every f-fiber $f^{-1}(\{f(p)\})$ is discrete.*

Proof. By the power series representation or the normal form, if f is nowhere locally constant, then the fibers are discrete. If f is somewhere locally constant, i.e., in a neighborhood of some $p \in D, f \equiv f(p) =: a$, we consider the interior Ω of the closed set $\{f = a\}$. Assume that the boundary $\partial\Omega$ in D is not empty and apply the normal form theorem at $p_o \in \partial\Omega$.[12] It follows that the f-fibers near p_0 are discrete, contrary to the definition of Ω. Thus $\partial\Omega = \emptyset$ and $\Omega = D$.   □

3. A brief introduction to complex analysis in higher dimensions

3.1 Complex differential forms and Dolbeault cohomology

For a domain D in $\mathbb{C}^n, n \geq 1$, let $\mathcal{A}^p(D), 0 \leq p \leq n$, denote the space of complex p-forms. Just as in the case of real forms we have the de Rham complex

$$0 \to \mathbb{C} \to \mathcal{A}(D) \to \mathcal{A}^1(D) \to \ldots \to \mathcal{A}^{2n}(D) \to 0$$

and the associated cohomology

$$H^k_{deR}(D, \mathbb{C}) := \frac{Ker(d : \mathcal{A}^k(D) \to \mathcal{A}^{k+1}(D))}{Im(d : \mathcal{A}^{k-1}(D) \to \mathcal{A}^k(D))}, \quad 1 \leq k \leq 2n.$$

Since the spaces $\mathcal{A}^k(D)$ and operators $d : \mathcal{A}^k(D) \to \mathcal{A}^{k+1}(D)$ are just the *complexifications* of the corresponding spaces of real forms and real operators which are defined by regarding $D \subset \mathbb{R}^{2n}$, $H^k_{deR}(D, \mathbb{C})$ is likewise just the complexification of the real de Rham cohomology and we have nothing new.

Let $(z_1, \ldots, z_n)$ be the standard complex vector space coordinates on $\mathbb{C}^n$. If $I = (i_1, \ldots, i_p)$ and $J = (j_1, \ldots, j_q)$ are multi-indices, let

$$dz_I \wedge d\bar{z}_J := dz_i, \wedge \ldots \wedge dz_{i_p} \wedge d\bar{z}_{j_1} \wedge \ldots d\bar{z}_{j_q}.$$

---

[12] Here it is really only necessary to use the power series representation.

Define $\omega \in \mathcal{A}^k(D)$ to be a (p, q)-form for $p, q \in \mathbb{N}^{\geq 0}$ with $p + q = k$ if it can be written

$$\omega = \sum_{|I|=p, |J|=q} a_{IJ} dz_I \wedge d\bar{z}_J.$$

Let $\mathcal{A}^{p,q}(D)$ denote the space of such forms. Of course $\mathcal{A}^k(D) = \bigoplus_{p+q=k} \mathcal{A}^{p,q}(D)$. This decomposition yields a splitting $d = \partial + \bar{\partial}$ for

$$d : \mathcal{A}^{p,q}(D) \to \mathcal{A}^{p+1,q}(D) \oplus \mathcal{A}^{p,q+1}(D),$$

where ∂, resp. $\bar{\partial}$, is defined by composing d with the projection on the first, resp. second, factor.

For example, if $\Omega^p(D) := Ker(\bar{\partial} : \mathcal{A}^{p,0}(D) \to \mathcal{A}^{p,1}(D))$ is defined to be the space of holomorphic p-forms, then we have the Dolbeault complex[13]

$$0 \to \Omega^p(D) \to \mathcal{A}^{p,0}(D) \to \mathcal{A}^{p,1}(D) \to \ldots \to \mathcal{A}^{p,n}(D) \to 0$$

and the associated cohomology

$$H^{p,q}_{Dolb}(D) := \frac{Ker(\bar{\partial} : \mathcal{A}^{p,q}(D) \to \mathcal{A}^{p,q+1}(D))}{Im(\bar{\partial} : \mathcal{A}^{p,q-1}(D) \to \mathcal{A}^{p,q}(D))}.$$

Remark. Unlike the case of domains in the complex plane, these Dolbeault spaces may not vanish. As we indicated before, one of the main goals in complex analysis is to determine complex geometric conditions which imply the vanishing of such groups. As a *negative example* it should be noted that $H^{0,1}_{Dolb}(\mathbb{C}^2 \setminus \{(0,0)\})$ is infinite-dimensional.

3.2 First properties of holomorphic functions and maps

The notion of holomorphic function was defined above, i.e., a *holomorphic* 0-form. Thus $f \in \mathcal{A}(D)$ is holomorphic if and only if $\bar{\partial}f = 0$, i.e.,

$$0 = \bar{\partial}f = \frac{\partial f}{\partial \bar{z}_1} d\bar{z}_1 + \ldots + \frac{\partial f}{\partial \bar{z}_n} d\bar{z}_n.$$

Equivalently, $f \in \mathcal{O}(D)$ if and only if it is a smooth function which is holomorphic in each variable, i.e.,

$$\frac{\partial f}{\partial \bar{z}_1} \equiv \ldots \equiv \frac{\partial f}{\partial \bar{z}_n} \equiv 0.$$

It is of course also important to consider *biholomorphic* coordinate changes in this several variable context. For this note first that the coordinates $(z_1, \ldots, z_n)$ define a nature complex structure J in the real tangent bundle TD: Using the real coordinates $(x_1, \ldots, x_n, y_1, \ldots, y_n)$, where $z_k = x_k + iy_k$, $1 \leq k \leq n$, one defines J so that the frame $(\frac{\partial}{\partial x_1}, \ldots, \frac{\partial}{\partial x_n}, \frac{\partial}{\partial y_1}, \ldots, \frac{\partial}{\partial y_n})$ is J-standard, i.e., $J(\frac{\partial}{\partial x_k}) := \frac{\partial}{\partial y_k}$ and $J(\frac{\partial}{\partial y_k}) := -\frac{\partial}{\partial x_k}$, $1 \leq k \leq n$.

Definition. A diffeomorphism $\varphi : U \to V$ between domains in $\mathbb{C}^n$ is said to be *biholomorphic* if φ_* is complex linear: $\varphi_* \circ J = J \circ \varphi_*$.

---

[13] One easily checks that $\partial^2 = 0$, $\bar{\partial}^2 = 0$ and $\partial\bar{\partial} = -\bar{\partial}\partial$.

Exercise. Show that the spaces $\mathcal{A}^{p,q}(D)$ and the operators ∂ and $\bar{\partial}$ are well-defined independently of a biholomorphic change of coordinates.

As in the 1-variable case, if $\varphi = (\varphi_1, \ldots, \varphi_n)$ is written in coordinates, then φ is holomorphic if and only if $\bar{\partial}\varphi_k = 0, 1 \leq k \leq n$.

Exercise. Let $\varphi : U \to V$ be a holomorphic map which is diffeomorphism. Show that its inverse $\varphi^{-1} : V \to U$ is holomorphic. In other words, holomorphic diffeomorphism between domains in $\mathbb{C}^n$ are biholomorphic.

We now turn to some of the basic properties of holomorphic functions of several complex variables. One of the few properties which can be formulated in exactly the same way as in the 1-variable case is the

Maximum Principle. *Let $f \in \mathcal{O}(D)$ be holomorphic function which is nowhere locally constant. Then, for all $p \in D$ and every neighborhood $U = U(p) \subset\subset D$, it follows that*

$$|f(p)| < |f|_U.$$

Proof. Note that if L is a small *complex* line segment, i.e., for $\lambda \in \mathbb{C}^n \setminus \{(0,0)\}$ fixed and $|z|$ sufficiently small, a disk through p defined by $z \mapsto p + \lambda z$, then $f|L$ is a holomorphic function of 1-variable. Thus the result follows from the 1-dimensional version of the maximum principle. $\square$

Let us attempt to develop some intuition for the level set or fiber $\{f = 0\}$ of a non-constant holomorphic function.

Proposition. *Suppose that f is holomorphic in a neighborhood of a point $p \in \mathbb{C}$ and assume that p is generic for f in the sense that $df(p) \neq 0$. Then there are holomorphic coordinates $(w_1, \ldots, w_n)$ in a neighborhood of p so that $\{f = f(p)\}$ agrees with the linear space $\{w_1 = 0\}$.*

Proof. Choose $f_2, \ldots, f_n$ to be holomorphic functions near p so that $df \wedge df_2 \wedge \ldots \wedge df_n(p) \neq 0$. [14] Then $\varphi = (f, f_2, \ldots, f_n)$ is biholomorphic near p and, after composing with an appropriate translation, we have the desired coordinates. $\square$

Remark. The above proposition shows that at the generic fibers of $f \in \mathcal{O}(D)$ in the case where $dim_{\mathbb{C}} D > 1$ are not discrete. This can in fact be shown for the *singular fibers* as well, i.e., those which contain points p where $df(p) = 0$.

One statement of an identity principle in higher dimensions is that a non-constant function has $dim_{\mathbb{R}}\{f = f(p)\} = 2n - 2$ in the topological sense. Of course sets of the form $\{f = f(p)\}$ have quite refined structure and such dimension statements are very rough.

Exercise. Use the 1-variable identity principle to prove the following *Weak Identity Principle*: Let D be a connected domain in $\mathbb{C}^n$, $f \in \mathcal{O}(D)$ and suppose that $\{f = 0\}$ has non-empty interior. Then $f \equiv 0$.

---

[14] This can be achieved by complex linear functions.

In several complex variables, domains which are obviously diffeomorphic may have completely different complex analytic properties. For example, the *polydisk* $\Delta :=$ $\{z : |z_i| < 1, 1 \le i \le n\}$ and the *ball* $B := \{z : |z_1|^2 + \ldots + |z_n|^2 < 1\}$ are not biholomorphically equivalent. Any two small generic perturbations of the ball are also not equivalent and the same goes for the polydisk. This is diametrically opposed to the 1-dimensional setting: *If D is a domain in $\mathbb{C}$ which is not $\mathbb{C}$ itself but which is topologically a cell, then D is biholomorphically equivalent to the unit disk.*

The polydisk is *the* domain which is most suitable for proving theorems that go one variable at a time. For example, using the 1-variable theorem one immediately derives the

Cauchy Integral Theorem. *Let $f \in O(\bar{\Delta})$. Then for $\xi \in \Delta$*

$$f(\xi) = \frac{1}{(2\pi i)^n} \int_{S(\partial\Delta)} \frac{f(z)}{z - \xi} dz.$$

The notation is explained as follows: $S(\partial\Delta) := \{|z_1| = \ldots = |z_n| = 1\}$ is the *Shilov-Boundary* of Δ. Note that it is a very small piece of the full boundary: It is an n-dimensional torus. The *integral kernel* $\frac{dz}{z-\xi}$ denotes the form $\frac{dz_1 \wedge \ldots \wedge dz_n}{(z_1-\xi_1)\cdot\ldots\cdot(z_n-\xi_n)}$.

Exercise. Show that if f is holomorphic in a neighborhood of $\bar{\Delta}$, then $|f|$ achieves its maximum on $S(\partial\Delta)$.

Just as in the 1-dimensional case this proves that $f \in \mathcal{O}(\bar{\Delta})$ has a convergent power series development

$$f(\xi) = \sum_{|I|=0}^{\infty} a_I \xi^I.$$

3.3 A continuation theorem

It should be underlined that the *singularities* of a function of the type $\frac{1}{z_1-\xi_1}$ are on a complex hyperplane and are by no means isolated as in the 1-variable case. One of the points where complex analysis in the higher-dimensional setting differs from its 1-variable relative is that *singularities* can only occur in certain geometric situations. The following is the first indicator of this phenomenon:

Hartogs' Lemma. *For $F := \Delta \cup A$, where $\Delta := \{z : |z_1| < 1 \text{ and } |z_i| < \epsilon \text{ for } 2 \le i \le n\}$ and $A := \{z : 1 - \delta < |z_1| < 1 \text{ and } |z_i| < 1 \text{ for } 1 \le i \le n\}$, for $0 < \epsilon, \delta < 1$ arbitrary, and $\hat{F} =:= \{|z_i| < 1 \text{ for } 1 \le i \le n\}$ the full polydisk, it follows that the restriction map $r : \mathcal{O}(\hat{F}) \to \mathcal{O}(F)$ is surjective, i.e. every function holomorphic on F can be uniquely continued to a function holomorphic on $\hat{F}$.*

Proof. Since we may approximate these domains from inside by domains of the same type, it is enough to consider $f \in \mathcal{O}(\bar{F})$ and show that it extends to $\hat{f} \in \mathcal{O}(\hat{F})$. For this simply define

$$\hat{f}(\xi) := \frac{1}{(2\pi i)^n} \int_{S(\partial\hat{F})} \frac{f(z)}{z - \xi} dz$$

which is clearly holomorphic on $\hat{F}$. For $\xi \in \Delta$ it follows

$$f(\xi) = \frac{1}{(2\pi i)^n} \int_{S(\partial \Delta)} \frac{f(z)}{z - \xi} dz.$$

But for $\xi \in \Delta$ there exists a homotopy of $S(\partial \hat{F})$ to $S(\partial \Delta)$ in a region where the closed form $\frac{f(z)}{z-\xi} dz$ is non-singular. Thus

$$\hat{f}(\xi) = \frac{1}{(2\pi i)^n} \int_{S(\partial \Delta)} \frac{f(z)}{z - \xi} = f(\xi) \text{ for } \xi \in \Delta.$$

The result then follows from the identity principle. $\qquad \square$

Remark. In words, the Hartogs' Lemma states that a holomorphic function cannot have singularities in the region between $\hat{F}$ and F. In particular, holomorphic functions in higher-dimensions do *not have isolated singularities*.

3.4 Two basic mapping theorems
Recall that a holomorphic mapping of a connected domain in the 1-variable case is either identically constant or an open mapping.

Example. Let $\varphi : \mathbb{C}^2 \to \mathbb{C}^2$ be defined by $(z, w) \mapsto (z, zw)$. Then $Im(\varphi) = \{(z, w) : z \neq 0\} \dot{\cup} \{(0, 0)\}$. In particular $Im(\varphi)$ contains no open neighborhood of $(0, 0)$.

Although $Im(\varphi)$ in the above example is not open, it is the *finite* union of sets which are quite reasonable. Of course the mapping φ in this case is algebraic and indeed the situation for holomorphic maps can be much more complicated. However, the image of a holomorphic map is no worse than a *countable* union of reasonable sets. We now make this precise.

The local models for algebraic varieties are *affine varieties*, i.e., subsets of $\mathbb{C}^n$ defined by polynomial equations, i.e.,

$$V(P_1, \ldots, P_m) := \{z \in \mathbb{C}^n : P_1(z) = \ldots = P_m(z) = 0\},$$

where $P_j \in \mathbb{C}[z_1, \ldots z_n]$ for $1 \leq j \leq m$. In this context an algebraic morphism $\varphi : X \to Y$ between affine varieties is just a polynomial map. Of course this notion can be made precise for varieties of a very general nature.

A *subvariety* V of an affine variety X is a closed subset defined by polynomial equations. A subset $Z \subset X$ is said to be *Zariski open in its closure* $\bar{Z}$ if

(i) $\bar{Z}$ is a subvariety of X

and

(ii) $V := \bar{Z} \setminus Z$ is a subvariety of $\bar{Z}$.

Here $\bar{Z}$ is the topological closure in the Euclidean topology of $\mathbb{C}^n$.

The sets of the type $Im(\varphi)$ in the above example are called *constructible*. A subset $Z \subset X$ of an algebraic variety X is said to be constructible if it contains a dense subset Z_1, which is Zariski-open in its closure in X such that $Z \setminus Z_1$ is constructible. Such sets are locally finite unions of subsets Z_k which are Zariski open in their closures. They fit together in a stratification based on dimension. The precise definition is given by recursion based on the dimension of the ambient algebraic variety X: any subset Z of a 0-dimensional variety is constructible. A subset Z of an n-dimensional variety is constructible if it contains a subset Z_1 which is Zariski open in its closure $\bar{Z}_1$ in X and $Z \cap (\bar{Z}_1 \setminus Z_1)$ is constructible in $\bar{Z}_1 \setminus Z_1$. Note that the notion is well-defined because $\bar{Z}_1 \setminus Z_1$ is a proper subvariety of lower dimension in X.

Theorem of Chevalley. *The image of an algebraic morphism is constructible.*

In the complex analytic setting a (closed) *analytic* subset A of a domain $D \subset \mathbb{C}^n$ is locally defined by finitely many holomorphic equations: For every $a \in A$ there exists an open neighborhood $U = U(a) \subset D$ and holomorphic functions $f_1, \ldots, f_m \in \mathcal{O}(U)$ such that

$$A \cap U = \{p \in U : f_1(p) = \ldots = f_m(p) = 0\}$$

The following is a rough description of the image of a holomorphic map: Let D be a domain in $\mathbb{C}^n$ and $\varphi : D \to \mathbb{C}^m$ a holomorphic map. Then there are countably many analytic subsets A_α of open domains $D_\alpha \subset \mathbb{C}^m$ such that

$$\varphi(D) = \bigcup_\alpha A_\alpha.$$

In closing we note that open mappings can be characterized.

Remmert's Theorem. *Let D be a domain in $\mathbb{C}^n$ and $\varphi : D \to \mathbb{C}^n$ a holomorphic map. Then φ is an open mapping if and only if all φ-fibers are discrete.*

Remark. Remmert's Theorem is much more general than the above *equi-dimensional* version. For example, if $D \subset \mathbb{C}^m$ with $m > n$, then φ is a open mapping if and only if the φ-fibers are all $(m - n)$-dimensional.

For further information on the elements of algebraic geometry in the context of algebraic groups see, e.g., [B] and [Hu1]. For the basics on complex geometric structures, in particular holomorphic mappings, see [GrRe] and [N]. Further references for complex analysis in several variables are, e.g., [GF], [GuRo], [KK], [Kr], [R], and [We].

4. Complex manifolds

A complex manifold X is a differentiable manifold which has a covering by coordinate charts $\{U_\alpha, z_\alpha\}$, where $z_\alpha : U_\alpha \to V_\alpha \subset \mathbb{C}^{n_\alpha}$ is such that the change of coordinate diffeomorphisms are biholomorphic. Given such a covering we equip X with the maximal covering compatible with it.

Example. An open subset $D \subset \mathbb{C}^n$ is clearly a complex manifold. More generally, let A be a (closed) analytic subset of D such that the locally defining functions have maximal rank, i.e.,

$$A \cap U = \{f_1 = \ldots = f_m = 0\}, \text{ where } f_1, \ldots, f_m \in \mathcal{O}(U)$$

and $df_1 \wedge \ldots \wedge df_m(q) \neq 0$ for all $q \in A$. Then A is a manifold which in appropriate coordinates can be locally defined as the simultaneous zero set of finitely many complex linear functions. Sets of this type are called *complex submanifolds* of D.

Exercise. Show that submanifolds in the sense of the above example have a covering by coordinate charts which are holomorphically compatible.

Let X be a complex manifold with a locally-finite covering by holomorphic coordinate charts $\{U_\alpha, z_\alpha\}$. If $z_\alpha = x_\alpha + iy_\alpha$, then we consider the associated frame for the top exterior power of the real cotangent bundle $\omega_\alpha = dx_\alpha^1 \wedge \ldots \wedge dx_\alpha^n \wedge dy_\alpha^1 \wedge \ldots \wedge dy_\alpha^n$. For simplicity assume that X is connected and $n := dim_{\mathbb{C}} X$. Let $\{\chi_\alpha\}$ be a partition of unity subordinate to the covering and define $\omega := \sum_\alpha \chi_\alpha \cdot \omega_\alpha$.

Now the holomorphic change of variables $\varphi_{\alpha\beta} : V_{\alpha\beta} \to V_{\beta\alpha}$ has Jacobian matrix $Jac_{\mathbb{R}}(\varphi_{\alpha\beta})$ in $Gl_n(\mathbb{C}) \subset GL_{2n}(\mathbb{R})^0$. Thus $J_{\alpha\beta} := det(Jac_{\mathbb{R}}(\varphi_{\alpha\beta})) > 0$.

Proposition. *A complex manifold and its submanifolds possess natural orientations.*

Proof. First, we show that ω above is an orientation form, i.e., it is nowhere vanishing. But this is clear, because on U_β, utilizing the change of coordinates, one calculates $\omega = \sum_{\alpha : U_{\alpha\beta} \neq \emptyset} \chi_\alpha \cdot J_{\alpha\beta} \omega_\beta$. If $Y \hookrightarrow X$ is a complex submanifold, then locally $Y = \{z_\alpha^1 = \ldots = z_\alpha^k = 0\}$ for an appropriate coordinate system $z_\alpha = (z_\alpha^1, \ldots, z_\alpha^n)$. The natural orientation for Y is defined similarly: $e_\alpha(Y) = dx_\alpha^{k+1} \wedge \ldots \wedge dx_\alpha^n \wedge dy_\alpha^{k+1} \wedge \ldots \wedge dy_\alpha^n$.   □

4.1 Almost complex structures

For a moment let us consider the underlying differentiable manifold M of a given complex manifold X.

As we have seen above, the holomorphic coordinates $(z_\alpha^1, \ldots, z_\alpha^n)$ define distinguished real coordinates $(x_\alpha^1, \ldots, x_\alpha^n, y_\alpha^1, \ldots, y_\alpha^n)$ and distinguished frames

$$e^\alpha := \left(\frac{\partial}{\partial x_\alpha^1}, \cdots \frac{\partial}{\partial x_\alpha^n}, \frac{\partial}{\partial y_\alpha^1}, \ldots, \frac{\partial}{\partial y_\alpha^n} \right)$$

for the tangent bundle $TM \to M$. Define the complex structure J_α in TM over U_α so that e^α is J_α-standard, i.e.,

$$J_\alpha \left(\frac{\partial}{\partial x_\alpha^i} \right) := \frac{\partial}{\partial y_\alpha^i} \text{ and } J_\alpha \left(\frac{\partial}{\partial y_\alpha^i} \right) := -\frac{\partial}{\partial x_\alpha^i} , \quad 1 \leq i \leq n.$$

Now the holomorphicity of the change of coordinates $\varphi_{\alpha\beta} : V_{\alpha\beta} \to V_{\beta\alpha}$ implies that $(\varphi_{\alpha\beta})_* \circ J_\alpha = J_\beta \circ (\varphi_{\alpha\beta})_*$. In other words, the bundle endomorphism $J \in End(TM)$ is well-defined by $J_\alpha := J|_{U_\alpha}$.

Definition. A complex structure J on the tangent bundle of a real manifold M is called an *almost complex structure* on M. The pair (M, J) is called an *almost complex manifold*.

Almost complex manifolds have recently become so important, e.g., in symplectic geometry, that in certain circles the word *almost* has been dropped.

As we have seen above, a complex manifold X has a uniquely associated almost complex structure J. Such complex structures have very special properties.

If (M, J) is an arbitrary almost complex manifold, then the complexified tangent bundle $TM^{\mathbb{C}} := TM \oplus_{\mathbb{R}} \mathbb{C}$ splits: $TM^{\mathbb{C}} = (TM)^{1,0} \oplus (TM)^{0,1}$, where $(TM)^{1,0}$, resp. $(TM)^{0,1}$, is the $+i$, resp $-i$, eigenspace of the induced complex linear endomorphism $J^{\mathbb{C}} : TM^{\mathbb{C}} \to TM^{\mathbb{C}}$. In the analogous way we have the eigenspace decomposition $T^*M^{\mathbb{C}} = (T^*M)^{1,0} \oplus (T^*M)^{0,1}$ for the induced complex structure on the dual bundle. This defines the bundle of (p, q)-forms $A^{p,q} := (\wedge^p(T^*M)^{1,0}) \wedge (\wedge^q(T^*M)^{0,1})$.

Thus the notion of a (p, q)-form is well-defined:

$$\mathcal{A}^{p,q}(M) := \Gamma(M, A^{p,q}).$$

The splitting $d = \partial + \bar{\partial} : \mathcal{A}(M) \to \mathcal{A}^1(M) = \mathcal{A}^{1,0}(M) \oplus \mathcal{A}^{0,1}(M)$ is defined, but for higher degree form there may be difficulties, e.g., the projections of $Im(d : \mathcal{A}^{p,q}(M) \to \mathcal{A}^{p+q+1}(M))$ onto the forms of bidegrees other than $(p + 1, q)$ and $(p, q + 1)$ may be non-trivial. This makes it difficult to discuss the $\bar{\partial}$-operator.

If (M, J) is the almost complex manifold associated to a complex manifold X, then via holomorphic coordinates one easily verifies that $Im(d : \mathcal{A}^{p,q}(M) \to \mathcal{A}^{p+q+1}(M))$ is contained in $\mathcal{A}^{p+1,q}(M) \oplus \mathcal{A}^{p,q+1}(M)$. Thus we have the splitting $d = \partial + \bar{\partial}$ for all degrees and, for example, the induced Dolbeault complex

$$0 \to \Omega^p(M) \to \mathcal{A}^{p,0}(M) \xrightarrow{\bar{\partial}} \mathcal{A}^{p,1}(M) \xrightarrow{\bar{\partial}} \ldots \xrightarrow{\bar{\partial}} \mathcal{A}^{p,n}(M) \to 0$$

and associated cohomology

$$H^{p,q}_{Dolb}(M) = \frac{Ker(\bar{\partial} : \mathcal{A}^{p,q}(M) \to \mathcal{A}^{p,q+1}(M))}{Im(\bar{\partial} : \mathcal{A}^{p,q-1}(M) \to \mathcal{A}^{p,q}(M))}.$$

Note that if $(z^1_\alpha, \ldots, z^n_\alpha)$ are local holomorphic coordinates, then $(\frac{\partial}{\partial z^1_\alpha}, \ldots, \frac{\partial}{\partial z^n_\alpha})$ is a frame for $TM^{1,0}$ over U_α. If $X, Y \in \Gamma(M, TM^{1,0})$, i.e., smooth complex vector fields of type $(1, 0)$, then

$$X = \sum a^i_\alpha \frac{\partial}{\partial z^i_\alpha} \text{ and } Y = \sum b^j_\alpha \frac{\partial}{\partial z^j_\alpha} \text{ on } U$$

and an explicit calculation shows that

$$[X, Y] = XY - YX = \sum_{i=1}^{n}(X(b^i_\alpha) - Y(a^i_\alpha))\frac{\partial}{\partial z^i_\alpha}.$$

In particular, $\Gamma(M, TM^{1,0})$ is a Lie subalgebra of the Lie algebra of complex vector fields $\Gamma(M, (TM)^{\mathbb{C}})$.

Let us consider the condition that $\Gamma(M, TM^{1,0})$ is closed under the $[\cdot, \cdot]$-operation at the level of the real fields.

Recall the canonical map $\Gamma(M,TM) \to \Gamma(M,TM^{1,0})$, $X \mapsto X^{\mathbb{C}} = \frac{1}{2}(X - iJX)$. This is complex linear in the sense that $(JX)^{\mathbb{C}} = iX^{\mathbb{C}}$. For $X,Y \in \Gamma(M,TM)$ we compute the condition that $[X^{\mathbb{C}}, Y^{\mathbb{C}}]$ is again a $(1,0)$-field:

$$[X - iJX, Y - iJY] = ([X,Y] - [JX,JY]) - i([JX,Y] + [X,JY])$$

Thus $[X^{\mathbb{C}}, Y^{\mathbb{C}}] \in \Gamma(M,TM^{1,0})$ if and only if

$$(I) \qquad\qquad J([X,Y] - [JX,JY]) = [JX,Y] + [X,JY].$$

If condition (I) is satisfied for all $X,Y \in \Gamma(M,TM) = Vect(M)$, then the almost complex stucture J is said to be *involutive*. The *Complex Frobenius Theorem* states that J is involutive if and only if it is the associated almost complex structure to a complex manifold structure, i.e, it is defined by holomorphic coordinates.

We have shown that the involutivity of J is a necessary condition for it to be associated to a complex manifold structure. Conversely, given J, it is a simple matter to find locally standard frames $(X_1, \dots, X_n, JX_1, \dots, JX_n)$ for TM. The goal is to find coordinates $(z_1, \dots, z_n)$ which yield such frames. This problem can be formulated as a system of differential equations where (I) is the appropriate *integrability* condition.

We close this section with some basic examples of complex manifolds.

Let M be a 2-dimensional oriented manifold and let $g \in \mathcal{R}(M)$ be a Riemannian metric. Then there exists a unique $J \in Iso(TM)$ so that $g(v,Jv) = 0$ and (v,Jv) is positively oriented. Clearly, $J^2 = -Id$. In the 2-dimensional case, every almost complex structure is involutive, because $TM^{1,0}$ has rank equal to one. Consequently, every orientable 2- dimensional manifold has a complex stucture.

4.2 Moduli: Remarks on surfaces

If M is a compact orientable surface, then $dim_{\mathbb{R}} H^1_{deR}(M) = 2g$, where $g = 0, 1, 2, \dots$ This number g is called the genus of M. It can be shown that g is the only differentiable invariant for orientable surfaces:

Theorem. *Suppose M_1 and M_2 are compact orientable smooth surfaces. Then the following are equivalent:*
  (i) *M_1 and M_2 are diffeomorphic*
  (ii) *M_1 and M_2 are homeomorphic*
  (iii) *M_1 and M_2 have the same genus.*

Let M be an oriented manifold and $\mathcal{J}(M)$ the space of positively oriented almost complex structures. In $\mathcal{J}(M)$ let $\mathcal{J}_I(M)$ be the integrable structures. The group $Diff^+(M)$ of orientation preserving diffeomorphisms acts on $\mathcal{J}(M)$ preserving $\mathcal{J}_I(M)$.

The moduli space of all integrable complex structures on M is defined as the quotient

$$\mathcal{M}(M) := \mathcal{J}_I(M)/Diff^+(M).$$

In general it is extremely difficult to say something about this space: For example, the action of $Diff^+(M)$ is often not proper!

However, if M is a compact surface, then quite a lot about this space is known. For example, for $g \geq 2$ it has a natural structure of a $(3g - 3)$-dimensional complex space.[15]

We consider briefly the case of $g = 1$. Let $H^+ := \{z \in \mathbb{C} : Im(z) > 0\}$ be the upper half plane. A point $\tau \in H^+$ determines a *lattice* $\Gamma_\tau := ((1, \tau))_{\mathbb{Z}} = \{n + m\tau : n, m \in \mathbb{Z}\}$. This in turn determines a torus $T_\tau := \mathbb{C}/\Gamma_\tau$. The quotient map $\mathbb{C} \to \mathbb{C}/\Gamma_\tau$ is locally a diffeomorphism. Thus there is a unique complex manifold structure on T_τ such that it is locally biholomorphic.

The group $Sl_2(\mathbb{R})$ acts on H^+ by $\tau \mapsto \frac{a\tau + b}{c\tau + d}$ for a matrix $\begin{pmatrix} a & b \\ b & d \end{pmatrix} \in Sl_2(\mathbb{R})$. One can show that T_{τ_1} and T_{τ_2} are biholomorphically equivalent if and only if there exists $g \in \Lambda := Sl_2(\mathbb{Z})$ with $g(\tau_1) = \tau_2$.

Using methods that are well beyond the scope of the present text, one can show that every compact 1-dimensional complex manifold of genus $g = 1$ can be holomorphically realized as a torus T_τ. Up to a choice of a base point, this identification is canonical.

Thus the moduli space of such surfaces can be identified with the quotient H^+/Λ. The action of Λ on H^+ is not quite fixed point free, but it is proper and $H^+ \to H^+/\Lambda$ has discrete fibers. In fact one can equip H^+/Λ with a natural complex manifold structure. In this structure $H^+/\Lambda \cong \mathbb{C}$.

Exercise. Determine the $\tau's$ with largest isotropy group Λ_τ. What does Λ_τ have to do with the torus T_τ?

4.3 Projective algebraic manifolds

We now consider (compact) *projective algebraic manifolds*. For this let V be a complex vector space and $\mathbb{P}(V)$ be the Grassmannian $Gr_1(V)$ of 1-dimensional complex subspaces of V. The discussion of complex Grassmannians, which is quite analogous to that in the real case, will be carried out in latter sections. Here, since only *complex lines* are involved, it is an easier matter.

Consider the free, proper $\mathbb{C}^*$-action on $V \setminus \{0\}$ which is given by scalar multiplication $v \mapsto \lambda v$. Since the set of complex lines through 0, i.e, 1-dimensional complex subspaces, corresponds to the closures of orbits of this action, it follows that the quotient $\mathbb{P}(V) := (V \setminus \{0\})/\mathbb{C}^*$ parameterizes the lines. We equip $\mathbb{P}(V)$ with the differentiable quotient structure associated to $\mathbb{P} : V \setminus \{0\} \to \mathbb{P}(V)$. This $Gl_\mathbb{C}(V)$-homogenous space is equivariantly identifiable with the closed $Gl_\mathbb{C}(V)$-orbit in the Grassmannian of real 2-planes in V (see Section IV.1).

For the complex manifold structure on $\mathbb{P}(V)$ we proceed as follows: Let $F \in V^*$ be a complex linear functional on V. The function F is not well-defined on $\mathbb{P}(V)$, but, since $F(\lambda v) = \lambda \cdot F(v)$, its zero set $\{[v] : F(v) = 0\}$ is well-defined. Let $U_F := \{F \neq 0\}$ be its complement.

---

[15] A complex space has a Zariski open subset of points where it is a complex manifold. Its geometric points are locally analytic subsets of domains in $\mathbb{C}^n$.

Let $A_F := \{v \in V : F(v) = 1\}$ and note that $\mathbb{P} : A_F \to U_F$ is a diffeomorphism. In fact the inverse map $\varphi_F : U_F \to A_F$ is given by $[v] \mapsto \frac{v}{F(v)}$. The quotient $\frac{v}{F(v)}$ is well-defined on U_F. We refer to $\{U_F, \varphi_F\}_{F \in V^*}$ as the standard coordinate charts on $\mathbb{P}(V)$.

Exercise. Let $F_1, F_2 \in V^*$ and show that the change of coordinates $\varphi_{12} : U_{12} \subset A_1 \to U_{21} \subset A_2$ is given by $v \mapsto \frac{v}{F_2(v)}$. (Here $A_1 = A_{F_1}$, $U_{12} = \varphi(U_{F_1} \cap U_{F_2})$ etc.) Thus $\mathbb{P}(V)$ has a natural complex structure – even algebraic!

It is quite often convenient to compute in $\mathbb{P}(\mathbb{C}^{n+1})$, which is denoted by $\mathbb{P}_n(\mathbb{C})$, $n = dim_{\mathbb{C}} \mathbb{P}_n(\mathbb{C})$. The equivalence class of $(z_0, \dots, z_n) \in \mathbb{C}^{n+1} \setminus \{0\}$ is denoted by the *homogeneous coordinate* $[z_0 : \dots : z_n]$. Here $[z_0 : \dots : z_n] = [w_0 : \dots : w_n]$ if and only if there exists $\lambda \in \mathbb{C}^*$ such that $\lambda \cdot z_i = w_i$, $0 \le i \le n$. The functionals z_i give us a finite covering of $\mathbb{P}_n(\mathbb{C})$, $U_i := \{z_i \ne 0\}$, $0 \le i \le n$. The coordinates $(\xi_0, \dots, \xi_{i-1}, \xi_{i+1}, \dots, \xi_N)$ in U_i are defined by $\xi_\alpha := \frac{z_\alpha}{z_i}$.

Observe that, if $P \in S^k(V^*) =: \mathbb{C}[V]_k$ is a homogeneous polynomical of degree k, then $V(P) := \{[v] : P(v) = 0\}$ is a well-defined closed subset of $\mathbb{P}(V)$. Now $P|_{A_F}$ defines $V(P)$ on the coordinate chart U_F and therefore $V(P)$ is locally defined by complex analytic – even algebraic equations. More generally, if $P_i \in \mathbb{C}[V]_{k_i}$, $1 \le i \le m$, are homogeneous polynomials, then

$$V(P_1, \dots, P_m) := \{[v] : P_1(v) = \dots = P_m(v)\}$$

is, by the same reasoning, an analytic subset of $\mathbb{P}(V)$. Since the defining equations are polynomial, one refers to such subsets as being *algebraic*.

Remark. One naturally considers *analytic subsets* of $\mathbb{P}(V)$ which are locally defined by *holomorphic* equations. It turns out, however, that every analytic subset is an algebraic subvariety $V(P_1, \dots, P_m)$ as above (Chow's Theorem).

Algebraic subvarieties of $\mathbb{P}(V)$ are not necessarily smooth complex submanifolds.

Example. Let $F \in V^*$ and $V_F = \{v \in V : F(v) = 0\}$ be the associated linear hyperplane in V. Let $\mathbb{P}(V) := U_F \dot\cup \mathbb{P}(V(F))$ be the decomposition of $\mathbb{P}(V)$ which is determined by a coordinate chart U_F. In this way $\mathbb{P}(V)$ can be regarded as a compactification of the affine space $A_F \cong U_F$ by adding the projective hyperplane $\mathbb{P}(V)$ at infinity.

Let $X \hookrightarrow \mathbb{P}(V(F))$ be any subvariety. For $v_0 \in U_F$ a base point, let $K(X)$ be the union of the projective lines[16] in $\mathbb{P}(V)$ through v_0 and any point of X. The set $K(X)$ is called the *projective cone* over X. It can be shown that, if X is non-linear, then $v_0 \in K(X)$ is a singular point.

Exercise. Suppose that $X := \{P_1 = \dots = P_m = 0\}$, where P_i, $1 \le i \le m$, is a homogeneous polynomial on $V(F)$. Derive the equations for the cone $K(X)$.

Remark. In order to check whether or not a given variety $V = V(P_1, \dots, P_m)$ is smooth at $p \in V$, one may analyze the dependencies among the differentials

---

[16] A projective line through $p, q \in \mathbb{P}(V)$ is the set $\{\zeta p + \eta q : [\zeta, \eta] \in \mathbb{P}_1(\mathbb{C})\}$.

$dP_1, \ldots, dP_m$ on the coordinate charts. On the chart U_i this amounts to setting the homogenous coordinate z_i equal to 1 and regarding P as a function of the remaining variables. After de-homogenizing in this way, the differential dP is computed as usual. To indicate that the situation is really quite concrete, we mention the *genus formula* for curves in a 2-dimensinal projective space:

Let V be a 3-dimensional complex vector space and $\mathbb{P}(V)$ the associated projective space. A 1-dimensional subvariety $C \subset \mathbb{P}(V)$ is called a (complex) *curve*. It can be shown that such curves can be defined by *one* polynomial equation, i.e., $C = V(P)$ for some homogeneous polynomial. Every curve is connected.

Let C be a curve in a 2-dimensional projective space $\mathbb{P}(V)$ and define $d(C) :=: deg(C)$ to be the minimal degree of a defining polynomial for C.

Genus Formula. *Let C be a smooth curve of degree d in a 2-dimensional projective space $\mathbb{P}(V)$. Then*

$$g(C) = \frac{(d-1)(d-2)}{2}.$$

Thus if $P \in \mathbb{C}[V]_{100}$ is a (generic) homogeneous polynomial of degree 100, then $C = \{P = 0\}$ is a smooth compact 1-dimensional complex manifold with $49 \cdot 99$ holes!

4.4 Complex Lie groups and their actions

A complex Lie group G is a (paracompact) complex manifold with holomorphic group multiplication $G \times G \to G$, $(g, h) \mapsto gh^{-1}$. Many such groups can be realized as subgroups of $Gl_n(\mathbb{C})$ which inherits its complex manifold structure from its realization as an open subset of the matrices: $Gl_n(\mathbb{C}) = \{A \in Mat(n \times n, \mathbb{C}) : det(A) \neq 0\}$.

If $G < Gl_n(\mathbb{C})$ is a complex submanifold, then we refer to it as a *linear complex group*.

Example. Let

$$U := \left\{ \begin{pmatrix} 1 & x & z \\ 0 & 1 & y \\ 0 & 0 & 1 \end{pmatrix} : x, y, z \in \mathbb{C} \right\}$$

be the (*nilpotent*) group of upper triangular matrices in $Gl_3(\mathbb{C})$. Then U, which is biholomorphically equivalent to $\mathbb{C}^3$ as a complex manifold, is clearly a linear complex group. Let the Gaussian integers $\Gamma := \mathbb{Z}[i]$ be embedded in U by

$$\Gamma \ni \xi \mapsto \begin{pmatrix} 1 & 0 & \xi \\ 0 & 1 & 0 \\ 0 & 0 & 1 \end{pmatrix}.$$

Then Γ is a discrete normal subgroup of U and $G := U/\Gamma$ is likewise a complex Lie group.[17] Note that G contains the *compact* complex torus $T = \mathbb{C}/\Gamma$, where

---

[17] We will discuss the complex structure on homogeneous spaces G/H of a complex Lie group in the sequel.

$\mathbb{C}$ is embedded in U by

$$\mathbb{C} \ni z \mapsto \begin{pmatrix} 1 & 0 & z \\ 0 & 1 & 0 \\ 0 & 0 & 1 \end{pmatrix}.$$

Remark. If X is a *compact* (connected) complex manifold, then it follows from the maximum principle that $\mathcal{O}(X) \cong \mathbb{C}$, i.e., X possesses only constant holomorphic functions.

Note that, since it is an open subset of $Mat(n \times n, \mathbb{C}) \cong \mathbb{C}^{n^2}$, given $p, q \in Gl_n(\mathbb{C})$ with $p \neq q$, there exists $f \in \mathcal{O}(Gl_n(\mathbb{C}))$ such that $f(p) \neq f(q)$. Consequently, any holomorphic mapping $\varphi : G \to Gl_n(\mathbb{C})$ of the above group G is constant on the torus T and in particular G is not realizable as a linear group.

Remark. Of course T itself is not realizable as a linear group, but we discussed the more complicated example of $G = U/\Gamma$ for the following reason: If we replace Γ by $\tilde{\Gamma} := \mathbb{Z}$, then the quotient $\mathbb{C}/\tilde{\Gamma} \cong \mathbb{C}^*$ is certainly realizable as a linear group. However, this is still *not* the case for $G = U/\tilde{\Gamma}$.

Many complex Lie groups do occur as closed linear subgroups of some $Gl_n(\mathbb{C})$. Quite often they arise as *algebraic* submanifolds. In this regard it is important to note that $Gl_n(\mathbb{C})$ can itself be realized as an affine variety, e.g., by the map $A \mapsto (A, \frac{1}{det A}) \in \mathbb{C}^{n^2} \times \mathbb{C}$. Hence, closed subgroups of $Gl_n(\mathbb{C})$ which are defined by (holomorphic) algebraic equations are referred to as *(affine) algebraic groups*.

Examples. The groups

$$SL_n(\mathbb{C}) = \{A \in Gl_n(\mathbb{C}) : det(A) = 1\},$$

$$O_n(\mathbb{C}) := \{A \in Gl_n(\mathbb{C}) : AA^t = Id\},$$

$$B_n := \left\{ A \in Gl_n(\mathbb{C}) : A = \begin{pmatrix} \lambda_1 & & * \\ & \ddots & \\ 0 & & \lambda_n \end{pmatrix} \right\},$$

$$N_n := \left\{ A \in GL_n(\mathbb{C}) : A = \begin{pmatrix} 1 & & * \\ & \ddots & \\ 0 & & 1 \end{pmatrix} \right\}$$

and

$$T_n := \left\{ A \in GL_n(\mathbb{C}) : A = \begin{pmatrix} \lambda_1 & & 0 \\ & \ddots & \\ 0 & & \lambda_n \end{pmatrix} \right\}$$

are examples of *complex* linear algebraic groups. The unitary group $U_n = \{A \in Gl_n(\mathbb{C}) : A\bar{A}^t = 0\}$ is not defined by holomorphic polynomial equations. It is of course an example of a *real* algebraic group.

If G is a complex Lie group and X is a complex manifold, then a G-action on X is said to be a *holomorphic action* if the action map $G \times X \to X$ is holomorphic.

The actions of a complex Lie group on itself by left- and right-multiplication and by conjugation are holomorphic.

A *holomorphic vector field* on a complex manifold X is a $(1,0)$-field

$$X \in \Gamma(X, TX^{1,0})$$

such that for any open subset $U \subset X$ and $f \in \mathcal{O}(U)$ it follows that $X(f) \in \mathcal{O}(U)$. If $(z_1, \ldots, z_n)$ are holomorphic coordinates on some open set $U \subset X$, then

$$X = \sum a_i \frac{\partial}{\partial z_i}, \quad \text{where } a_i \in \mathcal{O}(U), \quad 1 \le i \le n.$$

A local $\mathbb{C}$-action is defined in an analogous way to the notion of a local $\mathbb{R}$-action. In particular, if $g = g(z)$ is a holomorphic local $\mathbb{C}$-action and $f \in \mathcal{O}(U)$, then

$$X(f)(x) := \frac{d}{dz}\Big|_{z=0} f(g(z)(x))$$

defines a holomorphic vector field X. Conversely, to every holomorphic vector field there is a uniquely determined local $\mathbb{C}$-action which defines the field in this way.

Notation. Let $Vect_{\mathcal{O}}(X)$ denote the Lie algebra of holomorphic vector fields on the complex manifold X.

The above remarks show that elements of $Vect_{\mathcal{O}}(X)$ correspond to local 1-parameter group actions.

If X is compact, then the fields $X \in Vect_{\mathcal{O}}(X)$ yield *global* $\mathbb{C}$-actions. In fact in this case we have the following basic result.

Proposition. (Bochner-Montgomery) Let X be a compact complex manifold. Then, equipped with the compact-open topology, the group $G := Aut_{\mathcal{O}}(X)$ of holomorphic automorphisms of X is a complex Lie group. The natural action $G \times X \to X$ is holomorphic and $Lie(G) = Vect_{\mathcal{O}}(X)$. In particular, $Vect_{\mathcal{O}}(X)$ is finite-dimensional.

The last statement is a particular case of the important

Finiteness-Theorem (Cartan-Serre). *Let $E \to X$ be a holomorphic[18] vector bundle over a compact complex manifold X. Then the space of sections $\Gamma(X, E)$ is finite-dimensional.*

If G is an algebraic group and X an algebraic variety, e.g., an affine variety or a subvariety of some $\mathbb{P}(V)$, then a G-action on X is said to be *algebraic* if the action map is an algebraic morphism, i.e, its *graph* is an algebraic subvariety of the product space.

It follows from Chevalley's Theorem (see Section IV.4) that orbits of algebraic G-actions are Zariski open in their closures.

Let G be a complex Lie group. The complex structure J on TG defines a complex structure on $Vect_{\mathbb{R}}(G)$ which stabilizes both the spaces of left- and right-invariant

---

[18] This means that the transition maps $(x, v) \mapsto (x, g_{\alpha\beta}(x)(v))$ are holomorphic.

fields. These fields are indeed *real* vector fields, but they are defined by real 1-parameter subgroups of G which are acting by holomorphic transformations. Thus $Lie(G) =: (\mathfrak{g}, [\cdot, \cdot])$ comes equipped with a complex structure J such that $[\cdot, \cdot]$ is J-bilinear : $J[v, w] = [Jv, w] = [v, Jw]$ for all $v, w \in \mathfrak{g}$. Lie algebras of this type are called *complex* Lie algebras.

Using the identification $(\mathfrak{g}, J) \simeq \mathfrak{g}^{1,0} \hookrightarrow \mathfrak{g}^{\mathbb{C}}$, we regard $X^{\mathbb{C}} = \frac{1}{2}(X - iJX)$ as a holomorphic vector field on G. The same argumentation as in the case of real Lie groups shows that the fields $Z \in \mathfrak{g}^{1,0}$ are *globally* integrable and consequently we have the holomorphic exponential map $exp : \mathfrak{g}^{1,0} \to G$ which, since $exp_*(0) = Id$, is biholomorphic in a neighborhood of $0 \in \mathfrak{g}^{1,0}$. For brevity, if there is no confusion in notation, we replace $\mathfrak{g}^{1,0}$ by $\mathfrak{g}$.

Suppose G is a complex Lie group acting *properly* and holomorphically on a complex manifold X. Then we have the quotient $X \to X/G$. The space X/G has the natural structure of a complex space. Even in the simplest of cases it may not be smooth. Singularities quite often arise in the case of finite groups.

Example. Let $X = \mathbb{C}^2$ and let $G = \{\sigma, e\} \cong \mathbb{Z}_2$ act on X by $\sigma(z, w) = (-z, -w)$. The invariants of this *linear* action are generated by the functions z^2, zw, and w^2. The quotient is the subvariety $Y = X/G \hookrightarrow \mathbb{C}^3$ defined by the image of the map

$$F : \mathbb{C}^2 \to Y \hookrightarrow \mathbb{C}^3, (z, w) \mapsto (z^2, zw, w^2).$$

Thus $Y = \{(x_1, x_2, x_3) \in \mathbb{C}^3 : x_2^2 = x_1 x_3\}$. A simple check of the differential of the defining function $f = x_2^2 - x_1 x_3$ shows that Y is smooth at all of its points $y \neq y_0 = (0, 0, 0)$. But $F : \mathbb{C}^2 \backslash \{(0, 0)\} \to Y \backslash \{y_0\}$ is a $2 : 1$ covering map; in particular $Y \backslash \{y_0\}$ is not even simply-connected! If Y were smooth in a neighborhood of y_0, it would have neighborhoods which are (real) 4-dimensional cells and the complement of a point in such would certainly be simply-connected.

Proposition. *If the complex Lie group G is acting freely and properly on a complex manifold X, then there exists a unique structure of a complex manifold on the quotient X/G so that $X \to X/G$ is a holomorphic principal bundle.*

Proof. One definition of a holomorphic principal bundle is exactly what is assumed here, i.e., $X \to X/G$, where the complex Lie group is acting holomorphically, properly and freely on X. Here we observe that this fibration is locally trivial, because the slice Σ constructed in the real version of this result (see Section III.5) can be taken to be *any* sufficiently small complex submanifold of an open neighborhood $U = U(x)$ which is transversal to the orbit $G.x$ at x. The transition diffeomorphisms are therefore given by holomorphic maps $g_{\alpha\beta} : U \to G$. □

An important example of a proper, free holomorphic action is given by a closed complex subgroup H of a complex Lie group G. In this case the quotient G/H by the action of H by right-multiplication carries a unique complex structure so that $G \to G/H$ is an H-principal bundle. The action of G on G/H which is induced by left-multiplication is holomorphic.

Analogous to the situation for real Lie groups, if $G \times X \to X$ is a holomorphic action of a complex Lie group on a complex manifold, the isotropy subgroup G_x of a point is a closed complex subgroup and the canonical map $G/G_x \to G.x \hookrightarrow X$ is a G-equivariant holomorphic injection which identifies the homogeneous space G/G_x with the orbit $G.x$.

If G is a linear algebraic subgroup and H is a closed algebraic subgroup, then the homogeneous space $X = G/H$ carries the structure of an algebraic variety so that the natural G-action $G \times X \to X$ is algebraic. Conversely, we have the following basic fact.

Theorem. (Chevalley) *Let G be a linear algebraic group and H an algebraic subgroup.[19] Then there is an algebraic representation $\rho : G \to GL(V)$ on a complex vector space V and a point $x \in \mathbb{P}(V)$ such that $H = G_x$, i.e., the abstract homogeneous space G/H can be identified with the orbit $G.x$.*

4.5 Examples of complex homogeneous spaces

Projective spaces. Consider the natural transitive algebraic action of $G := Gl_{\mathbb{C}}(V)$ on $V\setminus\{0\}$. Let $v_0 \in V\setminus\{0\}$ be a base point and $H = G_{v_0}$. If $L := [v_0]$ is the line determined by v_0, then denote by P the isotropy group $G_{[v_0]}$. Set-theoretically it is clear that, since G acts transitively on $\mathbb{P}(V)$, we have the identificaiton $G/P = \mathbb{P}(V)$. In fact, this is an identification of algebraic varieties so that the induced G-action on $\mathbb{P}(V)$ is algebraic.

Exercise. Show that H is a normal subgroup of P so that $V\setminus\{0\} = G/H \to G/P$ is a $\mathbb{C}^*$-principal bundle.

Discrete Subgroups. A subgroup Γ in a Lie group G is called *discrete* whenever it is a discrete set in the induced topology. In this case we have the homogeneous space $X = G/\Gamma$. If G is a complex Lie group, then X is a complex manifold and the G-action is holomorphic. The quotient $G \to G/\Gamma$ is a holomorphic covering map. Discrete groups occur in important ways in many areas of mathematics, e.g., number theory, moduli space theory, etc. In certain situations one starts with a well-known complex Lie group G, e.g., $Sl_2(\mathbb{C}) := \{A \in Gl_2(\mathbb{C}) : det(A) = 1\}$ and a discrete subgroup Γ, and one ends up with a *very* mysterious complex homogeneous space G/Γ. For example, the following question is open.

Question. Let $G := Sl_2(\mathbb{C})$ and $\Gamma < G$ be a discrete subgroup so that $X = G/\Gamma$ is compact. Does X contain a smooth curve C as a complex submanifold with $g(C) \neq 1$?

Often discrete groups arise as something like the "integral" points in a Lie group G. For example, if $G = Sl_2(\mathbb{C})$, then we may consider the subgroup Γ consisting of the matrices whose coefficients are in the Gaußian integers $\mathbb{Z}[i]$. The complex manifold $X = G/\Gamma$ is really quite interesting.

---

[19] Algebraic subgroups are automatically closed.

Compact quotients can also be made in this way: Let $G := N_n$ be the nilpotent group of upper-triangular matrices and Γ the subgroup of matrices with entries in $\mathbb{Z}[i]$. Then G/Γ is compact.

Even the *abelian* case can be extremely interesting and occurs non-trivially in many areas of mathematics and physics, e.g., soliton theory. If $G = (\mathbb{C}^n, +)$ and Γ is a discrete subgroup so that $T = G/\Gamma$ is compact, then we refer to T as a *complex torus*. In this case $\Gamma = ((v_1, \dots, v_{2n}))_{\mathbb{Z}}$ is generated over the integers by $2n$ $\mathbb{R}$-linearly independent vectors. One can normalize the first n vectors to be the standard basis, i.e., $\Gamma = ((e_1, \dots, e_n, v_1, \dots, v_n))_{\mathbb{Z}}$, and regard the last n vectors as generating an $(n \times n)$-complex matrix Z. If T_1 and T_2 are defined by generically chosen matrices Z_1 and Z_2, then T_1 and T_2 are *not* biholomorphically equivalent. If $n > 1$, unlike the case $n = 1$, unless additional structures are imposed, there is no reasonable moduli space of all such tori.

Complex Grassmanians and Flag manifolds. Here we equip the Grassmanian $Gr_k(V)$ of k-dimensional complex subspaces of a complex vector space V with the structure of a complex algebraic homogeneous manifold. For brevity let $G = Gl_{\mathbb{C}}(V)$, W_0 be a base point in $Gr_k(V)$ and P_k the stabilizer of W_0, i.e., $P_k = \{g \in G : g(W_0) = W_0\}$. The group G clearly acts transitively on $Gr_k(V)$. For example, choose a basis $(e_1^0, \dots, e_k^0)$ for W_0 and, if W is some other complex k-plane, choose a basis $(e_1, \dots, e_k)$ for it and define $T \in Gl_{\mathbb{C}}(V)$ to be an arbitrary linear isomorphism which is an extension of the transformation which takes e_i^0 to $e_i, 1 \leq i \leq k$.

In matrix form, if $V = \mathbb{C}^n$, $(e_1, \dots, e_n)$ is the standard basis and

$$W_0 = ((e_1, \dots, e_k))_{\mathbb{C}},$$

then

$$P_k := \left\{ \left(\begin{array}{c|c} * & * \\ \hline 0 & * \end{array} \right) \right\}.$$

Since G acts transitivity on $Gr_k((V)$, we may define the complex algebraic manifold structure by the identification $Gr_k(V) = G/P_k$.

Let $h \in \mathcal{H}(V)$ be a Hermitian structure on V and $K := U(V, h)$ the corresponding unitary group. Recall that K is a maximal compact subgroup of G. Using the same argument as above, except that this time we choose unitary bases, i.e., orthonormal with respect to h, it follows that K acts transitively on $G_k(V)$ as well. Thus $Gr_k(V) = K/L$, where $L = K \cap P$.[20] In the representation $G_k(V) = K/L$ we do not see the complex structure, but we do see that $Gr_k(V)$ is compact.

To realize $Gr_k(V)$ as a G-orbit in some projective space, consider the space $\Lambda^k V$ with the linear G-action which is defined by extending the definition on pure wedge products, $g(v_1 \wedge \dots \wedge v_k) = g(v_1) \wedge \dots \wedge g(v_k)$, to the whole space.

---

[20] In matrices as above, $L = \left\{ \left(\begin{array}{cc} * & 0 \\ 0 & * \end{array} \right) \in K \right\} \cong U_k \times U_{n-k}$.

If $W_0 = ((e_1^0, \ldots, e_k^0))_{\mathbb{C}}$, then we consider $w_0 := e_1^0 \wedge \ldots \wedge e_k^0 \in \Lambda^k V$ and are in a situation which is completely analogous to that of projective space:

$$Gr_k(V) = G \cdot [w_0] \subset \mathbb{P}(\Lambda^k V).$$

This G-equivariant embedding of $Gr_k(V)$ is sometimes called the *Plücker embedding*.

Remark. Note that the center $Z_G \cong \mathbb{C}^*$ of $Gl_{\mathbb{C}}(V)$ acts trivially on $Gr_k(V)$ for all k. In fact it is exactly the ineffectivity of the G-action. Since $Gl_{\mathbb{C}}(V) = Sl_{\mathbb{C}}(V) \cdot Z_G$, one often replaces G by $S := Sl_{\mathbb{C}}(V)$.

A *flag* of length k of subspaces in a complex vector space V is an increasing sequence of complex subspaces

$$\{0\}) \subset V_1 \subset V_2 \subset \cdots \subset V_{k+1} = V$$

Associated to such a flag is the vector of codimensions $\mathfrak{a} = (a_1, \ldots, a_k)$, where $a_i := codim_{V_i} V_{i-1}$. Let $F_{\mathfrak{a}}$ denote the *flag manifold* of all flags with codimension vector $\mathfrak{a}$.

As before, let $G := Gl_{\mathbb{C}}(V)$ and $K := U(V, h)$ be the unitary subgroup with respect to a fixed Hermitian structure. Using the same type of basis arguments as we did in the case of Grassmannians, one easily shows that G and K act transitively on $F_{\mathfrak{a}}$. If F_0 is a base point in $F_{\mathfrak{a}}$ and $P_{\mathfrak{a}}$ its isotropy group in G, then $L_{\mathfrak{a}} := K \cap P_{\mathfrak{a}}$ is its K-isotropy.

The identification $F_{\mathfrak{a}} = G/P_{\mathfrak{a}}$ equips $F_{\mathfrak{a}}$ with the structure of a G-homogeneous algebraic variety. The fact that K acts transitively and $F_{\mathfrak{a}} = K/L_{\mathfrak{a}}$ shows that $F_{\mathfrak{a}}$ is compact.

Exercise. Compute the isotropy $P_{\mathfrak{a}}$ in the case $V = \mathbb{C}^n$, $V_1 = \{e_1, \ldots, e_{a_1}\}_{\mathbb{C}}$, $V_2 = \{e_1, \ldots, e_{a_1 + a_2}\}_{\mathbb{C}}, \ldots$.

5. Symplectic manifolds

5.1 Basic properties

A symplectic manifold is a pair (M, ω), where M is a differentiable manifold and ω is a *closed, non-degenerate* 2-form on M, i.e., $d\omega = 0$ and $(T_p M)^{\perp \omega_p} = \{0\}$ for all $p \in M$. Note that if (M, ω) is symplectic and $i : U \hookrightarrow M$ is an *open* subset, then $(U, i^*\omega)$ is likewise symplectic.

Example. A symplectic vector space (V, ω) is a symplectic manifold. The condition $d\omega = 0$ is automatically fulfilled for linear forms. For concrete examples recall the standard structure ω_{std} on $W = V \oplus V^*$: $\omega_{std}((v, f), (w, g)) = g(v) - f(w)$. If we write $\mathbb{R}^{2n} = \mathbb{R}^n \oplus (\mathbb{R}^n)^*$ and let $(q_1, \ldots, q_n, p_1, \ldots, p_n)$ be standard coordinates, then $\omega_{std} = \sum_{i=1}^n dq_j \wedge dp_j$.

The above concrete examples are special cases of an important general

Fact. The cotangent bundle $M = T^*Q$ of an arbitrary differentiable manifold Q has a canonical symplectic structure.

For this let $\pi : T^*Q \to Q$ be the standard projection. Denote a point in T^*Q by ν_p, where ν_p is a cotangent vector at the point $p \in Q$. If $t \in T_{\nu_p}(T^*M)$, then $\pi_*(t) \in T_pM$. Thus we define a 1-form θ on T^*Q by

$$\theta(\nu_p)(t) := \nu_p(\pi_*(t))$$

and let $\omega = -d\theta$.

It is useful to calculate θ in local coordinates. For this let $(q_1, \dots, q_n)$ be coordinates on Q. A cotangent vector $\nu_q \in T^*Q$ is uniquely described by

$$(*) \qquad \nu_q = p_1 dq_1|_q + \dots + p_n dq_n|_q.$$

Thus we regard $(q_1, \dots q_n, p_1, \dots, p_n)$ as coordinates in T^*Q over the coordinate chart $(q_1, \dots, q_n)$. Since θ is a 1-form which vanishes on the vertical fields $\frac{\partial}{\partial p_i}, 1 \leq i \leq n$, it follows that $\theta = f_1 dq_1 + \dots + f_n dq_n$, where f_i are functions to be determined, $1 \leq i \leq n$. Finally, if $\frac{\partial}{\partial q_i}$ is regarded at a point ν_q with coordinates (p, q), then ν_q has the form $(*)$ above and $\theta(\frac{\partial}{\partial q_i}) = p_i, 1 \leq i \leq n$. Thus

$$\theta = \sum p_i dq_i \text{ and } \omega_{std} = \sum dp_i \wedge dq_i.$$

In particular ω_{std} is non-degenerate and (T^*Q, ω_{std}) is a symplectic manifold.

Remark. (Liouville-Form) Let (M, ω) be a symplectic manifold and note that, since every tangent space has an induced symplectic structure, $dim_{\mathbb{R}} M = 2n$. The non-degeneracy of ω is equivalent to the fact that $\omega^n := \omega \wedge \dots \wedge \omega$ is nowhere zero. This form, or a normalized version such as $\frac{1}{n!}\omega^n$, is called the associated *Liouville form*. From time to time it will be regarded as a measure. In particular, the Liouville form defines a canonical associated orientation on M.

Corollary. *A symplectic manifold is orientable.*

In the 2-dimensional case the existence of an orientation is the only obstruction for a manifold to possess a symplectic structure.

Exercise. Let ω be an orientation form on M, where $dim_{\mathbb{R}} M = 2$. Show that (M, ω) is a symplectic manifold.

A far more serious obstruction is of cohomological nature: If (M, ω) is a compact symplectic manifold, then $[\omega] \neq 0 \in H^2_{deR}(M)$.

Proof. If $\omega = d\alpha$, then

$$0 < \int_M \omega^n = \int_M d\alpha \wedge \omega^{n-1} = \int_M d(\alpha \wedge \omega^{n-1}) = 0,$$

where the last step follows from Stokes' Theorem. $\qquad\qquad\square$

Consequently the only sphere which is symplectic is $M = S^2$.

Note that a symplectic manifold can also be equipped with the *opposite* structure which is obtained by replacing ω with $-\omega$. If (M_1, ω_1) and (M_2, ω_2) are symplectic, then $M := M_1 \times M_2$ has the natural product structure $\omega := \pi_1^*(\omega_1) + \pi_2^*(\omega_2)$, where

π_1 and π_2 are the projections. Sometimes it is convenient to equip M with the symplectic form $\pi_1^*(\omega_1) - \pi_2^*(\omega_2)$.

One origin of the study of sympletic geometry is *classical mechanics*. In this context physical laws are formulated in terms of *Hamiltonian Fields*.

To introduce this notion, let M a differentiable manifold, $X \in Vect(M)$ a vector field and let $g_X(t)$ denote the associated local $\mathbb{R}$-action, i.e., at the level of functions, $X(f) = \frac{d}{dt}\big|_{t=0} g_X^*(t)(f)$. This defining action has a natural extension to, e.g., the full exterior algebra $\mathcal{E}^*(M) = \oplus \mathcal{E}^k(M)$:

$$\mathcal{L}_X : \mathcal{E}^k(M) \to \mathcal{E}^k(M): \qquad \alpha \mapsto \frac{d}{dt}\Big|_{t=0} g_X^*(t)(\omega).$$

Of course there are other operators on $\mathcal{E}^*(M)$, e.g., $d : \mathcal{E}^*(M) \to \mathcal{E}^*(M)$ which *increases* degree by 1.

For $X \in Vect(M)$, *contraction* $i_X : \mathcal{E}^*(M) \to \mathcal{E}^*(M)$ is defined by $i_X \alpha = \alpha(X, \ldots, \cdot)$. This operator *lowers* degree by 1.

Cartan's formula, which can be proved by direct calculation, relates the three operations:

$$\mathcal{L}_X = d \circ i_X + i_X \circ d.$$

Now let (M, ω) be a symplectic manifold and notice that contraction with ω yields an isomorphism $Vect(M) \simeq \mathcal{E}^1(M)$.[21]

Since $d\omega = 0$, if we follow this with $d : \mathcal{E}^1(M) \to \mathcal{E}^2(M)$, we arrive at the *local orbit* of ω in $\mathcal{E}^2(M)$ by the action of $Vect(M) : \omega \mapsto \mathcal{L}_X\omega = d \circ i_X\omega$. The isotropy of this local action is regarded as the Lie algebra of *local symmetries* of ω. If we regard X as a smooth dynamical system, then such local symmetries have ω as a *constant of motion*, e.g., if ω is considered to measure area, then *area is a constant of motion*.

Notation. Let (M, ω) be a symplectic manifold. Then $Ham_{loc}(M) := \{X : \mathcal{L}_X\omega = 0\}$ denotes the space of (local) *Hamiltonian fields*.

If (M, ω) is a symplectic manifold, let $Diff_\omega(M)$ denote the group of diffeomorphisms of M which preserve ω, i.e., $Diff_\omega(M) := \{g \in Diff(M) : g^*(\omega) = \omega\}$. Elements of $Diff_\omega(M)$ are sometimes called *symplectomorphisms*.

Proposition. *If $X \in Ham_{loc}(M)$ comes from a globally defined $\mathbb{R}$-action $g_X(t)$, then $g_X(t) \in Diff_\omega(M)$ for all t.*

Proof. The curve $\omega(t) := g^*(t)(\omega)$ is in some appropriate infinite dimensional space where the rules of elementary calculus hold. From the group law and the fact that $\mathcal{L}_X\omega = 0$ it follows that $\frac{d\omega(t)}{dt} = 0$. $\qquad\qquad\square$

---

[21] The non-degeneracy of ω is equivalent to the fact that contraction establishes an isomorphism between the tangent and cotangent bundles.

Thus it of interest to produce globally integrable Hamiltonian fields.

Recall that $\mathcal{L}_X\omega = d \circ i_X\omega$. With the above discussion in mind, it is relevant to determine those fields X such that $i_X\omega$ is closed, i.e., *locally* $i_X\omega = dH$, where $H \in \mathcal{E}(U)$ is a smooth function on a neighborhood $U = U(x)$ of a given point with $H^1_{deR}(U) = \{0\}$.

Conversely, even globally there exists the natural map

$$\nu : \mathcal{E}(M) \to Ham_{loc}(M), \quad H \mapsto X_H,$$

where X_H is uniquely determined by $dH = i_{X_H}\omega$. Since $ker(\nu)$ just consists of the locally constant functions, one produces in this way numerous Hamiltonian fields.

Remark. If H has compact support, then so does the field X_H and consequently X_H is globally integrable. In this way we already observe that $Diff_\omega(M)$ is infinite-dimensional.

5.2 The Lie algebra of $Diff_\omega(M)$

It is possible to think about $Diff(M)$ as an infinite-dimensional manifold with compatible group stucture. For now we are only interested in the notion of a *smooth curve*: A curve $I = (-\epsilon, \epsilon) \to Diff(M), t \mapsto \varphi_t$, is said to be *smooth* if the induced map, $(x, t) \mapsto \varphi_t(x)$, is smooth.

Let $e \in Diff(M)$ denote the identity and $\mathcal{C}$ be the set of smooth curves φ with $\varphi(0) = e$ and let $\sim$ be the equivalence relation defined on $\mathcal{C}$ by

$$\varphi^1 \sim \varphi^2 : \iff (\varphi^1_t)^{-1} \circ \varphi^2_t = e + O(t^2).$$

This can be made precise in several ways. For example, if we embed M as a submanifold of a vector space V, then $(\varphi^1_t)^{-1} \circ \varphi^2_t$ can be regarded as an element of $C^\infty(M, V)$. Since the definition of $\sim$ only requires small t, this can also be considered pointwise in local coordinates.

The tangent space $T_e(Diff(M))$ is defined as $\mathcal{C}/\sim$. At this point there is no topology on this set.

Now let φ_t be a curve in $Diff(M)$ with $\varphi_0 = Id_M = e$. For each t we have the vector field X_t defined by $X_t \circ \varphi_t = \frac{d}{dt}\varphi_t$. This should be interpreted as follows:
$$X_{t_0}(f)(p) = \frac{d}{dt}\Big|_{t=t_0} f(\varphi_t(\varphi_{t_0}^{-1}(p))) \text{ for } f \text{ in } \mathcal{E}(M).$$

Let $N = M \times \mathbb{R}$ be the trivial family of manifolds over $\mathbb{R}$:

$$M = M_t := M \times \{t\} \hookrightarrow N.$$

The smooth curve φ_t defines a smooth map $\varphi : U \to N$ of an appropriately chosen neighborhood of $\{0\} \times N$ given by

(∗) $(s, (x, t)) \mapsto (\varphi_{t+s} \circ \varphi_t^{-1}(x), t + s).$

This defines a vector field on Z_φ on N. For t fixed, $(\pi_1)_*(Z_\varphi\big|_{M_t}) = X_t$ and $(\pi_2)_*(Z_\varphi) = \frac{\partial}{\partial t}.$

Notation. A vector field $Z \in Vect(M \times \mathbb{R})$ such that $(\pi_2)_*(Z)$ is well-defined and equal to $\frac{\partial}{\partial t}$ is called a *time dependent* vector field on M. The space of such fields is denoted by $Vect_t(M)$.

To a field $Z \in Vect_t(M)$ one has the local $\mathbb{R}$-action on $\mathbb{R} \times M$ which is given by $g_Z(s)(x,t) = (\psi(s,(x,t)), t + s)$. If $\varphi_t(x) := \psi(t,(x,0))$, then the condition of being a local $\mathbb{R}$-action implies that $\psi(s,(x,t)) = \varphi_{t+s} \circ \varphi_t^{-1}(x)$ as in $(*)$ above. This simply means that the integral curves of Z are the orbits $(\varphi_t(x), t)$ starting at the initial manifold M_0.

Exercise. Show that the map $\mathcal{C} \to Vect_t(M), \varphi \mapsto Z_\varphi$, is surjective and, for two curves φ^1 and φ^2 of diffeomorphisms,

$$X_0^1 = X_0^2 \iff \varphi^1 \sim \varphi^2 .$$

Corollary. *There is a natural identification $T_e(Diff(M)) = Vect(M)$.*

Corollary. *The tangent space $T_e(Diff_\omega(M))$ of the group of symplectic diffeomorphisms of a symplectic manifold (M, ω) can be naturally identified with $Ham_{loc}(M)$.*

Proof. If φ_t is a curve in $Diff_\omega(M)$, then $\frac{d}{dt}(\varphi_t^*(\omega)) = 0$. $\qquad\square$

Exercise. Show that $Ham_{loc}(M)$ is a Lie subalgebra of $Vect(M)$.

(Hint: Derive the general formula $\mathcal{L}_{[X,Y]} = \mathcal{L}_X \circ \mathcal{L}_Y - \mathcal{L}_Y \circ \mathcal{L}_X = [\mathcal{L}_X, \mathcal{L}_Y]$.)

Thus we refer to $Ham_{loc}(M)$ as the Lie algebra $Lie(Diff_\omega(M))$.

5.3 Time dependent fields: Moser's method

Let $t \mapsto \omega_t$ be a curve of forms on a smooth manifold M, $t \mapsto \varphi_t$ a curve of diffeomorphisms with $\varphi_0 = e$, $Z = Z_\varphi$ the associated *time dependent* vector field on $M \times \mathbb{R}$ and $t \mapsto X_t$ the associated curve of vector fields on M. Finally, let $i_t : M \hookrightarrow M \times \mathbb{R} =: N$, $x \mapsto (x, t)$, be the standard embedding.

If $\nu \in \mathcal{E}^*(N)$, then we have Cartan's formula

$$\mathcal{L}_Z \nu = i_Z \circ d\nu + d \circ i_Z \nu.$$

Let ν be the unique form defined by $i_t^* \nu = \omega_t$ and $i_{\frac{\partial}{\partial t}} \nu = 0$. Since

$$g_Z(s) \circ i_0 = i_s \circ \varphi_s \quad \text{and} \quad \frac{d}{dt}\Big|_{t=s} \Big((g_Z(t))^* \nu \Big) = (g_Z(s))^* (\mathcal{L}_Z \nu)$$

it follows from Cartan's formula for $\mathcal{L}_Z \nu$ that

$$\frac{d}{dt}(\varphi_t^* \omega_t) = \varphi_t^* \Big(d \circ i_{X_t} \omega_t + i_{X_t} \circ d\omega_t + \frac{d\omega_t}{dt} \Big).$$

We refer to this as the *Basic Identity for Time Dependent Fields*. Moser's Method provides an Ansatz for constructing a curve of diffeomorphisms φ_t so that

$$\frac{d}{dt}(\varphi_t^* \omega_t) = 0, \quad \text{i.e.,} \quad \varphi_t^* \omega_t = \omega_0.$$

In other words, one should try to realize a given curve of differential forms by a curve of diffeomorphisms which is constructed from a time dependent vector field.

Let (M, ω) be an arbitrary symplectic manifold and $x_0 \in M$ an arbitrary point. We may choose smooth coordinates $(q, p) = (q_1, \ldots, q_n, p_1, \ldots, p_n)$ so that x_0 corresponds to $0 \in \mathbb{R}^{2n}$ and $\omega(0) = \omega_{std}(0)$.

Lemma. *Let $U = U(0)$ be a neighborhood of $0 \in \mathbb{R}^n$ and ω_0, ω_1 be symplectic forms on U. Suppose that $\omega_0(0) = \omega_1(0)$. Then there is a diffeomorphism $\varphi : V \to W \subset U$ defined in a possibly smaller neighborhood $V = V(0)$ so that $\varphi(0) = 0$ and $\varphi^* \omega_1 = \omega_0$.*

Before giving the proof we state a basic consequence:

Corollary. (Darboux's Theorem) *Let (M, ω) be a symplectic manifold and $x_0 \in M$ an arbitrary point. Then there exists a neighborhood $V = V(x_0)$ and a diffeomorphism $\varphi : V \to W(0) \subset \mathbb{R}^{2n}$ such that $\varphi^*(\omega_{std}) = \omega$.*

In other words this states that local symplectic geometry is that of the standard model.

The proof of the Lemma is stronger than the statement: A curve φ_t of diffeomorphisms is produced so that $\varphi_t^* \omega_t = \omega_0$, where $\omega_t = (1 - t)\omega_0 + t\omega_1$, $0 \leq t \leq 1$.

For this note that $d\omega_t = 0$ and $\frac{d\omega_t}{dt} = \omega_1 - \omega_0 =: -\beta$ is likewise closed. Since the result is of a local nature, it may be assumed that $M = V = V(0)$ is a contractible, open neighborhood of $x_0 = 0$ in $\mathbb{R}^{2n}$. In particular, $\beta = d\alpha$, where α may be chosen with $\alpha(0) = 0$. Finally, define X_t by $i_{X_t}\omega_t = \alpha$. The basic identity implies that $\frac{d}{dt}(\varphi_t^* \omega_t) = 0$. – Of course the basic identity told us how to construct the forms α and β!

Finally, in order to make sure that the associated time dependent field Z can be integrated for $0 \leq t \leq 1$, we simply replace X_t by $\chi \cdot X_t$, where χ is a cut-off function with $\chi \equiv 1$ near x_0. Thus we have a curve $\varphi_t \in Diff(M)$ such that $\varphi_t^* \omega_t = \omega_0$ near x_0. Since $\alpha(0) = 0$, it follows that $X_t(0) = 0$, i.e., $\varphi_t(0) = 0$ for all $t \in [0, 1]$. The desired diffeomorphism is $\varphi := \varphi_1$.   □

Weinstein has shown how to use Moser's Methods in various important contexts in symplectic geometric. The following is typical in this regard.

Theorem. (Weinstein) *Let N be a closed Lagrangian submanifold[22] of a symplectic manifold (M, ω). Then there exists an open neighborhood $U = U(N)$ in M, an open neighborhood W of the zero-section in T^*N and a diffeomorphism $\varphi : U \to W$ such that φ maps N to be zero-section and $\varphi^*(\omega_{std}) = \omega$.*

---

[22] A submanifold N is Lagrangian if and only if $T_x N$ is Lagrangian in $(T_x M, \omega(x))$ for all $x \in N$.

5.4 The basic sequence

Let (M, ω) be a connected symplectic manifold. Define $\delta : Ham_{loc}(M) \to H^1_{deR}(M)$ by $X \mapsto [i_X \omega]$. This makes sense, because $\mathcal{L}_X \omega = 0$, $d\omega = 0$ and consequently $di_X \omega = 0$.

Recall that for a function $f \in \mathcal{E}(M)$ the associated field $\nu(f) = X_f \in Ham_{loc}(M)$ is defined by $df = i_{X_f} \omega$.

Finally, let $i : \mathbb{R} \hookrightarrow \mathcal{E}(M)$ be the canonical emdedding of the (locally) constant functions.

Proposition. *The basic sequence*

$$0 \to \mathbb{R} \to \mathcal{E}(M) \to Ham_{loc}(M) \to H^1_{deR}(M) \to 0$$

is exact.

Proof. To show that $Im(i) = ker(\nu)$ note that $\nu(f) = 0 \iff X_f = 0 \iff df = 0$. For $Im(\nu) = Ker(\delta)$ just recall that $X \in ker(\delta) \iff i_X \omega = df$, i.e. $X = X_f$. Finally, if $\xi = [\mu]$, then the non-degeneracy of ω implies that $\mu = i_X \omega$ for some $X \in Vect(M)$. Since $d\mu = 0$ and $d\omega = 0$, it follows from Cartan's formula that $\mathcal{L}_X \omega = 0$, i.e., $X \in Ham_{loc}(M)$. $\qquad\square$

Notation. Let $Ham(M)$ denote the image $\nu(\mathcal{E}(M))$, i.e., the space of local Hamiltonian fields which are defined by global functions.

The symplectic form ω yields a bilinear, alternating pairing

$$\{\,,\} : \mathcal{E}(M) \times \mathcal{E}(M) \to \mathcal{E}(M), \ \{f, g\} := \omega(X_f, X_g).$$

Exercise. Show that the degeneracy of this pairing, i.e., $\mathcal{E}(M)^{\perp\{\,\}} = \{f \in \mathcal{E}(M) : \{f, \cdot\} = 0\}$, just consists of the constant functions.

The $Diff(M)$ action on $\mathcal{E}^*(M)$ at the function level is defined by $\varphi(f) = f \circ \varphi^{-1}$. This determines the full action, since $\varphi(df) = d\varphi(f)$. The action of $Diff(M)$ on itself is given by conjugation. For example, if $g_X(t)$ is the local $\mathbb{R}$-action associated to a vector field X, then it is transformed by $\varphi \circ g_X(t) \circ \varphi^{-1}$. This is equivalent to the action on $Vect(M)$ which is defined by $X \mapsto \varphi_* \circ X \circ \varphi^{-1}$. In particular $\varphi(g_X(t)) = g_{\varphi(X)}(t)$.

The definition of the Lie bracket of fields in $Vect(M)$ has been chosen to be

$$[X, Y] = -\frac{d}{dt}\Big|_{t=0} g_X(t)(Y).$$

The Lie algebra structure $(\mathcal{E}(M), \{\cdot, \cdot\})$ is closely related to the compatibility of $\{\cdot, \cdot\}$ with $\nu : \mathcal{E}(M) \to Ham_{loc}(M)$. To see this it is necessary to carry out a calculation.

As usual define X_f by $df = i_{X_f} \omega$. Given $h \in \mathcal{E}(M)$, let $g_{X_h}(t)$ be the local $\mathbb{R}$-action associated to X_h. Then

$$d(g_{X_h}(t)(f)) = g_{X_h}(t)(df) = g_{X_h}(t)i_{X_f}\omega = i_{g_{X_h}(t)(X_f)}\omega.$$

Applying $\frac{d}{dt}|_{t=0}$, we have

$$dX_h(f) = -i_{[X_h, X_f]}\omega.$$

But $X_h(f) = \{f, h\}$. Thus

$$d\{f, h\} = i_{[X_f, X_h]}\omega.$$

In other words

(*) $$\nu(\{f, h\}) = [\nu(f), \nu(g)].$$

Proposition. *The pair $(\mathcal{E}(M), \{\cdot, \cdot\})$ is a Lie algebra, $Ham(M)$ is a Lie subalgebra of $(Ham_{loc}(M), [\cdot, \cdot])$ and $\nu : (\mathcal{E}(M), \{\cdot, \cdot\}) \twoheadrightarrow (Ham(M), [\cdot, \cdot])$ is a Lie algebra homomorphism.*

Proof. Note that the above calculation shows that

$$X_{f_1}(\omega(X_{f_2}, X_{f_3})) = \omega([X_{f_2}, X_{f_3}], X_{f_1}).$$

Thus the identity

$$d\omega(X_{f_1}, X_{f_2}, X_{f_3}) = 0$$

can be interpreted as the Jacobi-identity

(**) $$\{f_1, \{f_2, f_3\}\} + \{f_3, \{f_1, f_2\}\} + \{f_2, \{f_3, f_1\}\} = 0.$$

The identity (*) shows that ν is a Lie homomorphism. In fact, (*) already shows that the left-hand side of (**) is a constant.

Since the image of a Lie algebra morphism is automatically a Lie subalgebra, the proof is complete. $\qquad\square$

Remark. If $X, Y \in Ham_{loc}(M)$ and $f := \omega(X, Y)$, then $X_f = [X, Y]$. For this one applies Cartan's formula twice to obtain another basic formula:

$$\mathcal{L}_Y \circ i_X - i_Y \circ \mathcal{L}_X = d \circ i_Y \circ i_X.$$

One then uses the fact that $\mathcal{L}_Y i_X \omega = i_{\mathcal{L}_Y(X)}\omega = i_{[X,Y]}\omega$.

Of course the commutator $[Ham_{loc}(M), Ham_{loc}(M)]$ is an ideal in $Ham_{loc}(M)$. As a consequence of the above remark, we are able to give more precise information on the structure of $Ham_{loc}(M)$.

Corollary. *The ideal $[Ham_{loc}(M), Ham_{loc}(M)]$ is contained in $Ham(M)$.*

5.5 The classical Hamiltonian equations

For $H \in \mathcal{E}(M)$ one regards $\{H, \cdot\}$ as a first order linear differential operator, i.e., $\{H, f\} = -X_H(f)$.

Definition. A smooth Hamiltonian system is a triple (M, ω, H), where (M, ω) is a symplectic manifold and $H \in \mathcal{E}(M)$.

For Σ a subset of $\mathcal{E}(M)$, let $Z(\Sigma) := \{f \in \mathcal{E}(M) : \{H, f\} = 0 \ \forall H \in \Sigma\}$ be the centralizer of Σ in $\mathcal{E}(M)$.

If $\{H, f\} = 0$, then $X_H(f) = 0$, i.e., the function f is constant along the orbits of the local $\mathbb{R}$-action defined by X_H.

Notation. If $\{H, f\} = 0$, then f is said to be a *constant of motion* for the Hamiltonian system (M, ω, H).

The centralizer of H, i.e., $Z(H) = \{f : \{H, f\} = 0\}$ is the Lie algebra of constants of motion for (M, ω, H). The fact that $H \in Z(H)$ is sometimes referred to as *conservation of energy*.

Example. (Hamiltonian systems in the standard symplectic vector space)

Let $(\mathbb{R}^{2n}, \omega_{std})$ be the standard symplectic vector space. Since $\omega_{std} = \sum dp_i \wedge dq_i$, the definition $dH = i_{X_H} \omega$ means

$$X_H = \sum \frac{\partial H}{\partial q_i} \frac{\partial}{\partial p_i} - \frac{\partial H}{\partial p_i} \frac{\partial}{\partial q_i}.$$

The associated system of ordinary differential equations is given by

$$\dot{p}_i = \frac{\partial H}{\partial q_i}(q, p) \text{ and } \dot{q}_i = -\frac{\partial H}{\partial q_i}(q, p), \ 1 \leq i \leq n.$$

Notice that if J is the standard complex structure on $\mathbb{R}^{2n}$, i.e., $J(\frac{\partial}{\partial q_i}) = \frac{\partial}{\partial p_i}$ and $J(\frac{\partial}{\partial p_i}) = -\frac{\partial}{\partial q_i}$, then the differential equations can be written as

$$\dot{x} = -J \cdot \nabla H(x),$$

where the coordinate vector x is given by $x = (q_1, \ldots q_n, p_1, \ldots, p_n)$. The Hamiltonian field $X_H = -J\nabla H$ is often called the *skew-gradient*.

The fact that X_H occurs as a skew-gradient with respect to a complex structure is no coincidence.

Let $J \in \mathcal{J}_\omega(M)$ be an ω-compatible almost complex structure on a symplectic manifold (M, ω) and $h = g - i\omega$ the associated Hermitian metric. Recall that if (M, g) is a Riemannian manifold, then the gradient ∇H of $M \in \mathcal{E}(M)$ is defined by

$$g(\nabla H, Y) = dH(Y) \text{ for } Y \in Vect(M).$$

In this case, since the Hamiltonian field X_H is defined by $\omega(X_H, Y) = dH(Y)$ and $g(X, Y) = \omega(X, JY)$, it follows that

$$dH(Y) = \omega(X_H, Y) = g(JX_H, Y)$$

and consequently $X_H = -J\nabla H$.

Remark. Recall that the space $\mathcal{J}_\omega(M)$ is infinite-dimensional. Thus there is no difficulty coming up with a compatible almost complex structure. For certain applications it is in fact important to keep *all* such structures in mind.

6. Kähler manifolds

If (M, ω) is a symplectic manifold with an ω-compatible complex structure J which is integrable, i.e., it is associated to a complex manifold structure X on M, then (X, ω) is called a *complex Kähler manifold*. These manifolds are of central importance in complex geometry and provide a broad class of symplectic manifolds where the tools of complex analysis can be used.

6.1 Basic properties and examples

In general it is difficult to decide if a submanifold of a symplectic manifold is again symplectic. However, in the Kähler case, there is one obvious criterion.

Lemma. *Let (X, ω) be Kählerian and $\varphi : Y \to X$ be a holomorphic map which is everywhere of maximal rank. Then $(Y, \varphi^*(\omega))$ is Kählerian.*

Thus the complex submanifolds of a (perhaps) simple looking Kähler manifold may be a very interesting class of Kähler manifolds.

Example (Unitary vector spaces). Let V be a complex vector space equipped with a unitary norm $\rho : V \to \mathbb{R}^{\geq 0}$. The associated linear form, i.e., $\omega = -Im(h)$, where h is the Hermitian structure, is Kählerian. One can write ω as a $(1, 1)$-form in the following way: $\omega = \frac{i}{2} \partial \bar{\partial} \rho$.

Using classical methods of complex analysis, it is possible to prove

Kodaira's Lemma. *Let ω be a Kähler-form on the unit ball $\mathbb{B}_n := \{z : \|z\|^2 < 1\}$ in $\mathbb{C}^n$. Then there exists a potential function $\rho \in \mathcal{E}(\mathbb{B}_n)$ such that $\omega = \frac{i}{2} \partial \bar{\partial} \rho$.*

Remark. It is not essential that the domain in question is the unit ball, but it must satisfy certain complex analytic and topological conditions.

Starting with $\rho \in \mathcal{E}(X)$, one can always consider the form $\omega = \frac{i}{2} \partial \bar{\partial} \rho$. It is a closed real form, but it will not in general yield a Kähler metric: It is a Kähler metric, if and only if it is positive definite.

Notation. A real valued function $\rho \in \mathcal{E}(X)$ is said to *strictly pluri-subharmonic* whenever $\omega := \frac{i}{2} \partial \bar{\partial} \rho$ is a Kähler-form on X.

In local coordinates the Hermitian form of ω is given by

$$h_\varphi = \sum_{j,k} \frac{\partial^2 \varphi}{\partial z_j \partial \bar{z}_k} dz_j \otimes d\bar{z}_k.$$

The form is Kählerian if and only if the *Levi-matrix* $\left(\frac{\partial^2 \varphi}{\partial z_j \partial \bar{z}_k} \right)$ is positive-definite.[23]

It is quite a simple matter to find Kähler structures on complex manifolds X which have many holomorphic functions. For example, if

$$F : X \to \mathbb{C}^N, \quad x \mapsto (f_1(x), \dots, f_n(x)), \quad f_j \in \mathcal{O}(X),$$

has maximal rank at every point of X, then $\rho := \sum |f_j|^2$ is strictly pluri-subharmonic and $\omega := \frac{i}{2} \partial \bar{\partial} \rho$ is a Kähler-form.

A function is strictly pluri-subharmonic if it is *strictly subharmonic in all directions*. More precisely, we have the following elementary

Proposition. *A function $\rho : X \to \mathbb{R}$ is strictly pluri-subharmonic if and only if for every holomorphic map $\varphi : \Delta \to X$ of the unit disk $\Delta = \{z \in \mathbb{C} : |z| < 1\}$ which is everywhere of maximal rank the function $\rho_\varphi := \rho \circ \varphi$ is strictly subharmonic, i.e., $\Delta_{\rho_\varphi} > 0$.*

---

[23] This was Kähler's original definition.

Corollary. (Maximum Principle) *If X is a connected complex manifold, then a strictly pluri-subharmonic function $\rho : X \to \mathbb{R}$ attains its maximum nowhere.*

Thus, compact complex manifolds possess *no* strictly pluri-subharmonic functions and consequently in this case a Kähler form ω only has local potentials: $\omega = \frac{i}{2}\partial\bar{\partial}\rho_\alpha$ for $\rho_\alpha \in \mathcal{E}(U_\alpha)$, where $\mathcal{U} = \{U_\alpha\}$ is a covering.

Example. (Projective spaces) Let V be a complex vector space and $\mathbb{P}(V)$ the associated projective space. Instead of piecing together local potentials, we produce a Kähler form by *pushing down* a naturally defined form on $V\backslash\{0\}$.

We discuss this in a general context: Let $\pi : M \to N$ be a surjective differentiable map which is everywhere of maximal rank. A vector field $X \in Vect(M)$ is said to be *vertical* if $\pi_*(X(x)) = 0$ for all $x \in M$ or equivalently if the local $\mathbb{R}$-action $g_X(t)$ stabilizes all π-fibers. If $y \in N$, let $M_y := \pi^{-1}(y)$ and $i_y : M_y \to M$ be the natural injection.

Proposition. *If in addition to the above assumptions the π-fibers are connected, then*

$$\pi^*(\mathcal{E}^*(N)) = \{\alpha \in \mathcal{E}^*(M) : \mathcal{L}_X\alpha = 0 \text{ and } i_y^*\alpha = 0 \text{ for all } X \text{ vertical and } y \in N\}.$$

Proof. It is enough to check that a form which satisfies the conditions

(i) $i_y^*(\alpha) = 0, \forall y \in N$
(ii) $\mathcal{L}_X\alpha = 0, \forall X$ vertical

is a pull-back: $\alpha = \pi^*(\beta)$.

For this observe that if $Y \in Vect(N)$, $y \in N$ and $p \in \pi^{-1}(y)$, then condition (i) insures that $\beta(y)(Y) := \alpha(p)(\hat{Y})$ is well-defined independent of the choice of $\hat{Y} \in T_pM$.

If $p_1, p_2 \in \pi^{-1}(y)$ and $\hat{Y}_1$ and $\hat{Y}_2$ are lifts in the respective tangent spaces, then one can easily join p_1 to p_2 by a piece-wise smooth curve so that the smooth pieces are orbits of local $\mathbb{R}$-actions $g_{X_1}(t), g_{X_2}(t), \ldots, g_{X_n}(t)$. The result follows from (ii) by iteration of the result for the case where p_1 and p_2 are connected by one such orbit $g_X(t)$: Since $\mathcal{L}_X\alpha = 0$, it follows that $\alpha(p_1)(\hat{Y}_1) = \alpha(p_2)(g_*(\hat{Y}_1)) = \alpha(p_2)(\hat{Y}_2)$. $\square$

We now return to the concrete special case $\pi : V\backslash\{0\} \to \mathbb{P}(V)$ of the above proposition.

Exercise. Show that if $\alpha \in \mathcal{E}^*(V\backslash\{0\})$ is invariant with respect to the $\mathbb{C}^*$-action defined on $V\backslash\{0\}$ by scalar multiplication, then $\mathcal{L}_X\alpha = 0$ for all vectical fields. What is the relevant general result?

Consequently, the forms $\alpha \in \pi^*(\mathcal{E}^*(\mathbb{P}(V)))$ are those which are invariant and satisfy $i_y^*(\alpha) = 0$ for all $y \in \mathbb{P}(V)$.

Recall that the function $log|z|, z \in \mathbb{C}^*$, is *harmonic*, i.e., $\frac{\partial^2}{\partial z \partial \bar{z}}(log|z|) = 0$. Thus, $\rho : V\backslash\{0\} \to \mathbb{R}$, $v \mapsto log\|v\|^2$, defines a form $\omega := \frac{i}{2}\partial\bar{\partial}\rho$ which satisfies $\omega =$

$\pi^*(\omega_{FS})$ for a uniquely defined $(1,1)$-form ω_{FS} on $\mathbb{P}(V)$. It remains to show that the *Fubini-Study form* ω_{FS} is non-degenerate.

For this we note that ω_{FS} is invariant with respect to the unitary group $K = U(V, \| \cdot \|^2)$ and therefore we only need to check the positivity at one point. For example, if we choose $[v] \in \mathbb{P}(V)$ where $\|v\|^2 = 1$ and let A be the affine plane in V which is othogonal to the line $[v]$ in V at v, then, up to a constant, the pull-back of ω to A is the same as that of $\omega_{std} := \frac{i}{2}\partial\bar\partial\| \cdot \|^2$. Thus ω_{FS} is positive-definite.

We summarize this in the following

Proposition. *The Fubini-Study form ω_{FS} defines a unitary invariant Kähler metric on $\mathbb{P}(V)$.*

Hence, every smooth projective algebraic variety is Kählerian. Of course the Kähler metric on such a variety is not unique and one of the goals of complex differential geometry is to determine the existence or non-existence of Kähler metrics which satisfy additional natural conditions.

Another important way that Kähler manifolds occur is as quotient of complex manifolds by discrete groups.

Exercise. Let X be a complex manifold which is equipped with a free, proper, holomorphic action of a discrete group Γ. We have seen to the quotient X/Γ has a unique structure of a complex manifold so that the quotient map $X \to X/\Gamma$ is holomorphic. Show that a Γ-invariant Kähler-form on X defines a Kähler-form on X/Γ.

We mention two concrete cases of quotients. First, let $X = (\mathbb{C}^n, +)$ be the standard simply-connected abelian complex Lie group and Γ a discrete subgroup. It follows that $\Gamma = ((v_1, \dots, v_m))_{\mathbb{Z}}$, where $v_1, \dots, v_m$ are $\mathbb{R}$-independent. In particular, $m \leq 2n$ and $m = 2n$ if and only if $\mathbb{C}^n/\Gamma$ is compact. The form $\omega = \frac{i}{2}\sum dz_i \wedge d\bar z_i$ is Γ-invariant and induces a Kähler metric on the quotient $\mathbb{C}^n/\Gamma$. If $X_\Gamma = \mathbb{C}^n/\Gamma$ is compact, then for certain $\Gamma's$ it will be projective algebraic. However, for Γ generic this will not be the case.

As a second class of examples we would like to mention quotients of bounded domains $D \subset \mathbb{C}^n$ by discrete groups. Such domains possess natural Kähler metrics which are invariant for the full group $Aut_{\mathcal{O}}D$ of biholomorphic diffeomorphisms of D. Interesting discrete groups occur in a number of important concrete cases, e.g., for $D = \mathbb{B}_n = \{z \in \mathbb{C}^n, \|z\| < 1\}$ or the polydisk $D = \triangle_n := \{z \in \mathbb{C}^n : |z_i| < 1, 1 \leq i \leq n\}$. If such a quotient is compact, then it is projective algebraic.

6.2 Variation of complex structure

In this brief section we would like to briefly mention another way in which Kähler manifolds appear.

Let $\pi : X \to S$ be a surjective, proper, everywhere maximal rank, holomorphic map with connected fibers $X_s := \pi^{-1}(s)$. If we forget the complex structures which are involved, then this is a locally trivial differentiable fiber bundle. As we have seen, it is nevertheless possible that no two π-fibers are biholomorphically equivalent.

Given a complex manifold X_0 one would like to determine a maximal family $\pi : X \to S$ as above with $X_0 \cong \pi^{-1}(s_0)$ being the fiber over a base point. Optimally, this family should be without repetition, i.e., no two fibers should be biholomorphic. This later property is extremely difficult to obtain and thus one is satisfied with weaker properties and/or parameter spaces which parameterize complex manifolds with additional structure. – This is sometimes more natural and much easier.

In this context, interesting Kähler manifolds occur in at least two ways. First, starting with a known X_0, some or even most of the fibers X_s may be new.

Secondly, the parameter spaces or moduli spaces S may themselves be important Kähler manifolds. It should be underlined that a moduli space S quite often exits in situations where there is no family $X \to S$.

As a concrete example we consider the compact 2-dimensional complex manifold $X_0 := \{[z_0 : z_1 : z_2 : z_3] \in \mathbb{P}_3(\mathbb{C}) : z_0^4 + z_1^4 + z_2^4 + z_3^4 = 0\}$. Of course one can deform X_0 by changing the defining equation in $\mathbb{P}_3$. Trivial deformations arise by moving X_0 by linear transformations. Roughly speaking, the set of all such equations, modulo automorphisms, can be interpreted as a 19-dimensional family of so-called $K3$-surfaces.

However, there is a bigger, in a certain sense maximal, moduli space $\mathcal{M}$ which is 20-dimensional. Each manifold which is represented in this space is Kählerian and a dense set of them can not be found in a projective space.

IV. Symplectic manifolds with symmetry

1. Introduction to the moment map

Let (M, ω) be a symplectic manifold and $G \times M \to M$ a smooth action of a Lie group of symplectic diffeomorphisms, i.e. $g^*(\omega) = \omega$ for all $g \in G$. Note that if G is connected, this is equivalent to

$$\mathcal{L}_{X_\xi} \omega = 0 \quad \text{for all} \quad \xi \in Lie(G) = \mathfrak{g}.$$

Here $X_\xi \in Vect(M)$ is the field associated to the action of the 1-parameter group $exp(\xi t)$.

If $\mathcal{L}_{X_\xi} \omega = 0$, then $i_{X_\xi} \omega$ is closed and, at least locally, there is a smooth function μ^ξ such that

$$d\mu^\xi = i_{X_\xi} \omega.$$

If such a function exists globally on M, i.e., $\mu^\xi \in \mathcal{E}(M)$, then we refer to it as a *moment function* or *Hamiltonian* associated to X_ξ. The manifold M will always be assumed to be connected and therefore, if it exists, μ^ξ is unique up to an additive constant.

Recall that if $\varphi : M \to V$ is a map to a vector space and $\xi \in V^*$, then the ξth coordinate of φ is $\varphi^\xi := \xi \circ \varphi$. If $V = W^*$, then of course $\xi \in W$.

Definition. Let (M, ω) be a symplectic manifold equipped with a Lie group action $G \times M \to M$ of symplectic diffeomorphisms. A *moment map* $\mu : M \to Lie(G)^*$ is a smooth *equivariant* map which satisfies

$$(*) \qquad\qquad d\mu^\xi = i_{X_\xi}\omega \quad \text{for all} \ \ \xi \in Lie(G).$$

The group G acts on $\mathfrak{g}^*$ by the *coadjoint representation*, i.e., for $F \in \mathfrak{g}^*$,

$$Ad^*(g)(F)(v) = F(Ad(g^{-1})(v))$$

and equivariance means that $\mu(g(x)) = Ad^*(g)(\mu(x))$. If only $(*)$ is fulfilled, we refer to μ as a *non-equivariant moment map*.

Example. Let $(\mathbb{R}^{2n}, \omega_{std})$ be the standard symplectic vector space and $T := (S^1)^n$. Let T act by rotations in the usual linear way, i.e., if $z_\alpha = q_\alpha + ip_\alpha$ and $t = (e^{i\varphi_1}, \ldots, e^{i\varphi_n})$, then

$$t(z) = t(z_1, \ldots, z_n) = (e^{i\varphi_1}z_1, \ldots, e^{i\varphi_n}z_n).$$

In this complex notation ω_{std} is the linear Kähler form,

$$\omega_{std} = \frac{i}{2}(dz_1 \wedge d\bar{z}_1 + \ldots + dz_n \wedge d\bar{z}_n) = \frac{i}{2}\partial\bar\partial|z|^2 = r_1 dr_1 \wedge d\varphi_1 + \ldots + r_n dr_n \wedge d\varphi_n,$$

where the last expression has been written in polar coordinates.

Let $\mathfrak{t} = Lie(T) = ((\xi_1 \ldots, \xi_n))_\mathbb{R}$, where $X_{\xi_i} = \frac{\partial}{\partial\varphi_i}$. Thus, if $\xi = \sum a_j \xi_j$, then

$$i_{X_\xi}\omega = \sum_{j=1}^n a_j r_j dr_j = d\mu^\xi, \quad \text{where } \mu^\xi = \frac{1}{2}\sum a_j r_j^2.$$

Thus $\mu : \mathbb{R}^{2n} \to Lie(T)^*$ is given by $(p, q) \mapsto \sum_{j=1}^n \left(\frac{p_j^2 + q_j^2}{2}\right)\xi_j^*$.

Exercise. Show that μ is a moment map.

There may be topological obstrutions to the existence of global moment functions.

Example. Let Γ be a lattice of rank $2n$ in $\mathbb{R}^{2n}$ and $M := \mathbb{R}^{2n}/\Gamma$. Since ω_{std} is invariant under the full group of translations $G = (\mathbb{R}^{2n}, +)$, there is an induced symplectic form ω on M. Now, G acts on M and leaves ω invariant. For $\xi \in Lie(\mathbb{R}^{2n}) = \mathbb{R}^{2n}$, the vector field X_ξ is the constant field $\xi : X_\xi = \xi$. In concrete terms, if

$$\omega_{std} = \sum dp_j \wedge dq_j \quad \text{and} \quad \xi = \sum a_j \frac{\partial}{\partial q_j} + b_j \frac{\partial}{\partial p_j},$$

then $i_{X_\xi}\omega = \sum(b_j dq_j - a_j dp_j)$. Such a constant coefficient 1-form is globally exact on the torus M if and only if it is identically zero.

Lack of equivariance occurs in important physical situations.

Example. Let $M = \mathbb{R}^2$ be equipped with $\omega = dx \wedge dy$ and let $G = (\mathbb{R}^2, +)$ act by translations. It follows that $\{\frac{\partial}{\partial x}, \frac{\partial}{\partial y}\}$ is a basis of the induced vector fields. Now

$$i_{\frac{\partial}{\partial x}}\omega = dy \quad \text{and} \quad i_{\frac{\partial}{\partial y}}\omega = -dx.$$

Thus, up to additive constants, the global moment functions are $\mu^{\frac{\partial}{\partial x}} = y$ and $\mu^{\frac{\partial}{\partial y}} = -x$. Since G is abelian, equivariance would require that $\mu^{[\frac{\partial}{\partial x}, \frac{\partial}{\partial y}]} = \mu^0 = 0$. But $\omega(\frac{\partial}{\partial x}, \frac{\partial}{\partial y}) = 1$. Consequently there is no equivariant moment map for the group G.

If G is a Lie group acting on (M, ω) as a group of symplectic diffeomorphisms, then one may consider the following diagramm:

$$\mathcal{E}(M) \to Ham_{loc}(M)$$
$$\searrow \quad \uparrow$$
$$Lie(G)$$

Here the vertical map $\kappa : Lie(G) \to Ham_{loc}(M)$, $\xi \mapsto X_\xi$, always exists. The diagonal map should be defined by $\mu^*(\xi) = \mu^\xi$. It exists as a linear map if and only if $Im(\kappa) \subset Ham(M)$, i.e., if each field X_ξ has a global moment function: $i_{X_\xi}\omega = d\mu^\xi$.

Suppose that such global functions exist and let $\mu^* : Lie(G) \to \mathcal{E}(M), \xi \mapsto \mu^\xi$, be a linear map so that the diagram commutes, i.e., $\nu \circ \mu^* = \kappa$ or $d\mu^\xi = i_{X_\xi}\omega$ for all $\xi \in Lie(G)$. The map

$$\mu : M \to (Lie(G))^*$$

is equivariant if and only if

$$\mu(g_{X_\eta}(t)(x)) = Ad^*(exp(\eta t))(\mu(x))$$

for all $\eta \in Lie(G)$ and all $x \in M$, i.e.,

$$\mu^\xi(g_{X_\eta}(t)(x)) = Ad^*(exp(\eta))(\mu(x))(\xi)$$
$$= \mu(x)(Ad(exp(-\eta t))(\xi)) \quad \text{for all } \xi \in Lie(G).$$

Differentiating, we obtain

$$\frac{d}{dt}\Big|_{t=0}\mu^\xi(g_{X_\eta}(t)(x)) = d\mu^\xi(X_\eta)(x) = \omega(x)(X_\xi, X_\eta) = \{\mu^\xi, \mu^\eta\}(x)$$

and

$$\frac{d}{dt}\Big|_{t=0}\mu(x)(Ad(exp(-\eta))(\xi)) = \mu(x)([\xi, \eta]) = \mu([\xi, \eta])(x).$$

Thus, if G is a connected finite dimensional Lie group, $\mu : M \to (Lie(G))^*$ is equivariant if and only if the lifting map $\mu^* : Lie(G) \to \mathcal{E}(M)$ is a homomorphism of Lie algebras.

2. Central extensions

Assuming that $Im(\kappa : \mathfrak{g} \to Ham_{loc}(M)) \subset Ham(M)$, it is always possible to define a moment map for a larger group: Define $\tilde{\mathfrak{g}} := \nu^{-1}(\kappa(\mathfrak{g}))$. Let

$$\tilde{\kappa} := \nu : \tilde{\mathfrak{g}} \to Ham(M).$$

Then $\tilde{\mu}^* := Id$ is a Lie homomorphism so that

$$\mathcal{E}(M) \to Ham(M)$$
$$\searrow \qquad \uparrow$$
$$\tilde{\mathfrak{g}}$$

commutes.

By construction $\tilde{\mathfrak{g}}$ is a *central extension* by the 1-dimensional algebra $\mathbb{R}$ of $\kappa(\mathfrak{g})$. For convenience of notation assume that the G-action is *almost effective*, i.e., $I := \{g \in G : g(x) = x \ \forall x \in M\}$ is a discrete (normal) subgroup of G. Thus we may identify $\kappa(\mathfrak{g})$ with $\mathfrak{g}$ and we have the central extension

$$0 \to \mathfrak{z} \to \tilde{\mathfrak{g}} \to \mathfrak{g} \to 0 \quad \text{with} \quad \mathfrak{z} \cong \mathbb{R}.$$

Since we are dealing with finite-dimensional Lie algebras, there exists a unique simply-connected Lie group $\tilde{G}$ with $Lie(\tilde{G}) = \tilde{\mathfrak{g}}$ and $\tilde{G}$ contains a connected 1-dimensional central subgroup Z with $Lie(Z) = \mathfrak{z}$ an ideal in $\mathfrak{g}$.

Lemma. *The central subgroup Z is closed in $\tilde{G}$.*

Sketch of Proof. Let C be the connected component of the identity of the center of $\tilde{G}$. Since the center of $\tilde{G}$ is closed, C is closed and we have the principal bundle

$$(*) \qquad\qquad \{0\} \to C \to \tilde{G} \to \tilde{G}/C \to \{0\}.$$

It may be shown that $\pi_2(L) = \{0\}$ for any connected Lie group L. Thus an application of the homotopy sequence for the fiber bundle $(*)$ shows that $\{0\} = \pi_2(\tilde{G}/C) \cong \pi_1(C)$.

If C is a simply-connected abelian group, then $C \cong (\mathbb{R}^m, +)$ for some $m \in \mathbb{N}^{>0}$ and *any* 1-parameter subgroup $Z < C$ is closed. $\qquad\qquad\qquad\qquad\square$

For convenience assume that G is simply-connected. Therefore, if $\tilde{G}$ is the extension given by the above lemma, it follows that $\tilde{G}$ is a central extension of G by $Z \cong \mathbb{R}$: $\{0\} \to Z \to \tilde{G} \to G \to \{0\}$. Furthermore, there is a natural smooth acton of $\tilde{G}$ on M: $\tilde{g}(x) := q(\tilde{g})(x)$, where $q : \tilde{G} \to \tilde{G}/Z = G$ is the canonical quotient map.

Proposition. *The lifting $\tilde{\mu}^*$ defines a moment map for the action of $\tilde{G}$.*

Remark. Of course this proposition would seem to be formal, because, although the group has been enlarged, the part that was added does not act. However, this ineffectivity can be important.

Example. (Heisenberg algebra). Consider the $(\mathbb{R}, +)$-action on $M = (\mathbb{R}^2, dx \wedge dy)$ as above. The Lie algebra $\tilde{\mathfrak{g}} \subset \mathcal{E}(M)$ is generated by $y = \mu \frac{\partial}{\partial x}, -x = \mu \frac{\partial}{\partial y}$ and $1 = \omega(\frac{\partial}{\partial x}, \frac{\partial}{\partial y})$, i.e.

$$\tilde{\mathfrak{g}} = ((x, y, 1))_{\mathbb{R}} \quad , \text{where } \{x, y\} = 1.$$

Exercise. Show that the associated Lie group $\tilde{G}$ can be identified with the group of upper-triangular matrices

$$H_3 := \left\{ \begin{pmatrix} 1 & x & z \\ 0 & 1 & y \\ 0 & 0 & 1 \end{pmatrix} : x, y, z \in \mathbb{R} \right\},$$

where $Z = \{A \in H_3 : x = y = 0\}$.

It is useful to formalize the structure of central extensions. For this let V be a $\mathbb{R}$-vector space and $\mathfrak{g}$ a $\mathbb{R}$-Lie algebra. A *central extension* of $\mathfrak{g}$ by V is a Lie algebra $\tilde{\mathfrak{g}}$ which contains V as a central subalgebra such that $\mathfrak{g} \cong \tilde{\mathfrak{g}}/V$, i.e., an exact sequence of Lie algebras.

$$\{0\} \to V \to \tilde{\mathfrak{g}} \to \mathfrak{g} \to \{0\}.$$

Let $\sigma : \mathfrak{g} \to \tilde{\mathfrak{g}}$ be a splitting so that

$$\tilde{\mathfrak{g}} = V \oplus \sigma(\mathfrak{g})$$

as a direct sum of vector spaces.

We compute the Lie bracket in $\tilde{\mathfrak{g}}$ with respect to this splitting:

$$[v_1 + \sigma(a_1), v_2 + \sigma(a_2)]_{\tilde{\mathfrak{g}}} = [\sigma(a_1), \sigma(a_2)]_{\tilde{\mathfrak{g}}} = c(a_1, a_2) + \sigma([a_1, a_2]_{\mathfrak{g}}).$$

Thus the *cocycle*

$$c : \mathfrak{g} \times \mathfrak{g} \to V$$

measures the *deviation* between σ and a possible Lie algebra splitting.

Since $[\cdot, \cdot]_{\tilde{\mathfrak{g}}}$ is an alternating bilinear pairing, we regard $c \in \Lambda^2 \mathfrak{g}^* \otimes V$, as an alternating functional with values in V.

Exercise. Check that the Jacobi identity in $\tilde{\mathfrak{g}}$ is equivalent to

$$\delta c(a_1, a_2, a_3) := -c([a_1, a_2], a_3) + c([a_1, a_3], a_2) - c([a_2, a_3], a_1) = 0.$$

Here we regard δ as a linear map $\delta : \Lambda^2 \mathfrak{g}^* \otimes V \to \Lambda^3 \mathfrak{g}^* \otimes V$

Of course it is possible that $\tilde{\mathfrak{g}} = V \oplus \mathfrak{g}$ splits as a Lie algebra even though the cocycle is non-zero. In this case we would then have the following picture: $\tilde{\mathfrak{g}} = V \oplus \mathfrak{g}$ as a Lie algebra and there would exist a linear map $\tau : \mathfrak{g} \to V$ so that $\sigma = \tau + Id$. Computing brackets,

$$[\sigma(a_1), \sigma(a_2)]_{\tilde{\mathfrak{g}}} = [a_1, a_2]_{\mathfrak{g}} = \sigma([a_1, a_2]_{\mathfrak{g}}) - \tau([a_1, a_2]_{\mathfrak{g}}).$$

This can be summarized as follows:

Define $\delta : \Lambda^1 \mathfrak{g}^* \otimes V \to \Lambda^2 \mathfrak{g}^* \otimes V$ by $\delta\tau(a_1, a_2) := -\tau([a_1, a_2])$. Note that $\delta \circ \delta = 0$, where the second map was introduced in the above exercise. Let

$$H^2(\mathfrak{g}^*, V) := \frac{Ker(\delta : \Lambda^2 \mathfrak{g}^* \otimes V \to \Lambda^3 \mathfrak{g}^* \otimes V)}{Im(\delta : \Lambda^1 \mathfrak{g}^* \otimes V \to \Lambda^2 \mathfrak{g}^* \otimes V)}$$

The above discussion can therefore be summarized as follows.

Proposition. *The Lie algebra cohomology $H^2(\mathfrak{g}^*, V)$ parameterizes the central extensions*

$$\{0\} \to V \to \tilde{\mathfrak{g}} \to \mathfrak{g} \to \{0\}.$$

This means that if $c \in \xi \in H^2(\mathfrak{g}^*)$, then $[a_1, a_2]_{\tilde{\mathfrak{g}}} := c(a_1, a_2) + [a_1, a_2]_{\mathfrak{g}}$ defines a Lie algebra structure on $\tilde{\mathfrak{g}} = V \oplus \mathfrak{g}$ which is a central extension of the desired type. Furthermore, every extension occurs in this way and equivalence of the extensions is the same as cohomological equivalence.

Exercise. Prove this proposition.

Note that the *boundary mapping* δ can be defined in an analogous way at all levels:

$$\delta : \Lambda^k \mathfrak{g}^* \otimes V \to \Lambda^{k+1} \mathfrak{g}^* \otimes V,$$

$$\delta\alpha(\xi_1, \ldots, \xi_{n+1}) = \sum_{i<j} (-1)^{i+j} \alpha([\xi_i, \xi_j], \xi_1, \ldots \hat{\xi}_i, \ldots, \hat{\xi}_j, \ldots, \xi_{n+1}),$$

where a hat on a ξ means that it is to be omitted at that place. One easily checks that $\delta \circ \delta = 0$ and as a result we have

$$H^k(\mathfrak{g}^*) := \frac{Ker(\delta : \Lambda^k \mathfrak{g}^* \otimes V \to \Lambda^{k+1} \mathfrak{g}^* \otimes V)}{Im(\delta : \Lambda^{k-1} \mathfrak{g}^* \otimes V \to \Lambda^k \mathfrak{g}^* \otimes V)}, \quad k = 1, 2, \ldots$$

It is important to relate this cohomology to the de Rham cohomology of a Lie group G with $Lie(G) \cong \mathfrak{g}$. For this it is convenient to have a coordinate free description of $d : \mathcal{E}^k(M) \to \mathcal{E}^{k+1}(M)$:

$$d\alpha(X_1, \ldots, X_{k+1}) = \sum_{i=1}^{k+1} (-1)^i \alpha(X_1, \ldots, \hat{X}_i, \ldots, X_{k+1}) +$$

$$+ \sum_{i<j} (-1)^{i+j} \alpha([X_i, X_j], X_1, \ldots, \hat{X}_i, \ldots, \hat{X}_j, \ldots, X_{k+1}).$$

Now suppose $M = G$ is a Lie group, $\alpha \in \Lambda^k \mathfrak{g}^*$ is a left-invariant k-form and $X_i = \xi_i, 1 \le i \le k+1$, are left-invariant vector fields. Then $d\alpha(\xi_1, \ldots, \xi_{k+1}) = \delta\alpha(\xi_1, \ldots, \xi_{k+1})$. Thus there exists a natural map

$$deR : H^*(\mathfrak{g}^*) \to H^*_{deR}(G).$$

Proposition. *For a connected compact Lie group K the map deR is an isomorphism.*

Proof. Let dk be an invariant probability measure on K. Then the averaging map

$$A : \mathcal{E}^k(G) \to \Lambda^k \mathfrak{g}^*, \quad A(\alpha) = \int_{k \in K} k^*(\alpha) dk$$

defines an isomorphism at the level of cohomology which is in fact the inverse of deR. $\qquad\square$

Example. It can be proved that

$$H^1(\mathfrak{g}^*) = H^2(\mathfrak{g}^*) = \{0\}.$$

for any semi-simple Lie algebra $\mathfrak{g}$. For unitary Lie algebras this can be shown, for example, by using the above proposition. We do this here for $K = SU_n$.

Note that K acts transitively on the unit sphere $S^{2n-1} \subset \mathbb{C}^n$:

$$S^{2n-1} = K/L,$$

where L is the isotropy of the north pole so that we have a natural identification of L with SU_{n-1}. Hence we have the principal bundle $SU_{n-1} \to SU_n \to S^{2n-1}$ to which we may apply the homotopy sequence. Starting with $n = 2$, using the identification $SU_2 \cong S^3$ and computing recursively, we show

$$\pi_1(SU_n) = \pi_2(SU_n) = \{0\} \quad \text{for } n \geq 2.$$

In particular the de Rham groups vanish and $H^1(\mathfrak{g}^*) = H^2(\mathfrak{g}^*) = \{0\}$ for $\mathfrak{g} = \mathfrak{su}_n$, $n \geq 2$.

3. Existence and uniqueness of the moment map

Let us return to our main theme and consider the action of a Lie group G of symplectic diffeomorphisms on a symplectic manifold (M, ω).

As was explained above, provided $X_\xi \in Ham(M)$ for all $\xi \in \mathfrak{g}$, there always exists a moment map $\tilde{\mu} : M \to \tilde{\mathfrak{g}}^*$ for the central extension defined by the basic sequence: $0 \to \mathbb{R} \to \tilde{\mathfrak{g}} \to \mathfrak{g} \to 0.$[24] Furthermore, there exists a moment map $\mu : M \to \mathfrak{g}^*$ if and only if there exists a Lie algebra lifting $\mu^* : \mathfrak{g} \to \tilde{\mathfrak{g}}$, i.e., if and only if the central extension is trivial. This can be summarized as follows

Proposition. *Let G be a connected Lie group acting as a group of symplectic diffeomorphisms on a symplectic manifold (M, ω) such that $X_\xi \in Ham(M)$ for all $\xi \in \mathfrak{g}$. This action uniquely determines a cohomology class $[c] \in H^2(\mathfrak{g}^*, \mathbb{R})$ and there exists an equivariant moment map $\mu : M \to \mathfrak{g}^*$ if and only if $[c] = 0$.*

We now turn to the question of uniqueness of μ.

Suppose μ_1 and μ_2 are equivariant moment maps for a given G-action. Then there exists a constant $\alpha \in \mathfrak{g}^*$ such that $\mu_1 = \mu_2 + \alpha$. Of course the map

$$T_\alpha : \mathfrak{g}^* \to \mathfrak{g}^*, \quad \beta \mapsto \beta + \alpha,$$

must be equivariant and, conversely, equivariant translations act on moment maps.

Exercise. Show that T_α is equivariant if and only if $\alpha \in Fix(G)$.

Since G is acting on $\mathfrak{g}^*$ by the coadjoint representation, $Fix(G)$ is the annihilator $[\mathfrak{g}, \mathfrak{g}]^0 = \{\alpha \in \mathfrak{g}^* : \alpha|_{[\mathfrak{g},\mathfrak{g}]} = 0\}$ and consequently the lack of uniqueness, or equiv-

---

[24] Here, without loss of generality, we are assuming that G acts almost effectively.

alently the set of equivariant translations, is given by $(\mathfrak{g}/[\mathfrak{g},\mathfrak{g}])^*$. This is exactly the first Lie algebra cohomology group.

Corollary. *If the connected Lie group G is acting by symplectic diffeomorphisms on M with moment map $\mu : M \rightarrow \mathfrak{g}^*$, then μ is the unique moment map for the action if and only if $H^1(\mathfrak{g}) = \{0\}$, i.e., if and only if $\mathfrak{g} = [\mathfrak{g},\mathfrak{g}]$.*

Exercise. Compute the Lie algebra cohomology $H^k(\mathfrak{g})$, $k = 1, \ldots, dim\mathfrak{g}$, for an abelian Lie algebra.

4. Basic examples of the moment map

Many concrete cases of moment maps occur in the cotangent bundle context. For a discussion of this let N be an arbitrary differentiable manifold and $M := T^*N$. We equip M with the standard symplectic structure $\omega := -d\theta$, where θ is the canonical 1-form.

Here it is appropriate to consider the action of two infinite dimensional groups on M. First, note that the group $Diff(N)$ acts on M in a natrual way: For $g \in Diff(N)$ and $\alpha_q \in T_q^*N$ let $g(\alpha_q) := (g^*)^{-1}(\alpha_q)$. In this way the natural projection $\pi : M \rightarrow N$ is $Diff(N)$-equivariant.

Secondly, the additive group of 1-forms $\mathcal{E}^1(N)$ acts on T^*N by translation, i.e., if $\beta \in \mathcal{E}^1(N)$ and $\alpha_q \in T_f^*N$, then $\beta(\alpha_q) = \alpha_q + \beta(q)$. In this way we have an action on M by the smooth functions on the base:

$$\alpha_q \mapsto \alpha_q + df(q) \quad \text{for } f \in \mathcal{E}(N).$$

Exercise. Show that θ is $\mathcal{E}(N)$-invariant.

For $g \in Diff(M)$ it follows that

$$(g^*\theta)(\alpha_q)(t) = \theta(g(\alpha_q))(g_*(t)) = g(\alpha_q)(g_*(\pi_*(t))) =$$

$$= (g^*)^{-1}(\alpha_q)(g_*(\pi_*(t))) = \alpha_q(\pi_*(t)) = \theta(\alpha_q)(t).$$

Thus θ is also $Diff(N)$-invariant.

It is convenient to regard the simultaneous action of these groups. This can be formalized in terms of an action of their *semi-direct product*.

Notation. Let G be a group, A and B be subgroups and suppose that the map $A \times B \rightarrow G$, $(a,b) \mapsto a \cdot b$, is bijective. Then we say that G is the *set-theoretic product* of A and B. If A and B are *normal* subgroups of G, then they commute and group multiplication is given by

$$g_1 \cdot g_2 = (a_1 \cdot b_1) \cdot (a_2 \cdot b_2) = (a_1 \cdot a_2)(b_1 \cdot b_2).$$

In this case we say that G is the group theoretic product and write $G = A \times B$. If only A is normal in G, i.e., A is normalized by B, then

$$g_1 \cdot g_2 = (a_1 \cdot b_1)(a_2 \cdot b_2) = (a_1 \cdot b_1 a_2 b_1^{-1})(b_1 \cdot b_2).$$

Thus if we let $\rho : B \to Aut(A)$ be defined by conjugation, then group multiplication is given by

$$(a_1 \cdot b_1)(a_2 \cdot b_2) = (a_1 \cdot \rho(b_1)(a_2))(b_1 \cdot b_2).$$

In this case we say that G is the *semi-direct product* of A and B and write $G = A \rtimes B$. In this way it is indicated that A is normalized by B.

Finally, if A and B are arbitrary groups and $\rho : B \to Aut(A)$ is an arbitrary homomorphism, then we define $G = A \rtimes_\rho B$ by letting G be the set $A \times B$ with the group structure given by

$$(a_1, b_1) \cdot (a_2, b_2) = (a_1 \cdot \rho(b_1)(a_2), b_1 b_2).$$

Exercise. Check that $G = A \rtimes_\rho B$ is a group.

Let $Diff^0(N)$ act on $\mathcal{E}(N)$ in the natural way, i.e., $\rho(g)(f) = f \circ g^{-1}$ and define $G := \mathcal{E}(N) \rtimes_\rho Diff^0(M)$ to be the semi-direct product defined by ρ.

Exercise. Recall that the Lie algebra of $Diff^0(N)$ is $Vect(N)$ and that of $\mathcal{E}(N)$ is $\mathcal{E}(N)$ itself. Note that $Vect(N)$ acts on $\mathcal{E}(N)$ by $\tilde{\rho}(X)(f) = X(f)$. Show that

$$Lie(G) = \mathcal{E}(N) \rtimes_{\tilde{\rho}} Vect(N),$$

where the notion of *semi-direct product of Lie algebras* is defined in the analogous way to that of groups.

Note that G acts in an *affine way* on M:

$$(g, f)(\alpha_q) = g(\alpha_q) + df(g(q)) = (g^*)^{-1}(\alpha_q) + df_{g(q)}.$$

Exercise. Let V be a finite-dimensional vector space. Show that the Lie group $Aff(V)$ of affine transformations is a semi-direct product $Aff(V) = V \rtimes_\rho Gl(V)$, where V acts on itself by translations and $\rho(T)(v) = Tv$.

Since the group $G = \mathcal{E}(N) \rtimes Diff^0(N)$ is infinite-dimesnional, one should pay close attention to topological matters when dealing with actions. We shall prove the following statement: For $\xi \in Lie(G)$ and X_ξ the associated field on $M = T^*N$, the lifting defined by $\mu^\xi := -\theta(X_\xi)$ defines a moment function.

Without worrying about topologies, this only means $\mu^{[\xi,\eta]} = \{\mu^\xi, \mu^\eta\} = \omega(X_\xi, X_\eta)$.

Proof. Since θ is G-invariant, it is clear that

$$i_{X_\xi}\omega = i_{X_\xi} \circ d\theta = -di_{X_\xi}\theta = d\mu^\xi.$$

For the *equivariance* let $L(X) \in Vect(M)$ denote the canonical lifting of a vector field $X \in Vect(N)$, i.e., if $g_X(t)$ is the associated 1-parameter group on N, then $L(X)$ is defined by $(g_X(t)^*)^{-1}$.

Our goal is to show that

$$\omega(X_\eta, X_\xi) = -\theta([X_\xi, X_\eta])$$

for all $\xi, \eta \in Lie(G)$. Since $\omega = -d\theta$, this is equivalent to showing that

$$(*) \qquad X_\xi\theta(X_\eta) - X_\eta\theta(X_\xi) - \theta([X_\xi, X_\eta]) = \theta([X_\xi, X_\eta]).$$

Now $\theta(Z) = 0$ for Z any vertical field on T^*N and if $X_\eta = L(X)$, then $\theta(L(X)) = \pi^*(f)$ for some $f \in \mathcal{E}(N)$. Thus it is only necessary to check $(*)$ for $X_\xi = L(X)$ and $X_\eta = L(Y)$, where $X, Y \in Vect(N)$. We in fact show that $X_\xi \theta(X_\eta) = \theta([X_\xi, X_\eta])$.

Now $\theta(\alpha_q)(L(Y)) = \alpha_q(Y)$. Thus, if we define $f(\alpha_q) = \alpha_q(Y)$ and compute

$$L(X)(f)(\alpha_q) = \frac{d}{dt}\Big|_{t=0} f((g_X(t)^*)^{-1}(\alpha_q)) = \frac{d}{dt}\Big|_{t=0} \alpha_q(g_X(t)_*^{-1}(Y)) =$$

$$= \alpha_q([X, Y]) = \theta(\alpha_q)([L(X), L(Y)]).$$

Thus $(*)$ is valid. $\square$

Exercise. Let $N = S^2$ be embedded in $\mathbb{R}^3$ as the unit sphere so that it is an orbit of SO_3 in its standard representation. Let $V = ((x_1, x_2, x_3))_\mathbb{R}$ be the vector space of linear functions. Compute the moment map for the action of $G := V \rtimes SO_3$ on $M = T^*N$.

Let (X, ω) be a Kähler manifold so that ω has a potential: $\omega = \frac{i}{2}\partial\bar\partial\rho$.

Remark. A given Kähler form may not have a global potential, but in many applications it is enough to consider its restrictions to open sets where it does.

Now suppose that $G \times X \to X$ is an action of a connected real Lie group of holomorphic transformations which leave the *potential* invariant.

Remark. If $K \times X \to X$ is an action by a real compact Lie group with $k^*\omega = \omega$ for all $k \in K$, then ρ may be replaced by

$$A(\rho) = \int_{k \in K} k^*(\rho)dk$$

which is a K-invariant potential.

Let J be the complex structure on X.

Propostion. *The mapping* $\mu : X \to (Lie\, G)^*$ *defined by* $\mu^\xi := (JX_\xi)(\rho)$ *is an equivariant moment map.*

Proof. In real notation $\omega = dd^c\rho$, where $d^c\rho(V) = d\rho(JV) = (JV)(\rho)$. Thus $i_\xi\omega = d\mu^\xi$.

For the equivariance, note that

$$\omega(V, W) = (dd^c)\rho(V, W) = (V \circ JW - W \circ JV - (J[V, W]))\rho$$

$$= ([V, JW] - [W, JV] - J[V, W])\rho = -d^c\rho([V, W]).$$

This follows from the invariance of ρ and the integrability of the complex structure. $\square$

Exercise. Show that $d^c = \frac{\partial - \bar\partial}{2i}$.

5. The Poisson structure on (Lie G)* and on coadjoint orbits

A *Poisson structure* on a manifold M is a Lie algebra structure $\{\cdot,\cdot\}$ on $\mathcal{E}(M)$ so that the Leibnitz-rule is fullfilled, i.e.

$$\{f, gh\} = g\{f, h\} + h\{f, g\}.$$

This implies that $\{f, \cdot\}$ is a first order linear differential operator, i.e., a vector field X_f.

Example. If (M, ω) is a symplectic manifold, then $\{f, g\} := \omega(X_f, X_g)$ is a Poisson structure. If M is connected, we then have $Z(\mathcal{E}(M)) = \{f : \{f, \cdot\} = 0\} = \mathbb{R}$, i.e., the constant functions. We refer to such a Poisson structure as a symplectic structure, because if $Z(\mathcal{E}(M)) = \mathbb{R}$, then there is a unique symplectic struture ω on M which defines $\{\cdot, \cdot\} : \omega(X_f, X_g) := \{f, g\}$.

Exercise. Fill in the details of the example above.

In order to define the canonical Poisson structure on $(Lie\,G)^* = \mathfrak{g}^*$ we first note the structure at the level of linear functions is defined by the structure on $Lie\,G$: For linear functions $\ell_1, \ell_2 \in Lie\,G$ and $\alpha \in (Lie\,G)^*$

$$\{\ell_1, \ell_2\}(\alpha) := [\ell_1, \ell_2](\alpha) = \alpha([\ell_1, \ell_2]).$$

We extend this to the algebra $\mathcal{E}((Lie\,G)^*)$ as follows:

$$\{f_1, f_2\}(\alpha) := [df_1(\alpha), df_2(\alpha)](\alpha).$$

Here $df_1(\alpha)$ and $df_2(\alpha)$ are regarded as linear functions, i.e., $df_1(\alpha) \in Lie\,G$. The Leibnitz rule

$$d(fg) = f\,dg + g\,df$$

immediately implies the Leibnitz rule for $\{\cdot, \cdot\}$.

It remains to prove the *Jacobi-identity*. For this let

$$P = \oplus_{k=0}^{\infty} S^k(Lie\,G)$$

be the algebra of $\mathbb{R}$-valued polynomials. Here $S^k(Lie\,G)$ is the vector space of homogeneous polynomials of degree k. Let $P_n := \bigoplus_{k=0}^{n} S^k(Lie\,G)$.

Proposition. *The bilinear pairing $\{\cdot, \cdot\}$ on $\mathcal{E}((Lie\,G)^*)$ defines a Poisson structure.*

Proof. Since $\{\cdot, \cdot\}$ is local in both entries and P is dense in $\mathcal{E}(\mathfrak{g}^*)$ with respect ot the locally uniform C^1-topology, it is enough to prove that the Jacobi-identity is satisfied on P. We prove this by induction. Since $Lie\,G$ is a Lie algebra, it is enough to verify the induction step.

Assume that the Jacobi identity is fulfilled for any functions $f_1, f_2, f_3 \in P_n$. If $g_1, g_2, g_3 \in P_{n+1}$ are given, then they are linear combinations of products $h_1 f_1, h_2 f_2, h_3 f_3$, where $h_i, f_i \in P_n$. Applying the Leibnitz-rule to the terms of the form, e.g., $\{h_1 f_1, \{h_2 f_2, h_3 f_3\}\}$ yields the desired result. $\qquad\square$

For $\xi \in Lie\,G$ we now compute $\{\xi, \cdot\}$ as a differential operator. Applied to a linear function $\eta \in Lie\,G$

$$\{\xi, \eta\}(\alpha) = \alpha([\xi, \eta]) = \frac{d}{dt}\Big|_{t=0} Ad^*(exp(\xi t))(\alpha)(\eta) = \frac{d}{dt}\Big|_{t=0} \eta(Ad^*(exp(\xi t))(\alpha)).$$

Thus, since $\{\xi, f\}(\alpha) = [\xi, df(\alpha)](\alpha)$,

$$\{\xi, f\} = \frac{d}{dt}\Big|_{t=0} f \circ Ad^*(exp(\xi t)), \text{ i.e. } \{\xi, \cdot\} = X_\xi,$$

where X_ξ is the vector field on $(Lie\,G)^*$ associated to ξ by the coadjoint representation.

Let $\alpha \in \mathfrak{g}^*$, $f \in \mathcal{E}(\mathfrak{g}^*)$ and $df(\alpha) = \xi \in \mathfrak{g}$. Since $\{df(\alpha), \cdot\} = \{\xi, \cdot\} = X_\xi$ and the mapping $f \mapsto X_\xi(f)(\alpha)$ only depends on the restriction of f to the orbit $\mathfrak{o} = G.\alpha$, we may define the structure $\{\cdot, \cdot\}_\mathfrak{o}$ on $\mathfrak{o}$ as follows: Let $r_\mathfrak{o} : \mathcal{E}(\mathfrak{g}^*) \to \mathcal{E}(\mathfrak{o})$ be the canonical restriction map.

For $f, g \in \mathcal{E}(\mathfrak{o})$ let $\hat{f}, \hat{g} \in \mathcal{E}(\mathfrak{g}^*)$ be such that $r_\mathfrak{o}(\hat{f}) = f$ and $r_\mathfrak{o}(\hat{g}) = g$. Then

$$(*) \qquad\qquad \{f, g\}_\mathfrak{o} := r_\mathfrak{o}(\{\hat{f}, \hat{g}\})$$

is well-defined.

Remark. Of course $G.\alpha = \mathfrak{o}$ may not be closed, but it is locally a submanifold and that is the sense in which $(*)$ is meant.

Since $\{\xi, \cdot\} = X_\xi$ and $\{X_\xi(\alpha) : \xi \in \mathfrak{g}\} = T_\alpha \mathfrak{o}$, it follows that $\{,\}_\mathfrak{o}$ is non-degenerate, i.e., $\{,\}_\mathfrak{o}$ defines a *symplectic structure* $\omega_\mathfrak{o}$ on $\mathfrak{o}$.

Concretely,

$$\omega_\mathfrak{o}(\alpha)(X_\xi, X_\eta) = \{\xi, \eta\}(\alpha) = \alpha([\xi, \eta]).$$

In other words, if $v, w \in T_\alpha G.\alpha$ are defined by the 1-parameter groups $Ad^*(exp(\xi t))$ and $Ad^*(exp(\eta t))$, then

$$\omega_\mathfrak{o}(\alpha)(v, w) = \alpha([\xi, \eta]).$$

Remark. The restricted structures $\{\cdot, \cdot\}_\mathfrak{o}$ define, in a certain sense, the maximal symplectic submanifolds of $(\mathfrak{g}, \{\cdot, \cdot\})$. For example, the degeneracy $Z_{\{\cdot, \cdot\}} = \{f \in \mathcal{E}(M) : \{f, g\} = 0 \text{ for all } g \in \mathcal{E}(\mathfrak{g}^*)\}$ is just the algebra $\mathcal{E}(\mathfrak{g}^*)^G$ of G-invariant functions.

The G-action on a coadjoint G-orbit $\mathfrak{o}$ has an equivariant moment map, e.g., $\mu(\alpha) = \alpha$. To see this it is enough to check the Hamiltonian property $i_{X_\xi}\omega = d\mu^\xi$. But

$$d\mu^\xi(\alpha)(X_\eta) = \frac{d}{dt}\Big|_{t=0} \mu^\xi(Ad^*(exp(\eta t))\alpha) = \frac{d}{dt}\Big|_{t=0} Ad^*(exp(\eta t)\alpha)(\xi)$$

$$= \alpha\left(\frac{d}{dt}\Big|_{t=0} Ad(exp(-\eta t))(\xi)\right) = \alpha([\xi, \eta]).$$

6. The basic formula and some consequences

Let (M, ω) be a symplectic manifold equipped with an action $G \times M \to M$ of a Lie group of symplectic diffeomorphisms. Suppose that there exists a moment map $\mu : M \to \mathfrak{g}^*$. Then we have the following *basic formula*:

$$d\mu(x)(v)(\xi) = d\mu^\xi(x)(v) = \omega(x)(X_\xi, v)$$

for all $\xi \in \mathfrak{g}, x \in M$ and $v \in T_x M$.

Corollary 1. $Ker(d\mu(x)) = (T_x G.x)^{\perp_\omega}$.

Corollary 2. *Let $\mathfrak{g}_x$ denote the Lie algebra of the isotropy group G_x and $\mathfrak{g}_x^0 = \{\alpha \in \mathfrak{g}^* : \alpha(\xi) = 0 \ \forall \xi \in \mathfrak{g}_x\}$ its annihilator. Then $Im(d\mu(x)) = \mathfrak{g}_x^0$.*

Proof. The basic formula shows that $Im(d\mu(x)) \subset \mathfrak{g}_x^0$. Equality follows by computing dimensions via the formula for $Ker(d\mu(x))$. $\qquad\square$

Corollary 3. *If all G-orbits have the same dimension, then μ has constant rank. If all G-isotropy groups are discrete,[25] then $\mu : M \to \mathfrak{g}^*$ is an open mapping.*

Corollary 4. *If G acts transitively on M, then*

$$\mu : M = G/G_x \to G/G_{\mu(x)}$$

is a G-equivariant covering map onto a coadjoint orbit.

7. Moment maps associated to representations

Here we consider the Kählerian manifold $(\mathbb{P}(V), \omega_{FS})$, where ω_{FS} is the Fubini-Study metric associated to a unitary invariant metric h. As we have seen above, $H^2(\mathfrak{k}^*) = \{0\}$, where $K = SU(V, h)$.

Exercise. Show that $H^1(\mathfrak{k}^*) = \{0\}$, i.e., $\mathfrak{k} = [\mathfrak{k}, \mathfrak{k}]$.

Thus there exists a unique K-moment map $\mu : \mathbb{P}(V) \to \mathfrak{k}^*$. Now let G be a subgroup of K; in particular we have the restriction $\mu_G : \mathbb{P}(V) \to \mathfrak{g}^*$.

Remark. Uniqueness is no longer guaranteed, e.g., G might be abelian.

If $i : X \hookrightarrow \mathbb{P}(V)$ is a complex submanifold, then the pull-back $i^*(\omega)$ is a Kähler-form on X. Thus, if X is in addition G-stable, we have the moment map $\mu_G : X \to \mathfrak{g}^*$.

As a consequence of the above observations, it becomes relevant to compute the moment map $\mu : \mathbb{P}(V) \to \mathfrak{k}^* = \mathfrak{su}_{n+1}^*$, because the others are defined by one form or another of restriction.

Recall that

$$\pi^*(\omega_{FS}) = \frac{i}{2} \partial \bar\partial \log \rho, \quad \text{where } \pi : V \backslash \{0\} \to \mathbb{P}(V)$$

is the standard projection and $\rho = |\cdot|^2$ is a unitary invariant norm-function. If $\xi \in \mathfrak{k}$ and X_ξ is the associated vector field on $V \backslash \{0\}$, then

$$(*) \qquad\qquad \mu^\xi = (JX_\xi)(\log \rho)$$

---

[25] In this case one says that the G-action is *locally free*.

is invariant with respect to scalar multiplication and defines a global moment function on $\mathbb{P}(V)$.

Remark. Above we showed that $\mu : M \to \mathfrak{g}^*$ is equivariant if $\mu^\xi = (JX_\xi)(\rho)$. The proof did not require $\omega = \frac{i}{2}\partial\bar{\partial}\rho$ to be non-degenerate; only the fact that ρ is G-invariant was needed. Hence $\mu : V\backslash\{0\} \to \mathfrak{k}^*$ defined by $(*)$ is equivariant and, since it is invariant by scalar multiplication, defines a moment map by the same name $\mu : \mathbb{P}(V) \to \mathfrak{k}^*$.

Let us carry out a concrete calculation. If

$$K = U_{n+1} = \{A \in Gl_{n+1}(\mathbb{C}) : A\bar{A}^t = Id\},$$

then $\mathfrak{k} = \{A \in Mat(n \times n, \mathbb{C}) : A + \bar{A}^t = 0\}$. Let $z = (z_1, \ldots, z_{n+1})$ be the standard complex coordinates for $\mathbb{C}^{n+1}$ and $\langle , \rangle$ the standard Hermitian product: $\langle z, w \rangle = {}^tz\bar{w}$.

If $\rho = \log\langle z, z \rangle$, then

$$\mu^A(z) = (JX_A)(\log\langle z, z \rangle) = \frac{d}{dt}\Big|_{t=0} \log\langle z + t(iAz), z + t(iAz) \rangle = 2i\frac{\langle Az, z \rangle}{\langle z, z \rangle}.$$

V. Kählerian structures on coadjoint orbits of compact groups and associated representations

On a coadjoint orbit $M = K.\lambda$ of a compact group there is a unique integrable K-invariant complex structure $J : TM \to TM$ so that the canonical symplectic form ω_M is Kählerian. If λ has been chosen so that ω_M is integral, i.e., $\omega_M \in H^{1,1}(M, \mathbb{Z})$, then there exists a uniquely associated holomorphic K-homogeneous line bundle $E_\lambda \to M$ equipped with a natural Hermitian metric h with Chern form $c_1^h(E_\lambda) = \omega_M$. This bundle is very ample and yields an embedding $M \hookrightarrow \mathbb{P}(\Gamma(M, E_\lambda)^*)$. The K-representation on $\Gamma(M, E_\lambda)$ is irreducible and all irreducible K-representations occur in this way.

The essentially unique K-invariant Hermitian metric on $\Gamma(M, E_\lambda)$ yields a canonical Fubini Study structure $\omega_{FS} = \frac{1}{2\pi}dd^c\log\|\cdot\|^2$ on $\mathbb{P}(\Gamma(M, E_\lambda)^*)$ where M is realized as the unique complex K-orbit. The moment map $\mu : \mathbb{P}(\Gamma(M, E_\lambda)^*) \to \mathfrak{k}^*$ establishes a K-equivariant diffeomorphism between M embedded in the projective space and M embedded in $\mathfrak{k}^*$. In fact, if $v_{max} \in \Gamma(M, E_\lambda)^*$ is a maximal weight vector and $x_0 = [v_{max}] \in \mathbb{P}(\Gamma(M, E_\lambda)^*)$, then $\mu(x_0) = \lambda$.

The goal of this chapter is to give background information and detailed proofs for this "geometric quantization" of coadjoint K-orbits. This point of view was suggested by Kirillov, Kostant and Souriau (compare, e.g., [Kiri], [Kos] and [So]).

1. Generalities on compact groups

The main aim here is to give a first description of the coadjoint representation of a connected compact group K. Of course the center $Z = Z(K)$ acts trivially on $\mathfrak{k}^*$. Hence it is important that it can essentially be split off.

1.1 The splitting $K = Z^0.K_{ss}$

Regard the Lie algebra $\mathfrak{z}$ of the center Z as a K-invariant subspace of the adjoint representation $\mathfrak{k}$ and let $\mathfrak{k}_{ss}$ be a K-invariant complement, i.e., $\mathfrak{k} = \mathfrak{z} \oplus \mathfrak{k}_{ss}$. Since $Ad(K)(\mathfrak{k}_{ss}) \subset \mathfrak{k}_{ss}$, it follows that $\mathfrak{k}_{ss}$ is an ideal in $\mathfrak{k}$.

Note that this splitting procedure can be carried out for any ideal. It follows that $\mathfrak{k}_{ss} = \mathfrak{k}_{ss}^1 \oplus \ldots \oplus \mathfrak{k}_{ss}^m$ is a direct sum of ideals each of which only has trivial ideals. Furthermore, since the center has already been split off, none of these is Abelian, i.e., the summands are simple Lie algebras and $\mathfrak{k}_{ss}$ may be regarded as the "semisimple part of $\mathfrak{k}$".

Let K_{ss} be the unique connected (possibly not closed) subgroup of K with Lie algebra $\mathfrak{k}_{ss}$. Assuming K is connected, it immediately follows that $K = Z^0.K_{ss}$.

The essential point is the following

Proposition. *The semisimple part K_{ss} of a connected compact Lie group K is compact.*

Proof. The principle bundle $Z^0 \to K \to K_1 := K/Z^0$ realizes K_{ss} as a covering space of the compact group K_1 which also has Lie algebra $\mathfrak{k}_{ss}$. Note that $\mathfrak{k}_{ss} = [\mathfrak{k}_{ss}, \mathfrak{k}_{ss}]$, i.e.,
$$H_{deR}^1(K_1) = H^1(\mathfrak{k}^*) = (\mathfrak{k}/[\mathfrak{k}_{ss}, \mathfrak{k}_{ss}])^* = \{0\}.$$
Since the fundamental group $\pi_1(K_1)$ is Abelian, it follows that $\pi_1(K_1) \equiv H_1(K_1, \mathbb{Z})$ and the universal coefficient theorem implies that $\pi_1(K_1)$ consists only of torsion elements. Since K_1 is compact, $\pi_1(K_1)$ is therefore finite and consequently $K_{ss} \to K_1$ is a finite covering; in particular K_{ss} is also compact. $\square$

1.2 The existence of a faithful representation

If A is a connected Abelian group, then $exp : \mathfrak{a} \to A$ is a homomorphism from the additive group $(\mathfrak{a}, +)$ onto A. Thus, $A \cong \mathfrak{a}/\Gamma$, where Γ is a discrete additive subgroup. Hence, if T is a connected Abelian compact group, then it is a torus, $T = S^1 \times \ldots \times S^1$, and clearly possesses a faithful representation $\rho : T \to GL(V)$.

On the other hand, if $K = K_{ss}$ is semisimple, then $Ad : K \to GL(\mathfrak{k})$ has only a finite kernel. Thus, up to finite coverings, it is clear that $K = Z^0.K_{ss}$ has a faithful representation.

In fact it requires a bit of work to remove the "up to finite coverings" part of the above statement. One line of argumentation goes as follows: For any finite subset $\{k_1, \ldots, k_m\} \subset K$ there clearly exists a function $f \in C^\infty(K)$ such that $f(k_i) \neq f(k_j)$. If in addition the linear span $V = < K.f >$ of the orbit of f in $C^\infty(K)$ is finite dimensional, i.e., f is a "K-finite function", then the representation $\rho : K \to GL(V)$ separates the points of the given set. The following is therefore of central importance (see e.g. [BTD]).

Theorem. *The set of K-finite functions is a dense subset of $C^\infty(K)$.*

In fact, by using more refined arguments it can be shown that $\mathbb{R}$- analytic functions do the job.

Theorem. *A compact Lie group K admits a faithful $\mathbb{R}$-analytic representation.*

1.3 Rough slices
Given a K-action $K \times M \to M$ one is interested in constructing closed submanifolds $\Sigma \hookrightarrow M$ so that $K.\Sigma = M$. For example, if M is a representation space, then it is important to determine the smallest subspaces with this property. Here we prove an observation concerning the existence of such a "rough slice" which is of particular use in analyzing the coadjoint representation.

Notation. Let $K \times M \to M$ be a smooth action of a compact group on a connected manifold and $\Sigma \hookrightarrow M$ a closed submanifold. Let M_{gen} be the open subset of dimension theoretically generic orbits and $\Sigma_{gen} := \Sigma \cap M_{gen}$.

Definition. The action is said to be *generically transversal* to Σ if $\Sigma_{gen} \neq \phi$ and for all $x \in \Sigma_{gen}$ it follows that

$$T_x\Sigma + T_x K.x = T_x M.$$

Proposition. *If a smooth action $K \times M \to M$ of a compact group on a connected manifold is generically transversal to a closed submanifold Σ, then $K.\Sigma = M$.*

Proof. Clearly $K.\Sigma$ is closed with open complement U and interior $V \supset K.\Sigma_{gen}$. Thus the complement S of the disjoint union $U \dot\cup V$ is contained in a set $K.(M \setminus M_{gen})$ which is locally the union of finitely many submanifolds each being at least 2-codimensional. In particular, the Hausdorff codimension of S is at least two and consequently $M \setminus S$ is connected. Since $V \neq \phi$, it follows that $K.\Sigma = M$. $\square$

1.4 Maximal tori
Let K be a connected compact group acting via Ad on its Lie algebra $\mathfrak{k}$. A maximal connected Abelian subgroup $T < K$ is called a maximal torus. If there is no confusion, its Lie algebra $\mathfrak{t}$ is called a maximal toral subalgebra of $\mathfrak{k}$.

For $\xi \in \mathfrak{t}$ let $T_\xi = \overline{< exp(t\xi) >}_{t \in \mathbb{R}}$, i.e., the closure of the 1-parameter subgroup generated by ξ. It follows that the isotropy group K_ξ is the centralizer $Z_K(T_\xi)$. In particular, for ξ generic in the sense that $T_\xi = T$, $K_\xi = Z(T)$. Note that if $A < Z_K(T)$ is any 1-parameter subgroup, then the closure $\overline{A.T}$ is a torus which contains T. Hence, the maximality of T implies that $Z_K^0(T) = T$. Since connected solvable compact groups are Abelian, the same argument proves the following

Proposition. *The connected component $N_K(T)^0$ is T itself.*

At the level of Lie algebras the statement $\mathfrak{n} = \mathfrak{n}_\mathfrak{k}(\mathfrak{t}) = \mathfrak{t}$ is of great use, because for $\xi \in \mathfrak{t}, T_\xi K.\xi = ad(\mathfrak{k})(\xi)$. Since $\mathfrak{n} = \mathfrak{t}$, for $\xi \in \mathfrak{t}_{gen}$ it follows that $\{x \in \mathfrak{k} : ad(x)(\xi) \in \mathfrak{t}\} = \mathfrak{t}$, i.e., $K.\xi$ is transversal to $\mathfrak{t}$.

Corollary. *A maximal toral subalgebra $\mathfrak{t}$ is a rough slice for the adjoint representation of K; in particular $K.\mathfrak{t} = \mathfrak{k}$.*

At the group level this is equivalent to the statement that any two maximal tori are conjugate in K. In fact the same argument proves the direct translation at the group level.

Corollary. *A maximal torus T is a rough slice for the action of K on itself defined by conjugation, $k(x) := kxk^{-1}$; in particular $Int(K).T = K$.*

As a result, every $x \in K$ can be conjugated into a fixed maximal torus. The following is therefore a consequence of the surjectivity of $exp : \mathfrak{t} \to T$.

Corollary. *The exponential map $exp : \mathfrak{k} \to K$ of a connected compact Lie group is surjective.*

Since the normalizer $N = N_K(T)$ is a compact subgroup of K which has the same dimension as T, it follows that the *Weyl group* $\mathcal{W} = N/T$ is finite.

We regard $\mathcal{W}$ as acting on T by inner automorphisms and consider its induced action $\mathcal{W} \times \mathfrak{t} \to \mathfrak{t}$.

Proposition. *For $\xi \in \mathfrak{t}$ it follows that $K.\xi \cap \mathfrak{t} = \mathcal{W}.\xi$.*

Proof. For $\xi \in \mathfrak{t}_{gen}$ and $k(\xi) \in \mathfrak{t}$, it follows that $k(\xi) \in \mathfrak{t}_{gen}$ and $K_\xi^0 = K_{k(\xi)}^0 = T$. Thus $k \in N$ and the result follows.

For the proof in the case where $\xi \notin \mathfrak{t}_{gen}$, a local version of the "rough slice theorem" allows us to argue as follows: Let $\xi \in \mathfrak{t}$ and $k \in K$ be given with $k(\xi) \in \mathfrak{t}$. Take $\{\xi_n\} \subset \mathfrak{t}_{gen}$ with $\xi_n \to \xi$. Of course $k(\xi_n) \to k(\xi)$. By the above mentioned local version it follows that for $n >> 0$ there exist $\tilde{\xi}_n \in \mathfrak{t}$ with $\tilde{\xi}_n \to k(\xi)$ and $k_n \in K$ so that $k_n(\tilde{\xi}_n) = k(\xi_n)$. The sequence $\{k_n^{-1}k\}$ is contained in N. Furthermore, by the compactness of K, it may be assumed that $k_n \to k_0 \in K_{k(\xi)} < N$. It follows that $k \in N$ as well. $\qquad\square$

Remark. In §5 it is shown that the isotropy groups $K_\xi, \xi \in \mathfrak{t}$, are connected. In particular this is true for $\xi \in \mathfrak{t}_{gen}$. Consequently, $Z_K(T) = T$ and the action of $\mathcal{W}$ on $\mathfrak{t}$ is effective. $\qquad\square$

Exercise. Show that the set T of diagonal matrices in $K = U_n$ is a maximal torus. Compute the Weyl group $\mathcal{W} = N/T$. $\qquad\square$

2. Root decomposition for $\mathfrak{k}^{\mathbb{C}}$

2.1 Schur's Lemma

Let $G < Gl_{\mathbb{C}}(V)$ be a group of complex linear transformations and $End(V)^G$ the vector subspace of G-equivariant endomorphisms. For $T \in End(V)^G$ the spectral decomposition $V = \oplus V_\lambda, \lambda \in Spec(T)$, is G-invariant. Thus the following is immediate.

Proposition. *If the only G-invariant subspaces are $\{0\}$ and V itself, then*

$$End(V)^G = \mathbb{C}.Id.$$

A finite-dimensional representation $\rho : G \to GL_{\mathbb{C}}(V)$ is said to be irreducible if it only has the trivial invariant subspaces as above.

Corollary. A representation $\rho : G \to GL_{\mathbb{C}}(V)$ of an Abelian Lie group is irreducible if and only if $dim_{\mathbb{C}} V = 1$.

Proof. $\rho(G) \subset End(V)^G$. □

It is quite possible that G leaves a number of subspaces invariant, but nevertheless $End(V)^G = \mathbb{C}.Id$.

Exercise. Show that the group B of upper triangular matrices in $GL_n(\mathbb{C})$ satisfies $End(\mathbb{C}^n)^B = \mathbb{C}.Id$. □

On the other hand, for compact groups K there is a strong converse. For this let $\rho : K \to GL_{\mathbb{C}}(V)$ be a representation with an invariant subspace W. Since we may always average a given Hermitian structure to obtain an invariant one, it may be assumed that ρ is unitary. Thus the decomposition $V = W \oplus W^{\perp}$ is invariant. More generally we have the following

Proposition (Complete reducibility): *Let K be compact and $\rho : K \to GL_{\mathbb{C}}(V)$ a unitary representation. Then*

$$V = V_1 \oplus \ldots \oplus V_m$$

is an orthogonal direct sum of irreducible respresentations.

As a consequence one has Schur's Lemma for compact groups.

Corollary. *A representation $\rho : K \to GL_{\mathbb{C}}(V)$ is irreducible if and only if $End(V)^K = \mathbb{C}.Id$. If $\rho : K \to GL_{\mathbb{C}}(V)$ is an irreducible representation of an Abelian compact group, then V is 1-dimensional.*

Notation. For G a Lie group, $\mathfrak{X}(G)$ denotes the set of smooth 1-dimensional representations $\chi : G \to GL_{\mathbb{C}}(V) = \mathbb{C}^*$. If K is compact and $\chi \in \mathfrak{X}(K)$, then $\chi : K \to S^1$. If there is no confusion, such 1-dimensional representations will be called *characters*.

2.2 Characters of tori

Let T be a torus, i.e., a connected, compact Abelian Lie group. Since T is connected and Abelian, $exp : \mathfrak{t} \to T$ is a surjective homomorphism. Here $\mathfrak{t}$ is regarded as a Lie group with respect to addition. Consequently, there is the canonical identification $T \cong \mathfrak{t}/Ker(exp)$.

If $\chi \in \mathfrak{X}(T)$, then we have the following diagram:

$$
\begin{array}{ccc}
\mathfrak{t} & \xrightarrow{\lambda} & \mathbb{R} \\
exp \downarrow & & \downarrow e \\
T & \xrightarrow{\chi} & S^1
\end{array}
$$

where $e(x) = e^{2\pi i x}$ and $\lambda = \chi_*$ is the derivative of χ. Of course $\lambda(Ker(exp)) \subset \mathbb{Z}$. We let $\mathfrak{t}_{\mathbb{Z}}^*$ denote the set of such *integral functionals* and for $\lambda \in \mathfrak{t}_{\mathbb{Z}}^*$ let $\chi_{\lambda} = e^{2\pi i \lambda} = e \circ \lambda$. In this language it follows that $\mathfrak{X}(T)$ is naturally isomorphic to $\mathfrak{t}_{\mathbb{Z}}^*$.

Reformulating the results in the previous section for tori, we have the following

Proposition. *Let $\rho : T \to GL_{\mathbb{C}}(V)$ be a representation of a torus. Then there are uniquely determined functionals $\lambda_1, \ldots, \lambda_m \in \mathfrak{t}_{\mathbb{Z}}^*$ so that*

$$V = \bigoplus_{j=1}^{m} V_{\lambda_j},$$

where $V_{\lambda_j} = \{v \in V : t(v) = \chi_{\lambda_j}(t).v\} \neq (0)$.

2.3 Properties of root spaces

Let $\mathfrak{k}^{\mathbb{C}} = \mathfrak{k} \otimes_{\mathbb{R}} \mathbb{C}$ be the complexification of $\mathfrak{k}$ and consider $Ad : K \to GL_{\mathbb{C}}(\mathfrak{k}^{\mathbb{C}})$.

Remark. Formally we should write $Ad^{\mathbb{C}}$, because we are considering the transformations $Ad(k)$ as being extended by complex linearity to $\mathfrak{k}^{\mathbb{C}}$.

For clarity it is perhaps better to write $\mathfrak{g} = \mathfrak{k}^{\mathbb{C}}$ and let $\tau : \mathfrak{g} \to \mathfrak{g}$ be the anti-holomorphic Lie algebra involution which defines $\mathfrak{k}$ as its fixed points.

Now the induced T-action on $\mathfrak{g}$ can be diagonalized as in the above proposition:
$\mathfrak{g} = \mathfrak{g}_{\lambda_1} \oplus \ldots \oplus \mathfrak{g}_{\lambda_m}$.

Of course $\lambda = 0$ occurs, because T is Abelian. Furthermore, since $\mathfrak{t}$ is maximal, it follows that $\mathfrak{g}_0 = \mathfrak{t}^{\mathbb{C}}$. The remaining $\mathfrak{g}_\lambda$'s are called *root spaces* of $\mathfrak{g}$ with respect to T and are denoted by $\mathfrak{g}_\alpha, \alpha \in \Phi$. The functionals $\alpha \in \Phi$ are called *roots*.

In the general theory of root space decompositions for complex semisimple Lie algebras, the existence of the compact real form $\mathfrak{k}$ is shown only after a certain amount of significant work. Here we have the advantage that K is known to exist from the beginning.

For example, $\tau(\mathfrak{g}_\alpha) = \mathfrak{g}_{-\alpha}$ is immediate from the facts that $\tau|T = Id_T$ and its being anti-holomorphic; in particular

(1) $\alpha \in \Phi \iff -\alpha \in \Phi$.

Using the fact that T is acting as a group of Lie algebra automorphisms

$$\xi([x_\alpha, x_\beta]) = (\alpha(\xi) + \beta(\xi)) \cdot [x_\alpha, x_\beta]$$

for all $\xi \in \mathfrak{t}$, $x_\alpha \in \mathfrak{g}_\alpha$ and $x_\beta \in \mathfrak{g}_\beta$.

So, if $[x_\alpha, x_\beta] \neq 0$ and $\alpha + \beta \neq 0$, then

(2) $\alpha, \beta \in \Phi \implies \alpha + \beta \in \Phi$.

Now any system Φ satisfying (1) and (2) can be (non-uniquely) divided into two parts: $\Phi = \Phi^+ \dot\cup \Phi^-$, where $\Phi^- = -\Phi^+$ satisfying

(3) $\alpha, \beta \in \Phi^+$ and $[\mathfrak{g}_\alpha, \mathfrak{g}_\beta] \neq (0) \implies \alpha + \beta \in \Phi^+$.

It follows that

$$\mathfrak{u}^+ := \bigoplus_{\alpha \in \Phi^+} \mathfrak{g}_\alpha$$

(resp. $\mathfrak{u}^- := \tau(\mathfrak{u}^+)$) and $\mathfrak{b}^+ := \mathfrak{t}^{\mathbb{C}} \ltimes \mathfrak{u}^+$ (resp. $\mathfrak{b}^- := \tau(\mathfrak{b}^+)$) are complex Lie subalgebras of $\mathfrak{g}$. Here the notation $\mathfrak{t}^{\mathbb{C}} \ltimes \mathfrak{u}^+$ underlines the fact that $\mathfrak{t}$ normalizes $\mathfrak{u}^+$.

There are two basic properties of the root decomposition which we choose not to prove (see e.g. [Wal] or [Hu2]):

(4) $$\text{For all } \alpha \in \Phi, \ dim_{\mathbb{C}} \, \mathfrak{g}_\alpha = 1$$

and

(5) $$\text{For } \alpha \in \Phi, k\alpha \in \Phi \iff k = \pm 1.$$

2.4 The root SU_2's

Let $\alpha \in \Phi$ and $\chi_\alpha \in \mathfrak{X}(T)$ be the associated character. Define T_α to the connected component $Ker(\chi_\alpha)^0$ and $Z_\alpha(K) := Z_K(T_\alpha)^0$.

Since $T < Z_\alpha(K)$, the complexified Lie algebra $\mathfrak{z}_\alpha(\mathfrak{k})^{\mathbb{C}}$ is the direct sum of $\mathfrak{t}^{\mathbb{C}}$ and finitely many root algebras. Note that if $\mathfrak{g}_\beta < \mathfrak{z}_\alpha(\mathfrak{k})^{\mathbb{C}}$, then $\beta|\mathfrak{t}_\alpha \equiv 0$, i.e., $\beta = k\alpha$.

Thus, by (5) above it follows that

$$\mathfrak{z}_\alpha^{\mathbb{C}} = \mathfrak{t}^{\mathbb{C}} \oplus \mathfrak{g}_\alpha \oplus \mathfrak{g}_{-\alpha}.$$

It is more convenient to write this in another way. For this, note that $[\mathfrak{g}_\alpha, \mathfrak{g}_{-\alpha}] \neq (0)$ as otherwise $Z_\alpha(K)$ would be a compact solvable group, i.e., Abelian, contrary to the maximality of T.

Thus $\mathfrak{z}_\alpha^{\mathbb{C}} = \mathfrak{t}_\alpha^{\mathbb{C}} \oplus \mathfrak{s}_\alpha$, where $\mathfrak{s}_\alpha := \mathfrak{g}_{-\alpha} \oplus [\mathfrak{g}_{-\alpha}, \mathfrak{g}_\alpha] \oplus \mathfrak{g}_\alpha$ is σ-invariant and is therefore of the form $\mathfrak{s}_\alpha =: \mathfrak{k}_\alpha^{\mathbb{C}}$, i.e.,

$$\mathfrak{z}_\alpha = \mathfrak{t}_\alpha \oplus \mathfrak{k}_\alpha.$$

Now clearly T_α is the connected component of the center of $Z_\alpha(K)$ and thus $\mathfrak{k}_\alpha$ is the Lie algebra of its semisimple part $K_\alpha : Z_\alpha(K) = T_\alpha \cdot K_\alpha$ is the splitting.

It can easily be checked that SU_2 is the unique simply-connected, 3-dimensional, connected compact Lie group. Thus, although it might not be simply-connected $(SO_3 = SU_2/\{\pm Id\}$ is the other possiblility), we refer to K_α as the "root SU_2" associated to $\alpha \in \Phi$ and, if there is no confusion, write $SU_2(\alpha)$.

In this case $[\mathfrak{g}_{-\alpha}, \mathfrak{g}_\alpha]$ is σ-invariant and thus its real part is a maximal toral algebra in $su_2(\alpha)$. In this way $su_2(\alpha)$ can be identified with the matrix algebra

$$\mathfrak{su}_2 = \left(\left(\begin{pmatrix} i & 0 \\ 0 & -i \end{pmatrix}, \begin{pmatrix} 0 & 1 \\ -1 & 0 \end{pmatrix}, \begin{pmatrix} 0 & i \\ i & 0 \end{pmatrix} \right) \right)_{\mathbb{R}}.$$

The matrix $\begin{pmatrix} i & 0 \\ 0 & -i \end{pmatrix}$ is a basis of its maximal torus and the remaining two matrices form a basis for the σ-invariant subspace $\mathfrak{g}_{-\alpha} \oplus \mathfrak{g}_\alpha$ over $\mathbb{C}$. It is sometimes convenient to regard $\mathbb{C} \hookrightarrow su_2$ by

$$z = x + iy \mapsto x \begin{pmatrix} 0 & 1 \\ -1 & 0 \end{pmatrix} + y \begin{pmatrix} 0 & i \\ i & 0 \end{pmatrix} = \begin{pmatrix} 0 & z \\ -\bar{z} & 0 \end{pmatrix}. \qquad \square$$

3. Complexification of compact groups

3.1 Tori

Let $T = \mathfrak{t}/\Gamma$, $\Gamma := Ker(exp)$, be a compact n-dimensional torus. It follows that $\Gamma = < \gamma_1, \ldots, \gamma_n >_\mathbb{Z}$ is generated by n $\mathbb{R}$-independent vectors $\gamma_j \in \mathfrak{t}$. Each defines a linear functional $\lambda_j \in \mathfrak{t}_\mathbb{Z}^*$ by $x_1\gamma_1 + \ldots + x_n\gamma_n \mapsto x_j$ The map $\iota : T \to (S^1)^n$, $t \mapsto (\chi_{\lambda_1}(t), \ldots, \chi_{\lambda_n}(t))$ is a $\mathbb{R}$-analytic group isomorphism. Since $\mathfrak{t}_\mathbb{Z}^* = ((\lambda_1, \ldots, \lambda_n))_\mathbb{Z}$, it follows that every character $\chi \in \mathfrak{X}(T)$ can be uniquely written as a monomial $\chi = \chi_{\lambda_1}^{m_1} \cdot \ldots \cdot \chi_{\lambda_n}^{m_n}, m_j \in \mathbb{Z}$.

Now define a complexification $T^\mathbb{C}$ of T to be $(\mathbb{C}^*)^n$ along with the embedding $\iota : T \to (S^1)^n < (\mathbb{C}^*)^n$. Since the standard coordinate function z on S^1 extends to a holomorphic character on $\mathbb{C}^*$, we have the following interpretation of this information.

Proposition (Universality with respect to representations). *For every representation $\rho : T \to GL_\mathbb{C}(V)$ there exists a unique holomorphic representation, $\tau_\rho : T^\mathbb{C} \to GL_\mathbb{C}(V)$ such that $\rho = \tau_\rho \circ \iota$.*

Proof. The existence follows from the above remarks on extension of Laurent monomials to $T^\mathbb{C}$ and uniqueness is a consequence of the identity principle. □

Since $T^\mathbb{C}$ is itself a linear group, the universality property implies that the complexification $\iota : T \to T^\mathbb{C}$ is unique up to automorphisms of T.

Remark. As a hint of things to come we observe that the T-finite functions on T extend to Laurent monomials, i.e., algebraic geometric functions on the algebraic group $T^\mathbb{C}$.

3.2 Polar decomposition for $GL_n(\mathbb{C})$

The usual polar decomposition for $z \in \mathbb{C}^*$ can be written as

$$z = z\left(\sqrt{z\tau(z)}\right)^{-1}\left(\sqrt{z\tau(z)}\right) = e^{i\phi} \cdot r,$$

where $\tau : \mathbb{C}^* \to \mathbb{C}^*$, $z \mapsto \bar{z} = \bar{z}^t$, is the anti-holomorphic involution defining $U_1 = S^1 < \mathbb{C}^* = GL_1(\mathbb{C})$. In this way

$$GL_1(\mathbb{C}) = U_1 \times \mathbb{R}^{>0},$$

where $\mathbb{R}^{>0}$ should be regarded as the set of positive definite Hermitian matrices in $GL_1(\mathbb{C})$.

For the polar decomposition of $GL_n(\mathbb{C})$ let P be the set of Hermitian positive-definite matrices and $K := U_n$. Regard $M = GL_n(\mathbb{C})$ as a K-manifold given by left-multiplicaton $x \mapsto k(x) := k.x$.

Proposition. *At every $x \in P$ the orbit $K.x$ is transversal to P:*

$$T_x P \oplus T_x K.x = T_x M.$$

Proof. For $A \in \mathfrak{k}$ consider the orbit $e^{At}.H = H + AHt + O(t^2)$. Note that P is open in the vector space of Hermitian matrices. Thus this curve is tangent to P if and only if AH is Hermitian. But this would imply that $AH = -HA$; in particular if V_λ is the H-eigenspace for the eigenvalue λ, then $A.V_\lambda \subset V_{-\lambda}$. Since $H > 0$, it follows that $A \equiv 0$. $\qquad\square$

Corollary. *The natural map* $\lambda : K \times P \to GL_n(\mathbb{C}), (U, H) \mapsto U.H$, *is a surjective local diffeomorphism.*

Proof. Since P is closed in $GL_n(\mathbb{C})$ and K is compact, it follows that $Im(\lambda)$ is closed. The openness of $Im(\lambda)$ and the fact that λ is a local diffeomorphism follow immediately from the above Proposition. $\qquad\square$

In fact λ is also injective: Suppose $BH_1 = H_2$ with $B \in K$ and $H_j \in P, j = 1, 2$. Then the facts that B is unitary and the H_j's are Hermitian imply that $H_1 B^{-1} = H_2$. So $H_1^2 = H_2^2$ and, since both are positive-definite, $H_1 = H_2$.

In summary we have the following

Theorem (Polar decomposition for $GL_n(\mathbb{C})$). *The map*

$$\lambda : K \times P \to GL_n(\mathbb{C}), \quad (U, H) \mapsto U.H,$$

is an equivariant diffeomorphism which intertwines the diagonal action of K on $K \times P$ given by conjugation on both factors with conjugation on $GL_n(\mathbb{C})$.

Proof. It only remains to check the last point:

$$(kUk^{-1}, kHk^{-1}) \mapsto kUk^{-1}kHk^{-1} = kUHk^{-1}. \qquad\square$$

Remark. Since an arbitrary $H \in P$ is K-conjugate to a diagonal matrix, it follows that $exp : \mathfrak{p} \to P$ is surjective. Here $\mathfrak{p} := i\mathfrak{k}$ denotes the space of Hermitian matrices in $\mathfrak{gl}_n(\mathbb{C}) = End(\mathbb{C}^n)$. In fact:

Proposition. $exp : \mathfrak{p} \to P$ *is a diffeomorphism.*

One way of proving this is by "taking roots".

Lemma. *For every $n \in \mathbb{N}^{>0}$ the map $\varphi_n : P \to P, B \mapsto B^n$, is a diffeomorphism.*

Proof. Since every $B \in P$ is K-diagonalizable, it is enough to prove the maximal rank condition at points $B \in D$ in the diagonal matrices D in P. For $B + tH$ a line through B in $P, (B + tH)^n = B^n + ntBH + 0(t^2)$. Since $BH = 0$ only for $H = 0$, the map φ_n is of maximal rank.

Since $B \in D$ clearly has an n-th root, φ_n is clearly surjective.

For the injectivity note that $v \in \mathbb{C}^n$ is a λ-eigenvector for $A \in P$ if and only if it is a λ^n-eigenvector for A^n. Thus $A^n = B^n$ immediately implies that $A = B$. $\qquad\square$

The Proposition then follows from the fact that $exp \circ \psi_n = \varphi_n \circ exp$, where $\psi_n(x) = nx$, because by taking n-th roots one can push the necessary considerations to a neighborhood of $0 \in \mathfrak{p}$ where exp is a diffeomorphism. $\qquad\square$

Another formulation of the polar decomposition for $GL_n(\mathbb{C})$ is that the map

$$K \times \mathfrak{k} \to GL_n(\mathbb{C}), (k, \xi) \mapsto k \cdot exp(i\xi)$$

is a diffeomorphism.

3.3 Construction of $K^\mathbb{C}$

Now let K be a connected compact Lie group and $\iota : K \to GL_\mathbb{C}(V)$ a faithful unitary representation. Choosing a basis, we may assume that

$$\iota : K \hookrightarrow U_n < GL_n(\mathbb{C}).$$

Let $K^\mathbb{C}$ be the complex Lie subgroup of $GL_n(\mathbb{C})$ which is associated to the complex Lie algebra $\mathfrak{k}^\mathbb{C} < \mathfrak{gl}_n(\mathbb{C})$. Note that $K.exp(i\mathfrak{k})$ is a closed submanifold of $GL_n(\mathbb{C})$, because it is defined by restricting the polar decomposition of $GL_n(\mathbb{C})$. Since its dimension is the same as that of $K^\mathbb{C}$ and it is contained in $K^\mathbb{C}$, it follows that $K^\mathbb{C} = K \cdot exp(i\mathfrak{k})$ is a closed complex subgroup of $GL_n(\mathbb{C})$.

If K is not connected, then we construct $(K^0)^\mathbb{C}$ and define $K^\mathbb{C} = K.(K^0)^\mathbb{C}$. It is likewise a closed subgroup with polar decomposition.

Now given a representation $\rho : K \to GL_\mathbb{C}(V)$ we wish to construct the holomorphic group homomorphism $\tau_\rho : K^\mathbb{C} \to GL_\mathbb{C}(V)$ so that $\tau_\rho \circ \iota = \rho$.

Since ρ is defined near the identity $e \in K$ by the Lie algebra representation $\rho_* :$ $\mathfrak{k} \to \mathfrak{gl}_n(\mathbb{C})$, $exp : \mathfrak{k} \to K$ and $E : \mathfrak{gl}_n(\mathbb{C}) \to GL_n(\mathbb{C})$ and since all of these maps are $\mathbb{R}$-analytic, it follows that ρ is $\mathbb{R}$-analytic. Thus the map τ_ρ is automatically defined and is holomorphic on some neighborhood $U = U(K) \subset K^\mathbb{C}$.

We choose $U := Kexp(iV)$, where $V = V(0) \subset \mathfrak{k}$ is a sufficiently small open neighborhood. Note that

$$\lim_{n \to \infty} nV = \mathfrak{k}.$$

Thus, for every $g_0 = k_0 exp(i\xi_0) \in K^\mathbb{C}$ there exists a sufficiently small neighborhood $W = W(g_0)$ and $n \in \mathbb{N}^{>0}$ so that for all $kexp(i\xi) \in W$ it follows that $kexp(i\frac{\xi}{n}) \in U$.

Then we may define

$$\tau_\rho(kexp(i\xi)) = \tau_\rho(k)\tau_\rho(exp(i\frac{\xi}{n}))^n$$

It follows that if it is well-defined, then τ_ρ is holomorphic with $\tau_\rho \circ i = \rho$ following from the identity principle.

So it only remains to check that, if $exp(in_1\xi_1) = exp(in_2\xi_2)$, then the two definitions of τ_ρ agree. But in this case $n_1\xi_1 = n_2\xi_2$ and the question is reduced to 1-parameter groups, where the extension is uniquely given by the complex linear extension of ρ_* to $\mathbb{C}.\xi$:

$$\tau_\rho(exp(z\xi)) := e^{2\pi i z\rho_*(\xi)}$$

In summary, we have the following

Complexification Theorem. *The complexification $\iota : K \hookrightarrow K^\mathbb{C}$ satisfies the universality property with respect to representations. It is unique up to automorphisms* *of K. The group $K^{\mathbb{C}}$ is realized as a closed linear complex group and has polar decomposition $K^{\mathbb{C}} = K exp(i\mathfrak{k})$.*

Proof. Except for the uniqueness, which follows by universality from the fact that $K^{\mathbb{C}}$ is a linear group, all properties have been proved above.   □

Remark. In fact for any Lie group G there exists a complexification $\iota : G \to G^{\mathbb{C}}$ [Ho]. It is universal with respect to homomorphisms $\rho : G \to H$ into complex Lie groups. In the case where G is compact, it agrees with the complexification constructed above, but in general ι is not necessarily injective.   □

4. Algebraicity properties of complexifications of compact groups

4.1 Unipotent groups

Let $U^+ = U^+(GL_n(\mathbb{C}))$ be the subgroup of $GL_n(\mathbb{C})$ consisting of all unipotent upper-triangular matrices. Note that $U^+ = exp(\mathfrak{u}^+)$, where

$$\mathfrak{u}^+ = \bigoplus_{\lambda > 0} \mathfrak{g}_\lambda$$

for a standard basis of roots for the decomposition $\mathfrak{u}^- \oplus \mathfrak{t}^{\mathbb{C}} \oplus \mathfrak{u}^+ = \mathfrak{g} = \mathfrak{gl}_n(\mathbb{C})$.

Of course U^+ is a nilpotent group, i.e., its central series terminates at the identity. In this particular case the exponential map $E : \mathfrak{u}^+ \to U^+$, $\xi \mapsto Id + \xi + \frac{\xi^2}{2!} + \ldots + \frac{\xi^n}{n!}$, is a complex polynomial map with a polynomial inverse

$$E^{-1}(u) = log(u) = (1-u) - \frac{(1-u)^2}{2} + \frac{(1-u)^3}{3} - \ldots \pm \frac{(1-u)^n}{n}.$$

This observation is of course also valid for subgroups.

Notation. A complex Lie group N is called a unipotent matrix group if it can be realized as a subgroup of some $U^+(GL_n(\mathbb{C}))$.

Proposition. *The exponential map $exp : \mathfrak{n} \to N$ of a connected unipotent matrix group is a polynomial isomorphism; in particular $N \cong \mathbb{C}^m$ as a complex manifold.*

If $\rho : R \to GL_{\mathbb{C}}(V)$ is a representation of a connected solvable group, then, by the Flag Theorem, there is a basis of V so that $\rho(R) \subset B^+ = B^+(GL_n(\mathbb{C}))$, where B^+ denotes the group of all upper-triangular matrices.

Of course, even if R is Abelian, $\rho(R)$ might not land in U^+. The following is typical of the transcendental phenomena that can occur.

Example. Let $\mathbb{C}^* \times \mathbb{C}^*$ be regarded as the complexification of the standard maximal torus $S^1 \times S^1$ of U_2 and define

$$\rho : \mathbb{C} \to \mathbb{C}^* \times \mathbb{C}^*, z \mapsto (e^z, e^{iz}).$$

Then $\rho(\mathbb{C})$ is a closed subgroup of $\mathbb{C}^* \times \mathbb{C}^*$ with quotient $(\mathbb{C}^* \times \mathbb{C}^*)/\rho(\mathbb{C})$ a compact complex torus.   □

On the other hand, since the commutator subgroup of B^+ is U^+, if $\rho : R \to GL_\mathbb{C}(V)$ is a representation of a connected solvable group, then by choosing an appropriate basis $\rho(R') \subset U^+$.

Proposition. *Let $\rho : B^+ \to GL_\mathbb{C}(V)$ be a holomorphic representation. Then, equipped with the affine algebraic structure defined by $exp : \mathfrak{u}^+ \to U^+$, $\rho | U^+ : U^+ \to GL_\mathbb{C}(V)$ is an algebraic morphism onto an algebraic unipotent group.*

Proof. This follows immediately from the description $\rho = E \circ \rho_* \circ log$. ☐

4.2 Chevalley's Method and the algebraicity of $K^\mathbb{C}$

Recall Chevalley's Theorem: The image of a $\mathbb{C}$-algebraic morphism $\varphi : X \to Y$ is constructible; in particular $Im(\varphi)$ contains a dense set Ω which is Zariski open in its closure $\overline{\Omega} = \overline{Im(\varphi)}$.

As an example of an application, let G be a $\mathbb{C}$-algebraic group and $\rho : G \to GL_\mathbb{C}(V)$ an algebraic representation. All orbits of the induced action on $GL_\mathbb{C}(V)$ defined, e.g., by left-multiplication are abstractly the same: They are isomorphic to the image $\rho(G)$. Since G acts on the closure of any orbit $G.v$ and the complement $\overline{G.v} \backslash G.v$ is contained in the lower-dimensional algebraic set $\overline{\Omega} \backslash \Omega$, it follows that every such orbit is closed, i.e., $\rho(G)$ is a closed algebraic subgroup of $GL_\mathbb{C}(V)$.

We shall use this method of Chevalley to show, e.g., that the complexification $K^\mathbb{C}$ of a compact group possesses a canonical affine algebraic structure.

A key intermediate result is the

Algebraicity of holomorphic representations of $T^\mathbb{C}$. *Let $T^\mathbb{C}$ be equipped with the affine algebraic structure of $(\mathbb{C}^*)^n$ in any one of its constructions. Then every holomorphic representation $\rho : T^\mathbb{C} \to GL_\mathbb{C}(V)$ is algebraic.*

Proof. This follows immediately from the fact that in an appropriate basis

$$(\chi_1, \ldots, \chi_m) = \rho$$

and that the χ_j's are Laurent monomials on $(\mathbb{C}^*)^n$. ☐

Now let $K_{ss}^\mathbb{C} \hookrightarrow GL_\mathbb{C}(V)$ be embedded (as it was constructed) as a complex subgroup of $GL_\mathbb{C}(V)$. Let $\mathfrak{u}^- \oplus \mathfrak{t}^\mathbb{C} \oplus \mathfrak{u}^+$ be a root decomposition of $\mathfrak{g} = \mathfrak{k}^\mathbb{C}$ and B^+ (resp. U^+) be the complex Lie subgroup of $K^\mathbb{C}$ associated to $\mathfrak{b}^+ = \mathfrak{t}^\mathbb{C} \ltimes \mathfrak{u}^+$ (resp. $\mathfrak{u}^+$). By the flag theorem there is a basis of V so that B^+ is contained in the full group B_n^+ of upper-triangular matrices. Furthermore, since it is the commutator of B^+, U^+ is realized as a unipotent algebraic group.

Regard $B^+ = T^\mathbb{C} \cdot U^+$ and $T^\mathbb{C}$ as acting algebraically on $GL_\mathbb{C}(V)$ by left-multiplication. Thus $T^\mathbb{C} \times U^+ \to GL_\mathbb{C}(V), (t, u) \mapsto t \cdot u$, is an algebraic morphism with image the constructible set B^+. Chevalley's method then shows that B^+ itself is therefore embedded as a closed affine algebraic subgroup of $GL_\mathbb{C}(V)$.

Proposition. *The image of a holomorphic embedding $K^\mathbb{C} \hookrightarrow GL_\mathbb{C}(V)$ is a Zariski closed affine algebraic subgroup.*

Proof. We have just shown that the Borel subgroup B^+ associated to a system of positive roots is a closed affine algebraic subgroup of $GL_{\mathbb{C}}(V)$.

The unipotent subgroup U^- associated to $\mathfrak{u}^-$ is also affine algebraic. So the image $U^- \cdot B^+$ of the multiplication map is both constructible and open in $K_{ss}^{\mathbb{C}}$. Since the closure $\overline{U^- \cdot B^+}$ is $K_{ss}^{\mathbb{C}}$-invariant and irreducible, it follows that $K_{ss}^{\mathbb{C}} = \overline{U^- \cdot B^+}$ is affine algebraic.

Finally, since Z^0 is a torus, the same arguments as in the case for B^+ show that $K^{\mathbb{C}} = (Z^0)^{\mathbb{C}} \cdot K_{ss}^{\mathbb{C}}$ is likewise affine. $\qquad\square$

The same proof as above yields the analogous result for images of representations.

Zusatz. *The image of a holomorphic representation $\rho : K^{\mathbb{C}} \to GL_{\mathbb{C}}(V)$ is a closed, algebraic subgroup of $GL_{\mathbb{C}}(V)$.*

Proof. By the results in 4.1 $\rho|U^+$ and $\rho|U^-$ are algebraic. Since holomorphic representations of $T^{\mathbb{C}}$ are likewise, Chevalley's Method can be applied as above. $\qquad\square$

Corollary. *Let $K^{\mathbb{C}}$ be equipped with an affine algebraic structure so that group multiplication is a regular morphism and let $\rho : K^{\mathbb{C}} \to GL_{\mathbb{C}}(V)$ be a holomorphic representation. Then ρ is an algebraic morphism.*

Proof. Let $Z = Graph(\rho) \subset K^{\mathbb{C}} \times GL_{\mathbb{C}}(V)$. Since $Im(\rho)$ is algebraic, the diagonal $K^{\mathbb{C}}$-action is algebraic and it follows that the orbit Z is a closed algebraic subvariety. $\qquad\square$

Corollary. *The complex Lie group $K^{\mathbb{C}}$ possesses a unique structure as an affine algebraic group.*

5. Compact complex homogeneous spaces

For the representation theory associated to K-coadjoint orbits it is necessary to introduce K-invariant integrable complex structures so that such orbits can be regarded as complex homogeneous spaces. In the present section the relevant basic properties of such spaces will be proved.

5.1 Induced action of the complexification

Let X be a compact complex manifold. It follows that, equipped with the compact-open topology, the full group $Aut_{\mathcal{O}}X$ of holomorphic automorphisms is a complex Lie group and its natural action $Aut_{\mathcal{O}}X \times X \to X$ is holomorphic (see [Kob]).

Suppose now that K is a compact Lie group and $K \times X \to X$ is an action of holomorphic transformations. Then there is the canonically associated homomorphism $\varphi : K \to Aut_{\mathcal{O}}X$.

If $Aut_{\mathcal{O}}X$ were a linear group, then the universality statement proved in §3 would guarantee the existence of a holomorphic extension $\varphi^{\mathbb{C}} : K^{\mathbb{C}} \to Aut_{\mathcal{O}}X$ of this map. In fact, as we remarked in §3, any real Lie group K has a complexification $\iota : K \to K^{\mathbb{C}}$ which is universal with respect to homomorphisms with values in any complex Lie group.

In particular, if K acts transitively on X, then so does $G = K^{\mathbb{C}}$ and $X = G/H$ is complex homogeneous in the strongest possible sense.

5.2 Normalizer fibration

Let G be a complex Lie group, H a closed complex subgroup and $X = G/H$ the associated homogeneous space. Regard the Lie algebra $\mathfrak{h}$ as a point in the Grassmannian $Gr_k(\mathfrak{g})$ of complex subspaces of dimension $k := dim_{\mathbb{C}}\mathfrak{h}$.

The adjoint representation $Ad : G \to GL_{\mathbb{C}}(\mathfrak{g})$ yields a holomorphic action of G on $Gr_k(\mathfrak{h})$. The isotropy group $G_{\mathfrak{h}}$ is simply the normalizer

$$N := N_G(H^0) := \{g \in G : gH^0g^{-1} = H^0\}.$$

Of course $H^0 \triangleleft H$ and consequently $H < N$. Thus we have the picture

$$X = G/H \to G/N =: Y \hookrightarrow Gr_k(\mathfrak{g}) \hookrightarrow \mathbb{P}(\wedge^k \mathfrak{g}),$$

where all maps are holomorphic and G-equivariant. Given $((v_1,...,v_k))_{\mathbb{C}} \in Gr_k(\mathfrak{g})$, the embedding $Gr_k(\mathfrak{g}) \hookrightarrow \mathbb{P}(\wedge^k \mathfrak{g})$ arises by identifying $Gr_k(\mathfrak{g})$ with the $GL_{\mathbb{C}}(\mathfrak{g})$-orbit of $[v_1 \wedge ... \wedge v_k] \in \mathbb{P}(\wedge^k \mathfrak{g})$.

In particular, the base $Y = G/N$ of the *normalizer fibration* is an orbit in $\mathbb{P}(V)$ via a holomorphic G-representation $\rho : G \to GL_{\mathbb{C}}(V)$.

The fiber $F := N/H$ is also of a special type; in particular $H^0 \triangleleft N$ so that, after reducing by the ineffectivity, F is of the form L/Γ, where $L = N/H^0$ is a complex subgroup and $\Gamma = H/H^0$ a discrete subgroup. Note, however, that L may not be connected.

5.3 Compact orbits in projective space

Let G be a complex Lie group, $\rho : G \to GL_{\mathbb{C}}(V)$ a holomorphic representation and $X = G.x_0$ an orbit in $\mathbb{P}(V)$. If $H := G_{x_0}$, then X is biholomorphically G-equivariantly realizable as the complex homogeneous space $X = G/H$. Other than the fact that X is meromorphically separable, it is difficult to say more about general homogeneous manifolds of this type.

However, if X is compact, then such manifolds are in some sense completely understood. One of the main reasons for this is the

Fixed-Point Theorem. *If a projective variety $X \hookrightarrow \mathbb{P}(V)$ is stable with respect to a connected, solvable Lie group G of linear transformations, then the set of fixed points X^G is not empty.*

Proof (Induction on $dim\,X$ and $dim\,G$). Unless G is 1-dimensional, it possesses a normal connected Lie subgroup H with $0 < dim\,H < dim\,G$. By the induction assumption $X^H \neq \emptyset$. Since G acts on the linear span of X^H in $\mathbb{P}(V)$ with H in its ineffectivity, induction on the dimension of G handles this case.

If $dim\,G = 1$, then it is Abelian and Schur's Lemma guarantees the existence of a fixed point in $\mathbb{P}(V^*)$, i.e., a hyperplane $E \subset V$. Unless $dim\,X = 0$, in which case the result follows from the connectedness of G, the intersection $E \cap X$ is a non-empty, lower-dimensional variety and induction on $dim\,X$ may be applied. $\square$

Corollary. *A compact, complex orbit X of a connected solvable Lie group in $\mathbb{P}(V)$ consists of a single point. In particular, if G is a connected group of holomorphic automorphisms of a compact complex torus X, then a G-equivariant holomorphic map $F : X \to \mathbb{P}(V)$ is completely degenerate, i.e., $F(X) = (*)$.*

The *radical* $R = R_G$ of a Lie group G is the maximal, connected, solvable, normal subgroup. It is clearly closed and unique. Furthermore, the radical of the quotient $S = G/R$ is trivial.

Corollary. *Let X be a compact, complex orbit in $\mathbb{P}(V)$ of a Lie group G of linear transformations. Then R_G fixes X pointwise.*

Proof. If $I \triangleleft G$ and $I < J < G$, then I fixes every point of the homogeneous space G/J. $\qquad\square$

As a consequence, when studying compact complex G-orbits X in $\mathbb{P}(V)$ we may assume that $R_G = (e)$, i.e., that $G = S$ is semisimple. Furthermore, we may replace G by the stabilizer of X in the full group of linear transformations, e.g., we may assume that G is a complex Lie group. In summary, we may assume that $G = S$ is a complex semisimple Lie group which, by the results of the previous section, is acting algebraically.

As was indicated in §2, an analysis of root systems shows that S has a compact real form, i.e., $S = K^{\mathbb{C}}$ for a maximal compact subgroup K. Of course K is semisimple and, via the polar decomposition, is a strong deformation retract of S.

In §1 we noted that $\pi_1(K)$ is finite and consequently the same can be said of S.

Proposition. *Let X be a compact, complex orbit $G.x_0$ in $\mathbb{P}(V)$ of a connected Lie group G of linear transformations. Then there is a complex, semisimple linear algebraic group $S < GL_{\mathbb{C}}(V)$ so that $X = S.x_0$. The isotropy group $P = S_{x_o}$ is linear algebraic and has finitely many components. Furthermore, $\pi_1(X)$ is finite.*

Proof. Except for $\pi_1(X)$ being finite all statements have been proved above or are obvious. In order to prove that $\pi_1(X)$ is finite just apply the homotopy sequence to the bundle $P \to S \to S/P$ and use the facts that $\pi_1(S)$ is finite and P has only finitely many components. $\qquad\square$

In this setting the following theorem of Montgomery (see [Kob]) is of use.

Proposition. *Let G be a connected Lie group and $X = G/H$ a compact, homogeneous G-manifold with $\pi_1(X)$ finite. Then every maximal compact subgroup K of G acts transitively on X.*

Remark. Any two maximal compact subgroups of G are conjugate (see [Ho]). Thus the statement that every such group acts transitively is equivalent to any one doing so.

Corollary. *Let X be a compact, complex orbit in $\mathbb{P}(V)$ of a connected Lie group G of linear transformations with maximal compact subgroup $K < G$. Then K also acts transitively on X.*

5.4 Borel subgroups

It has been shown that compact, complex homogeneous spaces X which admit equivariant embeddings in some $\mathbb{P}(V)$ are of the form $X = S/P$, where $S = K^{\mathbb{C}}$ is a semisimple complex group and P is an algebraic subgroup. The Fixed-point Theorem implies that every connected, solvable subgroup of S fixes some point of X.

Definition. A maximal, connected solvable subgroup B of a complex Lie group G is called a Borel subgroup.

Remarks. It is easily checked that Borel groups are closed, complex subgroups. Obviously the radical of G is contained in every Borel subgroup.

Proposition. *Let S be a complex semisimple Lie group with a root decomposition $\mathfrak{u}^{-} \oplus \mathfrak{t}^{\mathbb{C}} \oplus \mathfrak{u}^{+}$. Then the connected subgroup B^{+} associated to the subalgebra $\mathfrak{b}^{+} := \mathfrak{t}^{\mathbb{C}} \ltimes \mathfrak{u}^{+}$ is a Borel subgroup of S.*

Proof. Since B^{+} is solvable, it is enough to show that it is maximal. For this note that if $\mathfrak{b}^{+}$ is properly contained in a complex Lie subalgebra $\mathfrak{h}$ of $\mathfrak{g}$, then $\mathfrak{h}$ must contain a negative root space $\mathfrak{g}_{-\lambda}$. Since $\mathfrak{g}_{-\lambda} \oplus [\mathfrak{g}_{-\lambda}, \mathfrak{g}_{\lambda}] \oplus \mathfrak{g}_{\lambda} \cong \mathfrak{sl}_{2}(\mathbb{C})$ is contained in $\mathfrak{h}$, it follows that $\mathfrak{h}$ is not solvable. $\qquad\square$

Since a conjugate of a Borel group is again a Borel group, if $X = S.x_0 = S/P$ is a closed (compact) orbit of a complex semisimple group in some $\mathbb{P}(V)$, then by the Fixed-point Theorem P contains some Borel group B.

Notation. A connected complex subgroup P of S which contains a Borel subgroup B is called a parabolic subgroup.

Example (Flag Manifolds). A flag F of subspaces of a (complex) vector space V is a sequence

$$(0) \subsetneq V_1 \subsetneq V_2 \subsetneq \ldots \subsetneq V_k \subsetneq V.$$

If $d_j := dim_{\mathbb{C}} V_j$ and $\mathfrak{d} := (d_1, \ldots, d_k)$, then $F_{\mathfrak{d}}(V)$ denotes the set of flags of this type. For $\mathfrak{d} = (1, \ldots, n-1), n := dim_{\mathbb{C}} V$, $F_{\mathfrak{d}}(V)$ is the set of full flags.

Note that $S = SL_n(\mathbb{C})$ acts in a natural way on any $F_{\mathfrak{d}}(\mathbb{C}^n)$. If $((e_1, \ldots, e_n))$ is the standard basis for $\mathbb{C}^n$, then $(0) \subset \langle e_1, \ldots, e_{d_1} \rangle_{\mathbb{C}} \subset \cdots \subset \langle e_1, \ldots, e_{d_k} \rangle \subset \mathbb{C}^n$ is called the standard flag.

Given any other basis $((v_1, \ldots, v_n))$, it is clear that there exists $T \in S$ which maps the standard flag to the flag $(0) \subset \langle v_1, \ldots, v_{d_1} \rangle_{\mathbb{C}} \subset \cdots \subset \langle v_1, \ldots, v_{d_k} \rangle \subset \mathbb{C}^n$ that it generates. In particular, S acts transitively on $F_{\mathfrak{d}}$ and $F_{\mathfrak{d}}$ can be identified with the complex homogeneous space $S/P_{\mathfrak{d}}$, where $P_{\mathfrak{d}}$ is the isotropy group of the standard flag.

We equip $F_{\mathfrak{d}}$ with the complex manifold structure of this homogeneous space.

Regard S as the complexification $K^{\mathbb{C}}$ of $K := SU_n$.

By choosing the basis $((v_1, \ldots, v_n))$ to be unitary and defining $T \in K$ by $Te_j = v_j$, one sees that T also maps the standard flag to the flag it generates. Thus K also acts transitively and $X = S/P_{\mathfrak{d}} = K/L_{\mathfrak{d}}$ is a compact complex manifold.

The isotropy group of the standard full flag is the Borel subgroup B^+ associated to $\mathfrak{b}^+ = \mathfrak{t}^{\mathbb{C}} \ltimes \mathfrak{u}^+$ of the standard root decomposition $\mathfrak{sl}_n(\mathbb{C}) = \mathfrak{u}^- \oplus \mathfrak{t}^{\mathbb{C}} \oplus \mathfrak{u}^+$.

Note that $P_{\mathfrak{d}} \supset B^+$ for all symbols $\mathfrak{d}$. Thus $F_{\mathfrak{d}}$ is at least a candidate to be an S-orbit in some $\mathbb{P}(V)$. $\qquad \square$

Exercise. Describe $P_{\mathfrak{d}}$ as a matrix group and show that the normalizer $N = N(P_{\mathfrak{d}})$ of $P_{\mathfrak{d}}$ in S is $P_{\mathfrak{d}}$ itself. Thus by the results in §5.2 the flag manifold $F_{\mathfrak{d}}$ is canonically embedded as an S-orbit in a certain projective space. As a consequence show that the Fixed-point Theorem implies the **Flag Theorem:** *If $\rho : G \to GL_{\mathbb{C}}(V)$ is a smooth representation of a connected solvable Lie group, then there exists a G-invariant full flag.*

Now consider an arbitrary connected semisimple complex Lie group S which, by the algebraicity results in §4, may be regarded as a linear algebraic subgroup of some $SL_n(\mathbb{C})$.

Proposition. *If $B < S$ is a Borel subgroup, then S/B is a compact projective algebraic manifold.*

Proof. Consider the algebraic S-action on the space F of full flags in $\mathbb{C}^n$. A minimal orbit $X = S.x_0$ is closed and is therefore compact. Since the $SL_n(\mathbb{C})$-isotropy at x_0 is solvable (It is conjugate to B^+!), it follows that S_{x_0} is likewise solvable.

Now F is canonically embedded as an $SL_n(\mathbb{C})$-orbit in some projective space. Thus the same can be said of X and it follows that S_{x_0} contains a Borel subgroup B of S. The maximality condition implies that $B = (S_{x_0})^0$.

Let $\tilde{X} := S/B$ and note that, since the map $\tilde{X} \to X$ is finite, $\tilde{X}$ is also projective algebraic. $\qquad \square$

Warning. It has not yet been shown that S/B is an orbit in a projective space. $\square$

Proposition. *Any two Borel subgroups of S are conjugate.*

Proof. Let $B_0 := (S_{x_0})^0$ above. Since any other Borel subgroup B_1 has a fixed point in $X = S/S_{x_0}$, it follows from the fact that it is connected that it is conjugate to a subgroup of B_0. The desired result follows from the maximality condition. $\qquad \square$

5.5 The Normalizer Theorem

If $X = S/P = S.x_0$ is a closed S-orbit in a projective space, then P is parabolic. It was observed above that the quotient $X = S/P$ of a semisimple complex group by a parabolic subgroup is compact projective algebraic manifold.

Here we prove

Borel's Normalizer Theorem. *For P a parabolic subgroup of a complex semisimple group S it follows that $N_S(P) = P$.*

Hence, by the results of §5.2, the converse statement holds: The quotient $X = S/P$ can be realized as a compact S-orbit in some projective space $\mathbb{P}(V)$.

We begin with a refinement of the topological observation in 5.3.

Proposition. *The fundamental group $X = S.x_0 = S/P$ of a compact S-orbit in a projective space is trivial: $\pi_1(X) = (1)$.*

Proof. Let B^+ be the Borel subgroup of S corresponding to $\mathfrak{b}^+ = \mathfrak{t}^\mathbb{C} \ltimes \mathfrak{u}^+$ of a root decomposition $\mathfrak{s} = \mathfrak{u}^- \oplus \mathfrak{t}^\mathbb{C} \oplus \mathfrak{u}^+$. Since any two Borel subgroups are conjugate, it may be assumed that $B^+ < P$.

Now the unipotent subgroup U^- is acting algebraically on X and $U^- \cdot x_0$ is Zariski open. Thus the desired result follows from the following description of U^--homogeneous spaces.   □

Lemma. *Let U be a unipotent algebraic group and H a closed complex subgroup with H/H^0 finite. Then $H = H^0$ and the homogeneous space U/H is algebraic geometrically isomorphic to some $\mathbb{C}^n$.*

Proof. In §4 it was noted that $exp : \mathfrak{u} \to U$ is an algebraic isomorphism. In particular, every element of U is in a unique 1-parameter subgroup. If $x \in H$, then $x^n \in H^0$ for some $n \in \mathbb{N}^{>0}$. Thus $x \in H^0$ and $H = H^0$.

Since $N(H)$ is an algebraic subgroup which properly contains H (nilpotent groups have non-trivial centers), induction can be applied to obtain an algebraic 1-co-dimensional subgroup $I \triangleleft U$ with $I > H$. The desired result follows by induction, because the orbits of any 1-dimensional subgroup $J \not\subset I$ trivialize the bundle $U/H \to U/I \cong \mathbb{C}$.   □

The normalizer theorem for Borel subgroups is an immediate consequence: It is clear that $Lie(N) = \mathfrak{b}$, because otherwise we could enlarge $\mathfrak{b}$ and contradict the maximality assumption. Thus $S/B \to S/N = S.x_0 \hookrightarrow \mathbb{P}(V)$ is a finite covering. Since the base is simply-connected, it follows that $N = B$.

To complete the proof of the normalizer theorem we consider the fibrations

$$S/B \to S/P \to S/N,$$

where $N := N(P^0)$. Of course the base is an S-orbit in a projective space and as a result is simply-connected; in particular, N is connected. Consequently, the group N/P^0 is a compact complex torus and therefore so is N/P.

Since the radical $R = R_N$ acts trivially on N/B (It is also a compact N-orbit in a projective space!), it follows that the torus N/P is the homogeneous space of a semisimple Lie group. But N/P is Abelian and therefore $N = P$.   □

5.6 Complex structures on coadjoint orbits

Let K be a compact, semisimple Lie group and for the discussion fix a maximal torus $T < K$. Recall that the adjoint representation of K satisfies $K.\mathfrak{t} = \mathfrak{t}$ and that if $k(\xi) = \eta$ for $\xi, \eta \in \mathfrak{t}$, then k is uniquely represented by an element $\sigma \in W = N_K(T)/T$ (see §1).

Since K is compact, real K-representations V and their duals V^* are isomorphic, because every such possesses a K-invariant inner product. Thus the above statements carry over to the coadjoint representation $\mathfrak{k}^*$. In particular, in discussing coadjoint orbits we may restrict to orbits $K.\lambda$ for $\lambda \in \mathfrak{t}^*$.

The isotropy K_λ is the centralizer $Z_K(T_\lambda)$ of the corresponding torus $T_\lambda < T$ which is obtained from the identification with the adjoint representation. For brevity of notation let $L := K_\lambda$, $M = K.\lambda = K/L$ and $\mathcal{J}(L)$ the set of K-invariant integrable complex structures on M.

The complexified tangent space $T_\lambda^{\mathbb{C}} M := T_\lambda M \otimes_{\mathbb{C}} \mathbb{R}$ is L-equivariantly identifiable with the L-representation space $\mathfrak{k}^{\mathbb{C}}/\mathfrak{l}^{\mathbb{C}}$ induced by the restriction of the adjoint representation of K to L. Splitting $\mathfrak{k}^{\mathbb{C}} = \mathfrak{l}^{\mathbb{C}} \oplus \mathfrak{m}^{\mathbb{C}}$ into L-invariant pieces, we identify $T_\lambda^{\mathbb{C}} M$ with $\mathfrak{m}^{\mathbb{C}}$.

A K-invariant almost complex structure $J : TM \to TM$ is uniquely defined by an L-equivariant linear map $J_\lambda \in End(T_\lambda M)^L$ with $J_\lambda^2 = -Id$. This yields an L-invariant splitting $\mathfrak{m}^{\mathbb{C}} = \mathfrak{m}^{0,1} \oplus \mathfrak{m}^{1,0}$ into the $(-i)$-and $(+i)$-eigenspaces of the complexified mapping $J_\lambda^{\mathbb{C}}$. Of course, $\overline{\mathfrak{m}^{0,1}} = \mathfrak{m}^{1,0}$ and conversely, such a splitting defines a K-invariant almost complex structure on M.

Such a structure is integrable if and only if the complexified group $K^{\mathbb{C}}$ acts holomorphically on M, i.e., $M = K^{\mathbb{C}}/P$ as a complex manifold, where $(K^{\mathbb{C}})_\lambda =: P$. In this case the real tangent space $T_\lambda M = \mathfrak{m}$ is identified with the complex L-representation space $\mathfrak{k}^{\mathbb{C}}/\mathfrak{p}$ where J_λ acts as multiplication by $+i$. Thus the $(-i)$-eigenspace $\mathfrak{m}^{0,1}$ is contained in $\mathfrak{p}$ and thus $\mathfrak{p} = \mathfrak{l}^{\mathbb{C}} \oplus \mathfrak{m}^{0,1}$.

Now the integrability condition is translated to the condition that $\mathfrak{p}$ is a complex Lie subalgebra of $\mathfrak{k}^{\mathbb{C}}$. Since $\mathfrak{t}^{\mathbb{C}} < \mathfrak{l}^{\mathbb{C}}$, $\mathfrak{l}^{\mathbb{C}}$ normalizes $\mathfrak{m}^{0,1}$, $\overline{\mathfrak{m}^{0,1}} = \mathfrak{m}^{1,0}$ with $\mathfrak{m}^{1,0} \oplus \mathfrak{p} = \mathfrak{k}^{\mathbb{C}}$ and $\overline{\mathfrak{l}^{\mathbb{C}}} = \mathfrak{l}^{\mathbb{C}}$, it is possible to give a simple interpretation of this condition.

First, since $\mathfrak{t}^{\mathbb{C}}$ normalizes $\mathfrak{m}^{0,1}$, it follows that the latter is a direct sum

$$\mathfrak{m}^{0,1} = \bigoplus_{\alpha \in \Phi_1} \mathfrak{g}_\alpha$$

of root spaces. The same is true of $\mathfrak{l}^{\mathbb{C}}$:

$$\mathfrak{l}^{\mathbb{C}} = \mathfrak{t}^{\mathbb{C}} \oplus \bigoplus_{\alpha \in \Phi_2} \mathfrak{g}_\alpha.$$

Utilizing the behavior with respect to complex conjugation, it follows that $-\Phi_2 = \Phi_2$ and $\{-\Phi_1\} \dot\cup \Phi_2 \dot\cup \Phi_1 = \Phi$, where the latter denotes the full set of roots. If Δ_2 is a system of positive roots for Φ_2, it follows that $\Delta_2 \dot\cup \Phi_1 =: \Delta$ is a system of positive roots for $\mathfrak{k}^{\mathbb{C}} = \mathfrak{g}$. In particular, $\mathfrak{p}$ is a Lie algebra of a parabolic subgroup and $\mathfrak{m}^{0,1}$ is distinguished as being the sum of those $\mathfrak{g}_\alpha$ such that $\mathfrak{g}_{-\alpha} \not\subset \mathfrak{p}$.

For $\mathfrak{g}_\alpha, \mathfrak{g}_\beta \subset \mathfrak{m}^{0,1}$ it follows that $\mathfrak{g}_{\alpha+\beta} \subset \mathfrak{p}$. If $\mathfrak{g}_{\alpha+\beta} \subset \mathfrak{l}^{\mathbb{C}}$, then $\mathfrak{g}_{-(\alpha+\beta)} \subset \mathfrak{l}^{\mathbb{C}}$ and , since $\mathfrak{l}^{\mathbb{C}}$ normalizes $\mathfrak{m}^{0,1}$, it would follow that $\mathfrak{g}_{-\alpha} = [\mathfrak{g}_{-(\alpha+\beta)}, \mathfrak{g}_\beta] \subset \mathfrak{m}^{0,1}$, which is not the case. Consequently, $\mathfrak{m}^{0,1}$ is a subalgebra of $\mathfrak{p}$.

Notice that $\mathfrak{m}^{0,1}$ is the maximal nilpotent ideal in $\mathfrak{p}$. Thus one refers to it as the nilradical. Due to its being a sum of root algebras, the associated Lie group is unipotent. Hence, for a finer description, we refer to it as the unipotent radical $\mathfrak{r}_u(\mathfrak{p})$.

Proposition. *The set $\mathcal{J}(L)$ of K-invariant integrable, complex structures on the coadjoint orbit $K.\lambda = K/L$, $\lambda \in \mathfrak{t}^*$, is in 1-1 correspondence with the unipotent radicals $\mathfrak{r}_u(\mathfrak{p})$ of parabolic algebras $\mathfrak{p} = \mathfrak{l}^{\mathbb{C}} \ltimes \mathfrak{r}_u(\mathfrak{p})$.*

Of course this statement is only a reformulation, but in fact it is possible to give a detailed description of such parabolic completions of $\mathfrak{l}^{\mathbb{C}}$. The case of $\mathfrak{l}^{\mathbb{C}} = \mathfrak{t}^{\mathbb{C}}$ is particularly simple.

For this let C be a connected component of the complement of

$$\bigcup_{\alpha \in \Phi} \{\alpha = 0\}$$

in $\mathfrak{t}$. Such sets are referred to as Weyl chambers. If $\alpha \in \Phi$, then $\alpha|_C$ is either positive or negative. Thus C defines a decomposition $\Phi = \Phi^+ \dot\cup \Phi^-$, where $\Phi^+ = \Phi^+(C) = \{\alpha \in \Phi : \alpha|_C > 0\}$ and $\Phi^- = -\Phi^+$. Clearly

$$\mathfrak{m}^{0,1} = \mathfrak{m}^{0,1}(C) := \bigoplus_{\alpha \in \Phi^+(C)} \mathfrak{g}_\alpha$$

is the unipotent radical of $\mathfrak{b} := \mathfrak{t}^{\mathbb{C}} \ltimes \mathfrak{m}^{0,1}$. In fact, since all Borel algebras occur in this way, it follows that they are in 1-1 correspondence with components C of this type.

The same type of considerations shed light on the general case. For this let T_λ be a given torus in T, $\Phi_\lambda := \{\alpha \in \Phi : \alpha|_{\mathfrak{t}_\lambda} \not\equiv 0\}$ and C_λ a connected component of the complement of

$$\bigcup_{\alpha \in \Phi_\lambda} \{\xi \in \mathfrak{t}_\lambda : \alpha|_{\mathfrak{t}_\lambda} = 0\}.$$

in $\mathfrak{t}_\lambda$.

Let $\Phi_\lambda^- := \{\alpha \in \Phi_\lambda : \alpha|_{C_\lambda} < 0\}$ (resp. $\Phi_\lambda^+ := \{\alpha \in \Phi_\lambda : \alpha|_{C_\lambda} > 0\}$). Of course $\Phi_\lambda = \Phi_\lambda^- \dot\cup \Phi_\lambda^+$.

Now let $\Psi_\lambda := \{\alpha \in \Phi : \alpha|_{C_\lambda} \leq 0\}$ and define the subalgebra

$$\mathfrak{p}_\lambda := \mathfrak{t}^{\mathbb{C}} \oplus \bigoplus_{\alpha \in \Psi_\lambda} \mathfrak{g}_\alpha.$$

Since $\mathfrak{l}^{\mathbb{C}} = \mathfrak{z}(\mathfrak{t}^{\mathbb{C}}_\lambda)$, it follows that $\mathfrak{l}^{\mathbb{C}} < \mathfrak{p}_\lambda$ and

$$\mathfrak{m}^{0,1} := \bigoplus_{\alpha \in \Phi^-_\lambda} \mathfrak{g}_\alpha.$$

Obviously $\mathfrak{m}^{0,1}$ is a Lie subalgebra of $\mathfrak{p}_\lambda$, because $\alpha|_{C_\lambda} < 0$ and $\beta|_{C_\lambda} < 0$ certainly implies $(\alpha + \beta)|_{C_\lambda} < 0$. The identical argument shows that $\mathfrak{m}^{0,1} \lhd \mathfrak{p}_\lambda$ and, since $\Phi = \Psi_\lambda \dot\cup \Phi^+_\lambda$, it follows that $\mathfrak{k}^{\mathbb{C}} = \mathfrak{m}^{0,1} \oplus \mathfrak{l}^{\mathbb{C}} \oplus \mathfrak{m}^{1,0}$ as required.

In summary, the integrable, invariant structures on coadjoint orbits can be classified as follows (see [BH]).

Proposition. *For $\lambda \in \mathfrak{t}^*$ and $L := K_\lambda = Z_K(T_\lambda)$ the set $\mathcal{J}(L)$ of the integrable, K-invariant, complex structures on the coadjoint orbit $K.\lambda = K/L$ is in 1-1 correspondence with the (restricted) Weyl chambers $P_\lambda \subset \mathfrak{t}_\lambda$. For $T_\lambda = T$ a maximal torus, $\mathcal{J}(T)$ can be identified with the set of Borel subgroups containing T.*

Corollary. *Coadjoint orbits of connected compact semisimple groups are simply-connected; in particular, the centralizer of a torus in such a group is connected.*

Proof. The above proposition shows that a coadjoint orbit $K.\lambda$ comes equipped with a complex structure so that, as a complex manifold, $K.\lambda = S/P = X$, where P is parabolic in S. In the previous section it was shown that $\pi_1(X) = 1$. □

Exercise. Let $K = SU_2$ and T be the standard torus of diagonal matrices. Show that $W = <w> \cong \mathbb{Z}_2$, where $w = -Id$ on $\mathfrak{t}$ and, since w can be represented in $N = N(T)$, show that it acts naturally on the coadjoint orbits $K.\lambda \cong S^2, \lambda \in \mathfrak{t}^*$, as the antipodal map. Equip $X = K.\lambda$ with some K-invariant, integrable complex structure and show that this can be identified with $\mathbb{P}_1(\mathbb{C})$. Show that the induced map $w : X \to X$ is anti-holomorphic. □

6. The root groups $SL_2(\alpha)$ and $H_2(G/P, \mathbb{Z})$

Let K be a connected, compact, semisimple Lie group and $S = K^{\mathbb{C}}$ its complexification. A root $\alpha \in \Phi$ is defined at the group level by $\chi_\alpha \in \mathfrak{X}(T)$, the 1-dimensional representation of T on $\mathfrak{g}_\alpha$. Of course this can be complexified to a 1-dimensional representation of $T^{\mathbb{C}}$. Using the same notation we regard $\chi_\alpha \in \mathfrak{X}(T^{\mathbb{C}})$ and $\alpha \in (\mathfrak{t}^{\mathbb{C}})^*_\mathbb{Z}$, the integral functional with respect to the complex exponential map $exp : \mathfrak{t}^{\mathbb{C}} \to T^{\mathbb{C}}$.

If $T_\alpha := (Ker(\chi_\alpha : T \to S^1))^0$, then $T^{\mathbb{C}}_\alpha = (Ker(\chi_\alpha : T^{\mathbb{C}} \to \mathbb{C}^*))^0$ and $Z_\alpha(S) := Z_S(T^{\mathbb{C}}_\alpha)^0$ is therefore the complexification $Z_\alpha(K)^{\mathbb{C}}$ of the compact group which was discussed in §2. From the splitting $Z_\alpha(K) = S_\alpha(K) \cdot T_\alpha$ we have the splitting of the complex groups: $Z_\alpha(S) = S_\alpha(K)^{\mathbb{C}} \cdot T^{\mathbb{C}}_\alpha$. Since its Lie algebra is isomorphic to $\mathfrak{sl}_2(\mathbb{C})$, we denote $S_\alpha(K)^{\mathbb{C}}$ by $SL_2(\alpha)$. Of course it is isomorphic to either $SL_2(\mathbb{C})$ or $SO_3(\mathbb{C}) = SL_2(\mathbb{C})/\{\pm Id\}$. If there is no confusion we write $Z_\alpha = SL_2(\alpha) \cdot T^{\mathbb{C}}_\alpha$.

Our goal here is to derive basic properties concerning these *root SL_2's* and show, e.g., how certain of their orbits in $X = S/P$ describe $H^2(X, \mathbb{Z})$.

6.1 The Ad-action of Z_α

Let $S = K^{\mathbb{C}}$ be a complex semisimple group, $\mathfrak{u}^- \oplus \mathfrak{t}^{\mathbb{C}} \oplus \mathfrak{u}^+$ be a decomposition of $Lie(S)$ determined by a choice of positive roots and $U^-, T^{\mathbb{C}}, U^+, B^+$, and B^- the associated (algebraic) subgroups of S.

Of course $\mathfrak{t} \cap \mathfrak{s}_\alpha(\mathfrak{k})$ is a maximal toral subalgebra of the compact form $\mathfrak{s}_\alpha(\mathfrak{k})$ and its Weyl group $\mathcal{W}_\alpha$ is generated by a single involution σ_α. Since σ_α normalizes the full maximal torus T (It centralizes T_α!), we may regard it as an element of the full Weyl group.

In the dual action on $\mathfrak{t}^*$, $\sigma_\alpha(\alpha) = -\alpha$ and therefore in $\mathfrak{t}$ it can be regarded as a reflection defined by the hyperplane $\{\alpha = 0\}$. Thus the special elements $\sigma_\alpha \in \mathcal{W}$ are often simply referred to as reflections. The Ad-action of Z_α and the action of σ_α on Φ as a permutation of root spaces are closely related.

To understand this action it is perhaps better to first think in more general terms.

Let $\rho : S \to GL_{\mathbb{C}}(V)$ be any finite-dimensional, holomorphic representation and

$$V = \bigoplus_{\lambda \in \mathfrak{t}^*_{\mathbb{Z}}} V_\lambda$$

be the decomposition into *isotypical components* of the associated $T^{\mathbb{C}}$-representation: $t(v) = X_\lambda(t).v$ for $t \in T^{\mathbb{C}}$ and $v \in V_\lambda$. At the Lie algebra level, i.e., at the level of $\rho_* : Lie(S) \to \mathfrak{gl}_{\mathbb{C}}(V) = End(V)$, $\xi(v) = 2\pi i \lambda(\xi).v$ for $\xi \in \mathfrak{t}^{\mathbb{C}}$.

If $\eta \in \mathfrak{g}_\alpha$, then, using the fact that ρ_* is a Lie algebra morphism and $[\xi, \eta] = 2\pi i \alpha(\xi) \cdot \eta$ for $\xi \in \mathfrak{t}^{\mathbb{C}}$, it follows that $\rho_*(\eta) : V_\lambda \to V_{\lambda+\alpha}$.

Now, following the notation of [Hu], let U_α denote the 1-dimensional *root subgroup* with Lie algebra $\mathfrak{g}_\alpha$; in particular, for a given root decomposition of $Lie(S)$, the group $U_\alpha < S_\alpha$ will be associated to the positive root. Let $U_{\sigma_\alpha} = \sigma_\alpha(U) \cap U$, where $U := U^+$.

The translation of the description $\rho_*(\eta) : V_\lambda \to V_{\lambda+\alpha}$ to the group level is of basic importance.

Proposition. *For a representation $\rho : S \to GL_{\mathbb{C}}(V)$ it follows that*

$$\rho(U_\alpha)(V_\lambda) \subset \bigoplus_{n>0} V_{\lambda+n\alpha}.$$

Proof. This follows immediately from the fact that $\rho \circ exp = E \circ \rho_*$ and consideration of the power series development $E(A) = e^A = Id + A + \frac{A^2}{2!} + \dots$. Since this is applied to the nilpotent matrix $\rho_*(\eta), \eta \in \mathfrak{g}_\alpha$, this sum is of course finite. □

Corollary. *For the root group Z_α*

$$Ad(Z_\alpha)(\mathfrak{g}_\beta) \subset \bigoplus_{m \in \mathbb{Z}} \mathfrak{g}_{\beta+m\alpha}.$$

Proof. Apply the above proposition to U_α and $U_{-\alpha}$. □

In particular $\sigma_\alpha(\mathfrak{g}_\beta) = \mathfrak{g}_{\beta+m\alpha}$ for some $m \in \mathbb{Z}$. If $\alpha, \beta > 0$ and $\beta \neq \alpha$, then it follows that $\beta + m\alpha > 0$.

Corollary. *For $\alpha > 0$ the reflection σ_α leaves $\Phi^+ \setminus \{\alpha\}$ invariant. At the level of groups*

$$U = U_\alpha \cdot U_{\sigma_\alpha}.$$

Proof. The Lie algebra statement shows that $U_{\sigma(\alpha)}$ is 1-codimensional and is therefore a normal subgroup of U. Thus $U_\alpha \cdot U_{\sigma(\alpha)}$ is an open subgroup of U and the result follows. □

6.2 The open cell in S/P and its boundary divisors

Let $X = S/P$ be a compact, projective algebraic homogeneous space, i.e., for an appropriate root decomposition we may assume that $P \supset B^-$. Note that the complex conjugate R_u^+ of the unipotent radical of P satisfies $B^+.x_0 = R_u^+.x_0$, where x_0 is the neutral point in X associated to P. In fact R_u^+ acts freely on this open orbit and thus it can be identified in a natural way with the unipotent group R_u^+ itself; in particular, it is identified with an affine cell $\Omega \cong \mathbb{C}^n$.

The boundary of this cell consists of finitely many 1-codimensional, irreducible complex hypersurfaces $D_1, \ldots D_r$. Of course they are not disjoint, but for $i \neq j$ the analytic set $D_i \cap D_j$ is lower-dimensional.

Standard topological methods (see, e.g., [RV]) lead to the following fact.

Proposition. *The homology classes $[D_1], \ldots, [D_r]$ freely generate $H_{2n-2}(X, \mathbb{Z})$.*

6.3 The root curves $\mathbb{P}_1(\alpha)$

For the moment let us restrict to the case of $X = S/B$, where $B = B^-$ is defined by a choice of positive roots. Here we are discussing the open cell $\Omega = U^+.x_0$. Note that $Z_\alpha.x_0 = S_\alpha.x_0$ can be identified with S_α/B_α^-, where B_α^- is the Borel subgroup of S_α defined by $T^{\mathbb{C}} \cap S_\alpha$ and $U_{-\alpha}$. In particular $S_\alpha.x_0$ is a copy of $\mathbb{P}_1(\mathbb{C})$ which we denote by $\mathbb{P}_1(\alpha)$. Note that, since it is compact, such a curve is not contained in the open cell. On the other hand, since $U_\alpha.x_0 \cong \mathbb{C}$, it follows that the boundary $\mathbb{P}_1(\alpha) \setminus \Omega$ consists of exactly one point which we denote by x_α. Of course $\{x_0, x_\alpha\}$ is the fixed point set of $T^{\mathbb{C}}$ in $\mathbb{P}_1(\alpha)$, $U_\alpha x_\alpha = x_\alpha$, $U_{-\alpha} x_\alpha = \mathbb{P}_1(\alpha) \setminus \{x_0\}$ and $\sigma_\alpha(x_0) = x_\alpha$.

In this discussion we have fixed a notion of positivity and in particular have a basis $\{\alpha_1, \ldots, \alpha_s\}$ of positive roots. For brevity let $x_i = x_{\alpha_i}$ denote the associated *points at ∞* in the $\mathbb{P}_1(\alpha_i)$'s.

By the results of the previous section $\sigma_\alpha(B^-) \cap U^+ = U_\alpha$ and therefore $U^+.x_i \cong \mathbb{C}^{n-1}$ are mutually disjoint cells with closures $D_1, \ldots, D_s$. The fact that these are the only boundary divisors, i.e., $s = r$, follows from the more refined results of the next section.

6.4 The Bruhat decomposition

Let $\sigma \in W$ be represented by $n \in N_K(T)$. The point $n(x_0) \in X = S/B^-$ is independent of the choice of this representative; in fact $nB^-n^{-1} =: \sigma(B^-)$ has $n(x_0) =: \sigma(x_0) =: x_\sigma$ as its unique fixed point. Since the isotropy S_{x_σ} is $\sigma(B^-)$ which contains $T^{\mathbb{C}}$, it follows that $B^+.x_\sigma = U^+.x_\sigma$ is an affine cell $\mathbb{C}^{n_\sigma}$ which is Zariski open in its closure in X.

The Bruhat decomposition of S amounts to the fact that the entire set of B^+-orbits in X is simply $\{B^+.x_\sigma : \sigma \in W\}$. This is a precise list in the sense that $B^+.x_\sigma = B^+.x_\tau$ if and only if $\sigma = \tau$. Of course this is fundamental for many topological and complex geometric considerations. For the proof we need two basic remarks.

The normalization property. *The set*

$$BWB = \bigcup_{\sigma \in W} B\sigma(B)$$

is invariant under multiplication by the root groups Z_α, i.e., $Z_\alpha BWB = BWB$.

Remark. Of course this implies, e.g., that the group generated by the Z_α's is contained in BWB. Since this is an open subgroup of S, it follows that $S = BWB$, i.e.,

$$S/B = \bigcup_{\sigma \in W} B.x_\sigma.$$

We refer to this as the normalization property, because it follows immediately from the fact that Z_α normalizes U_{σ_α}:

$$Z_\alpha B\sigma B = Z_\alpha U_{\sigma_\alpha} U_\alpha \sigma B = U_{\sigma_\alpha} Z_\alpha \sigma B = U_{\sigma_\alpha} U_\alpha \sigma B \cup U_{\sigma_\alpha} U_\alpha \sigma_\alpha \sigma B \subset BWB.$$

The fact that Z_α normalizes U_{σ_α} follows from the fact that U_{σ_α} is 1-codimensional in $\sigma_\alpha(U)$ and therefore is normalized by $U_{-\alpha}$. $\square$

The unicity statement follows from a

Fixed point property. *If $X = B/H$ is a B-homogeneous space with $H = H^0 > T^{\mathbb{C}}$, then $T^{\mathbb{C}}$ has a unique fixed point in X.*

Proof. (Induction on $\dim_{\mathbb{C}} X$). Since $T^{\mathbb{C}}$ acts non-trivially on X, the proof of the 1-dimensional case is immediate. Otherwise, note there is a $T^{\mathbb{C}}$-invariant 1-parameter subgroup $U_1 < U$ which normalizes H with $H_1 := U_1 H$ closed. By the induction assumption, $T^{\mathbb{C}}$ has exactly one fixed point in $X_1 := B/H_1$ and, since it acts non-trivially on the fiber of $X \to X_1$ over that point, it therefore has only one fixed point in X. $\square$

Bruhat Decomposition. *Let B be a Borel group containing a maximal torus $T^{\mathbb{C}}$ of a complex semisimple group $S = K^{\mathbb{C}}$. Then S is the disjoint union*

$$S = \bigcup_{\sigma \in W} B\sigma B,$$

where $W = N/T$ is the Weyl group associated to T.

Proof. It remains to prove the unicity: $B\sigma B = B\tau B \Leftrightarrow \sigma = \tau$. For this observe that since $N(B) = B$, if $n_\sigma B n_\sigma^{-1} = \sigma B = \tau B = n_\tau B n_\tau^{-1}$ as points in S/B, then $\sigma^{-1}\tau \in B$. But $K \cap B = T$ and consequently $\sigma = \tau$ in $N/T = W$. But, if $B\sigma B = B\tau B$, then $\sigma B = \tau B$ follows immediately from the fact that the B-orbit $B\sigma B = B\tau B$ contains exactly one $T^{\mathbb{C}}$-fixed point. $\square$

6.5 Divisors in S/B and Poincaré-duality

Let us set the notation as above: $X = S/B^-$, $x_0 \in X$ the neutral point in X which is the unique B^--fixed point, $\Omega := B^+.x_0$ the open cell and $X\backslash\Omega = D_1 \cup D_2 \cup \ldots \cup D_r$ the union of irreducible hypersurfaces. Notationally we identify such hypersurfaces with their associated divisors.

Since B^+ has only finitely many orbits, D_j is the closure of the B^+- orbit of a generic point in D_j. In summary:

Proposition. *Let* $\{\alpha_1, \ldots, \alpha_r\}$ *be a basis of simple roots associated to the choice of* B^+ *and* $x_j := \sigma_{\alpha_j} B^-$. *Then* $X\backslash\Omega = D_1 \cup \ldots \cup D_r$, *where* D_j *is the closure of the 1-codimensional cell* $B^+.x_j, j = 1, \ldots, r$.

Proof. It is enough to show that $B^+\sigma B^-$ is 1-codimensional only when σ is a simple reflection. However, since it contains $T^{\mathbb{C}}$, the 1-codimensional B^--isotropy algebra at $x_\sigma = \sigma(x_0)$ is a direct sum of root spaces and therefore cannot contain all of the simple root spaces $\mathfrak{g}_{-\alpha_i}$. Thus σ is the simple reflection determined by the simple (negative) root space which it does not contain. $\square$

The Bruhat decomposition gives $X = S/B^-$ the structure of the CW-complex with cells occurring only in even dimensions. Thus the full homology is visible with closures of B^--orbits being generators. Poincaré duality at the level of complex curves and divisors is particularly simple.

To see this consider the decomposition defined by the B^--orbits. The 1-dimensional orbits are simply $B^-.x_j$, where $x_j := \sigma_{\alpha_j}(x_0)$ as above. Notice that $\mathbb{C} \cong B^-.x_j \subset SL_2(\alpha_j).x_j = SL_2(\alpha_j).x_0 = \mathbb{P}_1(\alpha_j)$. Since $\sigma_{\alpha_j}(U^+)$ acts freely on its (open cell) orbit of x_j, contains $\sigma_{\alpha_j}(U_\alpha^+)$ which acts freely and transitively on $B^-.x_j$ and contains $\sigma_{\alpha_j}(U^+) \cap U^+$ which acts freely and transitively on $B^+.x_j$, one explictly sees that $\mathbb{P}_1(\alpha_j)$ intersects only D_j and that intersection is transversal at the smooth point $x_j \in D_j$.

Thus we have the following observation.

Proposition. *The homology classes of the root* $\mathbb{P}_1(\alpha_j)$'s *generate* $H_2(S/B, \mathbb{Z})$. *By the intersection pairing, the divisor classes* $[D_i]$ *generate the cohomology* $H^2(S/B, \mathbb{Z}), j = 1, \ldots, r$, *and* $D_i \cap \mathbb{P}_1(\alpha_j) = \delta_{ij}$.

Remark. The analogous results hold for $X = S/P^-$, where P^- is an arbitrary parabolic subgroup containing B^-. The appropriate parameterization of the simple root $\mathbb{P}_1(\alpha_j)$'s is given by $I = \{j : \mathfrak{g}_{-\alpha_j} < r_u(\mathfrak{p}^-)\} : H_2(X, \mathbb{Z})$ is generated by the classes of the $\mathbb{P}_1(\alpha_j)$'s and $H^2(X, \mathbb{Z})$ by the divisors $D_j = B^+.x_j, j \in I$. It is therefore convenient to use the notation $P = P_I^-$. $\square$

6.6 Elementary complex geometry of $X = S/P$

One measuring stick of the complex geometry of a complex manifold X is the Dolbeault complex, in particular the Dolbeault spaces $H^q(X, \Omega^p)$. Now these spaces are bimeromorphic invariants. Furthermore, since $R_u(P^+)$ acts freely, transitively and algebraically on the open cell $R_u(P^+).x_0 = B^+.x_0 \subset X = S/P^-$, it follows that the algebraic action map $R_u(P^+) \to R_u(P^+).x_0$ extends to birational map $\mathbb{P}_n(\mathbb{C}) \to X$. Thus we have the following fact.

Proposition. *For $X = S/P$ a compact, algebraic homogeneous space of a complex semisimple group S, with the exception of $H^0(X, \mathcal{O}) \cong \mathbb{C}$, it follows that $H^q(X, \Omega^p) = \{0\}$ for all p, q.*

This is of particular use for computing the group $Pic(X)$ of equivalence classes holomorphic line bundles on $X = S/P$.

Remark. Since X is projective algebraic, holomorphic line bundles carry unique algebraic structures, i.e., $Pic_{alg}(X) = Pic_{holo}(X)$.

Since $H^1(X, \mathcal{O}) = H^2(X, \mathcal{O}) = \{0\}$, it follows from the long exact sequence in cohomology associated to the exponential sequence $0 \to \mathbb{Z} \to \mathcal{O} \to \mathcal{O}^* \to 0$ that

$$c_1 : Pic(X) \cong H^1(X, \mathcal{O}^*) \to H^2(X, \mathbb{Z})$$

is an isomorphism.

Since the divisors $D_1, \dots, D_r$ in the complement of the open B^+-orbit generate $H^2(X, \mathbb{Z})$, we have the following description of the Picard group of $X = S/P$.

Proposition. *The map $\delta : \mathbb{Z}^r \to Pic(X)$, which associates to $m = (m_1, \dots, m_r)$ the line bundle $E(D)$ of the divisor $D = m_1 D_1 + \dots + m_r D_r$ is an isomorphism.*

If we use $D_1, \dots, D_r$ to identify $H^2(X, \mathbb{Z})$ with $\mathbb{Z}^r$, then the Chern class mapping $c_1 : Pic(X) \to H^2(X, \mathbb{Z}) \cong \mathbb{Z}^r$ is simply δ^{-1}. One would like to define $E \in Pic(X)$ as being positive (resp. semi-positive) if its Chern numbers m_j are all positive (resp. non-negative). Since the isomorphism δ depends on various choices, e.g., on the choice of T, it is better to define these notions in terms of intersections: E is said to be *numerically positive* (resp. *semi-positive*) if its restriction $E|_C$ to every complex curve $C \subset X$ is positive (resp. semi-positive).

To understand these notions it is useful to regard E as an S-homogeneous bundle $E = S \times_\chi \mathbb{C}$, where $X = S/P$ is as above and $\chi : P \to \mathbb{C}^*$ is a character.

Exercise. For $\hat{S} = Sl_2(\mathbb{C})$ and $\hat{P} = B^-$ the Borel group of lower-triangular matrices let $\chi : \hat{P} \to \mathbb{C}^*$ be defined by one of the generating characters of $T^{\mathbb{C}} = B^-/U^-$. Then $L := \hat{S} \times_\chi \mathbb{C}$ is one of the generating bundles of $Pic(X)$, $X = \hat{S}/\hat{P} \cong \mathbb{P}_1(\mathbb{C})$. Let $S = \hat{S}/(\pm Id) \cong SO_3(\mathbb{C})$. Show that the $\hat{S}$-action on X induces an S-action so that $X = S/P$, but that E *cannot* be regarded as an S-homogeneous bundle. $\square$

Of course complex semisimple groups have finite fundamental groups and therefore in many situations there is no serious price to pay when replacing S by its universal cover.

Proposition. *Every line bundle $E \in Pic(X)$ on a projective algebraic homogeneous space $X = S/P$ of a simply-connected, complex semisimple group is S-homogeneous: $E = S \times_\chi \mathbb{C}$.*

Proof. Since S is connected and $Pic(X) \cong H^2(X, \mathbb{Z})$ is discrete, S acts trivially on $Pic(X)$. Therefore every element $g \in S$ which acts as an automorphism $g : X \to X$ lifts (perhaps non-uniquely) to a holomorphic bundle map $\hat{g} : E \to E$.

Let $Aut_{\mathcal{O}}(E)$ denote the full group of these bundle transformations. Since the transformations in $Aut_{\mathcal{O}}(E)$ extend to the associated $\mathbb{P}_1$-bundle, it follows that it is a complex Lie group. Let $\hat{S}$ be a maximal semisimple part of the connected component of the pre-image of S in $Aut_{\mathcal{O}}(E)$ and $\pi : \hat{S} \to S$ the resulting surjective homomorphism. Due to the fact that $Ker(\pi)$ acts on the fibers of E which are complex lines, it follows that it is finite and is therefore an isomorphism, because $\hat{S}$ is simply-connected. The desired character is determined by the P-action on the E-fiber over the neutral point. $\square$

Accepting finite ineffectivity of its action on $X = S/P$, we assume from now on that $\pi_1(S) = \{1\}$ and therefore may regard $E \in Pic(X)$ as being S-homogeneous with associated holomorphic representation on the finite-dimensional vector space $\Gamma(X, E)$. Let $\varphi_E : X \to \mathbb{P}(\Gamma(X, E)^*), x \mapsto H(x) := \{\sigma \in \Gamma(X, E) : \sigma(x) = 0\}$.

Exercise. Identifying E with the P-quotient $S \times_\chi \mathbb{C} = (S \times \mathbb{C})/P$, where the P-action on $S \times \mathbb{C}$ is defined by $g(s, z) := (sg^{-1}, \chi(g) \cdot z)$, recall that $\Gamma(X, E)$ is naturally identified with $\{f \in \mathcal{O}(S) : f(gh^{-1}) = \chi(h)f(g) \forall h \in P\}$, i.e., the equivariant maps $f : S \to \mathbb{C}$. Show that $\varphi_E : X \to \mathbb{P}(\Gamma(X, E)^*)$ is equivariant, i.e., in the case where $\Gamma(X, E) \neq \{0\}$. $\square$

The equivariance of φ_E implies in particular that for every $x \in X$ there is a section $\sigma \in \Gamma(X, E)$ such that $\sigma(x) \neq 0$, i.e., φ_E is holomorphic and realizes X as a homogeneous fibration $X = S/P \to S/Q = S.\varphi_E(x_0) \hookrightarrow \mathbb{P}(\Gamma(X, E)^*)$. It follows from the simple connectivity of the base that the fiber $F = Q/P$ is connected and of course the pull-back of E to F is trivial.

Proposition. *A bundle $E \in Pic(X)$ is numerically semi-positive if and only if it possesses a section $s \not\equiv 0$. It is numerically positive if and only if it is very ample, i.e., φ_E is an embedding.*

Proof. If E has a non-trivial section, its restriction to complex curves in the fibers of φ_E are trivial. On the other hand, if $\varphi_E|C$ in non-constant, then $E|C$ has non-trivial sections.

Conversely, identifying $E = E(D)$ for $D = m_1 D_1 + \ldots + m_r D_r$, if E is semi-positive (resp. positive), then the intersection numbers m_j of D with the root $\mathbb{P}_1$'s are all non-negative (resp. positive). So if E is semi-positive, D defines a holomorphic section $s \neq 0$. If $\varphi_E : S/P \to S/Q \hookrightarrow \mathbb{P}(\Gamma(X, E)^*)$ is not an embedding, then some root $\mathbb{P}_1$ lies in the fiber Q/P and the corresponding intersection number is zero. Thus, if E is positive, then it is very ample. $\qquad\square$

Zusatz. *The bundle $E = E(D)$, $D = m_1 D_1 + \ldots + m_r D_r$, is semi-positive (resp. positive) if and only if $m_i \geq 0$ (resp. $m_i > 0$) for all i.*

The anti-canonical bundle $K^{-1} \in Pic(X)$ is a basic example of a very ample bundle on $X = S/P^-$. To analyze K^{-1}, for $((\xi_1, \ldots, \xi_m))_\mathbb{C}$ a basis of $\mathfrak{r}_u(\mathfrak{p}^+)$, let $V_1, \ldots, V_m \in Vect_\mathcal{O}(X)$ be associated fields on X and $\sigma := V_1 \wedge \ldots \wedge V_m$ the resulting section of K^{-1}. Since $R_u(P^+)$ operates freely and transitively on the open cell $\Omega := B^+.x_0$, it follows that σ is (the unique) P-invariant section of K^{-1} and $\{\sigma = 0\}$ is the union of the basic hypersurfaces $D_j, j = 1, \ldots r$.

Proposition. *The anti-canonical bundle on $X = S/P$ is the bundle $K^{-1} = E(D) \in Pic(X)$ associated to $D = 2D_1 + \ldots + 2D_r$; in particular it is very ample.*

Proof. Let $\mathbb{P}_1(\alpha_i)$ the simple root curves which occur, i.e., $\mathbb{P}_1(\alpha_i) \cdot D_j = \delta_{ij}$. The coefficients of $D(\sigma) = m_1 D_1 + \ldots + m_r D_r$ are simply the Chern numbers of the pull-backs of K^{-1} to each of the $\mathbb{P}_1(\alpha_i)$'s.

For a given simple root α_i let $\mathfrak{r}_i$ be the subalgebra in $\mathfrak{r}_u(\mathfrak{p})$ obtained by removing the root space $\mathfrak{g}_{\alpha_i}$ and $R_i := exp(\mathfrak{r}_i)$. Since R_i acts freely on $B^+.x_i$ as well as on the open cell and $\mathbb{P}_1(\alpha_i)$ is transversal to D_i at x_i, it follows that $\Omega_i := R_i.\mathbb{P}_1(\alpha_i)$ is a product neighborhood $\mathbb{C}^{m-1} \times \mathbb{P}_1(\alpha_i)$ of $\mathbb{P}(\alpha_i)$ in X. Thus the pull-back of K^{-1} to $\mathbb{P}_1(\alpha_i)$ is just the anti-canonical bundle of $\mathbb{P}_1(\alpha_i)$ and $m_i = 2$ for $i = 1, \ldots, r$. $\qquad\square$

6.7 Generators of the Weyl group

The simple reflections $\sigma_\alpha \in \mathcal{W}$ defined by the root SU_2's generate a subgroup $\mathcal{W}(\Phi)$ of $\mathcal{W}$.

Proposition. $\mathcal{W}(\Phi) = \mathcal{W}$.

Proof. Let $\sigma \in \mathcal{W}$ be given, B^- be regarded as the neutral point x_0 in $X = S/B^-$ and consider the point $\sigma B^- = n(x_0) = x_\sigma \in X$, where $n \in N$ is a representative of σ.

If $\sigma B^- = B^-$, then clearly $\sigma = Id \in \mathcal{W}$. Otherwise there exists $\alpha < 0$ so that the root group U_α is not contained in σB^-. In this case $U_\alpha.x_\sigma \cong \mathbb{C} \subset B^- \sigma B^-$ and we consider the point $\sigma_\alpha(x_\sigma)$ at infinity in the root $\mathbb{P}_1(\alpha)$, i.e., the unique point in $S_\alpha.x_\sigma \setminus U_\alpha.x_\sigma$. This point corresponds to $\sigma_\alpha \sigma B^-$ and its orbit $B^- \sigma_\alpha \sigma B^-$ is lower-dimensional than that of σB^-.

After finitely many steps $\sigma_1 \cdot \ldots \cdot \sigma_k \cdot \sigma B^-$ is the B^--fixed point and $\sigma = (\sigma_1 \cdot \ldots \cdot \sigma_k)^{-1}$ is a product of reflections. $\qquad\square$

Let us again consider the W-action on t and let C be a component of the complement of the union of the reflection hyperplanes $\{\alpha = 0\}, \alpha \in \Phi$. Recall that the bases of simple roots as well as the K-invariant integrable complex structures on K/T are in 1-1 correspondence with these components.

Proposition. *The closure $\overline{C}$ is an exact fundamental region for the W-action on t, i.e., every W-orbit in t intersects $\overline{C}$ in exactly one point.*

Proof. First, recall that for $\xi \in C$ the isotropy K_ξ is the maximal torus T. Hence, if $w(\xi) = \xi$, then $w = Id \in W$.

Furthermore, if $w(C) = C$, then, utilizing the convexity of C, the average

$$\frac{1}{n}(\xi + w(\xi) + \ldots + w^{n-1}(\xi)), \; n := \text{Order}(w)$$

of a point $\xi \in C$ is in C and is w-fixed. So in this case w is also necessarily the identity.

Since $W = W(\Phi)$, $w \in W$ acts on the set C of all such connected components. In fact for $C \in C$ fixed

$$\bigcup_{w \in W} w(\overline{C}) = t,$$

because otherwise the left-hand side cannot have 1-codimension boundary hypersurfaces, i.e., open pieces of reflection hyperplanes.

Finally, let $x_0 \in \overline{C}$ and $w \in W$ be such that $w(x_0) =: y_0 \in \overline{C}$. Then, by applying a sequence of reflections $w_1 \cdot \ldots \cdot w_k$ which fix y_0 we obtain an element $w_0 := w_1 \cdot \ldots \cdot w_k \cdot w \in W$ which stabilizes $\overline{C}$. Thus $w_0 = Id$ and, since $w_0(x_0) = y_0$, it follows that $x_0 = y_0$. In other words, every W-orbit intersects $\overline{C}$ in exactly one point and the proof is complete. $\square$

Exercise. It follows that W acts freely and transitively on the K-invariant, integrable complex structures on the generic coadjoint orbits $K.\lambda \cong K/T$. Discuss the analogous statement in the case where $K_\lambda = Z_K(T_\lambda)$. $\square$

7. Representations of complex semisimple groups

7.1 Irreducible respresentations: The associated complex K-orbit

Recall that a smooth representation $\rho : K \to GL_{\mathbb{C}}(V)$ of a compact group extends holomorphically to its complexification $S = K^{\mathbb{C}}$ and conversely a holomorphic representation of a complex semisimple group is always given in this way. Furthermore, if $\rho : S \to GL_{\mathbb{C}}(V)$ is such a representation, then the induced S-action on $\mathbb{P}(V)$ is algebraic and therefore the minimal dimensional S-orbits are compact homogeneous manifolds of the form $X = S/P$.

It has been shown in §5 that a compact form K of S acts transitively on such manifolds. Thus the closed S-orbits in $\mathbb{P}(V)$ can be described as the complex K-orbits.

Proposition. *A representation* $\rho : K \to GL_{\mathbb{C}}(V)$ *of a compact semisimple group is irreducible if and only if there is exactly one complex K-orbit in* $\mathbb{P}(V)$. *Or, equivalently, if and only if the complexification* $S = K^{\mathbb{C}}$ *has exactly one closed orbit.*

One direction of the proof is straightforward: If ρ is not irreducible and decomposes into a sum $\rho_1 \oplus \rho_2 : K \to GL_{\mathbb{C}}(V_1 \oplus V_2)$ of two representations, then $dim_{\mathbb{C}}V_j \geq 2$, $j = 1, 2$, and K has complex orbits in both $\mathbb{P}(V_1)$ and $\mathbb{P}(V_2)$. Therefore K has at least two complex orbits in $\mathbb{P}(V)$.

The converse direction requires a bit of preparation. For this, let $\rho : S \to GL_{\mathbb{C}}(V)$ be a holomorphic representation and $X = K.x_0 = S.x_0 = S/P$ be a complex K-orbit in $\mathbb{P}(V)$. As usual, normalize P to contain B^- which is defined by a maximal torus $T^{\mathbb{C}}$ and a basis of roots. Let $\pi : V \setminus \{0\} \to \mathbb{P}(V)$ and choose some $v_0 \in V \setminus \{0\}$ with $\pi(v_0) = x_0$. Regard V as an S-invariant open set of $\mathbb{P}(V \oplus \mathbb{C})$ and extend π to a map

$$\pi : H := \mathbb{P}(V \oplus \mathbb{C}) \setminus \{0\} \to \mathbb{P}(V),$$

where $0 := [0 \oplus 1] = \mathbb{P}(\{0\} \oplus \mathbb{C})$.

This is of course the hyperplane bundle.

Let $E := \pi^{-1}(X) \subset H$. Then E is a positive S-homogeneous bundle $E = S \times_\chi E_0$, where $E_0 \cong \mathbb{C}$ is the π-fiber through v_0 and $\chi : P \to GL_{\mathbb{C}}(E_0)$ is the dual character to that defined by the natural P-action. The following general observation is now of use.

Borel-Weil Theorem. *The canonical S-representation on the space of sections* $\Gamma(X, E)$ *of an S-homogeneous line bundle* $S \times_\chi E_0 = E$ *over* $X = S/P$ *is irreducible and either* $\Gamma(X, E) = \{0\}$ *or* $\mathbb{P}(\Gamma(X, E))$ *contains exactly one complex K-orbit.*

Proof. Since a complex K-orbit, or equivalently a closed S-orbit, contains a B^--fixed point which corresponds to a section $\sigma \in \Gamma(X, E)$ with $b(\sigma) = \chi(b).\sigma$ for some $\chi \in \mathfrak{X}(B)$, two closed S-orbits in V would yield U^--fixed sections $\sigma_i \in V$, $0 = 1, 2$, which in turn would lead to a U^--invariant meromorphic function $f = \sigma_1/\sigma_2$. Since U^- has an open orbit in X, it follows that f is constant and σ_1 is a multiple of σ_2, i.e., the two orbits are the same. $\square$

Corollary. *If* $\varphi : X \to \mathbb{P}(V)$ *is an equivariant holomorphic embedding and* $E = \varphi^*(H)$ *is the pull-back of the hyperplane bundle, then* $\Gamma(X, E) = \varphi^*(\Gamma(\mathbb{P}(V), H)) = \varphi^*(V^*)$.

Formulated in the embedded situation above, intersection with hyperplanes defines a surjective, equivariant, linear map $V^* \to \Gamma(X, E)$.

Corollary. *If* V *is irreducible, then* $\mathbb{P}(V^*) = \mathbb{P}(\Gamma(X, E))$ *contains exactly one complex K-orbit.*

It is now possible to give the desired description of an irreducible representation.

Theorem (Borel-Weil interpretation). *Given an irreducible, holomorphic representation* $\rho : S \to GL_\mathbb{C}(V)$ *there is a uniquely determined parabolic subgroup* $P > B^-$ *and a uniquely determined very ample S-homogeneous line bundle* $E = S \times_\chi E_0$ *over* $X = S/P$ *so that* $V \cong \Gamma(X, E)^*$. *The projective space* $\mathbb{P}(V)$ *contains* $X = S.x_0$ *as its unique complex K-orbit.*

Proof. It remains to prove the last statement. For this note that $V \cong V^*$ are anti-holomorphically isomorphic K-representations. Since the anti-holomorphic image of a complex submanifold of $\mathbb{P}(V)$ is a complex submanifold of $\mathbb{P}(V^*)$ and K has only one complex orbit in $\mathbb{P}(V^*)$, it has only one complex orbit in $\mathbb{P}(V)$ as well. $\square$

Remark. The equivalence of irreducibility and the existence of a unique B^--fixed point x_0 in $\mathbb{P}(V)$ is sometimes referred to as the theorem on the dominant weight. A vector v_{max} with $\pi(v_{max}) = x_0$ is called a maximal weight vector. It transforms by a character, $b(v_{max}) = \chi_\lambda(b) \cdot v_{max}$, and λ is called the dominant weight.

Exercise. Prove the theorem on the dominant weight by using the knowledge given in §6.1 about the actions of the root groups U_α^+ on v_{max}.

7.2 Remarks on the momentum geometry

We begin with an

Exercise. Show that if $\rho : K \to GL_\mathbb{C}(V)$ is an irreducible representation, then up to a constant multiple there is a unique K-invariant Hermitian structure $H : V \times V \to \mathbb{C}$ on V. $\square$

The exercise implies that the Fubini-Study Kählerian structure,
$$w_{FS} = (1/2\pi) \, dd^c log\| \cdot \|^2,$$
and the associated moment map are uniquely determined by ρ.

Let $X = S/P = S.x_0 = K.x_0 \hookrightarrow \mathbb{P}(V)$ be the unique complex K-orbit and set up the root notation so that $T^\mathbb{C} < P$ and $x_0 = \pi(v_{max})$, where $b(v_{max}) = \chi_\lambda(b) \cdot v_{max}$ defines the maximal weight $\lambda = \lambda_{max} \in \mathfrak{t}_\mathbb{Z}^*$. In a natural way $\mathfrak{t}^* \hookrightarrow \mathfrak{k}^*$ as the annihilator of the compact root spaces.

With respect to $T^\mathbb{C}$ the representation decomposes:
$$V = \bigoplus_{\lambda \in \mathfrak{t}_\mathbb{Z}^*} V_\lambda,$$
where $t(v) = \chi_\lambda(t).v$ for $v \in V_\lambda$ and $t \in T^\mathbb{C}$. The non-zero spaces V_λ are referred to as weight spaces and the associated functions $\lambda \in \mathfrak{t}_\mathbb{Z}^*$ as weights of the representation. This *eigenspace decomposition* is of course orthogonal with respect to the invariant unitary structure. Of course $V_{\lambda_{max}}$ is 1-dimensional.

Let us compute $\mu(v)$ for $v \in V_\lambda$. For this it is convenient to first recall our choice of the Fubini-Study form: $w_{FS} = \frac{1}{2\pi} dd^c (log\| \cdot \|^2)$, where $d^c := Im(\partial)$. Thus $d^c f(v) = -\frac{1}{2} J(v)(f)$ and
$$\mu_\xi(v) = \frac{1}{2\pi} \frac{Im < \rho_*(\xi).v, v >}{\|v\|^2}.$$

Secondly, recall that the associated representation ρ_* of the Lie algebra satisfies $\rho_*(\xi)(V_\lambda) \subset V_{\lambda+\alpha}$ for $\xi \in \mathfrak{g}_\alpha$. Thus the orthogonality of the above decomposition implies that, for $v \in V_\lambda$ and $\xi \in \mathfrak{g}_\alpha$, $\alpha \in \Phi$, $\mu_\xi(v) = 0$.

For $\xi \in \mathfrak{t}$,

$$\mu_\xi(v) = \frac{1}{2\pi} \frac{Im < 2\pi i \lambda(\xi).v, v >}{\|v\|^2} = \lambda(\xi).$$

Thus we have the following observation:

Proposition. *Let $\rho : K \to GL_\mathbb{C}(V)$ be an irreducible representation of a compact group. Fix a maximal torus $T < K$ and let V_λ be a weight space. It follows that $\mu|\mathbb{P}(V_\lambda) \equiv \lambda \in \mathfrak{t}_\mathbb{Z}^*.$*

Remark. Much is known about the image $\mu(\mathbb{P}(V))$, e.g., that its intersection with a closed Weyl chamber $\overline{P}$ is a convex polyhedron (see e.g. [GS] and [Kir] and the references therein). $\square$

7.3 Holomorphic line bundles on integral coadjoint orbits

Let $\lambda \in \mathfrak{t}_\mathbb{Z}^*$ and J be an integrable, K-invariant complex structure on the coadjoint orbit $K.\lambda$. Thus there is a canonically associated parabolic subgroup $\mathfrak{p} = \mathfrak{l}^\mathbb{C} \oplus \mathfrak{m}^{0,1}$ so that $K.\lambda = K/L$ is identified as a complex homogeneous space with $X = S/P$ for $S = K^\mathbb{C}$ a complex, semisimple group. The functional λ extends uniquely to an integral $\mathbb{C}$-linear functional $\lambda \in (\mathfrak{t}^\mathbb{C})_\mathbb{Z}^*$. Let $\chi_\lambda : T^\mathbb{C} \to \mathbb{C}^*$ denote the associated complex 1-dimensional representation of $T^\mathbb{C}$.

Of course $[\mathfrak{l}^\mathbb{C}, \mathfrak{l}^\mathbb{C}] \subset Ker(\lambda)$ and, since $\mathfrak{p} = \mathfrak{l}^\mathbb{C} \ltimes \mathfrak{m}^{0,1}$, χ_λ can be regarded as a 1-dimensional representation $\chi_\lambda : P \to \mathbb{C}^*$ by defining it to be zero on the unipotent radical $R_u(P) = exp(\mathfrak{m}^{0,1})$. Thus in a canonical way one has the associated S-homogeneous line bundle $E_\lambda := S \times_{\chi_\lambda} \mathbb{C}$ over X. It should be underlined that, since $\chi_\lambda(L) \subset S^1$, the bundle E_λ possesses a canonical unitary structure which is defined by the L-invariant function

$$N : S \times \mathbb{C} \to \mathbb{R}^{>0}, \ (s, z) \mapsto |z|^2.$$

We regard N as defining a Hermitian bundle metric h and a bundle norm $\| \cdot \|^2$ and recall that the associated Chern form (see [GH]) is

$$c_1^h(E_\lambda) = \frac{1}{2\pi} d^c d(log\| \cdot \|^2).$$

This is of course a deRham representative of the Chern class $c_1(E_\lambda) \in H^2(X, \mathbb{Z})$ which was previously interpreted as intersection with $D = m_1 D_1 + \ldots + m_r D_r$, where $E_\lambda = E(D)$. The coefficients m_j are therefore determined by integrating $c_1^h(E_\lambda)$ over the simple root curves $\mathbb{P}_1(\alpha_j)$, $j \in I$.

Up to a constant multiple $\| \cdot \|^2$ is the unique invariant bundle norm and therefore $\omega_\lambda := c_1^h(E_\lambda)$ is the unique K-invariant representative of the deRham class of $c_1(E_\lambda)$.

Regarding

$$\omega_\lambda = \frac{1}{2\pi} dd^c log \left(\frac{1}{\|\cdot\|^2} \right) = \frac{1}{2\pi} dd^c \tau$$

as a 2-form on the total space of the associated $\mathbb{C}^*$-principle bundle $\mathcal{P}_\lambda$, we define an equivariant moment map $\mu : \mathcal{P}_\lambda \to \mathfrak{k}^*$ in the usual way: $\mu_\xi(v) = d^c\tau(\xi_{\mathcal{P}_\lambda})$. Of course μ factors through the projection $\pi : \mathcal{P}_\lambda \to X$.

Proposition. *The image of the moment map $\mu : \mathcal{P}_\lambda \to \mathfrak{k}^*$ is the orbit $K.\lambda$ itself. If $v \in S \times_\chi \mathbb{C}^*$ is in the neutral fiber over the neutral point $x_0 \in K.\lambda = S/P$, then $\mu(v) = \lambda$. In particular, regarding ω_λ as a form on $\mathfrak{o} = K.\lambda$, it agrees with the canonical form $\omega_\mathfrak{o}$.*

Proof. It is enough to show that $\mu(v) = \lambda$, because the fact that $\omega_\lambda = \omega_\mathfrak{o}$ on $\mathfrak{o}$ then follows directly from the definition of μ.

The calculation of $\mu(v)$ is completely analogous to that in the previous section for a weight vector $v \in V_\lambda$.

First, if $\mathfrak{g}_{-\alpha} < \mathfrak{r}_u(\mathfrak{p})$ we show that for $\xi \in \mathfrak{g}_\alpha$ it follows that $\mu_\xi(v) = 0$. For this consider the complex curve $exp(\xi z).v$ and the *orbit* $exp(\mathfrak{m}_\alpha).v$ of the space of compact roots, i.e., where $\mathfrak{su}_2(\alpha) = \mathfrak{t}_\alpha \oplus \mathfrak{m}_\alpha$ is the usual splitting. Since the $SL_2(\alpha)$-isotropy at v, is, up to connected components, just U_α^-, it follows that these two manifolds share the same tangent space at v. Thus, for $\xi \in \mathfrak{g}_\alpha$, the real curve $exp(i\xi t).v$ is tangent to $K.v$ which is the level set of $\| \cdot \|^2$ through v. In particular, $d^c\rho(\xi_{\mathcal{P}_\lambda})(v) = 0$.

For $\xi \in \mathfrak{t}$, $exp(i\xi t).v$ is in the neutral fiber of $\mathcal{P}_\lambda$

$$\mu_\xi(v) = \frac{d}{dt}\bigg|_{t=0} \tau(exp(i\xi t).v) = \frac{d}{dt}\bigg|_{t=0} \tau([exp(i\xi t), z]),$$

where $[g, z]$ denotes an equivalence class in $S \times_\chi \mathbb{C}^*$.

Thus

$$\mu_\xi(v) = \frac{d}{dt}\bigg|_{t=0} \tau([e, \chi_\lambda(-i\xi t).z]) = -\frac{1}{4\pi}\frac{d}{dt}\bigg|_{t=0} log|e^{2\pi\lambda(\xi)t}z|^{-2} = \lambda(\xi). \qquad \square$$

Remark. It should be noted that in the first part of the above proof it was shown that tangent spaces of the orbits $exp(\mathfrak{g}_\alpha).v$, $\mathfrak{g}_{-\alpha} < \mathfrak{r}_u(\mathfrak{p})$, form a basis of the complex tangent space of the hypersurface $K.v$.

7.4 The Kählerian structure on a coadjoint orbit and the associated representation

Let $K.\lambda = K/L$ be the coadjoint orbit of a given functional $\lambda \in \mathfrak{t}_\mathbb{Z}^*$. As we have seen $\mathcal{J}(L) \neq \emptyset$, i.e., there exist integrable, K-invariant complex structures. Furthermore, every such is a complex K-orbit in the projective space $\mathbb{P}(V)$ of a unitary representation.

Proposition. *For every $\lambda \in \mathfrak{t}_\mathbb{Z}^*$ there is a unique K-invariant, integrable, complex structure J_λ so that the symplectic orbit structure is Kählerian.*

Proof. Note that the T-invariance of $J \in \mathcal{J}(L)$ implies that it must preserve the root $\mathfrak{su}_2$'s. Thus in each such there is a basis $< \xi_\alpha, \eta_\alpha >_\mathbb{R}$ of the compact roots so that $J\xi_\alpha = \eta_\alpha$ and $J\eta_\alpha = -\xi_\alpha$. Hence, ω_o is J-invariant:

$$\omega_o(\lambda)(\xi_{\mathfrak{k}*}, \eta_{\mathfrak{k}*}) = \lambda([\xi, \eta]) = \lambda([J\xi, J\eta]).$$

The existence of $J = J_\lambda$ so that $\omega_o > 0$ follows by applying the moment map to the complex K-orbit in $\mathbb{P}(V)$ mentioned above. Uniqueness is also clear, because, restricting to a root $\mathfrak{su}$, there is only a choice of sign which is then fixed by the positivity assumption. $\qquad\square$

Summarizing this chapter, we have the following

Theorem. *For $\lambda \in \mathfrak{t}^*_\mathbb{Z}$, the S-homogeneous holomorphic line bundle $E := E_\lambda := S \times_{\chi_\lambda} \mathbb{C}$ over $X = K.\lambda = S/P_\lambda$ associated to the unique complex structure J_λ having the property that the coadjoint symplectic form ω_o is Kählerian is very ample. The S-representation in the space of sections $V_\lambda = \Gamma(X, E)$ is irreducible, the image $\varphi_E(X)$ of X by the canonical equivariant map $\varphi_E : X = S/P \hookrightarrow \mathbb{P}(V^*_\lambda)$ is the unique complex K-orbit $S.[v_{max}] \subset \mathbb{P}(V^*_\lambda)$. The moment image $\mu(\varphi_E(X))$ defined by the canonical K-invariant unitary structure on V^*_λ is the orbit $K.\lambda$ in $\mathfrak{k}^*$ with $\mu([v_{max}]) = \lambda$.*

Literature

[B]   A. Borel, *Linear algebraic groups*, 2nd enlarged ed., Springer-Verlag, New York, 1991.

[BH]   A. Borel and F. Hirzebruch, Characteristic classes and homogeneous spaces. I., *Amer. J. Math.* 80 (1958), 458–538.

[BTD]   T. Bröcker and T. tom Dieck, *Representations of compact Lie groups*, Corrected reprint of the 1985 orig., Springer, New York, 1995.

[GF]   H. Grauert and K. Fritzsche, *Several complex variables*, Springer-Verlag, New York – Heidelberg – Berlin, 1976.

[GrRe]   H. Grauert and R. Remmert, *Coherent analytic sheaves*, Springer-Verlag, Berlin, 1984.

[GH]   P. Griffiths and J. Harris, *Principles of algebraic geometry*, 2nd ed., John Wiley Sons, New York, 1994.

[GS]   V. Guillemin and S. Sternberg, *Symplectic techniques in physics*, Reprinted with corrections, Cambridge University Press, Cambridge, 1990.

[GuRo]   R. C. Gunning and H. Rossi, *Analytic functions of several complex variables*, Prentice-Hall, Englewood Cliffs, N.J., 1965.

[Ho]   G. Hochschild, *The structure of Lie groups*, Holden-Day, San Francisco – London – Amsterdam, 1965.

[Hu1]   J. E. Humphreys, *Linear algebraic groups*, Corr. 2nd printing, Springer-Verlag, New York – Heidelberg – Berlin, 1981.

[Hu2]   J. E. Humphreys, *Introduction to Lie algebras and representation theory*, 3rd printing, rev., Springer-Verlag, New York – Heidelberg – Berlin, 1980.

[KK]   B. Kaup and L. Kaup, *Holomorphic functions of several variables. An introduction to the fundamental theory*, Walter de Gruyter, Berlin – New York, 1983.

[Kiri]   A.A. Kirillov, Geometric quantization, in: *Dynamical systems. IV. Symplectic geometry and its applications*, Eds. V.I. Arnol'd and S.P. Novikov, 137–172, Encycl. Math. Sci. 4, Springer, Berlin, 1990.

[Kirw] F. Kirwan, Convexity properties of the moment mapping III, *Invent. Math.* 77 (1984), 547–552.

[Kob]   S. Kobayashi, *Transformation groups in differential geometry*, Reprint of the 1972 ed., Springer-Verlag, Berlin, 1995.

[Kos]   B. Kostant, Quantization and unitary representations I: Prequantization, in: *Lectures in Modern Analysis and Applications III*, Lect. Notes Math. 170, 87–207, Springer, Berlin, 1970.

[Kr]   S. G. Krantz, *Function theory of several complex variables*, 2nd ed., Wadsworth and Brooks/Cole, Pacific Grove, CA, 1992.

[N]   R. Narasimhan, *Several complex variables*, The University of Chicago Press, Chicago-London, 1971.

[R]   M. Range, *Holomorphic functions and integral representations in several complex variables*, Springer-Verlag, New York, 1986.

[RV]   R. Remmert and T. van de Ven, Über holomorphe Abbildungen projektiv-algebraischer Mannigfaltigkeiten auf komplexe Räume, *Math. Ann.* 142 (1961), 453–486.

[So]   J.-M. Souriau, Quantification géométrique, *Comm. Math. Phys.* 1 (1966), 374–398.

[Sp]   M. Spivak, *Calculus on manifolds. A modern approach to classical theorems of advanced calculus*, W.A. Benjamin, New York – Amsterdam, 1965.

[Wal]   N. R. Wallach, *Harmonic analysis on homogeneous spaces*, Marcel Dekker, New York, 1973.

[War]   F. W. Warner, *Foundations of differentiable manifolds and Lie groups*, Reprint, Springer-Verlag, New York, 1983.

[We]   R. O. Wells, *Differential analysis on complex manifolds*, 2nd ed., Springer-Verlag, New York – Heidelberg – Berlin, 1980.

Mathematical Subject Classification (2000)
Primary: 32Mxx, 53Dxx.
Secondary: 14M15, 22E46, 32J25, 32M05, 32M10, 32Q15, 53D05, 53D20, 53D50.

Mathematisches Institut
Ruhr-Universität Bochum
Universitätsstraße 150
D-44780 Bochum
Germany
E-mail address: ahuck@cplx.ruhr-uni-bochum.de

Infinite-dimensional Groups and their Representations

Karl-Hermann Neeb

Introduction

In this paper we discuss some of the basic general notions and results which play a key role in the representation theory of infinite-dimensional Lie groups modeled over sequentially complete locally convex (s.c.l.c.) spaces. In the following each locally convex space will implicitly be assumed to be Hausdorff.

In the first section we review the basic facts on calculus in s.c.l.c. spaces. We choose the setup of s.c.l.c. spaces to ensure the existence of integrals of vector valued continuous functions on compact intervals which is the key to the Fundamental Theorem of Calculus. For the setting of Fréchet spaces these results can be found in [Ha82], but one readily notices that as soon as one has a Fundamental Theorem of Calculus the other results go through with the same proofs. The s.c.l.c. setting is also used in [Mi83]. Moreover, the setting of s.c.l.c. spaces is the natural general setting for holomorphic mappings between infinite-dimensional spaces (cf. [He89]). In particular we show that the usual notion of holomorphy is equivalent to being smooth with complex linear differential. In this section we also discuss Lie groups over s.c.l.c. spaces and how to define their Lie algebra. For the existence of an exponential function no general result is known, nevertheless in all known examples an exponential function seems to exist (cf. [Mi83, p. 1043]). Moreover the differential of the exponential function is given by the same formula as in the finite dimensional case ([Gr97]). A particularly interesting class of infinite-dimensional Lie groups are the direct limit Lie groups. For more details on such groups we refer to [NRW91], [NRW93], [NRW94] and [Gl99]. For more results on general s.c.l.c. Lie groups we refer to [Mi83] where one finds in particular a discussion of the class of "regular" Lie groups which is characterized by nice properties of the exponential function. A discussion of regular Lie groups in the "convenient setting" of [KM97a] can be found in [KM97b].

Section II consists of a collection of various results from functional analysis, in particular on dual spaces, which play a role in dealing with representations of infinite-dimensional groups. Since we are working with s.c.l.c. spaces, one has to make sure in many circumstances that the spaces obtained are in fact sequentially complete. This is where one needs some refined tools from functional analysis. In addition to completeness properties, we also discuss metrizability of dual spaces for certain natural topologies.

In Section III we show how the results from Section II can be used to define convenient spaces of smooth and holomorphic functions on infinite-dimensional manifolds in such a way that these spaces become s.c.l.c. We also analyze the natural actions of Fréchet Lie groups on these spaces which are naturally associated to smooth actions. In particular we show that a smooth action of a Fréchet semigroup S on a Fréchet manifold M induces a smooth action of S on $C^\infty(M, V)$ for every s.c.l.c. space V. We also derive a complex version of this result for holomorphic actions of complex semigroups on complex manifolds.

In Section IV these results are applied to define a derived representation of a representation (π, V) of an s.c.l.c. Lie group G on the subspace V^∞ of smooth vectors and to endow this space with a suitable complete locally convex topology inherited from $C^\infty(G, V)$ on which the action of G is smooth.

In the last Section V we then turn to a quite general setup for so called coherent state representations. Analytically these representations are characterized by the property that they can be realized in spaces of holomorphic sections of a homogeneous complex line bundle. On the geometric side this means that the action of G on the projective space of the dual space has a cyclic complex orbit. These concepts are well studied in the setting of Hilbert spaces and here we show that if one carefully distinguishes between the spaces and their duals, then one can generalize this correspondence to s.c.l.c. spaces.

I. Calculus in locally convex spaces

In this section we explain briefly how calculus works in s.c.l.c. spaces. The main point is that one uses the appropriate notion of differentiability which for the special case of Banach spaces differs from Fréchet differentiability but which is more convenient in the setup of s.c.l.c. spaces. Our basic reference will be [Ha82], where one finds detailed proofs for the case of Fréchet spaces. One readily observes that once one has the Fundamental Theorem of Calculus, then the proofs of the Fréchet case carry over to a more general setup where one still requires smooth maps to be continuous (cf. also [Mi83]). A different approach to differentiability in infinite-dimensional spaces in the framework of the so called convenient setting can be found in [FK88] and [KM97a]. A central feature of this approach is that smooth maps are no longer required to be continuous, but for calculus over Fréchet spaces one finds the same class of smooth maps described by Hamilton and Milnor. Another approach which also gives up the continuity of smooth maps and requires only continuity on compact sets is discussed by E. G. F. Thomas in [Th96].

It is also interesting to note that since the Cauchy Integral Formula plays a similar role for holomorphic functions as the Fundamental Theorem of Calculus does for differentiable functions, the setting of s.c.l.c. spaces also seems to be the appropriate one for holomorphic mappings between infinite-dimensional spaces.

We show in particular that these two concepts are related by the observation that the usual notion of holomorphy is equivalent to smoothness with complex linearity of the differential.

Then we turn to manifolds modeled over s.c.l.c. spaces. Due to the aforementioned relation between smooth and holomorphic functions, complex manifolds are special cases of real manifolds in any reasonable setting. One of our main objectives in this section is to discuss some of the most basic properties of Lie groups modeled over s.c.l.c. spaces. In particular we explain how to define their Lie algebra and the adjoint representation. A major difficulty of the s.c.l.c. setup which does not arise for Banach Lie groups is that one cannot guarantee a priori that they have any exponential function. Thus one is forced in many places to argue without using an exponential function.

1. Differentiable functions

Definition I.1. (a) Let X and Y be topological vector spaces, $U \subseteq X$ open and $f\colon U \to Y$ a continuous map. Then the *derivative of f at x in the direction of h* is defined as

$$df(x)(h) := \lim_{t \to 0} \frac{1}{t}\big(f(x + th) - f(x)\big)$$

whenever it exists. The function f is called *differentiable in x* if $df(x)(h)$ exists for all $h \in X$. It is called *continuously differentiable or C^1* if it is differentiable in all points of U and

$$df\colon U \times X \to Y, \quad (x, h) \mapsto df(x)(h)$$

is a continuous map.

(b) Higher derivatives are defined by

$$d^n f(x)(h_1, \ldots, h_n)$$
$$:= \lim_{t \to 0} \frac{1}{t}\big(d^{n-1}f(x + th_n)(h_1, \ldots, h_{n-1}) - d^{n-1}f(x)(h_1, \ldots, h_{n-1})\big).$$

The function f is called *n-times continuously differentiable or C^n* if

$$d^n f\colon U \times X^n \to Y, \quad (x, h_1, \ldots, h_n) \mapsto d^n f(x)(h_1, \ldots, h_n)$$

is a continuous map. We say that f is *smooth or C^∞* if it is C^n for all $n \in \mathbb{N}$.

(c) If X and Y are complex vector spaces, then the map f is called *holomorphic* if it is C^1 and for all $x \in U$ the map $df(x)\colon X \to Y$ is complex linear (cf. [Mi83, p. 1027]) □

We note that if X and Y are Banach spaces, then the strong notion of continuous differentiability is weaker than the usual notion of continuous differentiability in Banach spaces which requires that the map $x \mapsto df(x)$ is continuous with respect to the operator norm. We will discuss this point below (Example I.6 and Theorem I.7). We also note that the existence of linear maps which are not continuous shows that the continuity of f does not follow from the differentiability of f because each linear map $f\colon X \to Y$ is differentiable in the sense of Definition I.1(a).

So far we did not use any special property of the topological vector spaces involved. To be able to develop a calculus on topological vector spaces which has at least the most basic properties of calculus in finite dimensions, we will have to make the assumption that the vector spaces under consideration are sequentially complete locally convex (s.c.l.c.) spaces.

The main point in making this assumption is to be able to integrate continuous curves $\gamma\colon [a,b] \to X$ in the sense that there exists a unique element $y := \int_a^b \gamma(t)dt \in X$ with

$$\omega(y) = \int_a^b \langle \omega, \gamma(t) \rangle \; dt$$

for all continuous linear functionals ω on X (cf. [He89, Prop. 1.2.3]).

We recall that a locally convex space X is called *quasicomplete* if each closed bounded subset of X is complete as a uniform space. Since Cauchy sequences form bounded sets, it is clear that completeness implies quasicompleteness and that quasicompleteness implies sequential completeness. For the existence of integrals of continuous functions $\gamma\colon C \to X$, where C is a compact space, the quasicompleteness of X is the appropriate assumption (cf. [Bou59, §1, no. 2, Cor. de Prop. 5; no. 6]).

Now we recall the precise statements of the most fundamental facts.

Lemma I.2. *The following assertions hold:*
  (i) *If f is C^1 and $x \in U$, then $df(x)\colon X \to Y$ is a linear map, f is continuous, and if $x + th \in U$ holds for all $t \in [0,1]$, then*

$$f(x+h) = f(x) + \int_0^1 df(x+uh)(h) \; du.$$

  (ii) *If f is C^n, then the functions $(h_1,\dots,h_n) \mapsto d^n f(x)(h_1,\dots,h_n)$, $x \in U$, are symmetric n-linear maps.*

Proof. (i) The first part is [Ha82, Th. 3.2.5] and the integral representation is [Ha82, Th. 3.2.2]. To see that f is continuous, let p be a continuous seminorm on Y and $\varepsilon > 0$. Then there exists a balanced 0-neighborhood $U_1 \subseteq X$ with $x + U_1 \subseteq U$ and $p\big(df(x+uh)(h)\big) < \varepsilon$ for $u \in [0,1]$ and $h \in U_1$. Hence

$$p\big(f(x+h) - f(x)\big) \le \int_0^1 p\big(df(x+uh)(h)\big) \; du \le \varepsilon,$$

and thus f is continuous.
(ii) [Ha82, Th. 3.6.2]   □

Proposition I.3. (The chain rule) *If X, Y and Z are s.c.l.c. spaces, $U \subseteq X$ and $V \subseteq Y$ are open, and $f_1\colon U \to V$, $f_2\colon V \to Z$ are C^1, then $f_2 \circ f_1\colon U \to Z$ is C^1 with*

$$d(f_2 \circ f_1)(x) = df_2\big(f_1(x)\big) \circ df_1(x).$$

Proof. [Ha82, Th. 3.3.4]   □

Proposition I.4. *If X_1, X_2 and Y are s.c.l.c. spaces, $X = X_1 \times X_2$, $U \subseteq X$ is open, and $f \colon U \to Y$ is continuous, then the partial derivatives*

$$d_1 f(x_1, x_2)(h) := \lim_{t \to 0} \frac{1}{t} \big(f(x_1 + th, x_2) - f(x_1, x_2) \big)$$

and

$$d_2 f(x_1, x_2)(h) := \lim_{t \to 0} \frac{1}{t} \big(f(x_1, x_2 + th) - f(x_1, x_2) \big)$$

exist and are continuous if and only if df exists and is continuous. In that case we have

$$df(x_1, x_2)(h_1, h_2) = d_1 f(x_1, x_2)(h_1) + d_2 f(x_1, x_2)(h_2).$$

Proof. [Ha82, Th. 3.4.3]   □

Remark I.5. (a) If $f \colon X \to Y$ is a continuous linear map, then f is smooth with

$$df(x)(h) = f(h)$$

for all $x, h \in X$, and $d^n f = 0$ for $n \geq 2$.

(b) From (a) and Proposition I.4 it follows that a continuous k-linear map $m \colon X_1 \times \ldots \times X_k \to Y$ is continuously differentiable with

$$dm(x)(h_1, \ldots, h_k) = m(h_1, x_2, \ldots, x_k) + \cdots + m(x_1, \ldots, x_{k-1}, h_k).$$

Inductively one obtains that m is smooth with $d^{k+1} m = 0$.

(c) If $f \colon U \to Y$ is C^{n+1}, then Lemma I.2(ii) and Proposition I.4 imply that

$$d(d^n f)(x, h_1, \ldots, h_n)(y, k_1, \ldots, k_n) = d^{n+1} f(x)(h_1, \ldots, h_n, y)$$
$$+ d^n f(x)(k_1, h_2, \ldots, h_n) + \ldots + d^n f(x)(h_1, \ldots, h_{n-1}, k_n).$$

It follows in particular that, whenever f is C^n, then f is C^{n+1} if and only if $d^n f$ is C^1.

(d) If $f \colon U \to Y$ is holomorphic, then the finite-dimensional theory shows that for each $h \in X$ the function $U \to Y, x \mapsto df(x)(h)$ is holomorphic. Hence $d^2 f(x)$ is complex bilinear and therefore $d(df)$ is complex linear. Thus $df \colon U \times X \to Y$ is also holomorphic.   □

2. Differentiable functions on Banach spaces

In this subsection we discuss the relation between the notion of differentiability described in Definition I.1 and the notion of Fréchet differentiability in Banach spaces. In Example I.6 we will see that for maps between Banach spaces our C^1 concept differs from the concept of continuous Fréchet differentiability, and in Theorem I.7 we will show that smooth functions are also smooth in the Fréchet sense (the converse is obvious). For a more detailed discussion of several concepts of differentiability in Fréchet and Banach spaces we refer to [Ke74, p. 110].

Example I.6. Let $E := \{ f \in C(\mathbb{R}) \colon (\forall x \in \mathbb{R}) f(x+1) = f(x) \}$ denote the Banach space of 1-periodic continuous functions on $\mathbb{R}$ endowed with the norm $\| f \|_E :=$

$\sup\{|f(x)|: x \in \mathbb{R}\}$. Further let $F := \{f \in E \cap C^1(\mathbb{R}) : f' \in E\}$ be endowed with the norm $\|f\|_F := \|f\|_E + \|f'\|_E$. We consider the map

$$m: X := \mathbb{R} \times F \to E, \quad (x, f) \mapsto f(x + \cdot).$$

We claim that in the sense of Definition I.1(a) this map is C^1, but that $\tilde{d}m: X \to \mathcal{L}(X, E), x \mapsto (h \mapsto dm(x, h))$, where $\mathcal{L}(X, E)$ denotes the Banach space of all continuous operators from X to E, is not continuous, i.e., m is not C^1 in the Fréchet sense.

We first show that the differential of m is given by

$$dm(x, f)(y, h) = f'(x + \cdot)y + h(x + \cdot).$$

In fact, for $s \in \mathbb{R}$ and $t \neq 0$ we have

$$\frac{1}{t}(m(x + ty, f + th)(s) - m(x, f)(s)) - f'(x + s)y - h(x + s)$$

$$= \frac{1}{t}(f(x + ty + s) + th(x + ty + s) - f(x + s)) - f'(x + s)y - h(x + s)$$

$$= \frac{1}{t}(f(x + ty + s) - f(x + s)) - f'(x + s)y + h(x + ty + s) - h(x + s)$$

$$= \int_0^1 f'(x + s + uty)y \, du - f'(x + s)y + \int_0^1 h'(x + s + uty)ty \, du.$$

Now the facts that f' is uniformly continuous and that h' is bounded imply that this expression tends to 0 in E whenever $t \to 0$. This proves the formula for the differential of m.

Next we show that $dm: X \times X \to E$ is continuous. In fact, the continuity of $\mathbb{R} \times F \to E, (x, h) \mapsto h(x + \cdot)$ follows from

$$\|h(x + \cdot) - h_1(x_1 + \cdot)\|_E \leq \|h(x + \cdot) - h(x_1 + \cdot)\|_E + \|h(x_1 + \cdot) - h_1(x_1 + \cdot)\|_E$$
$$\leq \|h'\|_E |x - x_1| + \|h - h_1\|_E.$$

So it remains to see that $(x, f) \mapsto f'(x + \cdot)$ is also continuous. We have

$$\|f'(x + \cdot) - f_1'(x_1 + \cdot)\|_E \leq \|f'(x + \cdot) - f'(x_1 + \cdot)\|_E + \|f' - f_1'\|_E,$$

so that the asserted continuity follows from the uniform continuity of f'.

To see that $\tilde{d}m: X \to \mathcal{L}(X, E)$ is not continuous, we note that $d_2m(x, f)(h) = h(x + \cdot)$. If $\lambda_x.f = f(x + \cdot)$, then $x \neq x'$ implies that $\|\lambda_x - \lambda_{x'}\| = 2$. This shows that $(x, f) \mapsto d_2m(x, f) = \lambda_x$ is not continuous. $\square$

Theorem I.7. *Let X and Y be Banach spaces, $U \subseteq X$ open, and $f: U \to Y$ a map. Then the following assertions hold:*

(i) *If f is C^2, then it is C^1 in the Fréchet sense.*

(ii) *f is C^∞ if and only if it is C^∞ in the Fréchet sense.*

Proof. (i) Let us fix $x \in U$ and suppose that the open δ-ball $U_\delta(x)$ about x is contained in U. We write $d^2 f(x)(h)$ for the map $h_1 \mapsto d^2 f(x)(h, h_1)$ in $\mathcal{L}(X, Y)$. We claim that there exists an $\varepsilon \in \,]0, \delta[$ such that the set

$$M_\varepsilon := \left\{ \frac{1}{\sqrt{\|h\|}} d^2 f(x + h)(h) : 0 < \|h\| < \varepsilon \right\}$$

is bounded. Suppose that this is not the case. Then there exists a sequence $h_n \to 0$ such that $\|d^2 f(x + h_n)(h_n)\| \geq n\sqrt{\|h_n\|}$. For each $h_1 \in X$ we have

$$\frac{1}{\sqrt{\|h_n\|}} d^2 f(x + h_n)(h_n)(h_1) = d^2 f(x + h_n)\left(\frac{h_n}{\sqrt{\|h_n\|}}, h_1 \right) \to 0$$

because $d^2 f : U \times X^2 \to Y$ is continuous and $\frac{h_n}{\sqrt{\|h_n\|}} \to 0$. This contradicts the Banach-Steinhaus Theorem, and therefore one of the sets M_ε is bounded.

Now assume that $\|h\| < \varepsilon$ and that $\|d^2 f(x + h)(h)\| \leq C\sqrt{\|h\|}$ for $\|h\| < \varepsilon$. Then

$$\|\tilde{d}f(x + h) - \tilde{d}f(x)\| = \left\| \int_0^1 d^2 f(x + uh)(h)\, du \right\| \leq \int_0^1 \|d^2 f(x + uh)(uh)\| \frac{1}{u} du$$

$$\leq \int_0^1 C\sqrt{\|uh\|} \frac{1}{u} du = C\sqrt{\|h\|} \int_0^1 u^{-\frac{1}{2}} du = 2C\sqrt{\|h\|}.$$

We conclude that the map $\tilde{d}f : U \to \mathcal{L}(X, Y)$ is continuous.

Furthermore we have

$$\|f(x + h) - f(x) - \tilde{d}f(x)(h)\| = \left\| \int_0^1 \tilde{d}f(x + uh)(h) - \tilde{d}f(x)(h)\, du \right\|$$

$$\leq \sup\{\|\tilde{d}f(x + h_1) - \tilde{d}f(x)\| : \|h_1\| < \varepsilon\}\|h\|,$$

and, in view of the continuity of $x \mapsto \tilde{d}f(x)$, the expression on the right hand side is $o(\|h\|)$. This proves that f is C^1 in the Fréchet sense whenever it is C^2 in the sense of Definition I.1(a).

(ii) If f is C^∞ in the Fréchet sense, then it is trivially C^∞ in the sense of Definition I.1(a).

Suppose that f is C^∞. Then the map $df : U \times X \to Y$ is also C^∞, hence in particular C^2. Therefore (i) shows that the map

$$\tilde{d}(df) : U \times X \to \mathcal{L}(X^2, Y)$$

is continuous, hence in particular that $d^2 f : U \times X \to \mathcal{L}(X, Y)$ is continuous since $d^2 f(x)(h_1, h_2) = \tilde{d}(df)(x, h_1)(h_2, 0)$. Now

$$\tilde{d}f(x + h) - \tilde{d}f(x) - d^2 f(x)(h) = \int_0^1 d^2 f(x + uh)(h) - d^2 f(x)(h)\, du$$

implies that $d^2 f$ can be viewed as $d(\tilde{d}f)$. Iterating this argument, we conclude that the map $\tilde{d}f : U \to \mathcal{L}(X, Y)$ is smooth in the sense of Definition I.1. Now we can apply induction and obtain for all $n \in \mathbb{N}$ that the nth Fréchet derivative of f is smooth, and therefore that f is smooth in the Fréchet sense. $\square$

3. Holomorphic functions

In this subsection we clarify the relation between several concepts of holomorphy for functions between s.c.l.c. spaces.

Definition I.8. Let X be a complex vector space.

(a) A subset $U \subseteq X$ is called *finitely open* if for all finite-dimensional affine subspaces $F \subseteq X$ the set $F \cap U$ is open in F.

(b) Let V be a sequentially complete locally convex space. A function f on a finitely open subset $U \subseteq X$ is called *Gateaux holomorphic* ((G)-*holomorphic*) if for each finite-dimensional affine subspace $F \subseteq X$ the function $f|_{F \cap U}$ is (weakly) holomorphic on $F \cap U$ (cf. [He89, Th. 2.1.3]). We write $\mathcal{G}(U, V)$ for the space of (G)-holomorphic V-valued functions on U. Note that, in view of Hartog's Theorem, a function is (G)-holomorphic if the above criterion is satisfied for all affine complex lines $F \subseteq X$.

(c) Suppose that X is a locally convex space. A (G)-holomorphic function $f: U \to V$ is called *Fréchet holomorphic* ((F)-*holomorphic*) if for each continuous seminorm p on V the function $p \circ f$ is locally bounded. We recall from [He89, Prop. 2.4.2(a)] that this property is equivalent to the continuity of the function f. $\square$

If X is of countable dimension and we write $X = \bigcup_{n \in \mathbb{N}} X_n$ with $X_n \subseteq X_{n+1}$ and $\dim X_n < \infty$, then X carries a natural LF space structure which is the finest locally convex topology on X (cf. [Tr67, Ex. 13.1]). The open sets in this topology are exactly the finitely open sets ([He89, Prop. 2.3.2]). If $\dim X > \aleph_0$, then the topology defined by the finitely open sets is no longer a vector spaces topology and therefore does not coincide with the finest locally convex topology (cf. [He89, Rem. 2.3.3]).

The notion of (G)-holomorphy is the weakest possible notion of holomorphy in infinite-dimensional spaces. Unfortunately it has the drawback that in general it even does not imply continuity. In this sense the "nice" holomorphic functions are the (F)-holomorphic functions. Note that (F)-holomorphy is preserved by passing to locally uniform limits. The relations between (F)-holomorphy and weak holomorphy are clarified for "nice" spaces in the following result.

Proposition I.9. *For a function $f: U \to V$ from an open subset U of a locally convex space X to the s.c.l.c. space V the following assertions hold:*

  (i) *If X is metrizable, then f is (F)-holomorphic if and only if it is weakly (F)-holomorphic.*

  (ii) *If X is the inductive limit of locally convex spaces $(X_n)_{n \in \mathbb{N}}$ such that the origin in X_n has a neighborhood which is relatively compact in X_{n+1}, then*

  (a) *f is (F)-holomorphic if and only if it is weakly (F)-holomorphic.*

  (b) *f is continuous if and only if all the functions $f|_{U \cap X_n}$ are continuous for all $n \in \mathbb{N}$.*

(iii) *If X is Baire, $f \in \mathcal{G}(U, V)$, and there exists a sequence of continuous functions $f_n \colon U \to V$ converging pointwise to f, then f is continuous, i.e., (F)-holomorphic.*

Proof. (i), (ii)(a) [He89, Prop. 3.1.2]

(ii)(b) [He89, Prop. 1.5.1(b)]

(iii) [He89, Th. 2.4.4] □

Proposition I.10. *For a function $f \colon U \to V$ the following are equivalent:*
  (i) *f is holomorphic in the sense of Definition I.1(c).*
  (ii) *f is (F)-holomorphic.*
  (iii) *f is smooth with complex linear differentials $df(x)$, $x \in U$.*

Proof. (i) $\Rightarrow$ (ii): If f is complex differentiable in the sense of Definition I.1(c), then f is (G)-holomorphic (differentiable functions on open domains in the complex plane are holomorphic), and continuous (Lemma I.2(i)), hence (F)-holomorphic.

(ii) $\Rightarrow$ (iii): Suppose that f is (F)-holomorphic. We have to show that all its higher derivatives

$$d^n f \colon U \times E^n \to V, \quad (x, h_1, \dots, h_n) \mapsto d^n f(x)(h_1, \dots, h_n)$$

are continuous maps. It is clear that the (G)-holomorphy implies the (G)-holomorphy of $d^n f$ because a similar statement holds in finite dimensions. Moreover, the generalized Cauchy inequalities (cf. [He89, Th. 2.3.5]) imply that whenever f is locally bounded in the sense of Definition I.8(c), the same property is inherited by the functions

$$(x, h) \mapsto \hat{d}^n f(x, h) := d^n f(x)(h, \dots, h).$$

Next we use the formula

$$d^n f(x)(h_1, \dots, h_n) = \frac{1}{2^n n!} \sum_{\varepsilon \in \{1, -1\}^n} (\varepsilon_1 \cdots \varepsilon_n) \hat{d}^n f(x)(\varepsilon_1 h_1 + \dots + \varepsilon_n h_n)$$

(cf. [Na69, p.7]) to conclude that the function $d^n f$ is also locally bounded in the sense of Definition I.8(c), i.e., that $d^n f$ is (F)-holomorphic. It follows in particular that the functions $d^n f$ are continuous, hence that f is a smooth function.

(iii) $\Rightarrow$ (i): This is trivial since C^∞ implies C^1. □

The following result clarifies the concept of (F)-holomorphy in the Banach setting.

Proposition I.11. *If X and V are complex Banach spaces, $U \subseteq X$ a domain, and $f \colon U \to V$ a function. Then the following assertions hold:*
  (i) *If f is (F)-holomorphic, then f is complex Fréchet differentiable.*
  (ii) *The function f is (F)-holomorphic if and only if it is Fréchet differentiable at each point $x \in U$.*

Proof. (i) ([HP57, Th. 3.17.1]) If f is (F)-holomorphic, then Proposition I.10 shows that f is smooth, hence f is Fréchet smooth (Theorem I.7).

(ii) [He89, Cor. 3.1.4] □

4. Differentiable manifolds

Since we have a chain rule for differentiable maps between s.c.l.c. spaces, we can define smooth manifolds as one defines them in the finite-dimensional case (cf. [Ha82], [Mi83]). The underlying topological space is always required to be Hausdorff. Since locally convex spaces (which we always assume to be Hausdorff) are *regular* in the sense that each point has a neighborhood base consisting of closed sets, this property is inherited by manifolds modeled over these spaces (cf. [Mi83]). One also defines vector bundles and in particular the tangent bundle $TM \to M$ as usual.

Note that it is far more subtle to define a cotangent bundle because this requires an s.c.l.c. topology on the dual space of the underlying vector space and therefore depends on this topology. We will discuss topologies on the dual in Section II.

Let M and N be smooth manifolds modeled over s.c.l.c. spaces and $f\colon M \to N$ a smooth map. We write $Tf\colon TM \to TN$ for the corresponding map induced on the level of tangent vectors. Locally this map is given by

$$Tf(x, h) = \big(f(x), df(x)(h)\big),$$

where $df(p)\colon T_p(M) \to T_{f(p)}(N)$ denotes the differential of f in p. In view of Remark I.5(c), the tangent map Tf is also smooth if f is smooth. In the following we will always identify M with the zero section in TM. In this sense we have $Tf|_M = f$ with $Tf(M) \subseteq N \subseteq TN$.

A *vector field* on M is a smooth section of the tangent bundle $TM \to M$. We write $\mathcal{V}(M)$ for the space of all vector fields on M. If $f \in C^\infty(M)$ is a smooth function on M and $X \in \mathcal{V}(M)$, then we obtain a function on M via

$$(X.f)(p) := df(p)\big(X(p)\big).$$

Since locally $X(p) = (p, \tilde{X}(p))$, where $\tilde{X}$ is a smooth function, we have $X.f = df \circ X$. Therefore the smoothness of $X.f$ follows from the smoothness of the maps $df\colon TM \to \mathbb{C}$ and $X\colon M \to TM$.

Lemma I.12. *If $X, Y \in \mathcal{V}(M)$, then there exists a vector field $[X, Y] \in \mathcal{V}(M)$ which is uniquely determined by the property that on each open subset $U \subseteq M$ we have*

(1.1) $$[X, Y].f = X.(Y.f) - Y.(X.f)$$

for all $f \in C^\infty(U)$.

Proof. Locally the vector fields X and Y are given as $X(p) = (p, \tilde{X}(p))$ and $Y(p) = (p, \tilde{Y}(p))$. We define a vector field by

(1.2) $$[X, Y]\tilde{\ }(p) := d\tilde{Y}(p)\big(\tilde{X}(p)\big) - d\tilde{X}(p)\big(\tilde{Y}(p)\big).$$

Then the smoothness of the right hand side follows from the chain rule. The requirement that (1.1) holds on continuous linear functionals determines $[X, Y]\tilde{\ }$ uniquely. Since an easy calculation shows that (1.2) defines in fact a smooth vector field on M (cf. Lemma I.14 below), the assertion follows because locally (1.1) is a consequence of the chain rule. $\square$

Proposition I.13. $(\mathcal{V}(M), [\cdot, \cdot])$ *is a Lie algebra.*

Proof. The crucial part is to check the Jacobi identity. This follows from the observation that if $U \subseteq X$ is an open subset of an s.c.l.c. space, then the mapping

$$\Phi\colon \mathcal{V}(U) \to \mathrm{Der}\left(C^\infty(U)\right), \quad \Phi(X)(f) = X.f$$

is injective and satisfies $\Phi([X, Y]) = [\Phi(X), \Phi(Y)]$. Therefore the Jacobi identity in $\mathcal{V}(U)$ follows from the Jacobi identity in the associative algebra $\mathrm{End}\left(C^\infty(U)\right)$. $\square$

For the applications to Lie groups we will need the following lemma.

Lemma I.14. *Let M and N be smooth manifolds and $\phi\colon M \to N$ a smooth map. Suppose that $X_N, Y_N \in \mathcal{V}(N)$ and $X_M, Y_M \in \mathcal{V}(M)$ satisfy*

$$X_N\left(\phi(p)\right) = d\phi(p).X_M(p) \quad \text{and} \quad Y_N\left(\phi(p)\right) = d\phi(p).Y_M(p)$$

for all $p \in M$, i.e., $X_N \circ \phi = T\phi \circ X_M$ and $Y_N \circ \phi = T\phi \circ Y_M$. Then $[X_N, Y_N] \circ \phi = T\phi \circ [X_M, Y_M]$.

Proof. It suffices to perform a local calculation. Therefore we may w.l.o.g. assume that $M \subseteq F$ is open, where F is a s.c.l.c. space and that N is an s.c.l.c. space. Then

$$[X_N, Y_N]\widetilde{\,}(\phi(p)) = d\tilde{Y}_N\left(\phi(p)\right).\tilde{X}_N\left(\phi(p)\right) - d\tilde{X}_N\left(\phi(p)\right).\tilde{Y}_N\left(\phi(p)\right).$$

Next we note that our assumption implies that $\tilde{Y}_N \circ \phi = d\phi \circ (\mathrm{id}_F \times \tilde{Y}_M)$. Using the chain rule we obtain

$$d\tilde{Y}_N\left(\phi(p)\right)d\phi(p) = d(d\phi)\left(p, \tilde{Y}_M(p)\right) \circ \left(\mathrm{id}_F, d\tilde{Y}_M(p)\right)$$

which, in view of Remark I.5(c), leads to

$$\begin{aligned}
d\tilde{Y}_N\left(\phi(p)\right).\tilde{X}_N\left(\phi(p)\right) &= d\tilde{Y}_N\left(\phi(p)\right)d\phi(p).\tilde{X}_M(p) \\
&= d(d\phi)\left(p, \tilde{Y}_M(p)\right) \circ \left(\mathrm{id}_F, d\tilde{Y}_M(p)\right).\tilde{X}_M(p) \\
&= d^2\phi(p)\left(\tilde{Y}_M(p), \tilde{X}_M(p)\right) + d\phi(p)\left(d\tilde{Y}_M(p).\tilde{X}_M(p)\right).
\end{aligned}$$

Now the symmetry of the second derivative (Lemma I.2(ii)) implies that

$$[X_N, Y_N]\widetilde{\,}(\phi(p)) = d\phi(p)\left(d\tilde{Y}_M(p).\tilde{X}_M(p) - d\tilde{X}_M(p).\tilde{Y}_M(p)\right) = d\phi(p)\left([X_M, Y_M]\widetilde{\,}(p)\right).$$
$\square$

5. Infinite-dimensional Lie groups

In this subsection we consider *s.c.l.c. Lie groups*, i.e., Lie groups modeled over s.c.l.c. spaces. Basically we follow [Mi83]. Throughout this subsection G denotes such a Lie group, i.e., G is a smooth manifold which is a group such that multiplication and inversion are smooth maps. For $g \in G$ we write $\lambda_g \colon G \to G, x \mapsto gx$ for the left-multiplication with g and $\rho_g \colon G \to G, x \mapsto xg$ for the right-multiplication with g. Both are diffeomorphisms of G. Moreover, we write $m \colon G \times G \to G, (x, y) \mapsto xy$ for the multiplication map and $\eta \colon G \to G, x \mapsto x^{-1}$ for the Inversion.

Lemma I.15. *Let $\mathfrak{g} := T_1(G)$ denote the tangent space in the identity. Then the mapping*

$$\Phi \colon G \times \mathfrak{g} \to TG, \quad (g, X) \mapsto d\lambda_g(1).X$$

is a diffeomorphism.

Proof. First we note that for a product of two smooth manifolds M and N we have a canonical diffeomorphism $T(M \times N) \to TM \times TN$. Since the multiplication map $m \colon G \times G \to G$ is smooth, the same holds for its tangent map

$$Tm \colon T(G \times G) \cong TG \times TG \to TG.$$

In view of Proposition I.4, $dm(g, 1)(0, X) = d\lambda_g(1).X$. Therefore the smoothness of Φ follows from $\Phi(g, X) = Tm(g, X)$ for $(g, X) \in G \times T_1(G) \subseteq T(G) \times T(G)$ and the fact that the restriction of Tm to $G \times T_1(G) \subseteq TG \times TG$ is smooth.

To see that Φ^{-1} is also smooth, let $\pi \colon TG \to G$ denote the canonical projection. Then

$$\Phi^{-1} \colon TG \to G \times \mathfrak{g}, \quad v \mapsto \left(\pi(v), d\lambda_{\pi(v)^{-1}}\big(\pi(v)\big).v\right).$$

The maps

$$\alpha \colon TG \to TG \times TG, \quad v \mapsto \big(\pi(v), v\big) \in G \times TG$$

and $\tilde{m} \colon G \times G \to G, (g_1, g_2) \mapsto g_1^{-1} g_2$ are smooth by the chain rule. Now

$$T(\tilde{m}) \circ \alpha(v) = T(\tilde{m})\big(\pi(v), v\big) = d_2\tilde{m}\big(\pi(v), \pi(v)\big).v = d\lambda_{\pi(v)^{-1}}\big(\pi(v)\big).v$$

shows that Φ^{-1} is smooth.   □

The essential consequence of Lemma I.15 is that the tangent bundle of a Lie group is trivial, so that we can identify $\mathcal{V}(G)$ with $C^\infty(G, \mathfrak{g})$. We write $\mathcal{V}(G)^l \subseteq \mathcal{V}(G)$ for the subspace of *left invariant* vector fields, i.e., of those satisfying

(1.3) $$X(g) = d\lambda_g(1).X(1)$$

for all $g \in G$. These are the vector fields that correspond to constant functions $G \to \mathfrak{g}$. We see in particular that each left invariant vector field is smooth, so that the mapping

$$\mathcal{V}(G)^l \to \mathfrak{g}, \quad X \mapsto X(1)$$

is a bijection. Moreover, Lemma I.14 implies that for $X, Y \in \mathcal{V}(G)^l$ we have

$$[X, Y](g) = d\lambda_g(1).[X, Y](1),$$

i.e., that $[X, Y] \in \mathcal{V}(G)^l$. Thus there exists a unique Lie bracket on $\mathfrak{g}$ satisfying

$$[X, Y](\mathbf{1}) = [X(\mathbf{1}), Y(\mathbf{1})]$$

for all left invariant vector fields on G.

Definition I.16. The Lie algebra $(\mathfrak{g}, [\cdot, \cdot])$ is called *the Lie algebra of G*. $\square$

Definition I.17. Let G be a Lie group. Then for each $g \in G$ the map $I_g: G \to G, x \mapsto gxg^{-1}$, is a smooth automorphism, hence induces a continuous linear automorphism

$$\mathrm{Ad}(g) := dI_g(\mathbf{1}): \mathfrak{g} \to \mathfrak{g}.$$

We thus obtain an action $G \times \mathfrak{g} \to \mathfrak{g}, (g, X) \mapsto \mathrm{Ad}(g).X$ called the *adjoint action* of G on $\mathfrak{g}$. $\square$

Proposition I.18. *For a Lie group G the following assertions hold:*
  (i) *$dm(g_1, g_2)(X_1, X_2) = d\rho_{g_2}(g_1).X_1 + d\lambda_{g_1}(g_2).X_2$ and in particular we have $dm(\mathbf{1}, \mathbf{1})(X_1, X_2) = X_1 + X_2$.*
  (ii) *$d\eta(\mathbf{1}).X = -X$.*
  (iii) *The mapping $Tm: TG \times TG \to TG$ defines a Lie group structure on TG with identity element $\Phi(\mathbf{1}, 0)$ and inversion $T\eta$. More explicitly multiplication and inversion are given by*

$$\Phi(g_1, X_1) \cdot \Phi(g_2, X_2) = \Phi\big(g_1 g_2, \mathrm{Ad}(g_2)^{-1}.X_1 + X_2\big)$$

  and $\Phi(g, X)^{-1} = \Phi\big(g^{-1}, - \mathrm{Ad}(g).X\big)$.
  (iv) *If $X_l: G \to TG$ is a left invariant vector field with $X_l(\mathbf{1}) = X$, then $X_r: g \mapsto -X_l(g)^{-1}$ is a right-invariant vector field with $X_r(\mathbf{1}) = X$. The assignment $\mathfrak{g} \to \mathcal{V}(G)^r, X \mapsto X_r$ is an antiisomorphism of Lie algebras.*
  (v) *If $\sigma: G \times M \to M$ is a smooth action of G on the smooth manifold M, then $T\sigma: TG \times TM \to TM$ is a smooth action of TG on TM. The assignment*

$$\dot{\sigma}: \mathfrak{g} \to \mathcal{V}(M), \quad with \quad \dot{\sigma}(X)(p) := -d\sigma(\mathbf{1}, p)(X, 0)$$

  defines a homomorphism of Lie algebras.

Proof. (i) In view of Proposition I.4, we have

$$dm(g_1, g_2)(X_1, X_2) = d_1 m(g_1, g_2)(X_1) + d_2 m(g_1, g_2)(X_2)$$
$$= d\rho_{g_2}(g_1).X_1 + d\lambda_{g_1}(g_2).X_2.$$

(ii) From $m \circ (\mathrm{id}_G \times \eta) = \mathbf{1}$, we derive $0 = dm(\mathbf{1}, \mathbf{1})\big(X, d\eta(\mathbf{1}).X\big) = X + d\eta(\mathbf{1}).X$ and hence the assertion.

(iii) Let $\varepsilon: G \to \{\mathbf{1}\}$ denote the constant map and $u: \{\mathbf{1}\} \to G$ the group morphism representing the identity element. Then the group axioms for G are encoded in the relations $m \circ (m \times \mathrm{id}) = m \circ (\mathrm{id} \times m)$ (associativity), $m \circ (\eta \times \mathrm{id}) = m \circ (\mathrm{id} \times \eta) = \varepsilon$ (inversion), and $m \circ (u \times \mathrm{id}) = m \circ (\mathrm{id} \times u) = \mathrm{id}$ (unit element). Using the functorial properties of T, we see that these properties carry over to the corresponding maps on TG and show that TG is a Lie group with multiplication Tm, inversion $T\eta$, and unit element $\Phi(\mathbf{1}, 0)$.

To derive an explicit formula for the multiplication in terms of the trivialization described in Lemma I.15, using (i), we calculate

$$\Phi(g_1, X_1) \cdot \Phi(g_2, X_2) = dm(g_1, g_2)(d\lambda_{g_1}(1).X_1, d\lambda_{g_2}(1).X_2)$$
$$= d\rho_{g_2}(g_1)d\lambda_{g_1}(1).X_1 + d\lambda_{g_1}(g_2)d\lambda_{g_2}(1).X_2$$
$$= d\lambda_{g_1 g_2}(1)(d\lambda_{g_2}^{-1}(g_2)d\rho_{g_2}(1).X_1 + X_2)$$
$$= \Phi(g_1 g_2, \mathrm{Ad}(g_2)^{-1}.X_1 + X_2).$$

The formula for the inversion follows directly from this formula.

(iv) In view of (ii) above, we have

$$X_r(g) = -d\eta(g^{-1}).X_l(g^{-1}) = -d\eta(g^{-1})d\lambda_{g^{-1}}(1).X = -d\rho_g(1)d\eta(1).X = d\rho_g(1).X$$

and this proves the first part. The second part follows from Lemma I.14 which shows that

$$[X_r, Y_r](g) = d\eta(g^{-1}).[X_l, Y_l](g^{-1}) = d\eta(g^{-1}).[X, Y]_l(g^{-1}) = -[X, Y]_r(g).$$

(v) That $T\sigma$ defines an action of TG on TM follows in the same way as in (iii) above by applying T to the commutative diagrams defining a group action.

For the second part we pick $p \in M$ and write $\phi_p : G \to M, g \mapsto g.p$ for the smooth orbit map of p. Then the equivariance of ϕ_p means that $\phi_p \circ \rho_g = \phi_{g.p}$. From that we derive

$$-d\phi_p(g).X_r(g) = -d\phi_p(g)d\rho_g(1).X = -d\phi_{g.p}(1).X = \dot\sigma(X)(g.p).$$

Therefore Lemma I.14 and (iv) imply that

$$\dot\sigma([X, Y])(p) = -d\phi_p(1)[X, Y]_r(1) = d\phi_p(1)[X_r, Y_r](1) = [\dot\sigma(X), \dot\sigma(Y)](p). \qquad \square$$

Remark I.19. If S is an s.c.l.c. semigroup, i.e., a manifold modeled over an s.c.l.c. space which is endowed with a smooth semigroup multiplication $m : S \times S \to S$, then Proposition I.18(iii) and (v) also hold in the following sense. The mapping $Tm : TS \times TS \to TS$ is an s.c.l.c. semigroup structure on the tangent bundle TS, and if $\sigma : M \times S \to M$ is a smooth right action of S on the manifold M, then $T\sigma : TM \times TS \to TM$ is a smooth right action of TS on the tangent bundle TM.
$\square$

II. Dual spaces of locally convex spaces

In the next section we will have to deal with topologies on function spaces which play a crucial role in representation theory. In this section we discuss the basic properties of the relevant topologies on the dual space of a locally convex space. In particular we discuss completeness of the dual space, metrizability, and the properties of the corresponding evaluation map $\eta : X \to X''$ given by $\eta(x)(\alpha) = \alpha(x)$.

Let X' denote the space of continuous linear functionals on the locally convex space X, *the topological dual.* If X^* denotes the set of all linear functionals $X \to \mathbb{C}$, then $X' \subseteq X^*$ is a subspace. There are several natural locally convex topologies on the space X'. We write X'_σ (X'_γ, X'_c, X'_b) for the space X' endowed with the *weak-$*$-topology*, i.e., the topology of pointwise convergence (the topology of uniform convergence on compact convex, compact, bounded subsets of X). The space X'_b is called the *strong dual*. Note that we have the following continuous bijections:

$$X'_b \to X'_c \to X'_\gamma \to X'_\sigma.$$

Before we turn to a closer investigation of the various dual spaces of locally convex spaces, we introduce an important class of locally convex spaces.

Definition II.1. Let X be a vector space which can be written as $X = \bigcup_{n=1}^\infty X_n$, where $X_n \subseteq X_{n+1}$ are subspaces of X which are endowed with the structures of locally convex spaces in such a way that the inclusion mappings $X_n \to X_{n+1}$ are topological embeddings. Then we obtain a locally convex vector topology on X by defining a seminorm p on X to be continuous if and only if its restriction to all the subspaces X_n is continuous. We call X the *strict inductive limit* of the spaces $(X_n)_{n\in\mathbb{N}}$. If, in addition, the spaces X_n are Fréchet spaces, then X is called an *LF space*. $\qquad\square$

A locally convex space X is called *barreled* if all lower semicontinuous seminorms on X are continuous. Geometrically this property can be interpreted as follows. A closed convex balanced subset of X is called a *barrel* if it is absorbing. Then X is barreled if and only if all barrels are 0-neighborhoods (cf. [He89, p.11]). Baire spaces are always barreled ([He89, Prop. 1.4.1]).

Proposition II.2. *If X is a strict inductive limit of the spaces $(X_n)_{n\in\mathbb{N}}$, then the following assertions hold:*
  (i) $X_n \hookrightarrow X$ *is an embedding.*
  (ii) *A linear map $f \colon X \to Y$, where Y is a locally convex space, is continuous if and only if its restriction to each X_n is continuous.*
If, in addition, all the spaces X_n are complete, then:
 (iii) *Each X_n is closed in X and X is quasicomplete.*
 (iv) *Any bounded subset of X is contained in some X_n.*
  (v) *If the X_n are Baire spaces, then X is Baire if and only if $X = X_n$ holds for some $n \in \mathbb{N}$.*
 (vi) *If X is an LF space, then X is complete and barreled.*

Proof. (i) [He89, Prop. 1.5.2]

(ii) This follows directly from the description of the topology by continuous seminorms.

(iii),(iv) [He89, Prop. 1.5.3]

(v) First we recall from (iii) that the subspaces X_n are closed. If $X \neq X_n$ holds for all $n \in \mathbb{N}$, then no X_n has an interior point. Therefore $X = \bigcup_{n=1}^\infty X_n$ shows that this cannot happen if X is a Baire space. If, conversely, $X = X_n$ for some $n \in \mathbb{N}$, then (i) implies that X is a Baire space.

(vi) For the completeness of X we refer to [Tr67, Th. 13.1]. Let p be a lower semi-continuous seminorm on X. Then the restrictions $p|_{X_n}$ are lower semicontinuous, hence continuous because Fréchet spaces are Baire spaces and therefore barreled. Thus p is continuous, and this shows that X is barreled. $\qquad\square$

1. Metrizability

It is well known that for a normed space the strong dual space X'_b is a Banach space, hence that the category of Banach spaces is closed under taking dual spaces. This changes drastically for Fréchet spaces as we will see in Corollary II.7 below.

Definition II.3. Let X be a topological vector space. A subset $K \subseteq X$ is called *precompact* if for each 0-neighborhood $U \subseteq X$ there exists a finite subset $F \subseteq K$ with $K \subseteq F + U$. Note that if $\overline{X}$ denotes the completion of X ([Tr67, Th. 5.2]), then the precompactness of a subset $K \subseteq X$ is equivalent to the relative compactness of K as a subset of $\overline{X}$ (cf. [Tr67, Prop. 6.9]). $\qquad\square$

Lemma II.4. *If V is a locally convex space and $K \subseteq V$ is a precompact set, then $\overline{\mathrm{conv}}(K)$ is precompact. If, in addition, V is quasicomplete, then $\overline{\mathrm{conv}}(K)$ is compact.*

Proof. First we use [Tr67, Prop. 7.11] to see that $\mathrm{conv}(K)$ and hence also $C := \overline{\mathrm{conv}}(K)$ is precompact (cf. [Tr67, Def. 6.3]). Further each precompact set is bounded. In fact, let U be a balanced convex 0-neighborhood in X. Then there exists a finite set $F \subseteq X$ with $C \subseteq F + U$ and $F \subseteq nU$ holds for some $n \in \mathbb{N}$, hence $C \subseteq nU + U \subseteq (n+1)U$. If V is quasicomplete, then the fact that C is closed and bounded implies that C is complete and therefore compact because it is precompact. $\qquad\square$

For a subset B of a locally convex space we define its *polar*

$$\hat{B} := \{\alpha \in X' : (\forall x \in B)|\alpha(x)| \leq 1\}$$

and for $C \subseteq X'$ we put

$$\hat{C} := \{\alpha \in X : (\forall \alpha \in C)|\alpha(x)| \leq 1\}.$$

We recall the following basic properties of polar sets. They show in particular that the assignments $B \mapsto \hat{B}$ and $C \mapsto \hat{C}$ are mutually inverse bijections from the set of closed convex balanced subsets of X onto the set of weak-$*$-closed convex balanced subsets of X'.

Lemma II.5.
   (a) $B \subseteq \hat{C}$ *if and only if* $C \subseteq \hat{B}$.
   (b) $B \subseteq \hat{\hat{B}}$ *and* $\hat{\hat{B}}$ *is the balanced convex closure of* B.
   (c) $C \subseteq \hat{\hat{C}}$ *and* $\hat{\hat{C}}$ *is the balanced convex weak-$*$-closure of* C.

(d) *A closed convex balanced subset $B \subseteq X$ is a barrel if and only if $\hat{B}$ is weak-$*$-bounded.*

(e) *A subset $B \subseteq X$ is bounded if and only if $\hat{B}$ is absorbing.*

(f) *If $B \subseteq X$ is compact and convex, then $\hat{B}$ is compact.*

Proof. (a) is trivial and (b), (c) are consequences of the Bipolar Theorem.

(d) B is a barrel if and only if it is absorbing. In view of $B = \hat{\hat{B}}$ this means that the function
$$\eta(x): \hat{B} \to \mathbb{C}, \quad \alpha \mapsto \alpha(x)$$
is bounded for each $x \in X$. This in turn means that $\hat{B}$ is weak-$*$-bounded.

(e) According to [He89, Prop. 1.4.2], a subset $B \subseteq X$ is bounded if and only if it is bounded for the weak topology on X which in turn is equivalent to the boundedness of all continuous linear functionals on B, i.e., that $\hat{B}$ is absorbing.

(f) If $B \subseteq X$ is a compact convex set, then [Bou87, Ch. IV, §1, no. 1, Rem. 1] shows that $\hat{\hat{B}}$ is compact. In fact, it is closed and contained in the convex hull of the sets $\pm 2iB, \pm 2B$ which is compact. □

Proposition II.6. *Let X be a locally convex Baire space. Then the following assertions hold:*
  (i) *X'_b is metrizable if and only if X is normable.*
  (ii) *X'_c and X'_σ are metrizable if and only if $\dim X < \infty$.*

Proof. If X is finite-dimensional, then $X'_\sigma = X'_c = X'_b$ is metrizable, and if X is normable, then X'_b is a Banach space and in particular metrizable.

(a) Suppose that X'_b is metrizable. Then there exists a countable basis $(U_n)_{n \in \mathbb{N}}$ of 0-neighborhoods in X'_b. The sets $\hat{B} \subseteq X'_b$ for $B \subseteq X$ bounded form a neighborhood basis for 0. Hence there exist bounded sets $B_n \subseteq X$ with $\hat{B}_n \subseteq U_n$.

Let $C_n := \hat{\hat{B}}_n$. Then $\hat{C}_n = \hat{B}_n$ shows that C_n is bounded because $\hat{C}_n$ is absorbing (Lemma II.5(e)). Let $x \in X$. Then the evaluation functional
$$\eta(x): X'_b \to \mathbb{C}, f \mapsto f(x)$$
is continuous, i.e., $\{\hat{x}\} = \{f \in X': |f(x)| \leq 1\}$ is a 0-neighborhood in X'. Thus we find $n \in \mathbb{N}$ with $\hat{B}_n \subseteq \{\hat{x}\}$. Now the Bipolar Theorem implies that $x \in \{\hat{x}\} \subseteq \hat{\hat{B}}_n = C_n$ and therefore $X = \bigcup_{n \in \mathbb{N}} C_n$. Since the sets C_n are closed, the fact that X is a Baire space implies that one of the sets C_n has interior points. Hence $C_n - C_n$ is a bounded neighborhood of 0 in X, and therefore X is normable (cf. [He89, p.3]).

(b) Assume that X'_c is metrizable. Then the same argument as above shows that there exists a compact subset $K \subseteq X$ such that $C := \hat{\hat{K}}$ has interior points. Since C coincides with the closed balanced convex hull of K (Lemma II.5(b)), it is a precompact subset of X (Lemma II.4). Hence $C - C$ is a precompact 0-neighborhood. Therefore X is normable in such a way that the balls are precompact. Now the balls in the completion $\overline{X}$ of X are compact and therefore $\dim X \leq \dim \overline{X} < \infty$.

(c) If X'_σ is metrizable, then similar arguments as in (b) show that there exists a finite subset $F \subseteq X$ such that $\hat{F}$ has interior points. But since span F is closed, it follows that $\hat{F} \subseteq$ span F, whence $\dim X = \dim$ span $\hat{F} < \infty$. $\qquad\square$

Corollary II.7. *If X is a Fréchet space, then X'_c is a Fréchet space if and only if $\dim X < \infty$.* $\qquad\square$

2. Semireflexivity

We recall that for a locally convex space X we have several natural topologies on the dual space leading to the following continuous bijections:

$$X'_b \xrightarrow{\ \alpha\ } X'_c \xrightarrow{\ \beta\ } X'_\gamma \xrightarrow{\ \gamma\ } X'_\sigma$$

which induce weak-$*$-continuous injective maps

$$(X'_\sigma)' \xrightarrow{\ \gamma'\ } (X'_\gamma)' \xrightarrow{\ \beta'\ } (X'_c)' \xrightarrow{\ \alpha'\ } (X'_b)'.$$

We write $\eta_\sigma \colon X \to (X'_\sigma)'$ for the evaluation map, and $\eta_\gamma := \gamma' \circ \eta_\sigma$, $\eta_c := \beta' \circ \eta_\gamma$, and $\eta_b := \alpha' \circ \eta_c$. The space X is called *semireflexive* if the map η_b is surjective, hence a bijection. Note that all these maps are injective with a weak-$*$-dense range.

Theorem II.8. *For a locally convex space the following assertions hold:*
  (i) *The maps η_σ and η_γ are bijections.*
  (ii) *If X is quasicomplete, then η_c is a bijection.*
  (iii) *If X is semireflexive, then X is quasicomplete for the original topology and the weak topology.*

Proof. (i) We show that η_γ is surjective. Then η_σ is also surjective because γ' is injective.

If $C \subseteq X$ is a compact convex set, then $\hat{C}$ is compact (Lemma II.5(f)). Hence the topology on X'_γ coincides with the topology of uniform convergence on balanced compact convex sets. If C is a balanced compact convex set, then C is also weakly compact and hence $\eta_\gamma(C) \subseteq (X'_\gamma)'$ is weak-$*$-compact. Each $\alpha \in (X'_\gamma)'$ is bounded on some set $\hat{C} \subseteq X'$, hence contained in some set of the type $\widehat{\eta_\gamma(C)} = \eta\eta_\gamma(\hat{C}) \subseteq \eta_\gamma(X)$ (Bipolar Theorem). This proves that $\eta_\gamma(X) = (X'_\gamma)'$.

(ii) If X is quasicomplete and $C \subseteq X$ is compact, then $\overline{\mathrm{conv}(C)}$ is compact (Lemma II.4). Therefore the mapping $\beta \colon X'_c \to X'_\gamma$ is a homeomorphism, i.e., $X'_c = X'_\gamma$. Since η_γ is bijective according to (i), the surjectivity of $\eta_c = \beta' \circ \eta_\gamma$ follows.

(iii) (cf. [He89, Th. 1.1.2(e)]) Let $C \subseteq X$ be closed balanced convex and bounded. Then C is also weakly closed, and therefore $\eta_b(C) \subseteq \eta_b(X) = (X'_b)'$ is a weak-$*$-closed convex balanced subset. Since $\widehat{\eta_b(C)} = \hat{C} \subseteq X'_b$ is a 0-neighborhood, the set $\eta_b(C)$ is weak-$*$-compact (Banach-Alaoglu Theorem). Hence C is weakly compact.

Now let $B \subseteq X$ be closed and bounded. Then its closed balanced convex hull C is also bounded, hence weakly compact and therefore in particular weakly complete. Further each Cauchy net in B for the original topology is a weak Cauchy net, hence converges weakly in B and therefore also in the strong topology because the closed convex neighborhoods of a point in X are also weakly closed. □

Proposition II.9. *Let X be a locally convex space.*

(i) *A subset $K \subseteq X'$ is equicontinuous if and only if its polar $\hat{K} \subseteq X$ is a 0-neighborhood in X.*

(ii) *If K is equicontinuous, then*

    (a) *K is weak-$*$-relatively compact.*

    (b) *K is relatively compact in X'_c.*

    (c) *K is strongly bounded.*

Furthermore (a), (b) or (c) implies that K is weak-$*$-bounded, i.e., $\hat{K} \subseteq X$ is a barrel. These properties are all equivalent if and only if X is barreled.

(iii) *If X is barrelled, then the following properties are equivalent for $K \subseteq X'$:*

    (a) *K is equicontinuous.*

    (b) *K is bounded for one of the topologies X'_σ, X'_γ, X'_c or X'_b.*

    (c) *K is relatively compact for one of the topologies X'_σ, X'_γ or X'_c.*

Proof. (i) This is more or less the definition of equicontinuity (cf. [Tr67, Prop. 32.7]).

(ii) ([He89, Th. 1.4.4]) If K is equicontinuous, then its balanced convex closure in the weak-$*$-topology of K has the same polar set $\hat{K} \subseteq X$ (Lemma II.5(c)). So we may w.l.o.g. assume that $K = \hat{\hat{K}}$. Since $\hat{K}$ is a 0-neighborhood in X, the weak-$*$-compactness of $K = \hat{\hat{K}}$ follows from the Banach-Alaoglu Theorem. Now the topology of compact convergence and the weak-$*$-topology coincide on K ([Tr67, Prop. 32.5]), so that K is also compact in X'_c. If $B \subseteq X$ is bounded, then there exists $n \in \mathbb{N}$ with $B \subseteq n\hat{K}$, i.e., $K \subseteq n\hat{B}$. Hence K is strongly bounded. It is clear that (a), (b) or (c) implies that K is weak-$*$-bounded.

The equivalence of the stated properties is equivalent to the assertion that if K is weakly bounded then K is equicontinuous, i.e., that the barrel $\hat{K}$ is a 0-neighborhood (Lemma II.5(d)). This is true if X is barreled, and if, conversely, X is not barreled and $B \subseteq X$ is a barrel which is not a 0-neighborhood, then its polar $\hat{B} \subseteq X'$ is weakly bounded but not equicontinuous.

(iii)(a) $\Rightarrow$ (b): If K is equicontinuous, then (ii) implies that K is bounded in X'_b, hence also in the spaces X'_σ, X'_γ and X'_c.

(b) $\Rightarrow$ (c): If (b) holds, then K is in particular bounded in X'_σ, i.e., weak-$*$-bounded. Hence (ii) shows that it is also relatively compact in X'_c. Thus it is also compact as a subset of X'_γ and X'_σ.

(c) $\Rightarrow$ (a): If K is relatively compact for one of the topologies X'_σ, X'_γ or X'_c, then it is in particular weak-$*$-relatively compact, hence weak-$*$-bounded. As we have seen in the preceding argument, this implies that K is equicontinuous. □

Lemma II.10. *For a locally convex space X the following assertions hold:*

(i) *The mapping $\eta_c \colon X \to (X'_c)'_c$ is an open map onto $\eta_c(X)$.*

(ii) *The mapping $\eta_b \colon X \to (X'_b)'_b$ is an open map onto $\eta_b(X)$.*

(iii) *If X is barreled, then the maps $\eta_c\colon X \to (X'_c)'_c$ and $\eta_b\colon X \to (X'_b)'_b$ are embeddings.*

Proof. (i) If $U \subseteq X$ is a closed convex balanced 0-neighborhood, then $\hat{U} \subseteq X'_c$ is closed and equicontinuous, hence compact in X'_c (Proposition II.9(ii)(b)). Therefore $\hat{\hat{U}} \subseteq (X'_c)'_c$ is a 0-neighborhood with $\hat{\hat{U}} \cap \eta_c(X) = \eta_c(U)$ (Bipolar Theorem). Thus η_c is open onto $\eta_c(X)$.

(ii) For a closed convex balanced 0-neighborhood $U \subseteq X$ the polar set $\hat{U} \subseteq X'$ is equicontinuous and therefore strongly bounded (Proposition II.9(ii)(c)). Thus $\hat{\hat{U}} \subseteq (X'_b)'_b$ is a 0-neighborhood with $\hat{\hat{U}} \cap \eta_b(X) = \eta_b(U)$. Therefore η_b is open onto $\eta_b(X)$.

(iii) Suppose that X is barreled. If $K \subseteq X'_c$ is compact or $K \subseteq X'_b$, then it is equicontinuous (Proposition II.9(iii)), and therefore $\hat{K} \subseteq X$ is a 0-neighborhood. Hence $\eta_c\colon X \to (X'_c)'_c$ and $\eta_b\colon X \to (X'_b)'_b$ are continuous maps. In view of (i) and (ii), this means that both are embeddings. $\qquad\square$

Theorem II.11. (Reflexivity criterion for the c-topologies) *If X is a quasicomplete barreled space, then $\eta_c\colon X \to (X'_c)'_c$ is an isomorphism of topological vector spaces. This holds in particular if X is an LF space.*

Proof. Since X is quasicomplete, the surjectivity of η_c follows from Theorem II.8(ii). If, in addition, X is barreled, then Lemma II.10(iii) shows that η_c is an isomorphism of topological vector spaces.

To see that the assertion holds for LF spaces, we recall from Proposition II.2(vi) that they are complete and barreled. $\qquad\square$

3. Completeness properties of the dual space

Now we turn to the question whether a dual space X' is complete with respect to a given topology. The following lemma is the topological background for the completeness criteria.

Proposition II.12. (i) *Let X be a topological space satisfying the first axiom of countability and V be a (sequentially) complete locally convex space. Then the space $C(X,V)_c$ of continuous maps $X \to V$ is a (sequentially) complete locally convex space with respect to the topology of uniform convergence on compact subsets of X.*

(ii) *If X is an LF space and V is a (sequentially) complete locally convex space, then the space $\mathcal{L}(X,V)_c$ of continuous linear maps endowed with the topology of uniform convergence on compact subsets of X is a (sequentially) complete locally convex space.*

(iii) *If X is a Baire space and V is an s.c.l.c. space, then the space $\mathcal{L}(X,V)$ is sequentially complete with respect to any topology of uniform convergence on a system of subsets of X whose union is X.*

Proof. (i) That $C(X,V)_c$ is a locally convex space follows from the fact that its topology is defined by the seminorms

$$p_K(f) := \sup\{p(f(x)) : x \in K\},$$

where $K \subseteq X$ is a compact subset and $p : V \to \mathbb{R}^+$ is a continuous seminorm.

Let $\mathcal{F}$ be a Cauchy-Filter in $C(X,V)_c$. Since V is complete, $\mathcal{F}$ converges pointwise to a function $f : X \to V$. We claim that $\mathcal{F}$ converges uniformly on each compact subset K of X. In fact, let p be a continuous seminorm on V and $\varepsilon > 0$. Then there exists $F \in \mathcal{F}$ with $p_K(g - h) \leq \varepsilon$ for all $g, h \in F$. Since $f(x) \in \overline{F(x)}$ holds for all $x \in K$, we conclude that $p_K(g - f) \leq \varepsilon$ for all $g \in F$. Hence $\mathcal{F} \to f$ holds uniformly on each compact subset $K \subseteq X$ and thus f is continuous on each compact subset of X.

If $(x_n)_{n \in \mathbb{N}}$ with $x_n \to x$ is a convergent sequence in X, then the set $\{x\} \cup \{x_n : n \in \mathbb{N}\}$ is compact. Since f is continuous on this set, it is continuous by our assumption on the space X. This proves that $C(X,V)_c$ is complete.

If V is sequentially complete, then similar arguments show that each Cauchy sequence in $C(X,V)_c$ converges, hence that $C(X,V)_c$ is sequentially complete.

(ii) ([Tr67, Cor. 32.2.4, p.345]) First we note that Fréchet spaces satisfy the assumption of (i). So let $(X_n)_{n \in \mathbb{N}}$ be a defining sequence for the topology on X. That $\mathcal{L}(X,V)_c$ is locally convex follows as in (i). If $\mathcal{F}$ is a Cauchy filter in $\mathcal{L}(X,V)_c$, then we see as in (i) that $\mathcal{F}$ converges pointwise to some function $f : X \to V$. Then f must be linear, and, in view of (i), f is continuous on each of the subspaces X_n, hence is continuous on X. This proves that $\mathcal{L}(X,V)_c$ is complete. If V is sequentially complete, then we see by a similar argument that $\mathcal{L}(X,V)_c$ is sequentially complete.

(iii) If $(f_n)_{n \in \mathbb{N}}$ is a Cauchy sequence in $\mathcal{L}(X,V)$ for the topology of uniform convergence on a system $\mathcal{S}$ of subsets of X whose union is X, then the sequential completeness of V implies that f_n converges pointwise to a linear function $f : X \to V$. It follows in particular that f is (G)-holomorphic. Therefore the continuity of f follows from Proposition I.9(iii). Since (f_n) is a Cauchy sequence for the topology of uniform convergence on the sets in $\mathcal{S}$, we see that $f_n \to f$ holds uniformly on sets in $\mathcal{S}$. This proves that $\mathcal{L}(X,V)$ is sequentially complete with respect to the topology of uniform convergence on sets in $\mathcal{S}$. $\square$

Corollary II.13. (a) *If X is an LF space, then X'_c is a complete locally convex space.*

(b) *If X is a Baire space, then X'_σ, X'_γ, X'_c, and X'_b are sequentially complete.* $\square$

Note that in general one cannot expect that the dual X' is complete with respect to the topology of pointwise convergence. With respect to this topology the embedding $X'_\sigma \hookrightarrow X^*$ is a dense embedding if X^* carries the topology of pointwise convergence. Therefore X'_σ is not complete unless $X' = X^*$, i.e., each linear functional on X is continuous. This holds in particular for the finest locally convex topology on X, i.e., the topology for which all seminorms are continuous, and also for the weak topology defined by X^*.

Lemma II.14. *If X'_σ is quasicomplete, then the same holds for X'_γ, X'_c and X'_b.*

Proof. If $B \subseteq X'$ is closed and bounded for one of the topologies X'_γ, X'_c or X'_b, then B is also weak-$*$-bounded. Let $\mathcal{F}$ be a Cauchy filter in B. Then $\mathcal{F}$ converges to some element α in the weak-$*$-closure of B. Then $\mathcal{F}$ also converges to α in the original topology, and we see that $\alpha \in B$. This shows that B is complete, i.e., that X'_γ, X'_c and X'_b are quasicomplete. $\qquad\square$

Proposition II.15. *If X is barreled or semireflexive, then the spaces X'_σ, X'_γ, X'_c, and X'_b are quasicomplete.*

Proof. First we assume that X is barreled. In view of Lemma II.14, it suffices to show that X'_σ is quasicomplete. Let $B \subseteq X'_\sigma$ be closed and bounded. Then $\hat{B}$ is a barrel (Lemma II.5(d)), hence a 0-neighborhood, and therefore Proposition II.9(ii) shows that B is weak-$*$-compact, hence in particular weak-$*$-complete.

If X is semireflexive, then X'_b is also semireflexive and therefore weakly quasicomplete ([He89, Th. 1.1.2(d)(e)] and Theorem II.8). Further $\eta_b(X) = (X'_b)'$, so that the weak topology on X'_b coincides with the weak-$*$-topology. Thus X'_σ is quasicomplete. $\qquad\square$

To clarify the relation between the assumptions in Proposition II.15, we note that a barreled space need not be semireflexive because there exist Banach spaces which are not reflexive. On the other hand one would not expect that the semireflexivity has strong implications for the topology on X because it only means that the map η_b is surjective. Nevertheless the following lemma shows that it has consequences for the strong dual.

Lemma II.16. *If X is semireflexive, then the strong dual X'_b is barreled. Furthermore the maps*

$$\tilde{\eta}_b \colon X'_b \to ((X'_b)'_b)'_b \quad and \quad \tilde{\eta}_c \colon X'_b \to ((X'_b)'_c)'_c$$

are topological isomorphisms.

Proof. Let $C \subseteq X'_b$ be a barrel. Then C is convex and closed in X'_b, hence also weakly closed. Thus $\eta_b(X) = (X'_b)'$ shows that C is also weak-$*$-closed, and the Bipolar Theorem gives $\hat{\hat{C}} = C$. But $\hat{C} \subseteq X$ is weakly bounded (Lemma II.5(e)), and so $\hat{C}$ is bounded which in turn implies that $C = \hat{\hat{C}}$ is a 0-neighborhood in X'_b. This proves that X'_b is barreled.

Moreover X'_b is semireflexive and quasicomplete ([He89, Th. 1.1.2(d)(e)]), so Theorem II.11 implies that $\tilde{\eta}_c$ is an isomorphism. Since X'_b is semireflexive and barreled, the assertion about $\tilde{\eta}_b$ follows from Lemma II.10(iii). $\qquad\square$

III. Topologies on function spaces

To construct and analyze representations of infinite-dimensional Lie groups and semigroups one often has to consider representations in spaces of smooth functions on G. So one has to endow these function spaces with a suitable (sequentially) complete locally convex topology. The importance of these spaces comes from the fact that for smooth representations a dense subspace of the representation space V can be embedded in $C^\infty(G, V)$.

First we discuss the space $C^\infty(M, V)$ of smooth functions on M with values in an s.c.l.c. space V and show that this space carries a natural s.c.l.c. topology which is, roughly stated, the topology of uniform convergence of all derivatives on compact sets. The main point here is to use the appropriate interpretation of the higher derivatives that permits inductive arguments. We also show that smooth Lie group actions lead to smooth actions on the corresponding spaces of smooth functions.

Next we show that smooth mappings between open subsets of s.c.l.c. spaces induce smooth mappings on the level of function spaces. This result is crucial to show that groups of the type $C^\infty(M, G)$, M a compact manifold and G a finite dimensional Lie group are in fact Lie groups modeled over Fréchet spaces in the sense specified in Section I (cf. [Ne99]).

Finally we turn to the space of holomorphic functions on a complex manifold M over a Baire s.c.l.c. space with values in a s.c.l.c. space V and show that it is sequentially complete with respect to the topology of uniform convergence on compact subsets and that holomorphic semigroup actions lead to holomorphic actions on the corresponding spaces of holomorphic functions. Here the assumption that M is modeled on a Baire space, an assumption which is in particular satisfied for Fréchet spaces, is crucial for the sequential completeness of the space of holomorphic functions on M.

1. The space $C^\infty(M, V)$

Let V be a (sequentially) complete locally convex space. If M is a smooth Fréchet manifold, then we write $C^\infty(M, V)_c$ for the space $C^\infty(M, V)$ endowed with the topology of compact convergence. This topology on $C^\infty(M, V)$ need not be complete. Nevertheless, the space $C(M, V)_c$ is (sequentially) complete by Proposition II.12(i).

For $f \in C^\infty(M, V)$ we obtain a smooth function $df \colon T(M) \to V$, where we identify $T_v(V)$ with V in each point $v \in V$, and inductively we get smooth functions $d^{(n)}f \colon T^{(n)}(M) \to V$. Thus we obtain an embedding

$$C^\infty(M, V) \to \prod_{n=0}^{\infty} C^\infty\big(T^{(n)}(M), V\big)_c.$$

We endow $C^\infty(M, V)$ with the topology induced by the product topology via this embedding (cf. [Th95]). Note that if $M = X$ is a vector space, then $X'_c \to C^\infty(X, \mathbb{C})$ is a topological embedding.

Proposition III.1. *If M is a Fréchet manifold and V is a (sequentially) complete locally convex space, then the space $C^\infty(M, V)$ is a (sequentially) complete locally convex space.*

Proof. Let $(f_i)_{i \in I}$ be a Cauchy net in $C^\infty(M, V)$. Then Proposition II.12(i) implies the existence of continuous functions $F_n \colon T^{(n)}(M) \to V$ such that $d^{(n)} f_i \to F_n$ holds uniformly on each compact subset of $T^{(n)}(M)$.

Next we show that $f \in C^1(M, V)$. To do this, we may w.l.o.g. assume that M is an open subset of a Fréchet space X. Then the uniform convergence of $df_i \to F_1$ on compact sets implies for each sufficiently small $t \neq 0$ that

$$\frac{1}{t}\big(f(x+th) - f(x)\big) = \lim_I \frac{1}{t}\big(f_i(x+th) - f_i(x)\big) = \lim_I \int_0^1 df_i(x+uth)(h)\ du$$

$$= \int_0^1 F_1(x+uth)(h)\ du.$$

Now the continuity of F_1 leads to

$$\lim_{t \to 0} \frac{1}{t}\big(f(x+th) - f(x)\big) = \lim_{t \to 0} \int_0^1 F_1(x+uth)(h)\ du = \int_0^1 F_1(x)(h)\ du = F_1(x)(h).$$

This proves that $f \in C^1(M, V)$ with $df = F_1$. By induction we now obtain $f \in C^n(M, V)$ and $d^{(n)} f = F_n$. Thus $f \in C^\infty(M, V)$ and $f_i \to f$ holds in $C^\infty(M, V)$.
□

Before we proceed, we need a topological lemma.

Lemma III.2. *Let M and N be Hausdorff spaces and V a locally convex space. Then the following assertions hold:*
  (i) *For $f \in C(M \times N, V)$ the map*

$$M \to C(N, V)_c, \quad x \mapsto \big(y \mapsto f(x, y)\big)$$

  is continuous.
  (ii) *If $\alpha \colon M \to N$ is continuous, then the map*

$$\alpha^* \colon C(N, V)_c \to C(M, V)_c, \quad f \mapsto f \circ \alpha$$

  is continuous.
  (iii) *Let S be a metrizable topological semigroup which acts continuously on M from the right. Then the action*

$$S \times C(M, V)_c \to C(M, V)_c, \quad (s, \phi) \mapsto \big(x \mapsto \phi(x.s)\big)$$

  is continuous.

Proof. (i) First we recall that the topology on $C(N, V)$ coincides with the compact open topology (cf. [Bou71, §3, no. 4, Th. 10]). Let $K \subseteq N$ be compact and $U \subseteq V$ be open. We write $W(K, U) := \{h \in C(N, V) \colon h(K) \subseteq U\}$ for the corresponding fundamental open subset of $C(N, Y)_c$. Suppose that $f_x \colon y \mapsto f(x, y)$ is contained in $W(K, U)$, i.e., $\{x\} \times K \subseteq f^{-1}(U)$. Since $f^{-1}(U)$ is an open subset of $M \times N$ and $\{x\} \times K \subseteq M \times N$ is compact, there exists an open neighborhood $O \subseteq M$ of

x such that $O \times K \subseteq f^{-1}(U)$. This means that $x \in O \subseteq \{p \in M : f_p \in W(K, U)\}$ which proves the assertion.

(ii) Let $K \subseteq M$ be compact, p a continuous seminorm on V, and $p_K(f) :=$ $\sup\{p(f(x)) : x \in K\}$ the corresponding seminorm on $C(M, V)_c$. These seminorms define the topology on this space. Now the set $\alpha(K)$ is compact and $p_K(\alpha^* f) \leq$ $p_{\alpha(K)}(f)$ shows that the seminorms $p_K \circ \alpha^*$ are continuous for each choice of p and K, hence that α^* is continuous.

(iii) Let $s_n \to s$, $f_i \to f$ in $C(M, V)_c$, $K \subseteq M$ a compact subset, and p a continuous seminorm on V. Then the closure $\tilde{K}$ of the set $\bigcup_{n=1}^\infty K.s_n$ is compact because it is the image of the compact set $K \times \{s, s_n : n \in \mathbb{N}\}$ under the action map. For $x \in K$ we have

$$p\big((s_n.f_i)(x) - (s.f)(x)\big) = p\big(f_i(x.s_n) - f(x.s)\big)$$
$$\leq p\big(f_i(x.s_n) - f(x.s_n)\big) + p\big(f(x.s_n) - f(x.s)\big)$$
$$\leq p_{\tilde{K}}(f_i - f) + p\big(f(x.s_n) - f(x.s)\big).$$

Therefore the uniform continuity of f on $\tilde{K}$ implies that $p_K(s_n.f_i - s.f) \to 0$. Hence $s_n.f_i \to s.f$ in $C(M, V)_c$. Thus the action of S on $C(M, V)_c$ is continuous. $\qquad \square$

In the following lemma the assumption that M is Fréchet is made to insure that the space $C^\infty(M, V)$ is sequentially complete (Proposition III.1), a property needed to make calculus work (cf. Section I).

Lemma III.3. (i) *Let* $\alpha : M \to N$ *be a smooth map between Fréchet manifolds. Then the linear map*

$$\alpha^* : C^\infty(N, V) \to C^\infty(M, V), \quad f \mapsto f \circ \alpha$$

is continuous.

(ii) *Let* M *be a Fréchet manifold and* $\pi_M : TM \to M$ *the canonical projection. Then the assignment*

$$C^\infty(M, V) \to C^\infty(TM, TV) \cong C^\infty(TM, V)^2, \quad f \mapsto Tf = (f \circ \pi_M, df)$$

is an embedding of locally convex spaces.

Proof. (i) (cf. [Th95, Prop. 3]) For $f \in C^\infty(N, V)$ we have $d(f \circ \alpha) = df \circ T\alpha$ and inductively $d^{(n)}(f \circ \alpha) = d^{(n)} f \circ T^{(n)} \alpha$. Therefore the continuity of α^* follows from Lemma III.2(ii).

(ii) Since $d^{(n)} df = d^{(n+1)} f$ for $n \in \mathbb{N}$, it is clear that the map $C^\infty(M, V) \to$ $C^\infty(TM, V), f \mapsto df$ is continuous. Since $C^\infty(M, V) \to C^\infty(TM, V), f \mapsto f \circ \pi_M$ is continuous according to (i), we see that $f \mapsto Tf$ is continuous.

If $\alpha : M \to TM$ is the natural embedding as the 0-section, then $(f \circ \pi_M) \circ \alpha =$ f. Therefore (i) shows that the inverse $Tf \to f$ is also continuous. This proves that $f \mapsto Tf$ is an embedding. $\qquad \square$

In many applications the following theorem is a very efficient tool.

Theorem III.4. *Let M and N be Fréchet manifolds, $f \in C^\infty(M \times N, V)$, and $f_x(y) := f(x, y)$. Then the map*

$$\Phi : M \to C^\infty(N, V), \quad x \mapsto f_x$$

is smooth.

Proof. We prove the theorem in several steps. First we note that w.l.o.g. we may assume that M is an open subset of a Fréchet space X.

Claim 1: Φ is continuous. We have $(d^{(n)} f_x)(y) = d^{(n)} f(x, y)$. Therefore

$$T^{(n)}(M \times N) \cong T^{(n)} M \times T^{(n)} N$$

and Lemma III.2(i) show that

$$M \to C(T^{(n)} N, V)_c, \quad x \mapsto d^{(n)} f_x$$

is continuous. In view of the definition of the topology on $C^\infty(N, V)$, this proves that Φ is continuous.

Claim 2: The map

$$\Psi : M \times X \to C^\infty(N, V), \quad (x, h) \mapsto \big(y \mapsto d_1 f(x, y)(h)\big)$$

is continuous. This follows from Claim 1 and the fact that $d_1 f \in C^\infty(M \times X \times N, V)$ (cf. Lemma I.5(c)).

Claim 3: Φ is C^1 with $d\Phi(x)(h) = \Psi(x, h)$. First we note that for a sufficiently small $\varepsilon > 0$ the map

$$]-\varepsilon, \varepsilon[\times [0, 1] \times M \times X \to C^\infty(N, V), \quad (t, u, x, h) \mapsto \Psi(x + uth, h)$$

is continuous by Claim 2. Therefore

$$]-\varepsilon, \varepsilon[\times M \times X \to C^\infty(N, V), \quad (t, x, h) \mapsto \int_0^1 \Psi(x + uth, h)\, du$$

is continuous and so

$$\lim_{t \to 0} \frac{1}{t}\big(\Phi(x + th) - \Phi(x)\big) = \lim_{t \to 0} \int_0^1 \Psi(x + uth, h)\, du = \int_0^1 \Psi(x, h)\, du = \Psi(x, h).$$

Thus $d\Phi(x)(h) = \Psi(x, h)$, and the continuity of Ψ implies that Φ is C^1.

Claim 4: Φ is smooth. Since $\Psi(x, h)(y) = d_1 f(x, y)(h)$ and

$$d_1 f \in C^\infty(M \times X \times N, V),$$

Claim 3 implies that $\Psi \in C^1$, hence that $\Phi \in C^2$. Proceeding inductively, we see that Φ is C^∞. $\qquad\square$

In the following we call a Fréchet manifold S endowed with a smooth associative multiplication $S \times S \to S$ a *Fréchet semigroup*.

Theorem III.5. *If M is a Fréchet manifold and the Fréchet semigroup S acts smoothly on M via $\sigma : M \times S \to M$, then the action map $\tilde{\sigma} : S \times C^\infty(M, V) \to C^\infty(M, V)$ given by $(s.f)(x) := f(x.s)$ is smooth.*

Proof. The partial derivative $d_2\tilde{\sigma}$ with respect to the second argument is given by

$$d_2\tilde{\sigma}(f,s)(h) = s.h = \tilde{\sigma}(s,h)$$

because the linear mappings $f \mapsto s.f$ are continuous (Lemma III.3). To see that this maping is continuous means to show that the action of S on $C^\infty(M,V)$ is continuous. We recall that we have defined the topology on $C^\infty(M,V)$ via the embedding

$$C^\infty(M,V) \to \prod_{n=0}^{\infty} C^\infty\big(T^{(n)}(M),V\big)_c.$$

Therefore it suffices to prove the continuity of the action map for S on the spaces

$$C^\infty\big(T^{(n)}(M),V\big)_c.$$

This action comes from the action of S on the manifold $T^{(n)}(M)$. The natural map

$$T^{(n)}\sigma \colon T^{(n)}(M \times S) \to T^{(n)}(M)$$

is smooth. Comparing with the injection

$$T^{(n)}(M) \times S \hookrightarrow T^{(n)}(M) \times T^{(n)}(S) \cong T^{(n)}(M \times S),$$

we see that the action of S on $T^{(n)}(M)$ is smooth and in particular continuous. So the continuity of the action of S on $C^\infty\big(T^{(n)}(M),V\big)_c$ follows from Lemma III.2(iii).

Now we turn to the first partial derivative $d_1\tilde{\sigma}$. We write $\pi_S \colon TS \to S$ and $\pi_M \colon TM \to M$ for the canonical projections, $\phi_x \colon S \to M$, $s \mapsto x.s$ for the orbit map of $x \in M$, and $\rho_s \colon M \to M$, $x \mapsto x.s$ for the translation maps on M. For each f the smoothness of the map $s \mapsto s.f$ follows from the smoothness of the function $(s,x) \mapsto f(x.s) = (f \circ \sigma)(x,s)$ on $S \times M$ and Theorem III.4 which also implies that $d_1\tilde{\sigma}(s,f).v = d_2(f \circ \sigma)(x,s).v$. To see that the partial derivative

$$d_1\tilde{\sigma} \colon TS \times C^\infty(M,V) \to C^\infty(M,V)$$

is continuous, we will use the embedding $C^\infty(M,V) \to C^\infty(TM,TV)$, $f \mapsto Tf$ from Lemma III.3(ii). According to Remark I.19, the smooth action $\sigma \colon M \times S \to M$ induces a smooth right action $T\sigma \colon TM \times TS \to TM$ so that the first part of the proof shows that the induced action map

$$TS \times C^\infty(TM,V) \to C^\infty(TM,V)$$

is continuous. If $\alpha \colon M \to TM$ is the 0-section, then we conclude with Lemma III.3(i) that the map

$$(v,f) \mapsto (v,T.f) \mapsto v.Tf = Tf \circ T\sigma(\cdot,v) = T(f \circ \sigma)(\cdot,v)$$
$$\mapsto T(f \circ \sigma)(\cdot,v) \circ \alpha \mapsto d(f \circ \sigma)(\cdot,v) \circ \alpha$$

from $TS \times C^\infty(M,V) \to C^\infty(M,V)$ is continuous. Now

$$d(f \circ \sigma)(\cdot,v) \circ \alpha(x) = d(f \circ \sigma)(x,v) = d_2(f \circ \sigma)\big(x,\pi(v)\big).v = d_1\tilde{\sigma}\big(\pi(v),f\big).v$$

shows that $d_1\tilde{\sigma}$ is continuous.

We have shown that $d_1\tilde{\sigma}$ and $d_2\tilde{\sigma}$ are continuous, so that Proposition I.4 implies that $d\tilde{\sigma}$ exists and is continuous, i.e., $\tilde{\sigma} \in C^1$ with

$$d\tilde{\sigma}(s, f)(v, h) = d_1\tilde{\sigma}(\pi(v), f).v + s.h.$$

The fact that $\tilde{\sigma}$ is C^1 implies in particular that $d_2\tilde{\sigma}$ is C^1 and since $d_1\tilde{\sigma}$ comes from the smooth action of TS on $C^\infty(TM, V)$, we conclude that this action is a C^1 map. But then $\tilde{\sigma}$ is C^2. Proceeding inductively we see that $\tilde{\sigma}$ is a smooth map. $\square$

2. Smooth mappings between function spaces

In the preceding subsection we have seen how to topologise the space $C^\infty(M, V)$ of smooth functions on a Fréchet manifold M with values in an s.c.l.c. space. Let X and Y be s.c.l.c. spaces, $U \subseteq X$ an open subset, M a compact manifold, and $f: M \times U \to Y$ a smooth map. Then $C^\infty(M, U)$ is an open subset of the s.c.l.c. space $C^\infty(M, X)$, and

$$f_*: C^\infty(M, U) \to C^\infty(M, Y), \quad \gamma \mapsto f \circ (\mathrm{id}_M, \gamma)$$

is a well defined map. We will show that this map is smooth. First we consider a purely topological situation:

Lemma III.6. *If M is a topological space and $f: M \times U \to Y$ continuous, then the mapping*

$$f_*: C(M, U)_c \to C(M, Y)_c, \quad \gamma \mapsto f \circ (\mathrm{id}_M, \gamma)$$

is continuous.

Proof. First we recall that the topology of uniform convergence coincides with the compact open topology (cf. [Bou71, §3, no. 4, Th. 10]). Let $K \subseteq M$ be compact and $V \subseteq Y$ be open. We write $W(K, V) := \{h \in C(M, Y): h(K) \subseteq V\}$ for the corresponding fundamental open subset of $C(M, Y)_c$. Then

$$f_*^{-1}(W(K, V)) = \{\gamma \in C(M, U): (\mathrm{id}_M, \gamma)(K) \subseteq f^{-1}(V)\}.$$

To see that this set is open in the compact open topology, let γ_0 be contained in this set and choose for each $x \in K$ a compact neighborhood K_x of x in K and an open neighborhood $U_x \subseteq U$ of $\gamma_0(x)$ such that $\gamma_0(K_x) \subseteq U_x$ and $K_x \times U_x \subseteq f^{-1}(V)$. Then we find finitely many points $x_1, \ldots, x_n \in K$ such that the K_{x_j} cover K. Now each $\gamma \in C(M, U)$ with $\gamma(K_{x_j}) \subseteq U_{x_j}$ satisfies $(\mathrm{id}_M, \gamma)(K_{x_j}) \subseteq K_{x_j} \times U_{x_j} \subseteq f^{-1}(V)$. Hence

$$\bigcap_{j=1}^n W(K_{x_j}, U_{x_j}) \subseteq (f_*)^{-1}(W(K, V))$$

proves the continuity of f_*. $\square$

Proposition III.7. *The map*

$$f_*: C^\infty(M, U) \to C^\infty(M, Y), \quad \gamma \mapsto f \circ (\mathrm{id}_M, \gamma)$$

is smooth.

Proof. First we show that f_* is continuous. For $\gamma \in C^\infty(M, X)$ the mapping $T\gamma \colon T(M) \to T(X) \cong X \times X$ can be split as $T\gamma(v_p) = (\gamma(p), d\gamma(p).v_p)$, where $d\gamma \in C^\infty(T(M), X)$. Inductively we obtain $d^{(n)}\gamma \in C^\infty(T^{(n)}M, X)$. In this sense $C^\infty(M, X)$ carries the topology induced by the embedding

$$C^\infty(M, X) \hookrightarrow \prod_{n=0}^\infty C^\infty\left(T^{(n)}(M), X\right)_c,$$

where the spaces on the right hand side carry the topology of uniform convergence on compact sets. We have

$$T(f_*\gamma) = T\left(f \circ (\mathrm{id}_M, \gamma)\right) = Tf \circ \left(\mathrm{id}_{TM}, T\gamma\right)$$

and thus $d(f_*\gamma) = df \circ \left(\mathrm{id}_{TM}, T\gamma\right)$. Inductively we obtain

(3.1) $$d^{(n)}(f_*\gamma) = d^{(n)}f \circ \left(\mathrm{id}_{T^{(n)}M}, T^{(n)}\gamma\right).$$

In view of Lemma III.6, this shows that the maps $\gamma \to d^{(n)}(f_*\gamma)$ are continuous. We conclude that f_* is continuous.

Next we calculate the derivative of f_*. For each $x \in M$ we have

$$\lim_{h \to 0} \frac{1}{h}\left(f\left(x, (\gamma + h\eta)(x)\right) - f\left(x, \gamma(x)\right)\right)$$

$$= \lim_{h \to 0} \int_0^1 d_2 f\left(x, (\gamma + uh\eta)(x)\right)(\eta(x))\, dx = df_2\left(x, \gamma(x)\right)(\eta(x)),$$

where, in view of the continuity of the integrand, the limit on the left hand side exists uniformly on compact subsets of M. In view of (3.1), the same argument applies to the higher derivatives $d^{(n)}f_*$. So we see that $(df_*)(\gamma, \eta)$ exists and equals $d_2 f \circ (\mathrm{id}_M, \gamma, \eta) \in C^\infty(M, Y)$. This means that $d(f_*) = (d_2 f)_* \colon C^\infty(M, TU) \to C^\infty(M, Y)$. Using the first part of our proof, we now see that $d(f_*)$ is continuous, i.e., f_* is C^1. Since the map $d(f_*)$ can be written as $(d_2 f)_*$, it has the same structure as f_*, and iteration of the argument shows that f_* is smooth. $\quad\square$

Corollary III.8. *If $f \colon U \to Y$ is a smooth map, then*

$$f_* \colon C^\infty(M, U) \to C^\infty(M, Y), \quad \gamma \mapsto f \circ \gamma$$

is smooth.

Proof. Put $\tilde{f}(x, y) := f(y)$ and apply Proposition III.7. $\quad\square$

3. Applications to groups of continuous mappings

Remark III.9. (a) If F is an s.c.l.c. space and X a compact metric space, then $C(X, F)_c$ is an s.c.l.c. space with respect to the topology of uniform convergence (Propositition II.12(a)).

(b) If $U \subseteq F$ is an open subset, then $C(X, U)$ is an open subset of $C(X, F)_c$. Now let $U_j \subseteq F_j$, $j = 1, 2$, be open subsets of s.c.l.c. spaces and $\phi \colon U_1 \to U_2$ a smooth map. We consider the map

$$\phi_X \colon C(X, U_1) \to C(X, U_2), \quad \gamma \mapsto \phi \circ \gamma.$$

Then ϕ_X is smooth. The continuity follows from Lemma III.6. For each $x \in X$ and $\gamma, \eta \in C(X, F_1)$ we have

$$\lim_{t \to 0} \frac{\phi(\gamma(x) + t\eta(x)) - \phi(\gamma(x))}{t} = \lim_{t \to 0} \int_0^1 d\phi(\gamma(x) + st\eta(x)).\eta(x)\, ds$$

$$= d\phi(\gamma(x)).\eta(x).$$

Since the integrand is continuous in $[0, 1]^2 \times X$, the limit exists uniformly in X, hence in the space $C(X, F_2)$. Therefore $d\phi_X(\gamma)(\eta)$ exists. Since $d\phi \colon TU_1 \cong U_1 \times F_1 \to F_2$ is a continuous map, the first part of the proof shows that

$$d\phi_X \colon C(X, TU_1) \cong C(X, U_1) \times C(X, F_1) \to C(X, F_2)$$

is continuous, so that ϕ_X is C^1. Iterating this argument shows that ϕ_X is C^∞. $\square$

Proposition III.10. *If G is a Lie group and X is a compact metric space, then $C(X, G)_c$ is a Lie group with Lie algebra $C(X, \mathfrak{g})_c$.*

Proof. We use Remark III.9(b) to see that the inversion and multiplication in the canonical local charts are smooth. The remainder is a routine verification. $\square$

4. Spaces of holomorphic functions

In this subsection we turn to spaces of holomorphic functions. In particular we show that holomorphic actions of complex Fréchet semigroups lead to holomorphic actions on the corresponding spaces of holomorphic functions, and that the inclusion $\mathrm{Hol}(M, V) \to C^\infty(M, V)$ is an embedding if $\mathrm{Hol}(M, V)$ carries the topology of uniform convergence on compact subsets. For refined investigations on topologies on spaces of holomorphic functions between Banach spaces we refer to [Na69].

In the following a *Baire manifold* is a manifold modeled over a s.c.l.c. Baire space.

Theorem III.11. *For a complex Baire manifold M the following assertions hold:*

(i) *If V is an s.c.l.c. space, then $\mathrm{Hol}(M, V)$ is s.c.l.c. with respect to the topology of uniform convergence on compact sets.*

(ii) *If, in addition, M is Fréchet and V is complete, then $\mathrm{Hol}(M, V)$ is complete.*

Proof. (i) Let $(f_n)_{n \in \mathbb{N}}$ be a Cauchy sequence in $\mathrm{Hol}(M, V)$. Since V is sequentially complete, this sequence converges uniformly on compact subsets of M to a function $f \colon M \to V$ (see the proof of Proposition II.12). It remains to show that f is holomorphic. For that we may w.l.o.g. assume that M is an open subset of a Baire space X. Since (G)-holomorphy is equivalent to weak (G)-holomorphy ([He89, Th. 2.1.3]), and for each $\alpha \in V'$ the function $\alpha \circ f \colon M \to \mathbb{C}$ is holomorphic on the intersection with each finite dimensional affine subspace, we see that $f \in \mathcal{G}(M, V)$. Now Proposition I.9(iii) implies that f is continuous, hence that f is (F)-holomorphic and therefore holomorphic (Proposition I.10).

(ii) (cf. [He71, p.79]) In view of Proposition II.12(i), it suffices to show that $\mathrm{Hol}(M, V)$ is closed in $C(M, V)_c$ because the latter space is complete. Suppose that $f_i \to f$, where f is continuous and the functions $f_i \colon M \to V$ are holomorphic. We have to show that f is holomorphic and, as in (i), we may w.l.o.g. assume that M is an open subset of a Fréchet space X. An argument similar to that in (i) implies that f is (G)-holomorphic, but then the continuity of f shows that $f \in \mathrm{Hol}(M, V)$. $\qquad\square$

Corollary III.12. *Let M and N be complex manifolds, where M is Fréchet. We write $\mathrm{Hol}(M, N)_c$ for the set of holomorphic maps $M \to N$ endowed with the compact open topology. Then the subspace $\mathrm{Hol}(M, N)_c$ is closed in $C(M, N)_c$.*

Proof. Since M is Fréchet, it is first countable, and therefore $C(M, N)_c$ is a complete uniform space. Now let $f \in C(M, N)$ and assume that $f_i \to f$ holds for $f_i \in \mathrm{Hol}(M, N)$ uniformly on compact subsets of M. We have to show that f is holomorphic. This is a local property, so that we may assume that M is an open subset of a Fréchet space F. In view of the continuity of f, it suffices to show that f is Gateaux-holomorphic, so that we may even assume that M is one-dimensional, hence locally compact (Proposition I.9). Let $x_0 \in M$ and fix a compact neighborhood K of x_0 and an open neighborhood $U \subseteq N$ of $f(x_0)$ which is diffeomorphic to an open subset of an s.c.l.c. space V. Then we may w.l.o.g. assume that $f_i(K) \subseteq U$ holds for all i, so that the same argument as in the proof of Theorem III.11(i) shows that f is holomorphic in a neighborhood of x_0. $\qquad\square$

In the following the assumption that the manifolds under consideration are Baire is made to ensure that the spaces $\mathrm{Hol}(M, V)$ are sequentially complete (Theorem III.11(i)).

Proposition III.13. *Let M and N be complex Baire manifolds, $f \colon M \times N \to V$ holomorphic, and $f_x(y) := f(x, y)$. Then the map*

$$\Phi \colon M \to \mathrm{Hol}(N, V), \quad x \mapsto f_x$$

is holomorphic.

Proof. First the continuity of the map Φ follows from Lemma III.2(i). Next we note that we may w.l.o.g. assume that M is an open subset of a Baire space X.

Claim 1: The map

$$\Psi \colon M \times X \to \mathrm{Hol}(N, V), \quad (x, h) \mapsto \big(y \mapsto d_1 f(x, y)(h)\big)$$

is continuous. This follows from Lemma III.2(i) and the fact that

$$d_1 f \in \mathrm{Hol}(M \times X \times N, V)$$

(Remark I.5(d)).

Claim 2: Φ is C^1 with $d\Phi(x)(h) = \Psi(x, h)$. This is proved exactly as the corresponding assertion in the proof of Theorem III.4.

This shows that Φ is C^1 with complex linear differentials, i.e., that Φ is holomorphic. $\qquad\square$

Theorem III.14. *Let M be a complex Baire manifold, S a complex Fréchet semigroup, and $M \times S \to M$ a holomorphic right action. Then the action*

$$S \times \mathrm{Hol}(M,V) \to \mathrm{Hol}(M,V)$$

with $\big(\pi(s).f\big)(x) = f(x.s)$ is holomorphic.

Proof. According to Lemma III.2(iii), the action of S on $\mathrm{Hol}(M,V) \subseteq C(M,V)_c$ is continuous.

For each $s \in S$ the map $\mathrm{Hol}(M,V) \to \mathrm{Hol}(M,V), f \mapsto s.f$ is continuous linear, hence holomorphic. Now let $f \in \mathrm{Hol}(M,V)$. Then the function defined by $\tilde{f}(s,x) \mapsto f(x.s)$ is in $\mathrm{Hol}(S \times M,V)$. Hence the holomorphy of $S \to \mathrm{Hol}(M,V), s \mapsto s.f = \tilde{f}_s$ follows from Proposition III.13. This proves that the action map is partially holomorphic in each argument. Now [He89, Prop. 2.3.8] implies that the action map is (G)-holomorphic, and finally the continuity implies that it is (F)-holomorphic, i.e., holomorphic (Proposition I.10). $\qquad\square$

We have already seen in Proposition I.10 that holomorphic functions are in particular smooth, i.e., that $\mathrm{Hol}(M,V) \subseteq C^\infty(M,V)$ holds for each complex manifold M. We have endowed the space $\mathrm{Hol}(M,V)$ with the topology of compact convergence which could be coarser than the topology induced from $C^\infty(M,V)$ but it turns out that on $\mathrm{Hol}(M,V)$ the latter topology coincides with the original one.

Proposition III.15. *If M is manifold modeled over a s.c.l.c. space, then the inclusion $\mathrm{Hol}(M,V) \hookrightarrow C^\infty(M,V)$ is an embedding of locally convex spaces.*

Proof. It is clear that the topology $\mathrm{Hol}(M,V)$ inherits from $C^\infty(M,V)$ is finer than the original one. Therefore it suffices to show that the inclusion map is continuous. If f is holomorphic, then $df : TM \to V$ is also holomorphic. Therefore it suffices to show that $\mathrm{Hol}(M,V) \to \mathrm{Hol}(TM,V), f \mapsto df$ is a continuous map. Then the assertion follows by induction.

Since each compact subset of TM is the union of finitely many pieces lying in coordinate neighborhoods, we may w.l.o.g. assume that M is an open subset of the s.c.l.c. space X. Let $x \in M$ and $h \in X$ with $x + zh \in M$ whenever $|z| \le 1$. Then

$$df(x)(h) = \frac{1}{2\pi} \int_0^{2\pi} e^{-i\theta} f(x + e^{i\theta}h) \, d\theta.$$

For each continuous seminorm p on V we therefore have

$$p\big(df(x)(h)\big) \le \sup_{|z|=1} p\big(f(x + zh)\big).$$

Let $K \subseteq TM \cong M \times X$ be a compact subset and w.l.o.g. $K = K_1 \times K_2$ with $K_1 \subseteq M$ and $K_2 \subseteq X$ compact and balanced. Then we find a balanced 0-neighborhood $V \subseteq X$ with $K_1 + V \subseteq M$ and $n \in \mathbb{N}$ with $K_2 \subseteq nV$. This means that for $(x, h) \in K$ we have $x + z\frac{h}{n} \in M$ whenever $|z| \leq 1$. Hence

$$p\big(df(x)(h)\big) = np\big(df(x)(\frac{h}{n})\big) \leq n \sup_{h \in \frac{1}{n}K_2} p\big(f(x + h)\big),$$

i.e., $p_K(df) \leq np_{K_1 + \frac{1}{n}K_2}(f)$. Since the set $K_1 + \frac{1}{n}K_2$ is compact, convergence in $\mathrm{Hol}(M, V)$ implies uniform convergence on this set, hence uniform convergence of df on K. This completes the proof.   □

One of the main features of the representation theory of finite-dimensional Lie groups is that they have an exponential function which makes it possible to translate analytic problems on a Lie group G to algebraic problems on $\mathfrak{g}$ without loosing too much information. This works in particular quite well for representations with analytic or holomorphic orbit mappings. To obtain a suitable generalization to the infinite-dimensional setting, let us say that a smooth function $\exp\colon \mathfrak{g} \to G$ is an *exponential function for G* if for each $X \in \mathfrak{g}$ the curve $\gamma_X\colon t \mapsto \exp(tX)$ is an integral curve of the corresponding left invariant vector field $\tilde{X} \in \mathcal{V}(G)$. Further we say that a Lie group G modeled over the s.c.l.c. space $\mathfrak{g}$ has a *good exponential function* if the closure $\overline{\exp(\mathfrak{g})}$ of the exponential image contains a neighborhood of the identity. If G is complex, we require, in addition, that the exponential function $\exp\colon \mathfrak{g} \to G$ is a holomorphic map. For a discussion of the exponential function for the class of regular Lie groups we refer to [KM97a]. We write $D_l(G) \subseteq \mathrm{End}\,\big(C^\infty(G)\big)$ for the unital algebra of all operators on $C^\infty(G)$ generated by the action of the left invariant vector fields. An element $D \in D_l(G)$ is called a *left invariant* differential operator on G.

Lemma III.16. (a) (Identity Theorem for Holomorphic Functions) *If M is connected and two functions $f, f' \in \mathrm{Hol}(M, V)$ coincide on a non-empty open subset of M, then $f = f'$.*

(b) *If G is a connected complex Lie group with a good exponential function and $f \in \mathrm{Hol}(G, V)$ with $(D.f)(1) = 0$ for all $D \in D_l(G)$, then $f = 0$.*

Proof. (a) Since V is locally convex, the linear functionals on V separate the points, and so we may w.l.o.g. assume that $V = \mathbb{C}$. Let

$$D := \{x \in M\colon f(x) = f'(x)\}.$$

Then D is a closed subset of M which contains an open subset.

Since M is connected, it suffices to show that the interior D^0 of D is closed, i.e., that each point $x \in \overline{D^0}$ belongs to D^0. Choosing a local chart around x, we may w.l.o.g. assume that M is an open convex subset of the s.c.l.c. space X. Pick $y \in D^0$ and $x \in M$. Then we consider the holomorphic map $\phi\colon \mathbb{C} \to X, z \mapsto x + z(y - x)$ and note that $f \circ \phi$ and $f' \circ \phi$ are holomorphic functions on $\phi^{-1}(M)$ which coincide on an open neighborhood of y, hence also in 0 because $[0, 1] \subseteq \phi^{-1}(M)$. Thus $f(x) = f'(x)$, and therefore $D = M$ which completes the proof.

(b) For each $X \in \mathfrak{g}$ we obtain a holomorphic function $F \colon \mathbb{C} \to V, z \mapsto f(\exp zX)$. Inductively our assumption implies that

$$0 = (\tilde{X}^n . f)(1) = F^{(n)}(0).$$

Since F is holomorphic, we conclude that $F = 0$ and hence that $f\vert_{\exp \mathfrak{g}} = 0$. The assumption that G has a good exponential function now implies that f vanishes on a neighborhood of $\mathbf{1}$ and by (a) also on G. $\qquad \square$

IV. Representations of infinite-dimensional groups

Let V be an s.c.l.c. space and G a Lie group modeled over a s.c.l.c. space. In this section we will apply the results of Section III to define a derived representation of a representation (π, V) of G on the subspace V^∞ of smooth vectors and to endow this space with a suitable complete locally convex topology inherited from $C^\infty(G, V)$ on which the action of G is smooth. For many purposes it is irrelevant that G is a group and it will suffice to assume that it is an s.c.l.c. semigroup, i.e., a manifold modeled over an s.c.l.c. space with a smooth semigroup multiplication.

Definition IV.1. Let V be an s.c.l.c. space and S an s.c.l.c. semigroup.

(a) A *representation* (V, π) of S is a continuous action $S \times V \to V$ such that the mappings $\pi(s) \colon v \mapsto s.v$ are linear and π denotes the corresponding homomorphism $\pi \colon S \to \mathcal{L}(V)$.

(b) If (V, π) is a representation of S, then a vector $v \in V$ is called *smooth* if the orbit map $S \to V, s \mapsto \pi(s).v$ is smooth. We write V^∞ for the subspace of smooth vectors. $\qquad \square$

Let (V, π) be a representation of the s.c.l.c. Lie group G, $v \in V^\infty$ and $\phi_v \colon G \to V, g \mapsto \pi(g).v$, denote the corresponding orbit map. Then $d\phi_v(\mathbf{1}) \colon \mathfrak{g} \cong T_\mathbf{1}(G) \to V \cong T_v(V)$ is a continuous linear map. We define

$$d\pi(X).v := X.v := d\phi_v(\mathbf{1}).X.$$

Lemma IV.2. *The prescription* $\mathfrak{g} \times V^\infty \to V^\infty$ *defines a representation of* $\mathfrak{g}$ *on* V^∞.

Proof. First we show that for $X \in \mathfrak{g}$ and $v \in V^\infty$ the element $X.v \in V$ is in fact contained in V^∞.

For $g \in G$ we have $\pi(g) \circ \phi_v = \phi_v \circ \lambda_g$ because the orbit map ϕ_v is equivariant with respect to left multiplications. Hence the chain rule implies

$$\pi(g) d\phi_v(\mathbf{1}).X = d\phi_v(g) d\lambda_g(\mathbf{1}).X.$$

Let $X_l \in \mathcal{V}(G)$ denote the left invariant vector field with $X_l(\mathbf{1}) = X$. Then the preceding calculation shows that

(4.1) $$g \mapsto \pi(g)(X.v) = d\phi_v(g).X_l(g)$$

is smooth since the map

$$T(\phi_v) \circ X_l \colon G \to TV \cong V \times V, \quad g \mapsto \big(\pi(g).v, d\phi_v(g).X_l(g)\big)$$

is smooth. This proves that $X.v \in V^\infty$.

It remains to show that $d\pi \colon \mathfrak{g} \to \mathrm{End}(V^\infty)$ is a homomorphism of Lie algebras. For $v \in V^\infty$ we obtain a map

$$\Phi_v \colon V' \to C^\infty(G), \quad \omega \mapsto \big(g \mapsto \langle \omega, \pi(g).v \rangle\big).$$

For $X \in \mathfrak{g}$, the corresponding left invariant vector field X_l, and $\omega \in V'$ the chain rule and (4.1) show that

$$\big(X_l.\Phi_v(\omega)\big)(g) = \langle \omega, d\phi_v(g).X_l(g) \rangle = \langle \omega, \pi(g).(X.v) \rangle = \Phi_{X.v}(\omega)(g),$$

i.e., $X_l \circ \Phi_v = \Phi_{X.v}$. Therefore

$$\Phi_{[X,Y].v} = [X_l, Y_l] \circ \Phi_v = X_l \circ \Phi_{Y.v} - Y_l \circ \Phi_{X.v} = \Phi_{X.(Y.v)} - \Phi_{Y.(X.v)} = \Phi_{X.(Y.v) - Y.(X.v)}.$$

Evaluating this at $g = 1$ we obtain $\omega([X,Y].v) = \omega\big(X.(Y.v) - Y.(X.v)\big)$ for all $\omega \in V'$ and, since the continuous linear functionals on V separate the points, $[X,Y].v = X.(Y.v) - Y.(X.v)$. $\square$

Remark IV.3. If G is finite-dimensional, then Gårding's Theorem (cf. [Wa72, Prop. 4.4.1.1]) shows that V^∞ is a dense subspace of V. Another important fact on smooth vectors is Harish-Chandra's Theorem ([Wa72, Th. 4.4.2.1]) saying that if G is finite-dimensional and compact, $\hat{G}$ is the set of equivalence classes of irreducible representations, and $P(\delta) \colon V \to V$ the projection onto the isotypical component of type δ, then for each $v \in V^\infty$ the Fourier series

$$v = \sum_{\delta \in \hat{G}} P(\delta).v$$

converges in V. $\square$

Lemma IV.4. *Let X be a topological space, S a metrizable topological semigroup acting continuously from the right on X, and V a (sequentially) complete locally convex space.*

  (i) *If, in addition, X satisfies the first axiom of countability, then $C(X,V)_c$ is a (sequentially) complete locally convex space and we obtain a representation of S on this space by $(s.f)(x) := f(x.s)$.*

  (ii) *If (π, V) is a representation of the s.c.l.c. group G, then the action $g.\alpha := \alpha \circ \pi(g^{-1})$ on the dual space V'_c is continuous. If, in addition, V is an LF-space, then we obtain a representation of G on V'_c.*

Proof. (i) The completeness follows from Proposition II.12(i), and the continuity of the action from Lemma III.2(iii).

(ii) Since V'_c is endowed with the topology of uniform convergence on compact subsets of V, Lemma III.2(iii) implies that the action of G on the space $V'_c \subseteq C(V, \mathbb{C})_c$ is continuous. If, in addition, V is an LF-space, then V'_c is complete by Corollary II.13, and we thus obtain a representation of G on this space. $\square$

Next we discuss an appropriate topology on the space V^∞ of smooth vectors. The key tool is Theorem III.5.

Proposition IV.5. *Let (π, V) be a continuous representation of the Fréchet semigroup S with identity element 1 on V and $V^\infty \subseteq V$ the space of smooth vectors. Via the map $v \mapsto \phi_v \colon s \mapsto \pi(s).v$ we obtain a linear embedding $V^\infty \hookrightarrow C^\infty(S, V)$ which we use to define a locally convex topology on V^∞. Then the natural action of S on V^∞ defines a representation of S on V^∞ for which the action map $S \times V^\infty \to V^\infty$ is smooth.*

Proof. For $v \in V$ and $s, t \in S$ we have $\phi_v(st) = \pi(st).v = \pi(s).(\pi(t).v) = \pi(s).\phi_v(t)$, i.e., $\phi_v \colon S \to V$ is equivariant. If, conversely, $\phi \colon S \to V$ is a smooth equivariant map, then $\phi(s) = s.\phi(1)$ shows that $\phi(1) \in V^\infty$. Thus

$$V^\infty \cong C^\infty(S, V)^S = \{f \in C^\infty(S, V)) \colon (\forall s, t \in S) f(st) = \pi(s).f(t)\}$$

is a closed subspace of $C^\infty(S, V)$ because the representation of S on V is continuous, hence V^∞ is a complete locally convex space because S is Fréchet (Proposition III.1).

In view of Theorem III.5, the action map

$$S \times C^\infty(S, V) \to C^\infty(S, V), \quad (s, f) \mapsto s.f$$

with $(s.f)(x) = f(xs)$ is smooth. Since

$$(s.\phi_v)(x) = \phi_v(xs) = \pi(xs).v = \pi(x).(\pi(s).v) = \phi_{\pi(s).v}(x),$$

this implies that the action of S on V^∞ is also smooth. □

Corollary IV.6. *If G is a Fréchet Lie group and (π, V) a continuous representation of G, then the action map*

$$\mathfrak{g} \times V^\infty \to V^\infty, \quad (X, v) \mapsto d\pi(X).v$$

is continuous.

Proof. If $\sigma \colon G \times V^\infty \to V^\infty$ denotes the action map, then $d\pi(X).v = d_1\sigma(1, v)(X)$, so that the asserted continuity follows from $\sigma \in C^1$ (Proposition IV.5). □

Remark IV.7. (a) Note that Corollary IV.6 implies in particular that the operators

$$d\pi(X) \colon V^\infty \to V^\infty$$

are continuous, hence that $\mathfrak{g}$ acts naturally on the dual space $V^{-\infty} := (V^\infty)'$ of continuous linear functionals on V^∞ by $(X.\alpha)(v) = -\alpha(X.v)$.

(b) With respect to the natural topology on V^∞ the inclusion map $V^\infty \to V$ is continuous because the evaluation map $C^\infty(G, V) \to V, f \mapsto f(1)$ is continuous. □

Example IV.8. Let G be a Lie group and $\mathrm{Ad} \colon G \to \mathrm{Aut}(\mathfrak{g})$ the adjoint representation. Then Ad is a representation of G on $\mathfrak{g}$ with a smooth action map.

In fact, since the action map can be written as $\mathrm{Ad}(g).X = dI_g(1).X = d\Phi(g,1)(0,X)$, where $\Phi(g,x) = gxg^{-1}$, it is a restriction of the smooth map $T\Phi \colon T(G \times G) \to TG$, hence a smooth map. Thus the adjoint action of G is a representation in the sense of Definition IV.1 with $\mathfrak{g}^\infty = \mathfrak{g}$. Using Taylor expansions up to a certain order, one can show that the derived action $d\,\mathrm{Ad} = \mathrm{ad}$ is given by $\mathrm{ad}(X).Y = [X,Y]$. We refer to [Mi83, Sect. 5] for the details.

We give a direct proof for the case where G has enough smooth functions such that the representation $\mathfrak{g} \to \mathrm{Der}\left(C^\infty(G)\right)$ is injective. It follows in particular from the results in [Th95] that this is true if $\mathfrak{g}$ is a nuclear LF space.

Let $f \in C^\infty(G)$, $g \in G$, and $X \in \mathfrak{g}$. We write π for the natural representation of G on $C^\infty(G)$ given by $\left(\pi(g).f\right)(x) = f(g^{-1}.x)$. Passing to the derivative of the smooth map

$$\psi \colon G \to C^\infty(G), \quad h \mapsto \pi(g)\pi(h)\pi(g^{-1}).f = \pi(ghg^{-1}).f$$

yields

$$\pi(g)d\pi(Y)\pi(g^{-1}).f = d\pi\left(\mathrm{Ad}(g).Y\right).f.$$

In view of the smoothness of the map ψ, we see that we can take the derivative with respect to g in $\mathbf{1}$, and since f is arbitrary, we get

$$d\pi([X,Y]) = d\pi(X)d\pi(Y) - d\pi(Y)d\pi(X) = d\pi\left(d\,\mathrm{Ad}(X).Y\right).$$

If $d\pi$ is injective, then $d\,\mathrm{Ad} = \mathrm{ad}$ follows.

The above argument can be generalized to the setting where one only considers germs of smooth functions in $\mathbf{1}$. Then one does not have to worry about the existence of enough smooth function, and one can still show that the derivative of the map $G \to \mathfrak{g}, g \mapsto \mathrm{Ad}(g).X$ is $\mathrm{ad}(\cdot).X$ for every $X \in \mathfrak{g}$. $\qquad\square$

In the next proposition we record an important application of the Identity Theorem for Holomorphic Functions (Lemma III.16(a)) to representation theory.

Proposition IV.9. *Let G be a connected complex Lie group with a good exponential function $\exp \colon \mathfrak{g} \to G$ and (π, V) a representation of G such that all orbit maps $G \to V, g \mapsto \pi(g).v$ are holomorphic. Then the following assertions hold:*

(i) *If $F \subseteq V$ is a subspace which is invariant under $\mathfrak{g}$, then its closure is invariant under G.*

(ii) *If $v \in V$ is annihilated by $\mathfrak{g}$, then v is fixed by G.*

Proof. (i) Let $\alpha \in F^\perp \subseteq V'$ be a continuous linear functional vanishing on F. For $v \in F$ we consider the function $f_v \colon G \to \mathbb{C}, g \mapsto \alpha(g.v)$, i.e., $f_v = \alpha \circ \phi_v$, where ϕ_v is the orbit map. Then the calculation in the proof of Lemma IV.2 shows that for each $X \in \mathfrak{g}$ and the associated left invariant vector field X_l we have

$$(X_l.f_v)(g) = df_v(g).X_l(g) = \langle \alpha, d\phi_v(g).X_l(g)\rangle = \langle \alpha, \pi(g)X.v\rangle = f_{X.v}(g),$$

i.e., $X_l.f_v = f_{X.v}$.

For $g = 1$ we now obtain $(X_l.f_v)(1) = \alpha(X.v) = 0$. In view of $X.v \in F$, we can apply this argument inductively and thus obtain $(D.f_v)(1) = 0$ for all $D \in D_l(G)$. Now Lemma III.16(b) implies that $f_v = 0$, hence that $\pi(G).v \subseteq \ker \alpha$ for all $\alpha \in F^\perp$. Next we use the Hahn-Banach Theorem to see that $\overline{F} = (F^\perp)^\perp$ from which we obtain $\pi(G).v \subseteq \overline{F}$. Since G acts by continuous operators on V, we conclude that $\overline{F}$ is invariant under G.

(ii) As above, we consider the function $f_v: g \mapsto \alpha(g.v) - \alpha(v)$ but now with an arbitrary element $\alpha \in V'$. Taking derivatives, we see that $X_l.f_v = f_{X.v} = 0$ for all $X \in \mathfrak{g}$ and therefore $(D_l(G).f_v)(1) = 0$ because $f_v(1) = 0$. So Lemma III.16(b) implies that $f_v = 0$, hence that $\alpha(g.v) = \alpha(v)$ for all $\alpha \in V'$ and $g \in G$. Since V' separates the points of V, the group G fixes v. $\qquad \square$

V. Generalized coherent state representations

In this section we describe a general setup for so called coherent state representations. Analytically these representations are characterized by the property that they can be realized in spaces of holomorphic sections of a homogeneous complex line bundle. On the geometric side this means that the action of G on the projective space of the dual space has a cyclic orbit which is a complex manifold. These concepts are well studied in the setting of Hilbert spaces (cf. [Li95]) and here we show that if one carefully distinguishes between the spaces and their duals, then one can generalize this correspondence to general s.c.l.c. spaces.

In the first subsection we describe how to construct a natural complex line bundle on the projection space $\mathbb{P}(V)$ of an s.c.l.c. space. In the second subsection we then turn to group representations and show in particular that for finite-dimensional Lie groups the representations of G in an LF space which are generalized coherent state representations are precisely those on subspaces of the space of holomorphic sections of a homogeneous complex line bundle.

1. The line bundle over the projective space of a topological vector space

In this section V denotes an s.c.l.c. space and $\mathbb{P}(V)$ its projective space. We write $[v]$ for the element of $\mathbb{P}(V)$ which corresponds to the one-dimensional subspace generated by $v \in V \setminus \{0\}$. Furthermore we write $\mathrm{GL}(V)$ for the group of continuously invertible linear operators on V and V' for the topological dual of V.

Lemma V.1. *The group* $\mathrm{GL}(V)$ *acts transitively on*
  (i) $V \setminus \{0\}$,
  (ii) $\mathbb{P}(V)$,
  (iii) $V' \setminus \{0\}$, *and*
  (iv) $\mathbb{P}(V')$.

Proof. (i) Let $v, w \in V \setminus \{0\}$. If v and w are linearly dependent, then there exists $\lambda \in \mathbb{C}^\times \subseteq \mathrm{GL}(V)$ with $w = \lambda v$. We now assume that v and w are linearly independent. Since V is locally convex, there exists a continuous linear functional $\alpha \in V'$ with $\alpha(v + w) = 0$ and $\alpha(v - w) = 1$, i.e., $\alpha(v) = -\alpha(w) = \frac{1}{2}$. Then

$$\Phi(x) := x - 2\alpha(x)(v - w)$$

is a continuous reflection in the hyperplane $\ker \alpha$ satisfying $\Phi(v) = w$ and $\Phi^{-1} = \Phi$. It follows in particular that $\Phi \in GL(V)$.

(ii) This is an immediate consequence of (i).

(iii) We endow V' with the weak-$*$-topology. If $\alpha, \beta \in V' \setminus \{0\}$ are linearly independent, then there exists $x \in V$ with $(\alpha + \beta)(x) = 0$ and $(\alpha - \beta)(x) = 1$. Therefore the same argument as in (i) works in this case.

(iv) This is a direct consequence of (iii). $\square$

Proposition V.2. *The space $\mathbb{P}(V)$ carries the structure of a complex manifold modeled over closed hyperplanes of V. The charts are given by $(U_\alpha, \phi_\alpha)_{\alpha \in V' \setminus \{0\}}$, where*

$$(5.1) \quad U_\alpha = \{[v] \in \mathbb{P}(V) : \alpha(v) \neq 0\} \quad and \quad \phi_\alpha : U_\alpha \to \ker \alpha, \quad [v] \mapsto \frac{v}{\alpha(v)} - v_\alpha,$$

where $v_\alpha \in V$ is chosen with $\alpha(v_\alpha) = 1$.

Proof. First we note that the condition defining U_α makes sense because either α vanishes on the one-dimensional space $\mathbb{C}v$ or $\alpha(w) \neq 0$ holds for all $w \in \mathbb{C}v \setminus \{0\}$. According to Lemma V.1, for two different non-zero continuous functionals their kernels are isomorphic as topological vector spaces because they are conjugate under the group $GL(V)$. Since these kernels are precisely the closed hyperplanes of V, we also see that two such hyperplanes are isomorphic.

Next we note that the inverse of ϕ_α is given by

$$\phi_\alpha^{-1} : \ker \alpha \to U_\alpha, \quad v \mapsto [v + v_\alpha].$$

For $[v] \in U_\alpha \cap U_\beta$ and $w := \phi_\beta([v])$ we have

$$\phi_\alpha \circ \phi_\beta^{-1}(w) = \frac{w + v_\beta}{\alpha(w + v_\beta)} - v_\alpha$$

which is a holomorphic map of an open subset of $\ker \beta$ to $\ker \alpha$. Hence the atlas given by the above charts defines on $\mathbb{P}(V)$ the structure of a complex manifold. $\square$

We put $U_{\alpha\beta} := U_\alpha \cap U_\beta$ for $\alpha, \beta \in V' \setminus \{0\}$. We define functions

$$g_{\beta\alpha} : U_{\alpha\beta} \to \mathbb{C}^\times, \quad [v] \mapsto \frac{\alpha(v)}{\beta(v)}$$

and note that these functions satisfy $g_{\gamma\beta}([v]) \cdot g_{\beta\alpha}([v]) = g_{\gamma\alpha}([v])$ on $U_\alpha \cap U_\beta \cap U_\gamma$, i.e., the functions $g_{\alpha\beta}$ form a *system of transition functions* in the sense of [Hu94, Def. 5.2.4]. Next we construct a holomorphic line bundle $p : L_V \to \mathbb{P}(V)$ as follows. On the disjoint union

$$\tilde{L}_V := \bigcup_{0 \neq \alpha \in V'} U_\alpha \times \mathbb{C} \times \{\alpha\}$$

we define an equivalence relation by

$$([v], z, \alpha) \sim ([v], g_{\beta\alpha}([v])z, \beta) = \left([v], \tfrac{\alpha(v)}{\beta(v)} z, \beta\right).$$

Proposition V.3. *The space* $L_V := \tilde{L}_V / \sim$ *carries the structure of a complex line bundle over* $\mathbb{P}(V)$ *with projection*

$$q \colon L_V \to \mathbb{P}(V), \quad [[v], z, \alpha] \mapsto [v].$$

Proof. It is clear that L_V inherits the structure of a complex manifold because the transition functions are holomorphic and the sets $U_\alpha \times \mathbb{C} \times \{\alpha\}$ carry natural complex manifold structures.

The subset $q^{-1}(U_\alpha)$ is biholomorphically equivalent to $\ker \alpha \times \mathbb{C}$, where the charts are given by

$$\psi_\alpha \colon q^{-1}(U_\alpha) \to \ker \alpha \times \mathbb{C}, \quad [[v], z, \alpha] \mapsto (\phi_\alpha([v]), z).$$

Note that for these coordinate charts we have

$$\psi_\beta \circ \psi_\alpha^{-1}(v, z) = \psi_\beta\Big([[v + v_\alpha], z, \alpha]\Big) = \psi_\beta\Big([[v + v_\alpha], g_{\beta\alpha}([v + v_\alpha])z, \beta]\Big)$$

$$= \psi_\beta([[v + v_\alpha], \tfrac{z}{\beta(v+v_\alpha)}, \beta]) = \big(\phi_\beta \circ \phi_\alpha^{-1}(v), \tfrac{z}{\beta(v+v_\alpha)}\big).$$

Since this map is holomorphic, we obtain another proof for the fact that L_V is a complex manifold. Moreover, the fact that this map is linear in the second argument shows that L_V is a holomorphic vector bundle with fiber $\mathbb{C}$, i.e., a holomorphic line bundle. $\qquad\square$

Theorem V.4. *The assignment*

(5.2) $$s_\alpha([v]) := [[v], \tfrac{\alpha(v)}{\beta(v)}, \beta], \quad [v] \in U_\beta$$

yields a topological isomorphism $\eta \colon V_c' \to \Gamma(L_V)_c$, *where* $\Gamma(L_V)_c$ *denotes the space of holomorphic sections of* L_V *endowed with the topology of uniform convergence on compact subsets of* $\mathbb{P}(V)$.

Proof. First let $\alpha \in V'$. Then

$$[[v], \tfrac{\alpha(v)}{\beta(v)}, \beta] = [[v], g_{\gamma\beta}([v]) \tfrac{\alpha(v)}{\beta(v)}, \gamma] = [[v], \tfrac{\alpha(v)}{\gamma(v)}, \gamma]$$

so that (5.2) defines in fact a section $\eta(\alpha)$ of L_V which is holomorphic. Now we show that the so obtained map $\eta \colon V' \to \Gamma(L_V)$ is a bijection. The subset

$$L_V^\times := \{[[v], z, \alpha] \colon z \neq 0, [v] \in \mathbb{P}(V), 0 \neq \alpha \in V'\},$$

of L_V is the complement of the zero section in L_V. We have a natural map

$$j \colon V \setminus \{0\} \to L_V^\times, \quad v \mapsto [[v], \tfrac{1}{\alpha(v)}, \alpha]$$

for $[v] \in U_\alpha$. For $[v] \in U_{\alpha\beta}$ we have

$$[[v], \tfrac{1}{\alpha(v)}, \alpha] = [[v], g_{\beta\alpha}([v]) \tfrac{1}{\alpha(v)}, \beta] = [[v], \tfrac{1}{\beta(v)}, \beta].$$

The inverse of this map is given by

$$j^{-1} \colon L_V^\times \to V, \quad [[v], z, \alpha] \mapsto \tfrac{v}{z\alpha(v)},$$

for $[v] \in U_\alpha$, where we have to note that the expression on the right hand side is well defined because

$$\frac{v}{z\alpha(v)} = \frac{v}{g_{\beta\alpha}([v])z\beta(v)}.$$

Now let $s \in \Gamma(L_V)$ be a holomorphic section. Then we obtain a holomorphic function $\tilde{s}\colon L_V^\times \to \mathbb{C}$ with $s(p(x)) = \tilde{s}(x) \cdot x$. Note that $\tilde{s}(\lambda x) = \frac{1}{\lambda}\tilde{s}(x)$. Therefore the function $\hat{s} := \tilde{s} \circ j\colon V \setminus \{0\} \to \mathbb{C}$ is holomorphic and satisfies $\hat{s}(\lambda x) = \lambda\hat{s}(x)$ for all $\lambda \in \mathbb{C}^\times$. We claim that $\hat{s}$ is the restriction of a continuous linear functional. If V is one-dimensional, then $\mathbb{P}(V)$ consists of one point and there is nothing to show. Let $W \subseteq V$ be a two-dimensional subspace. Then the restriction f of $\hat{s}$ to $W \setminus \{0\}$ is a holomorphic function satisfying

(5.3) $f(\lambda v) = \lambda f(v), \quad 0 \neq v \in W, \lambda \in \mathbb{C}^\times.$

Since $\{0\}$ is an isolated singularity of this function, Hartog's Theorem shows that f extends holomorphically to W. Now the Taylor expansion in the origin and (5.3) imply that f is linear. Thus the extension of $\hat{s}$ by $\hat{s}(0) := 0$ yields a linear functional $\hat{s}$ on V. If $\hat{s} \neq 0$, then $\ker \hat{s}$ is a complex hyperplane with the property that $(V \setminus \{0\}) \cap \ker \hat{s}$ is closed. Hence $\ker \hat{s}$ is closed and therefore $\hat{s}$ is continuous. Thus for each holomorphic section s there exists a continuous linear functional $\alpha \in V'$ such that

$$s([v]) = \tilde{s}\big([[v], z, \beta]\big) \cdot [[v], z, \beta] = \alpha\big(j^{-1}([[v], z, \beta])\big) \cdot [[v], z, \beta]$$
$$= \frac{\alpha(v)}{z\beta(v)} \cdot [[v], z, \beta] = [[v], \tfrac{\alpha(v)}{\beta(v)}, \beta],$$

i.e., $s = s_\alpha$. This completes the proof of the bijectivity of η.

Now we show that η also is a topological isomorphism. We may w.l.o.g. assume that $V \neq \{0\}$. First we observe that the topology on V'_c coincides with the topology of uniform convergence on all compact subsets $C \subseteq V$ for which there exists a linear functional $\beta \in V'$ with $\inf \Re\beta(C) \geq 1$. In fact, if $C \subseteq V$ is a compact subset, then we pick $x \in V$ with $\Re\beta(x) > \max\big(1, 1 - \inf \Re\beta(C)\big)$. Then $\inf \Re\beta(C + x) = \inf \Re\beta(C) + \Re\beta(x) > 1$, and the uniform convergence on $C + x$ and x implies the uniform convergence on $C = (C + x) - x$. On the other hand, a covering argument using that the quotient map $p\colon V \setminus \{0\} \to \mathbb{P}(V), v \mapsto [v]$ is open and has local sections shows that every compact subset of $\mathbb{P}(V)$ is a finite union of compact subsets lying in some open subset $U_\beta, \beta \in V' \setminus \{0\}$.

Now let $C \subseteq V$ be a compact subset with $\inf \Re\beta(C) > 1$. Then $p(C) \subseteq \mathbb{P}(V)$ is a compact subset of $p(\{v \in V\colon \beta(v) \neq 0\}) = U_\beta$ and we have $\eta(\alpha)([v]) = [[v], \tfrac{\alpha(v)}{\beta(v)}, \beta]$ for $[v] \in U_\beta$. In view of $\inf |\beta(C)| > 1$, this formula implies that a net $(\alpha_j)_{j \in J}$ in V' converges uniformly on C if and only if the net $(\eta(\alpha_j))_{j \in J}$ of holomorphic sections of L_V converges uniformly on $p(C)$. Therefore η is a topological isomorphism $V'_c \to \Gamma(L_V)_c$. $\square$

2. Applications to representation theory

Definition V.5. A continuous representation (π, V) of G on an s.c.l.c. space V is called a *generalized coherent state representation* (GCS representation for short) if there exists $v \in V \setminus \{0\}$ such that

(1) v is cyclic,
(2) the homogeneous space $G/G_{[v]}$, where $G_{[v]} = \{g \in G: g.[v] = [v]\}$ carries the structure of a complex homogeneous space modeled over a Fréchet space such that the natural map $\eta: G/G_{[v]} \to \mathbb{P}(V), gG_{[v]} \mapsto g.[v]$ is holomorphic.

A vector $v \in V \setminus \{0\}$ satisfying (1) and (2) is called a *GCS vector*. □

If $p: L \to M$ is a holomorphic line bundle over a Fréchet manifold M, then we endow the space $\Gamma(L)$ of holomorphic sections with the compact open topology which turns it into a complete locally convex space (cf. Theorem III.11). If V is a topological vector space, then we write V'_c for the topological dual of V endowed with the topology of uniform convergence on the compact subsets of V (cf. Section II).

Proposition V.6. *If (π, V) is a generalized coherent state representation, then the contragredient representation (π', V'_c) can be injected continuously into the natural representation of G on the space $\Gamma(L)$ of holomorphic sections of a holomorphic line bundle $p: L \to M$.*

Proof. Let $v \in V$ be a GCS vector and $M := G/G_{[v]}$. Then M carries the structure of a complex manifold such that the inclusion map

$$\eta: M \to \mathbb{P}(V), \quad gG_{[v]} \mapsto g.[v]$$

is holomorphic. Let $L_V \to \mathbb{P}(V)$ denote the line bundle from Proposition V.3. Then the pull back $L := \eta^* L_V$ is a holomorphic line bundle over M and thus we obtain a natural map

$$\psi: V' \cong \Gamma(L_V) \to \Gamma(L).$$

We claim that ψ is injective. So let $\alpha \in V'$ and suppose that $\psi(s_\alpha) = 0$. This means that the section s_α vanishes on $\eta(M) \subseteq \mathbb{P}(V)$. For $\beta \in V' \setminus \{0\}$ and $[w] \in U_\beta \subseteq \mathbb{P}(V)$ we have

$$(5.4) \qquad\qquad s_\alpha([w]) := \left[[w], \tfrac{\alpha(w)}{\beta(w)}, \beta\right].$$

Hence s_α vanishes in $[w]$ if and only if $\alpha(w) = 0$. Therefore α vanishes on $G.v$, and the fact that v is cyclic implies that $\alpha = 0$, i.e., that ψ is injective.

To see that ψ is continuous, let $K \subseteq M$ be a compact subset. Then there exists a compact subset $C \subseteq V \setminus \{0\}$ with $\eta(K) = [C]$. Now convergence in V'_c implies uniform convergence on C, hence (5.4) shows that the corresponding sections converge uniformly on $K \subseteq M$. This proves that ψ is continuous. □

Lemma V.7. *Let $p: L \to M$ be a holomorphic line bundle, M a complex Fréchet manifold, and $V \subseteq \Gamma(L)$ a closed subspace with the property that for each $x \in M$ the exists a holomorphic section $s \in V$ with $s(x) \neq 0$. Then the following assertions hold:*

(i) *The system $U_s := \{x \in M : s(x) \neq 0\}$, $s \in V \setminus \{0\}$, and the transition functions*

$$g_{ts} \colon U_s \cap U_t \to \mathbb{C}^\times, \quad x \mapsto \frac{s(x)}{t(x)}$$

define a line bundle over M which is isomorphic to L.

(ii) *Assume that V is a Fréchet space. For $x \in L^\times$ we define a holomorphic map $\gamma \colon L^\times \to V'_c$ by $s(p(x)) = \gamma(x)(s) \cdot x$. Then $\gamma(L^\times) \subseteq V'_c \setminus \{0\}$, and we obtain a holomorphic map*

$$\overline{\gamma} \colon M \to \mathbb{P}(V'_c), \quad p(x) \mapsto [\gamma(x)].$$

Furthermore the pull-back line bundle $\overline{\gamma}^* L_{V'_c}$ is isomorphic to L.

Proof. (i) We construct a holomorphic line bundle $q \colon E \to M$ as $\tilde{E}/\sim$, where

$$\tilde{E} := \bigcup_{0 \neq s \in V} U_s \times \mathbb{C} \times \{s\}$$

and

$$(x, z, s) \sim (x, g_{ts}(x)z, t) = \left(x, \tfrac{s(x)}{t(x)} z, t\right).$$

Then the projection $q \colon E \to M$ is given by $q([x, z, s]) = x$. To see that this bundle is isomorphic to L, we define a holomorphic mapping

$$\Phi \colon E \to L, \quad [x, z, s] \mapsto z \cdot s(x) \quad \text{for} \quad x \in U_s.$$

To see that Φ is well defined, we note that for $x \in U_s \cap U_t$ we have $[x, z, s] = \left[x, \tfrac{s(x)}{t(x)} z, t\right]$ and

$$z \cdot s(x) = \frac{s(x)}{t(x)} z \cdot t(x).$$

Hence Φ is a well defined holomorphic bundle map with $p \circ \Phi = q$.

Moreover, if $\Phi([x, z, s]) = \Phi([x', z', s'])$, then $x = p(\Phi(x)) = x' \in U_s \cap U_{s'}$, and $z \cdot s(x) = z' \cdot s'(x)$, i.e., $z' = \tfrac{s(x)}{s'(x)} z$. Hence Φ is bijective. Moreover, for $y \in p^{-1}(U_s)$ we have

$$\Phi^{-1}(y) = \left[p(y), \frac{y}{s(p(y))}, s\right],$$

which shows that $\Phi^{-1} \colon L \to E$ is also holomorphic.

(ii) First we note that $V \to \mathbb{C}, s \mapsto \gamma(x)(s)$ is continuous, so that $\gamma(V) \subseteq V'$. We claim that γ is holomorphic. Since by assumption V is a Fréchet space, Corollary II.13 shows that V'_c is a complete locally convex space, and that the natural map $\eta_V \colon V \to (V'_c)'_c$ is surjective (Theorem II.8(ii)). Therefore each continuous linear functional on V'_c is given by evaluation in an element $s \in V$, and for each such s the mapping $x \mapsto \gamma(x)(s)$ is a holomorphic function on $L^\times$. This proves that γ is weakly holomorphic, hence that γ is holomorphic because V'_c is sequentially complete and M is Fréchet (Proposition I.9).

Since, by assumption, for each $x \in M$ there exists an $s \in V$ with $s(x) \neq 0$, we have $\gamma(L^\times) \subseteq V_c' \setminus \{0\}$. Moreover we have $\gamma(\lambda x) = \lambda^{-1}\gamma(x)$ for $\lambda \in \mathbb{C}^\times$, so that γ factors to a holomorphic map

$$\overline{\gamma}\colon M \to \mathbb{P}(V_c'), \quad p(x) \mapsto [\gamma(x)].$$

Let $E := \overline{\gamma}^* L_{V_c'}$ denote the pull-back line bundle with projection $q\colon E \to M$. Then $\overline{\gamma} \circ q = p_{V_c'} \circ \gamma$, and since the bundle $L_{V_c'}$ is defined by the transition functions

$$g_{\beta\alpha}([v]) = \frac{\alpha(v)}{\beta(v)} \quad \text{for} \quad \alpha(v), \beta(v) \neq 0, \alpha, \beta \in (V_c')',$$

the bundle E is defined by the transition functions

$$g_{\beta\alpha}(p(x)) = \frac{\alpha(\gamma(x))}{\beta(\gamma(x))} \quad \text{for} \quad \alpha(\gamma(x)), \beta(\gamma(x)) \neq 0.$$

Using $\eta_c(V) = (V_c')_c'$ (Theorem II.8(ii)), we write $\alpha = \eta_V(s)$ and $\beta = \eta_V(t)$ to obtain

$$g_{\beta\alpha}(p(x)) = \frac{\gamma(x)(s)}{\gamma(x)(t)} = \frac{s(p(x))}{t(p(x))} = g_{ts}(p(x))$$

for $p(x) \in U_s \cap U_t$. Therefore (i) shows that the holomorphic line bundle E is isomorphic to L. $\qquad\square$

For the remainder of this section we will restrict our attention to finite-dimensional Lie groups because we will need the differential geometric machinery describing complex structures and holomorphic sections in terms of the underlying real structure of the manifold.

Lemma V.8. *Let G be a finite-dimensional Lie group, H a closed subgroup, and suppose that the homogeneous space G/H is a complex manifold in such a way that G acts by holomorphic maps. Suppose further that M is a not necessarily finite-dimensional complex manifold on which G acts by holomorphic maps. If $\gamma\colon G/H \to M$ is a holomorphic equivariant map, $x_0 := \gamma(\mathbf{1}H)$, and G_{x_0} is the stabilizer of x_0, then $H \subseteq G_{x_0}$ and the homogeneous space G/G_{x_0} carries a unique complex structure such the quotient map $G/H \to G/G_{x_0}, gH \mapsto gG_{x_0}$ and the induced map $\overline{\gamma}\colon G/G_{x_0} \to M, gG_{x_0} \mapsto g.x_0$ are holomorphic.*

Proof. Let $\sigma\colon G \times M \to M$ denote the action of G on the complex manifold M and write $\mathcal{V}_{\mathrm{hol}}(M) \subseteq \mathcal{V}(M)$ for the Lie algebra of holomorphic vector fields on M. Then

$$\dot{\sigma}\colon \mathfrak{g} \to \mathcal{V}_{\mathrm{hol}}(M), \quad X \mapsto \big(p \mapsto -d\sigma(\mathbf{1}, p)(X, 0)\big)$$

is a homomorphism of Lie algebras. In fact, this follows easily from a local computation in coordinate charts.

We conclude that $\dot{\sigma}$ extends to a $\mathbb{C}$-linear homomorphism $\mathfrak{g}_\mathbb{C} \to \mathcal{V}_{\mathrm{hol}}(M)$ which we also denote by $\dot{\sigma}$. As the formula for the Lie bracket in local coordinates shows, the subspace

$$\mathfrak{a} := \{\mathcal{X} \in \mathcal{V}_{\mathrm{hol}}(M)\colon \mathcal{X}(x_0) = 0\}$$

is a Lie subalgebra of $\mathcal{V}_{\mathrm{hol}}(M)$. Hence $\mathfrak{b} := \dot{\sigma}^{-1}(\mathfrak{a})$ is a complex subalgebra of $\mathfrak{g}_{\mathbb{C}}$. Moreover $\mathfrak{g}_{x_0} = \mathfrak{b} \cap \mathfrak{g}$ according to the fact that the G-orbit is an equivariant image of the finite-dimensional homogeneous manifold G/H. This can also be written as $\mathfrak{b} \cap \overline{\mathfrak{b}} = (\mathfrak{g}_{x_0})_{\mathbb{C}}$ for the complex conjugation $X \mapsto \overline{X}$ on $\mathfrak{g}_{\mathbb{C}}$. Further it is easy to see that $\mathrm{Ad}(G_{x_0}).\mathfrak{b} = \mathfrak{b}$.

The holomorphy of γ now implies that $d\gamma(\mathbf{1}H) \colon T_{\mathbf{1}H}(G/H) = \dot{\sigma}(\mathfrak{g})(x_0)$ is a complex subspace of $T_{x_0}(M)$. This means that $\dot{\sigma}(\mathfrak{g})(x_0) = \dot{\sigma}(\mathfrak{g}_{\mathbb{C}})(x_0)$ which shows that

$$\mathfrak{g}_{\mathbb{C}} = \mathfrak{g} + \mathfrak{b}.$$

Thus we find for each $X \in \mathfrak{g}_{\mathbb{C}}$ an element $Y \in \mathfrak{g}$ and $Z \in \mathfrak{b}$ with $X = Y + Z$. Hence $X - \overline{X} = Z - \overline{Z} \in \mathfrak{b} + \overline{\mathfrak{b}}$, and therefore $i\mathfrak{g} \subseteq \mathfrak{b} + \overline{\mathfrak{b}}$ which in turns gives

$$\mathfrak{g}_{\mathbb{C}} = \mathfrak{g} + i\mathfrak{g} \subseteq i(\mathfrak{b} + \overline{\mathfrak{b}}) + \mathfrak{b} + \overline{\mathfrak{b}} = \mathfrak{b} + \overline{\mathfrak{b}}.$$

This completes the proof of

$$\mathrm{Ad}(G_{x_0}).\mathfrak{b} = \mathfrak{b}, \quad \mathfrak{b} \cap \overline{\mathfrak{b}} = (\mathfrak{g}_{x_0})_{\mathbb{C}}, \quad \text{and} \quad \mathfrak{b} + \overline{\mathfrak{b}} = \mathfrak{g}_{\mathbb{C}},$$

which, according to [Ki76, p. 203], is equivalent to the existence of a complex structure on G/G_{x_0} such that G acts by holomorphic mappings. More explicitly, this complex structure can be described by identifying the tangent space $T_{\mathbf{1}G_{x_0}}(G/G_{x_0}) \cong \mathfrak{g}/\mathfrak{g}_{x_0}$ with the complex vector space $\mathfrak{g}_{\mathbb{C}}/\mathfrak{b}$. From this description of the complex structure it follows that the canonical maps $G/H \to G/G_{x_0}$ and $G/G_{x_0} \to M$ are holomorphic because they are G-equivariant, smooth, and their differentials are complex linear in the base point. This completes the proof. $\square$

Proposition V.9. *Suppose that G is finite-dimensional and L is a holomorphic G-homogeneous line bundle. Then G acts on the Fréchet space $\Gamma(L)$ by $(g.s)(x) := g.s(g^{-1}.x)$. Let $\{0\} \neq V \subseteq \Gamma(L)$ be a closed invariant subspace. Then the representation of G on V_c' is a GCS representation.*

Proof. First we note that V inherits the structure of a Fréchet space. We claim that V satisfies the assumptions of Lemma V.7. Let $x \in M$. Since $V \neq \{0\}$, there exists $s \in V \setminus \{0\}$. Pick $y \in M$ with $s(y) \neq 0$. Then there exists $g \in G$ with $g.y = x$, and we see that $(g.s)(x) = g.s(y) \neq 0$. This means that V satisfies the assumptions of Lemma V.7, and thus $L \cong \overline{\gamma}^* L_{V_c'}$ holds for the natural holomorphic map $\overline{\gamma} \colon M \to \mathbb{P}(V_c')$.

Moreover

$$\big(g.\gamma(x)\big)(s) \cdot x = \gamma(x)(g^{-1}.s) \cdot x = (g^{-1}.s)\big(p(x)\big) = g^{-1}.s\big(g.p(x)\big)$$
$$= g^{-1}.\gamma\big(g.p(x)\big)(s) \cdot (g.x) = \gamma\big(g.p(x)\big)(s) \cdot x,$$

shows that $\gamma \colon L^{\times} \to V_c' \setminus \{0\}$ is G-equivariant and hence that $\overline{\gamma}$ is G-equivariant.

Pick $x_0 \in M$ and let $\overline{\gamma}(x_0) = [\alpha_0]$. Then the G-homogeneous space $G/G_{[\alpha_0]} \cong \overline{\gamma}(M)$ inherits the structure of a complex manifold because $\overline{\gamma}$ is holomorphic (Lemma V.8). Moreover, the natural map $G/G_{[\alpha_0]} \to \mathbb{P}(V_c')$ is obtained by factorization of $\overline{\gamma}$ and therefore holomorphic. So, in view Definition V.5, it remains to prove that $\alpha_0 \in V_c'$ is a cyclic vector.

In fact, if α_0 is not cyclic, then $V \cong (V_c')'$, and the Hahn-Banach Theorem imply the existence of $0 \neq s \in V$ vanishing on $G.\alpha_0$. This means that the section s of $\Gamma(L)$ vanishes on $G.x_0 = M$, contradicting $s \neq 0$. This completes the proof. $\square$

Theorem V.10. *If G is finite-dimensional, then a non-zero continuous representation (π, V) of G, where V is an LF space is a generalized coherent state representation if and only if the contragredient representation permits a continuous equivariant injection into $\Gamma(L)$ for a homogeneous line bundle $p: L \to M$.*

Proof. If (π, V) is a GCS representation, then Proposition V.6 shows that the contragredient representation permits a continuous equivariant injection into $\Gamma(L)$ for a homogeneous line bundle L.

Suppose, conversely, that $\psi: V_c' \to \Gamma(L)$ is a continuous equivariant injection. In view of Proposition V.9, the representation of G on $\Gamma(L)_c'$ is a GCS representation because this space contains $\psi(V_c')$, hence is non-zero. The adjoint map $\psi': \Gamma(L)_c' \to (V_c')_c' \cong V$ is continuous and G-equivariant. Let $\alpha_0 \in \Gamma(L)_c'$ be a GCS vector. We claim that $\psi'(\alpha_0)$ is a GCS vector in V.

First we show that it is cyclic. In fact, if it is not cyclic, then there exists a non-zero $\beta \in V'$ vanishing on $G.\psi'(\alpha_0) = \psi'(G.\alpha_0)$, i.e., $\psi(\beta)$ vanishes on $G.\alpha_0$, and thus $\psi(\beta) = \{0\}$ because α_0 is cyclic, contradicting the injectivity of ψ. Thus $\psi'(\alpha_0)$ is cyclic, and it follows in particular that $\psi'(\alpha_0) \neq 0$.

Now the fact that the natural map

$$\mathbb{P}(\Gamma(L)_c') \setminus \psi(V_c')^\perp \to \mathbb{P}(V), \quad [\alpha] \mapsto [\psi'(\alpha)]$$

is holomorphic and G-equivariant implies that $G/G_{[\psi'(\alpha_0)]}$ is a complex homogeneous G-space such that the natural map $G/G_{[\psi'(\alpha_0)]} \to \mathbb{P}(V)$ is holomorphic (Lemma V.8). This proves that (π, V) is a GCS representation. $\square$

References

[Bou59] Bourbaki, N., "Intégration," Chap. 6, Hermann, Paris, 1959

[Bou71] —, "Topologie Générale," Chap. 10, Hermann, Paris, 1971

[Bou87] —, "Topological Vector Spaces," Chap. 1-5, Springer-Verlag, 1987

[FK88] Fröhlicher, A., and A. Kriegl, "Linear Spaces and Differentiation Theory," J. Wiley, Interscience, 1988

[Gl99] Glöckner, H., *Direct limit Lie groups and manifolds*, Kyoto J. Math., to appear

[Gr97] Grabowski, J., *Derivative of the exponential mapping for infinite-dimensional Lie groups*, Preprint, 1997

[Ha82] Hamilton, R., *The inverse function theorem of Nash and Moser*, Bull. Amer. Math. Soc. **7** (1982), 65–222

[He71] Hervé, M., "Analytic and Plurisubharmonic Functions," Lecture Notes in Mathematics **198**, Springer Verlag, Berlin, 1971

[He89] —, "Analyticity in Infinite-dimensional Spaces," de Gruyter, Berlin, 1989

[HP57] Hille, E., and R.S. Phillips, "Functional Analysis and Semigroups," Amer. Math. Soc. Colloquium Publications **XXXI**, Providence, Rhode Island, 1957

[Hu94] Husemoller, D., "Fibre Bundles," Graduate Texts in Math., Springer, New York, 1994

[Ka85] Kac, V. G., "Infinite-dimensional Groups with Applications," Mathematical Sciences Research Institute Publications **4**, Springer-Verlag, Berlin, Heidelberg, New York, 1985

[Ke74] Keller, H. H., "Differential Calculus in Locally Convex Spaces," Lecture Notes in Math. **417**, Springer Verlag, 1974

[Ki76] Kirillov, A. A., "Elements of the Theory of Representations," Grundlehren der mathematischen Wissenschaften **220**, Springer-Verlag, Berlin, Heidelberg, 1976

[KM97a] Kriegl, A., and P. W. Michor, "The Convenient Setting of Global Analysis," Amer. Math. Soc., Providence R. I., 1997

[KM97b] —, *Regular infinite-dimensional Lie groups*, Journal of Lie Theory **7** (1997), 61–99

[Li95] Lisiecki, W., *Coherent state representations. A survey*, Reports on Math. Phys. **35** (1995), 327–358

[Mi89] Mickelsson, J., "Current algebras and groups," Plenum Press, New York, 1989

[Mi83] Milnor, J., *Remarks on infinite-dimensional Lie groups*, Proc. Summer School on Quantum Gravity, Ed. B. DeWitt, Les Houches, 1983

[Na69] Nachbin, L., "Topology on Spaces of Holomorphic Mappings," Ergebnisse der Math. **47**, Springer Verlag, Berlin, 1969

[NRW91] Natarajan, L., Rodriquez-Carrington, E., and J. A. Wolf, *Differentiable structure for direct limit groups*, Letters in Mathematical Physics **23** (1991), 99–109

[NRW93] —, *Locally convex Lie groups*, Nova Journal of Algebra and Geometry **2:1** (1993), 59–87

[NRW94] —, *New classes of infinite-dimensional Lie groups*, Proceedings of Symposia in Pure Math. **56:2** (1994), 59–87

[Ne99] Neeb, K.-H., *Borel-Weil theory for loop groups*, in this volume

[PS86]   Pressley, A., and G. Segal, "Loop Groups," Oxford University Press, Oxford, 1986

[Th95]   Thomas, E. G. F., *Vector fields as derivations on nuclear manifolds*, Math. Nachr. **176** (1995), 277–286

[Th96]   —, *Calculus on locally convex spaces*, Preprint **W-9604**, Univ. of Groningen, 1996

[Tr67]   Treves, F., "Topological Vector Spaces, distributions, and kernels," Academic Press, New York, 1967

[Wa72]   Warner, G., "Harmonic Analysis on Semisimple Lie Groups I," Springer, Berlin, Heidelberg, New York, 1972

Mathematical Subject Classification (2000)
Primary: 22E65.
Secondary: 58B25, 46G25, 81R10.

Technische Universität Darmstadt
Fachbereich Mathematik
Schlossgartenstrasse 7
D-64289 Darmstadt
Germany
E-mail address: neeb@mathematik.tu-darmstadt.de

Borel-Weil Theory for Loop Groups

Karl-Hermann Neeb

Introduction

Let K be a compact Lie group and $LK := C^\infty(\mathbb{S}^1, K)$ the group of smooth loops with values in K. This is a group under pointwise multiplication and it carries the structure of a Lie group modeled over the Fréchet space $L\mathfrak{k} := C^\infty(\mathbb{S}^1, \mathfrak{k})$ of smooth loops with values in the Lie algebra of $\mathfrak{k}$. These notes grew out of a reworking of the proof of the Borel-Weil theory for loop groups as it is presented in the book of Pressley and Segal ([PS86]). Our main objective is to develop the techniques which are relevant for this theory in the setting of the Fréchet groups of smooth loops from an analytic point of view. We will describe an intrinsic construction of the irreducible positive energy representations which does not refer to embeddings of loop groups into infinite-dimensional classical Banach Lie groups as in [GW84] and [Ner83]. For an algebraic version of a Borel-Weil Theorem for general Kac-Moody groups, considered as algebraic groups of infinite type, we refer to [Ka85b, p. 192]. Generalizations of Bott-Borel-Weil theory to direct limits of Lie groups are discussed in [NRW99]. A realization of the spin representation of the group $O(\infty, \mathbb{C})$ in a Fréchet space of holomorphic sections is constructed by Neretin in [Ner87].

In Section I we recall some basic notation from the theory of compact groups and their Lie algebras. In Section II we briefly discuss the loop group LK, where K is a compact group, its complexification and certain central extensions. We then turn to the root decomposition of the loop algebra $L\mathfrak{k}$ in Section III, and in Section IV we discuss some general aspects of the representation theory of loop groups. Section V is dedicated to some rather general constructions of representations of involutive semigroups on pre-Hilbert spaces. We include this material to make the construction of a certain (pre-)Hilbert space of holomorphic functions in Section VI more transparent. The heart of this article is Section VI, where we explain how the finite-dimensional Borel-Weil theory can be extended to loop groups. The basic idea is that the irreducible representations of positive energy (maybe after passing to a dense subspace) can be realized as the holomorphic sections of a certain line bundle over a complex homogeneous space of LK. Then the parametrization of these representations is derived from a characterization of those line bundles which have non-zero holomorphic sections. Finally we discuss in Section VII how these constructions can be used to analyze more general positive energy representations.

I. Compact groups

In this section we collect the basic material concerning compact groups that we will need in the following. In particular we introduce the notation that will be used later on.

In the following we will write $\mathbb{S}^1 := \mathbb{R}/2\pi\mathbb{Z}$ for the circle as a real smooth manifold. If the circle is considered as a group, we will write $\mathbb{T} := \mathbb{R}/2\pi\mathbb{Z}$.

Let $\mathfrak{k}$ be a compact Lie algebra and $\mathfrak{t} \subseteq \mathfrak{k}$ be a Cartan subalgebra. Then the set $\operatorname{ad}\mathfrak{t}$ is commutative and diagonalizable on the complexification $\mathfrak{k}_\mathbb{C}$ of $\mathfrak{k}$. Thus we obtain the *root decomposition*

$$\mathfrak{k}_\mathbb{C} = \mathfrak{t}_\mathbb{C} \oplus \bigoplus_{\alpha \in \Delta_\mathfrak{k}} \mathfrak{k}_\mathbb{C}^\alpha,$$

where

$$\mathfrak{k}_\mathbb{C}^\alpha = \{X \in \mathfrak{k}_\mathbb{C} : (\forall Y \in \mathfrak{t})[Y, X] = i\alpha(Y)X\} \quad \text{and} \quad \Delta_\mathfrak{k} = \{\alpha \in \mathfrak{t}^* \setminus \{0\} : \mathfrak{k}_\mathbb{C}^\alpha \neq \{0\}\}.$$

Let K be a connected Lie group with Lie algebra $\mathfrak{k}$ and $T := \exp\mathfrak{t}$ the subgroup corresponding to $\mathfrak{t}$. If K is compact, then T is a maximal torus in K. We identify the *character group* $\hat{T} := \operatorname{Hom}(T, \mathbb{T})$ with a subset of $\mathfrak{t}^*$ by associating to each continuous character $\chi : T \to \mathbb{T}$ the linear functional $\lambda = -id\chi(1) \in \mathfrak{t}^*$ satisfying

$$(1.1) \qquad\qquad \chi(\exp X) = e^{i\lambda(X)}$$

for all $X \in \mathfrak{t}$. Since the Lie algebra $\mathfrak{t}$ of T can be identified with the set of one-parameter subgroups $\mathbb{R} \to T$ we can in particular identify the group $\check{T} := \operatorname{Hom}(\mathbb{T}, T)$ with a subset of $\mathfrak{t}$ by associating to each $\gamma \in \check{T}$ the uniquely determined element $X_\gamma \in \mathfrak{t}$ with

$$\gamma(\theta) = \exp(\theta X_\gamma)$$

for all $\theta \in \mathbb{T} \cong \mathbb{R}/(2\pi\mathbb{Z})$. In this sense we obtain the identification

$$\check{T} \cong \{X \in \mathfrak{t} : \exp(2\pi X) = 1\}.$$

An important tool which will also be crucial in the infinite-dimensional setting are certain subalgebras of $\mathfrak{k}$ which are isomorphic to the three-dimensional compact Lie algebra $\mathfrak{su}(2) \cong \mathfrak{so}(3, \mathbb{R})$. Let $\alpha \in \Delta_\mathfrak{k}$ and $X \mapsto \overline{X}$ denote complex conjugation on $\mathfrak{k}_\mathbb{C}$. Then we choose for each $\alpha \in \Delta_\mathfrak{k}$ an element $e_\alpha \in \mathfrak{k}_\mathbb{C}^\alpha$ in such a way that $e_{-\alpha} = \overline{e_\alpha}$ and put $ih_\alpha := [e_\alpha, e_{-\alpha}]$. Here $h_\alpha \in \mathfrak{t} \cap [\mathfrak{k}_\mathbb{C}^\alpha, \mathfrak{k}_\mathbb{C}^{-\alpha}]$ is uniquely determined by the requirement that $\alpha(h_\alpha) = 2$ and called the *coroot* associated to α. We then have

$$[h_\alpha, e_{\pm\alpha}] = \pm 2ie_{\pm\alpha}.$$

We call $(e_\alpha, e_{-\alpha}, h_\alpha)$ a *basic $\mathfrak{su}_2$-triple*. This is justified by the fact that

$$\mathfrak{k}(\alpha) := \mathbb{R}h_\alpha + \mathbb{R}(e_\alpha + e_{-\alpha}) + \mathbb{R}i(e_\alpha - e_{-\alpha})$$

is a subalgebra of $\mathfrak{k}$ isomorphic to $\mathfrak{su}(2)$. For $\mathfrak{k} = \mathfrak{su}(2)$ the corresponding basis elements are given by

$$e_\alpha = \begin{pmatrix} 0 & 1 \\ 0 & 0 \end{pmatrix}, \quad e_{-\alpha} = \begin{pmatrix} 0 & 0 \\ -1 & 0 \end{pmatrix}, \quad \text{and} \quad h_\alpha = \begin{pmatrix} i & 0 \\ 0 & -i \end{pmatrix}.$$

Lemma I.1. *If κ is an invariant $\mathbb{C}$-bilinear form on $\mathfrak{k}_{\mathbb{C}}$, then $2\kappa(e_\alpha, e_{-\alpha}) = \kappa(h_\alpha, h_\alpha)$.*

Proof. The invariance of κ implies that

$$2\kappa(e_\alpha, e_{-\alpha}) = \kappa(-i[h_\alpha, e_\alpha], e_{-\alpha}) = -i\kappa(h_\alpha, [e_\alpha, e_{-\alpha}]) = \kappa(h_\alpha, h_\alpha). \qquad \square$$

II. Loop groups and their central extensions

Before we turn to the specific case of the loop group of a compact group, we first study the Fréchet-Lie group structure on the group $C^\infty(M, G)$, where G is a finite-dimensional Lie group and M is a compact manifold.

1. Groups of smooth maps

Theorem II.1. *If M is a compact manifold, then the group*

$$\mathcal{D}(M, G) := C^\infty(M, G)$$

is a Fréchet-Lie group and its Lie algebra is $\mathcal{D}(M, \mathfrak{g}) := C^\infty(M, \mathfrak{g})$ endowed with its natural Fréchet structure. The exponential function

$$\exp \colon \mathcal{D}(M, \mathfrak{g}) \to \mathcal{D}(M, G), \quad \gamma \mapsto \exp \circ \gamma$$

is a smooth map which maps a 0-neighborhood in $\mathcal{D}(M, \mathfrak{g})$ diffeomorphically onto a 0-neighborhood in the group $\mathcal{D}(M, G)$, and we have the Trotter-Product-Formula:

$$\exp(\gamma + \eta) = \lim_{n \to \infty} \left(\exp\left(\tfrac{1}{n}\gamma\right) \exp\left(\tfrac{1}{n}\eta\right) \right)^n.$$

If, in addition, G is a complex Lie group, then $\mathcal{D}(M, G)$ is a complex Lie group and the exponential function is holomorphic.

Proof. First we note that the spaces $\mathcal{D}(M, \mathfrak{g})$ are nuclear Fréchet spaces (cf. [Alb93, Sect. 1.2]). The topology defined on these spaces is the same as the topology described in [Ne00a, Sect. II].

Next we have to describe the manifold structure on $\mathcal{D}(M, G)$. In [Alb93] it is shown how $\mathcal{D}(M, G)$ can be given the structure of a topological group such that the exponential function $\exp \colon \mathcal{D}(M, \mathfrak{g}) \to \mathcal{D}(M, G)$ is a local homeomorphism. Hence $\mathcal{D}(M, G)$ carries the structure of a topological manifold modeled over the complete locally convex space $\mathcal{D}(M, \mathfrak{g})$.

Let $U \subseteq \mathfrak{g}$ be an open convex 0-neighborhood such that the exponential function $\exp \colon U \to \exp(U) \subseteq G$ is a diffeomorphism and the Campbell-Hausdorff series

$$X * Y = X + Y + \tfrac{1}{2}[X, Y] + \cdots$$

converges for $X, Y \in U$. Let further $V \subseteq U$ be a symmetric 0-neighborhood with $V * V \subseteq U$. Then we obtain charts

$$\phi_g \colon \mathcal{D}(M, V) \to \mathcal{D}(M, G), \quad \gamma \mapsto g \cdot (\exp \circ \gamma).$$

Suppose that two chart neighborhoods $g \exp \left(\mathcal{D}(M, V) \right)$ and $h \exp \left(\mathcal{D}(M, V) \right)$ intersect. Then

$$g^{-1} h \in \exp \left(\mathcal{D}(M, V) \right) \exp \left(\mathcal{D}(M, V) \right)^{-1}$$
$$= \exp \left(\mathcal{D}(M, V) * \mathcal{D}(M, V) \right) \subseteq \exp \left(\mathcal{D}(M, V * V) \right)$$
$$\subseteq \exp \left(\mathcal{D}(M, U) \right).$$

Let $g^{-1} h = \exp \circ \alpha$, $\alpha \in \mathcal{D}(M, U)$. Then

$$\phi_g^{-1} \phi_h(\eta) = (\exp|_U)^{-1} (g^{-1} h \exp \circ \eta) = (\exp|_U)^{-1} \left(\exp \circ \alpha \cdot \exp \circ \eta \right)$$
$$= (\exp|_U)^{-1} \left(\exp \circ (\alpha * \eta) \right) = \alpha * \eta.$$

The mapping

$$\lambda_\alpha^* \colon M \times V \to U, \quad (p, X) \mapsto \alpha(p) * X$$

is smooth. Therefore Corollary III.8 in [Ne00a] shows that $\eta \mapsto \alpha * \eta, \mathcal{D}(M, V) \to \mathcal{D}(M, U)$ is smooth. This proves that the charts ϕ_g, $g \in \mathcal{D}(M, G)$, have smooth transition functions, hence define the structure of a smooth manifold on $\mathcal{D}(M, G)$.

Applying [Ne00a, Cor. III.8] to the Campbell-Hausdorff multiplication

$$V \times V \to U, \quad (X, Y) \mapsto X * Y = X + Y + \tfrac{1}{2}[X, Y] + \dots,$$

we see that the map

$$\mathcal{D}(M, V) \times \mathcal{D}(M, V) \to \mathcal{D}(M, U), \quad (\alpha, \beta) \mapsto \alpha * \beta$$

is smooth. Moreover, $\alpha \mapsto -\alpha$ is a smooth involution of $\mathcal{D}(M, U)$. From that it easily follows that multiplication and inversion in $\mathcal{D}(M, G)$ are smooth mappings, i.e., that $\mathcal{D}(M, G)$ is a Lie group.

To see that the exponential function is smooth, let $\gamma \in \mathcal{D}(M, \mathfrak{g})$. Then there exists an open 0-neighborhood $C \subseteq \mathfrak{g}$ such that $\exp \left(-\gamma(p) \right) \exp \left(\gamma(p) + X \right) \in \exp(U)$ for all $X \in C$. We put

$$f \colon M \times C \to \mathfrak{g}, \quad (p, X) \mapsto (\exp|_U)^{-1} \left(\exp \left(-\gamma(p) \right) \exp \left(\gamma(p) + X \right) \right)$$

For $\eta \in \mathcal{D}(M, C)$ we now have

$$\phi_{\exp \circ \gamma}^{-1} \left(\exp \circ (\gamma + \eta) \right) = f \circ (\mathrm{id}_M, \eta).$$

Hence the smoothness of this map follows from [Ne00a, Prop. III.7], and we conclude that the exponential function is smooth.

To prove the Trotter-Product-Formula, we consider the analytic map

$$f \colon U \times U \times]-1, 1[\to \mathfrak{g}, \quad (X, Y, t) \mapsto \begin{cases} \dfrac{1}{t}(tX * tY) \text{ for } t \neq 0 \\ X + Y \text{ for } t = 0. \end{cases}$$

Note that the analyticity follows from the fact that if $X * Y = \sum_{n=1}^{\infty} h_n(X, Y)$ denotes the expansion of the Campbell-Hausdorff product into homogeneous terms, then

$$f(X, Y, t) = \sum_{n=1}^{\infty} t^{2n-1} h_n(X, Y).$$

Using [Ne00a, Cor. III.8], we see that the map

$$C^{\infty}(M, U) \times C^{\infty}(M, U) \times C^{\infty}(M,]-1, 1[) \to C^{\infty}(M, \mathfrak{g}), \quad (\alpha, \beta, \gamma) \mapsto f \circ (\alpha, \beta, \gamma)$$

is smooth and therefore in particular continuous. Fixing α and β, we conclude that

(2.1) $$\alpha + \beta = \lim_{n \to \infty} f \circ (\alpha, \beta, \tfrac{1}{n})$$

holds in $C^{\infty}(M, \mathfrak{g})$. Now the Trotter-Product-Formula is proved as follows. For $\gamma, \eta \in C^{\infty}(M, \mathfrak{g})$ we use the compactness of M to find $m \in \mathbb{N}$ with $\frac{1}{m}\gamma, \frac{1}{m}\eta \in C^{\infty}(M, U)$. Then

$$\exp(\gamma + \eta) = \exp\left(\tfrac{1}{m}\gamma + \tfrac{1}{m}\eta\right)^m,$$

and we can apply (2.1) to the right hand side.

If, in addition, G is a complex Lie group, then transition functions, multiplication and inversion are smooth maps with complex linear differentials, i.e., holomorphic (cf. [Ne00a, Sect. I]). Thus $\mathcal{D}(M, G)$ is a complex Lie group with holomorphic exponential function. $\qquad \Box$

Remark II.2. The curves $\mathbb{R} \to \mathcal{D}(M, G), t \mapsto \exp(t\gamma)$ are one-parameter groups and the curves $t \mapsto g \exp(t\gamma)$ are integral curves of the left invariant vector field corresponding to the element $\gamma \in \mathcal{D}(M, \mathfrak{g})$. $\qquad \Box$

Corollary II.3. *If G is a connected finite-dimensional Lie group, then the loop group $LG := C^{\infty}(\mathbb{S}^1, G)$ is a Fréchet-Lie group and $L\mathfrak{g} := C^{\infty}(\mathbb{S}^1, \mathfrak{g})$ its Lie algebra. If, in addition, G is a complex Lie group, then the same holds for LG.* $\qquad \Box$

Remark II.4. The groups described in Theorem II.1 are special cases of the groups $\mathcal{D}(M, G)$ discussed in [Alb93], where G is a connected Lie group and M a not necessarily compact manifold. Here $\mathcal{D}(M, G) \subseteq C^{\infty}(M, G)$ denotes the subgroup of those smooth functions having compact support in the sense that the set $\{m \in M : f(m) \neq 1\}$ has compact closure in M. This group is topologized as a topological group locally homeomorphic to the space $\mathcal{D}(M, \mathfrak{g})$ of $\mathfrak{g}$-valued test functions which is a nuclear LF space.

The representation theory of these groups seems to be much more involved if M is non-compact because in this case it is not clear whether $\mathcal{D}(M, G)$ is a K-space, i.e., functions on $\mathcal{D}(M, G)$ are continuous if they are continuous on all compact subsets. This makes it harder to put complete topologies on function spaces on these groups (cf. [Ne00a, Sect. III]). $\qquad \Box$

2. Central extensions of loop groups

In this section K denotes a compact connected Lie group. We write $LK=C^\infty(\mathbb{S}^1,K)$ for the associated *loop group*, i.e., the group of all smooth maps with values in K, where the group structure is given by pointwise multiplication. First we discuss this group as an infinite-dimensional Lie group modeled on a Fréchet space in the sense of Section I in [Ne00a].

As before, we identify $\mathbb{S}^1$ with $\mathbb{R}/2\pi\mathbb{Z}$, hence identify functions on $\mathbb{S}^1$ with 2π-periodic functions on $\mathbb{R}$.

To each K-invariant symmetric bilinear form κ on $\mathfrak{k}$ we associate the skew-symmetric bilinear form on $L\mathfrak{k}$ given by

$$\omega(\xi,\eta) := \frac{1}{2\pi} \int_0^{2\pi} \kappa\big(\xi(\theta),\eta'(\theta)\big)\,d\theta.$$

If κ is $\mathfrak{k}$-invariant, then ω is a cocycle, hence defines a central extension

(2.2) $$\{0\} \to \mathbb{R} \to \tilde{L}\mathfrak{k} \to L\mathfrak{k} \to \{0\},$$

where $\tilde{L}\mathfrak{k} = L\mathfrak{k} \oplus \mathbb{R}$ is endowed with the bracket

$$[(\xi,s),(\eta,t)] = \big([\xi,\eta],\omega(\xi,\eta)\big)$$

(cf. [PS86, p. 39]). If K is simply connected, then a corresponding central extension

(2.3) $$\{1\} \to \mathbb{T}_c \cong \mathbb{T} \longrightarrow \tilde{L}K \stackrel{q}{\longrightarrow} LK \to \{1\}$$

exists if and only if

(2.4) $$\kappa(h_\alpha,h_\alpha) \in 2\mathbb{Z}$$

(cf. Section I for the notation) is satisfied for all roots $\alpha \in \Delta_\mathfrak{k}$ (see Theorem II.5 below).

For a discussion of the case where K is not simply connected we refer to [PS86, p.55].

The group $\mathbb{T}_r := \mathbb{T}$ acts by rotating loops $\gamma \in LK$ via

$$(R_\theta.\gamma)(t) = \gamma(t-\theta).$$

Since the cocycle ω on $L\mathfrak{k}$ is invariant under this action, this action lifts to an action by automorphisms on $\tilde{L}K$ and we obtain a semidirect product group $\hat{L}K := \mathbb{T}_r \ltimes \tilde{L}K$ which is a central extension of the semidirect product $L_rK := \mathbb{T}_r \ltimes LK$. (Corollary II.20 below). Its Lie algebra $\hat{L}\mathfrak{k}$ is a semidirect sum $\mathfrak{t}_r \ltimes \tilde{L}\mathfrak{k}$, where $\mathfrak{t}_r$ denotes the Lie algebra of the group $\mathbb{T}_r$.

We identify the group K with the subgroup of constant loops in LK and accordingly $\mathfrak{k}$ with a Lie subalgebra of $L\mathfrak{k}$. The triviality of the cocycle ω on $\mathfrak{k}$ implies that the Lie algebra $\mathfrak{k}$ lifts to a subalgebra (also denoted $\mathfrak{k}$) of $\tilde{L}\mathfrak{k}$. If K is simply connected, then we conclude that there exists a continuous homomorphism $K \to \tilde{L}K$ which is injective because q maps the image isomorphically to the group K of constant loops. Thus we can identify K with a subgroup of the central extension $\tilde{L}K$.

If $\mathfrak{k}_{\mathbb{C}}$ is the complexification of $\mathfrak{k}$ and κ denotes the complex bilinear extension of the bilinear form on $\mathfrak{k}$, then the complex bilinear extension of ω to $L\mathfrak{k}_{\mathbb{C}}$ is given by the same formula and defines a central extension

$$(2.5) \qquad \{0\} \to (\mathfrak{t}_c)_{\mathbb{C}} \cong \mathbb{C} \to \tilde{L}\mathfrak{k}_{\mathbb{C}} \to L\mathfrak{k}_{\mathbb{C}} \to \{0\}.$$

If κ satisfies condition (2.4) and K is simply connected, then we use Theorem II.5 to see that we also obtain a central extension on the level of the complex groups, where the homomorphisms are holomorphic:

$$(2.6) \qquad \{1\} \to (\mathbb{T}_c)_{\mathbb{C}} \cong \mathbb{C}^{\times} \xrightarrow{\quad} \tilde{L}K_{\mathbb{C}} \xrightarrow{\quad q \quad} LK_{\mathbb{C}} \to \{1\}$$

(cf. also [PS86, p.90] and the remarks at the end of Section II.3 in [Wu00]). By the same argument as above, we see that $K_{\mathbb{C}}$ can be realized as a subgroup of $\tilde{L}K_{\mathbb{C}}$.

3. Appendix IIa: Central extensions and semidirect products

In this appendix we use the results on central extensions of general infinite-dimensional Lie groups in [Ne00b] to derive the existence of the central extensions $\tilde{L}K$ and $\tilde{L}K_{\mathbb{C}}$, and also the existence of the semidirect product Lie group $\hat{L}K \cong \mathbb{T}_r \ltimes \tilde{L}K$ discussed above.

Theorem II.5. *If K is a compact simple simply connected Lie group and $\kappa \colon \mathfrak{k} \times \mathfrak{k} \to \mathbb{R}$ an invariant symmetric bilinear form with $\kappa(h_\alpha, h_\alpha) \in 2\mathbb{Z}$ for all $\alpha \in \Delta_{\mathfrak{k}}$, then the 2-cocycle*

$$\omega(\xi, \eta) := \frac{1}{2\pi} \int_0^{2\pi} \kappa(\xi(t), \eta'(t)) \, dt$$

of the Lie algebra $L\mathfrak{k}$ corresponds to a central group extension

$$\mathbb{T} \cong \mathbb{R}/2\pi\mathbb{Z} \hookrightarrow \tilde{L}K \twoheadrightarrow LK.$$

In this case we also obtain a central extension of complex Lie groups

$$\mathbb{T}_{\mathbb{C}} \cong \mathbb{C}^{\times} \hookrightarrow \tilde{L}K_{\mathbb{C}} \twoheadrightarrow LK_{\mathbb{C}}.$$

Proof. Since the second homotopy group of a finite-dimensional Lie group K vanishes, we have

$$\pi_0(K) = \pi_1(K) = \pi_2(K) = \{1\} \quad \text{and} \quad \pi_3(K) \cong \mathbb{Z}$$

([Mi95, Ths. 3.7 and 3.9]). With $\Omega K := \{\gamma \in LK \colon \gamma(0) = 1\}$ we have a semidirect decomposition $LK \cong \Omega K \rtimes K$, so that

$$\pi_k(\Omega K) \cong \pi_{k+1}(K), \quad k \in \mathbb{N}_0$$

(cf. [Br93, Cor. VII.4.4]) leads to

$$\pi_k(LK) = \pi_k(\Omega K \rtimes K) \cong \pi_k(\Omega K) \times \pi_k(K) \cong \pi_{k+1}(K) \times \pi_k(K),$$

so that LK is connected and simply connected with $\pi_2(LK) \cong \mathbb{Z}$. Therefore we have to show that the left invariant 2-form Ω on LK with $\Omega_1 = \omega$ satisfies

$$\int_{S^2} \gamma^* \Omega \in 2\pi\mathbb{Z},$$

where $\gamma\colon \mathbb{S}^2 \to LK$ is a smooth representative of a generator of $\pi_2(LK)$ (cf. [Ne00b, Ths. IV.12(b) and V.7]). Since the inclusion $K \hookrightarrow K_{\mathbb{C}}$ is a homotopy equivalence, the same holds for the inclusion $LK \hookrightarrow LK_{\mathbb{C}}$, so that the condition for the existence of a central extension of the complex group $LK_{\mathbb{C}}$ is the same.

Next we prove the proposition for the special case $K = \mathrm{SU}(2) \cong \mathbb{S}^3$. On the Lie algebra $\mathfrak{k} = \mathfrak{su}(2)$ we consider the operator norm corresponding to the complex euclidean norm on $\mathbb{C}^2$. On $\mathfrak{su}(2)$ this norm is given by the scalar product $\frac{1}{2}\kappa(x,y) := -\frac{1}{2}\operatorname{tr}(xy)$ (note that the unit balls in $\mathfrak{su}(2)$ are invariant under the adjoint action, hence euclidean balls). Let $B_3 \subseteq \mathfrak{su}(2)$ denote the closed unit ball and idenitfy $\mathbb{S}^2$ with ∂B_3. We consider the smooth map

$$\gamma\colon \mathbb{S}^2 \to L\,\mathrm{SU}(2), \quad \gamma(x)(t) := e^{tx} = \exp(tx).$$

It is easy to see that $\|x\| = 1$ for $x \in \mathfrak{su}(2)$ leads to $e^{2\pi x} = \mathbf{1}$. The map $\mathbb{S}^3 \to \mathrm{SU}(2)$ corresponding to γ is induced by $B_3 \to \mathrm{SU}(2), x \mapsto e^{2\pi x}$, which maps ∂B_3 to $\mathbf{1}$, hence leads to a map $\mathbb{S}^3 \to \mathrm{SU}(2)$. It is not hard to verify (counting inverse images of points) that this map is of degree 2, hence its homotopy class corresponds to the element $2 \in \mathbb{Z} \cong \pi_3(\mathrm{SU}(2))$. Now the isomorphism $\pi_3(\mathrm{SU}(2)) \cong \pi_2(L\,\mathrm{SU}(2))$ from above implies that $[\gamma] = 2 \in \mathbb{Z} \cong \pi_2(L\,\mathrm{SU}(2))$. We claim that the left invariant 2-form Ω on $L\,\mathrm{SU}(2)$ satisfies

$$\int_{\mathbb{S}^2} \gamma^*\Omega = 4\pi.$$

To facilitate this computation, we first observe that the Lie algebra cocycle ω on $L\mathfrak{su}(2)$ is $\mathrm{Ad}(\mathrm{SU}(2))$-invariant, so that Ω on $L\,\mathrm{SU}(2)$ is invariant under conjugation by $\mathrm{SU}(2)$. For $g \in \mathrm{SU}(2)$ we further have $\gamma(\mathrm{Ad}(g).x) = g\gamma(x)g^{-1}$, showing that $\gamma^*\Omega$ is an invariant 2-form on $\mathbb{S}^2$, hence determined by the value in a fixed element x_0.

In $\mathfrak{su}(2)$ there exist elements y_0, z_0 with the brackets

$$[x_0, y_0] = 2z_0, \quad [x_0, z_0] = -2y_0 \quad \text{and} \quad [y_0, z_0] = 2x_0$$

and such that (x_0, y_0, z_0) is an orthonormal basis with respect to κ. We have to evaluate

$$(\gamma^*\Omega)(x_0)(y_0, z_0) = \Omega(\gamma(x_0))(d\gamma(x_0).y_0, d\gamma(x_0).z_0).$$

Writing $\lambda_g(x) := gx$ for group elements, the element

$$d\gamma(x_0).y_0 \in T_{\gamma(x_0)}(L\,\mathrm{SU}(2))$$

is given by

$$(d\gamma(x_0).y_0)(t) = d\exp(tx_0)(ty_0) = d\lambda_{\exp(tx_0)}(\mathbf{1})f(\operatorname{ad} tx_0).ty_0,$$

where f is the holomorphic function given by $f(z) = \frac{1-e^{-z}}{z}$. With

$$\xi(t) := f(\operatorname{ad} tx_0).ty_0 \quad \text{and} \quad \eta(t) := f(\operatorname{ad} tx_0).tz_0$$

we therefore obtain

$$(\gamma^*\Omega)(x_0)(y_0, z_0) = \omega(\xi, \eta).$$

We compute

$$\xi(t) = f(\operatorname{ad} tx_0).ty_0 = \frac{1 - e^{-\operatorname{ad} tx_0}}{\operatorname{ad} tx_0}.ty_0$$

$$= \frac{1 - \cosh(\operatorname{ad} tx_0)}{\operatorname{ad} tx_0}.ty_0 + \frac{\sinh(\operatorname{ad} tx_0)}{\operatorname{ad} tx_0}.ty_0$$

$$= \frac{\cos(2t) - 1}{2} z_0 + \frac{\sin(2t)}{2} y_0.$$

Applying $\frac{1}{2} \operatorname{ad} x_0$, we also get

$$\eta(t) = f(\operatorname{ad} tx_0).tz_0 = \frac{1 - \cos(2t)}{2} y_0 + \frac{\sin(2t)}{2} z_0,$$

and hence

$$\eta'(t) = \sin(2t)y_0 + \cos(2t)z_0.$$

This leads to

$$\kappa(\xi(t), \eta'(t)) = \sin^2(2t) + \left(\cos^2(2t) - \cos(2t) \right) = 1 - \cos(2t).$$

Integration of this expression leads to

$$\omega(\xi, \eta) = \frac{1}{2\pi} \int_0^{2\pi} \kappa(\xi(t), \eta'(t))\, dt = 1.$$

This proves that $\gamma^*\Omega$ is the volume form on $\mathbb{S}^2$, and hence that

$$\int_{\mathbb{S}^2} \gamma^*\Omega = \operatorname{vol}(\mathbb{S}^2) = 4\pi,$$

which, in view of $[\gamma] = 2 \in \mathbb{Z} \cong \pi_2(LK)$, we had to show.

Let $\alpha \in \Delta_{\mathfrak{t}}$ be a long root and $\mathfrak{k}(\alpha) \subseteq \mathfrak{k}$ the corresponding $\mathfrak{su}(2)$-subalgebra (cf. Section I). Then the corresponding homomorphism inclusion $SU(2) \cong \mathbb{S}^3 \to K$ represents a generator of $\pi_3(K)$ ([Bo58]), so that the corresponding map

$$\gamma_\alpha : \mathbb{S}^2 \to LK$$

represents twice a generator of $\pi_2(LK)$. Therefore the preceding calculation shows that for κ on $\mathfrak{k}$ as above, we get

$$\int_{\mathbb{S}^2} \gamma_\alpha^*\Omega = 2\pi \cdot \kappa(h_\alpha, h_\alpha) \in 4\pi\mathbb{Z}$$

if and only if $\kappa(h_\alpha, h_\alpha) \in 2\mathbb{Z}$ holds for one and hence for all long roots α.

If $\beta \in \Delta_k$ is a short root, then h_β is a long coroot, so that $\kappa(h_\beta, h_\beta) \in \mathbb{N}\kappa(h_\alpha, h_\alpha) \subseteq 2\mathbb{Z}$. Therefore the requirement that $\kappa(h_\alpha, h_\alpha) \in 2\mathbb{Z}$ holds for all long roots is equivalent to the same requirement for all roots. $\square$

4. Appendix IIb: Smoothness of group actions

We consider the rotation action $\mathbb{T}_r \times LK \to LK$. In view of Theorem III.5 in [Ne00a], the action of $\mathbb{T}_r$ on $L\mathfrak{k}$ is smooth. Now we also show that the action on LK is smooth.

Lemma II.6. *Let M be a compact manifold, K a finite-dimensional Lie group and $G \times M \to M$ a smooth Lie group action on M. Then the action of G on $C^\infty(M, K)$ given by $(g.f)(x) := f(g^{-1}.x)$ is smooth.*

Proof. Step 1: In view of Theorem III.5 in [Ne00a], the action of G on the Fréchet space $C^\infty(M, \mathfrak{k})$ is smooth. Since the exponential function of $C^\infty(M, K)$ is a local diffeomorphism around 0, we obtain an open 1-neighborhood $U \subseteq K$ such that the action of G on $C^\infty(M, U)$ is smooth.

Step 2: Fix $f \in C^\infty(M, K)$ and write $\sigma: G \times M \to M$ for the smooth action map. We claim that $G \to C^\infty(M, K), g \mapsto g.f$ is a smooth map. Since each $g \in G$ acts by a diffeomorphism on $C^\infty(M, K)$, it suffices to verify smoothness in a neighborhood of 1. The smoothness of the left multiplications on $C^\infty(M, K)$ shows that it even suffices to show that the map $g \mapsto f^{-1} \cdot (g.f)$ is smooth in a neighborhood of 1. Since M is compact, f is uniformly continuous, and there exists an open neighborhood U of 1 in G and an open neighborhood V of 0 in $\mathfrak{k}$ such that $\exp|_{V_\mathfrak{k}}: V_\mathfrak{k} \to \exp(V_\mathfrak{k})$ is a diffeomorphism and $f(x)^{-1} f(g^{-1}.x) \in \exp(V_\mathfrak{k})$ holds for all $g \in U$ and $x \in M$. Now it suffices to see that the map

$$U \to C^\infty(\mathfrak{k}), \quad g \mapsto (\exp|_{V_\mathfrak{k}})^{-1}(f(x)^{-1} f(g^{-1}.x))$$

is smooth, which follows from Theorem III.4 in [Ne00a]. This completes the proof of Step 2.

Step 3: For $f \in C^\infty(M, K)$ the set $f \cdot C^\infty(M, U)$ is an open neighborhood of f, and the arguments above show that the map

$$G \times \left(f \cdot C^\infty(M, U)\right) \to C^\infty(M, K), \quad (g.f, x) \mapsto (g.f) \cdot (g.x)$$

is smooth. This eventually proves that the action of G on $C^\infty(M, K)$ is smooth.
$\square$

Corollary II.7. *Let G act on $C^\infty(M, K)$ by a smooth action of M as above. Then the semidirect product group $C^\infty(M, K) \rtimes G$ is a Lie group with Lie algebra $C^\infty(M, \mathfrak{k}) \rtimes \mathfrak{g}$.*

Proof. The preceding lemma implies that the inversion and the multiplication of this group are smooth maps, showing that the product manifold $C^\infty(M, K) \times G$ is a Lie group with the natural semidirect product structure.
$\square$

Corollary II.8. *The natural rotation action of $\mathbb{T}_r$ on LK is smooth, so that we obtain a semidirect product Lie group $L_r K := \mathbb{T}_r \rtimes LK$.*
$\square$

5. Appendix IIc: Lifting automorphisms to central extensions

In the following we will use the concept of an infinite-dimensional Lie group described in detail in [Ne00a]. In this context central extensions of Lie groups are always assumed to have a smooth local section. Let $Z \hookrightarrow \hat{G} \twoheadrightarrow G$ be a central extension of the connected Lie group by the connected abelian group Z which can be written as $\mathfrak{z}/\pi_1(Z)$. This means that Z is a quotient of a sequentially complete locally convex space $\mathfrak{z}$ modulo a discrete subgroup which can then be identified with $\pi_1(Z)$. Since the quotient map $q \colon \hat{G} \to G$ has a smooth local section, the corresponding Lie algebra homomorphism $\hat{\mathfrak{g}} \to \mathfrak{g}$ has a continuous linear section, hence is defined by a continuous cocycle $\omega \in Z^2_c(\mathfrak{g}, \mathfrak{z})$ in the sense that

$$\hat{\mathfrak{g}} \cong \mathfrak{g} \oplus_\omega \mathfrak{z}$$

with the bracket $[(x, z), (x', z')] = ([x, x'], \omega(x, x'))$. From [Ne00b] we recall the period homomorphism $\mathrm{per}_\omega \colon \pi_2(G) \to \mathfrak{z}$ of ω which on smooth representatives $\gamma \colon \mathbb{S}^2 \to G$ of elements of $\pi_2(G)$ is given by $\mathrm{per}_\omega([\gamma]) = \int_{\mathbb{S}^2} \gamma^* \Omega$, where Ω is the $\mathfrak{z}$-valued left invariant 2-form on G with $\Omega_1 = \omega$ ([Ne00b, Th. IV.12]).

We recall from [Ne00b, Prop. IV.2] that central Lie group extensions as above can always be written as

$$\hat{G} \cong G \times_f Z,$$

where $f \in Z^2_s(G, Z)$, the group cocycles $f \colon G \times G \to Z$ which are smooth in a neighborhood of $(\mathbf{1}, \mathbf{1})$. Two such cocycles f_1, f_2 define equivalent extensions if and only if their difference is of the form $h(gg')h(g)^{-1}h(g')^{-1}$, where $h \colon G \to Z$ is smooth in an identity neighborhood. The abelian group all these functions is called $B^2_s(G, Z)$, and the quotient group $H^2_s(G, Z) := Z^2_s(G, Z)/B^2_s(G, Z)$ now parametrizes the equivalence classes of central Z-extensions of G with smooth local sections ([Ne00b, Remark IV.4]). On the Lie algebra level we likewise have the space $Z^2_c(\mathfrak{g}, \mathfrak{z})$ of continuous linear 2-cocycles $\omega \colon \mathfrak{g} \times \mathfrak{g} \to \mathfrak{z}$, the subspace $B^2_c(\mathfrak{g}, \mathfrak{z})$ of coboundaries, i.e., the cocycles of the form $(x, x') \mapsto \alpha([x, x'])$, where $\alpha \colon \mathfrak{g} \to \mathfrak{z}$ is a continuous linear map. The quotient space $H^2_c(\mathfrak{g}, \mathfrak{z}) := Z^2_c(\mathfrak{g}, \mathfrak{z})/B^2_c(\mathfrak{g}, \mathfrak{z})$ classifies the central $\mathfrak{z}$-extensions of $\mathfrak{g}$ with continuous linear sections. For more details on all that we refer to [Ne00b].

For a Lie group G we write $\mathrm{Aut}(G)$ for the group of Lie group automorphisms of G. We also define

$$\mathrm{Aut}_Z(\hat{G}) := \{\gamma \in \mathrm{Aut}(\hat{G}) \colon f|_Z = \mathrm{id}_Z\}.$$

Then we have a natural homomorphism

$$\eta \colon \mathrm{Aut}_Z(\hat{G}) \to \mathrm{Aut}(G), \quad \eta(\gamma)(q(g)) = q(\gamma(g)),$$

where $q \colon \hat{G} \to G$ is the quotient map of the central extension.

Lemma II.9. *To each $f \in \mathrm{Hom}(G, Z)$ we assign the element of $\mathrm{Aut}_Z(\hat{G})$ given by $\hat{f}(g) := gf(q(g))$. Then*

$$\ker \eta = \{\hat{f} \colon f \in \mathrm{Hom}(G, Z)\} \cong \mathrm{Hom}(G, Z).$$

Proof. For each $f \in \operatorname{Hom}(G, Z)$ we have

$$\hat{f}(g_1 g_2) = g_1 g_2 f(q(g_1) q(g_2)) = g_1 f(q(g_1)) g_2 f(q(g_2)) = \hat{f}(g_1) \hat{f}(g_2),$$

showing that $\hat{f} \in \operatorname{Aut}_Z(G)$. It is clear that $\hat{f} \in \ker \eta$.

If, conversely, $\gamma \in \ker \eta$, then

$$f \colon G \to Z, \quad q(g) \mapsto \gamma(g) g^{-1}$$

is well defined, and it is easy to verify that f is a group homomorphism with $\gamma = \hat{f}$.
$\square$

Corollary II.10. *If G has no nontrivial homomorphisms into abelian groups, then η is injective.*
$\square$

In view of [Ne00b, Cor. III.20], the preceding condition is equivalent to the density of the commutator algebra $[\mathfrak{g}, \mathfrak{g}]$ in $\mathfrak{g}$.

Lemma II.11. *Let $f \in Z_s^2(G, Z)$ be a group cocycle which is smooth in a neighborhood of $(1, 1)$ and for which $\hat{G}$, as an abstract group, not as a manifold, is isomorphic to the group $G \times_f Z$ with the multiplication*

$$(g, z)(g', z') = (gg', zz' f(g, g'))$$

and the map $G \to \hat{G}, g \mapsto (g, 1)$ is smooth in an identity neighborhood. Then $\gamma \in \operatorname{Aut}(G)$ is contained in the image of η if and only if the cocycle $\gamma^* f := f \circ (\gamma, \gamma)$ is equivalent to f, i.e., there exists a function $h \colon G \to Z$, smooth on an identity neighborhood, such that

$$(2.7) \qquad (\gamma^* f)(g, g') f(g, g')^{-1} = h(gg') h(g)^{-1} h(g')^{-1}, \quad g, g' \in G.$$

Proof. Let us first assume that $\gamma = \eta(\hat{\gamma})$. Writing $\hat{G}$ as $G \times_f Z$, the automorphism $\hat{\gamma}$ has the form

$$\hat{\gamma}(g, z) = (\gamma(g), z h(g)),$$

where $h \colon G \to Z$ is a function which is smooth in an identity neighborhood. The condition that $\hat{\gamma}$ is an automorphism of $\hat{G}$ is equivalent to the relation (2.7).

If, conversely, (2.7) is satisfied, then the formula above defines an element $\hat{\gamma} \in \operatorname{Aut}_Z(\hat{G})$ which is smooth in an identity neighborhood, and since $\hat{G}$ is connected, it is smooth on $\hat{G}$, hence an isomorphism of Lie groups.
$\square$

Lemma II.12. *If $\gamma \in \operatorname{Aut}(G)$ is contained in the range of η, then there exists a continuous linear map $\alpha \colon \mathfrak{g} \to \mathfrak{z}$ such that $(\gamma^* \omega)(x, y) := \omega(\gamma.x, \gamma.y)$ satisfies*

$$(2.8) \qquad (\gamma^* \omega - \omega)(x, y) = \alpha([x, y]),$$

i.e., $[\gamma^* \omega] = [\omega]$ in $H_c^2(\mathfrak{g}, \mathfrak{z})$. If G is simply connected, then the preceding condition is also sufficient for γ to be in the range of η.

Proof. If $\gamma = \eta(\hat{\gamma})$, then $\hat{\gamma}$ induces an automorphism of the Lie algebra $\hat{\mathfrak{g}} \cong \mathfrak{g} \oplus_\omega \mathfrak{z}$ with the bracket

$$[(x, z), (x', z')] = ([x, x'], \omega(x, x')).$$

Hence

$$\hat{\gamma}.(x, z) = (\gamma.x, z + \alpha(x)),$$

where $\alpha \colon \mathfrak{g} \to \mathfrak{z}$ is a continuous linear map. The relation (2.8) means that such a map is a Lie algebra automorphism.

If, conversely, (2.8) is satisfied by γ, then we use the exact sequence for Lie group extensions to see that the natural map

$$\mathrm{Ext}(G, Z) \cong H_s^2(G, Z) \to \mathrm{Ext}(\mathfrak{g}, \mathfrak{z}) \cong H_c^2(\mathfrak{g}, \mathfrak{z})$$

is injective. Moreover, it is equivariant with respect to the action of $\mathrm{Aut}(G)$ on $H_s^2(G, Z)$, resp., $H_c^2(\mathfrak{g}, \mathfrak{z})$. Therefore $[\gamma^* \omega] = [\omega]$ in $H_c^2(\mathfrak{g}, \mathfrak{z})$ implies that $[\gamma^* f] = [f]$ in $H_s^2(G, Z)$ if $f \in Z_s^2(G, Z)$ represents $\hat{G}$ as in Lemma II.11. $\qquad\square$

Corollary II.13. *If G is simply connected and $\omega \in Z_c^2(\mathfrak{g}, \mathfrak{z})$ is a Lie algebra cocycle corresponding to the Lie algebra extension $\mathfrak{z} \hookrightarrow \hat{\mathfrak{g}} \twoheadrightarrow \mathfrak{g}$, then an automorphism $\gamma \in \mathrm{Aut}(G)$ lifts to an automorphism $\hat{\gamma} \in \mathrm{Aut}_Z(\hat{G})$ if and only if $[\gamma^* \omega] = [\omega]$, i.e., if the corresponding automorphism of $\mathfrak{g}$ lifts to an automorphism of $\hat{\mathfrak{g}}$ fixing $\mathfrak{z}$ pointwise.*

Proof. This is a direct consequence of Lemma II.12. $\qquad\square$

If G is not simply connected, then it has non-trivial central Z-extensions with trivial Lie algebra extension. These are discussed in the following lemma.

Lemma II.14. *If $\hat{G}$ is of the form*

$$\hat{G} = (\tilde{G} \times Z)/\Gamma(\phi^{-1}),$$

where $q_G \colon \tilde{G} \to G$ is the universal covering group of G, $\pi_1(G) \cong \ker q_G$ is identified with a subgroup of $\tilde{G}$, $\phi \colon \pi_1(G) \to Z$ is a homomorphism, and

$$\Gamma(\phi^{-1}) := \{(d, \phi(d)^{-1}) \colon d \in \pi_1(G)\}$$

the graph of ϕ^{-1} (pointwise inverse), then an automorphism $\gamma \in \mathrm{Aut}(G)$ is in the range of η if and only if $(\phi \circ \pi_1(\gamma)) \cdot \phi^{-1}$ extends to a smooth homomorphism $\tilde{G} \to Z$.

Proof. Let $\tilde{\gamma}$ be the canonical lift of γ to $\tilde{G}$ (cf. [Ne00b, Lemma II.3]). The canonical map $\tilde{G} \times Z \to \hat{G}$ is a covering, and $\tilde{G} \times \mathfrak{z}$ is the universal covering group of $\hat{G}$. Therefore, if $\gamma = \eta(\hat{\gamma})$, the automorphism $\hat{\gamma}$ also lifts to some automorphism $\tilde{\gamma}$ of $\tilde{G} \times Z$ preserving the subgroup $\Gamma(\phi^{-1})$. Then $\tilde{\gamma}$ is of the form

$$\tilde{\gamma}(g, z) = (\tilde{\gamma}_0(g), z f(g)),$$

with $f \in \mathrm{Hom}(\tilde{G}, Z)$. The condition that $\tilde{\gamma}$ preserves $\Gamma(\phi^{-1})$ means that f extends $\phi \cdot (\phi \circ \pi_1(\gamma))^{-1}$. If, conversely, this condition is satisfied, then the above formula yields an automorpism $\tilde{\gamma}$ on $\tilde{G} \times Z$ preserving $\Gamma(\phi^{-1})$ and hence factoring to the quotient group $\hat{G}$. $\qquad\square$

With Lemma II.14 one can easily construct examples showing that in Corollary II.13 the assumption that G is simply connected is crucial.

Example II.15. (a) We consider $G = \mathrm{SL}(2, \mathbb{R})$ with the automorphism $\gamma(g) = JgJ$ for $J = \begin{pmatrix} 1 & 0 \\ 0 & -1 \end{pmatrix}$. Then $\pi_1(G) \cong \pi_1(\mathrm{SO}(2, \mathbb{R})) \cong \mathbb{Z}$ and $\pi_1(\gamma) = -\mathrm{id}_\mathbb{Z}$. On the other hand $\tilde{G}$ is perfect, so that $\mathrm{Hom}(\tilde{G}, Z)$ is trivial for every abelian group. Therefore $\phi \in \mathrm{Hom}(\mathbb{Z}, Z)$ satisfies the condition from Lemma II.14 if and only if $\phi(d)^2 = 1$ for all $d \in \mathbb{Z}$, i.e., $2\mathbb{Z} \subseteq \ker \phi$.

If $Z = \mathbb{T}$ and $\phi \colon \mathbb{Z} \to \mathbb{T}$ is injective, then we obtain a central $\mathbb{T}$-extension $\hat{G}$ of G whose corresponding Lie algebra extension is trivial, but γ does not lift to an element of $\mathrm{Aut}_Z(\hat{G})$.

(b) Let

$$\hat{\mathfrak{g}} := \left\{ \begin{pmatrix} 0 & a_{12} & a_{13} & a_{14} \\ 0 & 0 & a_{23} & a_{24} \\ 0 & 0 & 0 & a_{34} \\ 0 & 0 & 0 & 0 \end{pmatrix} \in \mathfrak{gl}(4, \mathbb{R}) \colon a_{jk} \in \mathbb{R} \right\}.$$

Then $\hat{\mathfrak{g}}$ contains the ideals

$$\hat{\mathfrak{g}}_1 := \left\{ \begin{pmatrix} 0 & 0 & a_{13} & a_{14} \\ 0 & 0 & 0 & a_{24} \\ 0 & 0 & 0 & 0 \\ 0 & 0 & 0 & 0 \end{pmatrix} \right\} = [\hat{\mathfrak{g}}, \hat{\mathfrak{g}}] \quad \text{and} \quad \hat{\mathfrak{g}}_2 := \left\{ \begin{pmatrix} 0 & 0 & 0 & a_{14} \\ 0 & 0 & 0 & 0 \\ 0 & 0 & 0 & 0 \\ 0 & 0 & 0 & 0 \end{pmatrix} \right\} = \mathfrak{z}(\hat{\mathfrak{g}})$$

satisfying $[\hat{\mathfrak{g}}_1, \hat{\mathfrak{g}}] \subseteq \mathfrak{z}(\hat{\mathfrak{g}})$. We define $\mathfrak{g} := \hat{\mathfrak{g}}/\mathfrak{z}(\hat{\mathfrak{g}})$ and consider the central extension $\mathfrak{z}(\mathfrak{g}) \hookrightarrow \hat{\mathfrak{g}} \twoheadrightarrow \mathfrak{g}$. The Lie algebra $\mathfrak{g}$ is a 2-step nilpotent Lie algebra, and $\mathrm{ad}\,\mathfrak{g} \cong \mathbb{R}^3$ is abelian. On the other hand, the adjoint action of $\hat{\mathfrak{g}}$ on $\hat{\mathfrak{g}}$ factors through an action of $\mathfrak{g}$ on $\hat{\mathfrak{g}}$, where the image $\mathrm{ad}_{\hat{\mathfrak{g}}}\,\mathfrak{g}$ of this Lie algebra is isomorphic to $\mathfrak{g}$. This means that the action of $\mathrm{ad}\,\mathfrak{g} \cong \mathbb{R}^3$ on $\mathfrak{g}$ does not lift to an action of the same Lie algebra on $\hat{\mathfrak{g}}$ because the central extension

$$\mathrm{Hom}(\mathfrak{g}, \mathfrak{z}(\mathfrak{g})) \hookrightarrow \mathrm{Hom}(\mathfrak{g}, \mathfrak{z}(\mathfrak{g})) + \mathrm{ad}_{\hat{\mathfrak{g}}}(\mathfrak{g}) \twoheadrightarrow \mathrm{ad}\,\mathfrak{g}$$

is non-trivial. From this one obtains an example of an $\mathbb{R}^3$-action on a simply connected group G which does not lift to an action on a central extension $\hat{G}$, even though the action of every element can be lifted. □

6. Appendix IId: Lifting automorphic group actions to central extensions

In the preceding subsection we have lifted automorphisms of G to automorphisms of $\hat{G}$. Now we consider automorphic actions of groups R on G and want to lift those to actions on $\hat{G}$.

Lemma II.16. *Let* $Z^\sharp := \mathfrak{z}/\mathrm{im}(\mathrm{per}_\omega)$. *Then there exists a central Lie group extension*

$$Z^\sharp \hookrightarrow G^\sharp \xrightarrow{q^\sharp} \tilde{G}$$

corresponding to the cocycle ω, *and* $G^\sharp$ *is a universal covering group of* $\hat{G}$.

Proof. In view of [Ne00b, Th. V.7], $\mathrm{im}(\mathrm{per}_\omega)$ is a subgroup of the discrete group $\pi_1(Z) \subseteq \mathfrak{z}$, so that $Z^\sharp := \mathfrak{z}/\mathrm{im}(\mathrm{per}_\omega)$ is a covering group of $Z \cong \mathfrak{z}/\pi_1(Z)$. The relation $\pi_2(G) \cong \pi_2(\tilde{G})$ and the criterion Theorem V.7 in [Ne00b] imply the existence of a central extension

$$Z^\sharp \hookrightarrow G^\sharp \xrightarrow{q^\sharp} \tilde{G}$$

corresponding to the cocycle ω. Now $\pi_2(Z) = \{\mathbf{1}\}$ and the exact homotopy sequence of the bundle $G^\sharp \to \tilde{G}$ lead to an exact sequence

$$\pi_2(G^\sharp) \to \pi_2(\tilde{G}) \xrightarrow{\delta} \pi_1(Z^\sharp) \twoheadrightarrow \pi_1(G^\sharp).$$

Since $\delta = -\mathrm{per}_\omega$ ([Ne00b, Prop. VII.7]), it is surjective, which implies that $G^\sharp$ is simply connected. On the other hand, the construction of $G^\sharp$ implies that $G^\sharp$ and $\hat{G}$ are locally isomorphic ([Ne00b, Lemmas IV.8, V.8]), so that $G^\sharp$ is the universal covering group of $\hat{G}$.   ☐

We assume that we have a smooth automorphic action of the Lie group R on G, which leads to a semidirect product Lie group $G \rtimes R$. We are looking for sufficient conditions to lift the smooth action of R on G to a smooth action on $\hat{G}$ which apply in particular to the rotation action of $\mathbb{T}_r$ on LK, where K is a compact simple simply connected Lie group.

Lemma II.17. *The action of R on G lifts to a smooth action of R on the simply connected covering group $\tilde{G}$ of G.*

Proof. Since each automorphism of G lift in a unique fashion to an automorphism of $\tilde{G}$, the action of R on G directly leads to an action of R on $\tilde{G}$. That the action map is smooth follows easily by using local sections of the universal covering map $q_G : \tilde{G} \to G$.   ☐

Theorem II.18. *(Lifting Theorem) Let $\sigma_G : R \times G \to G$ be a smooth automorphic action of the Lie group R on the connected Lie group G. Assume that G is simply connected and that there exists a smooth function $\alpha : R \times \mathfrak{g} \to \mathfrak{z}$ with*

(2.9) $$r^* \omega - \omega = \alpha(r, [\cdot, \cdot]), \quad r \in R$$

and the cocycle condition

(2.10) $$\alpha(r_1 r_2, x) = \alpha(r_2, x) + \alpha(r_1, r_2.x), \quad r_1, r_2 \in R, x \in \mathfrak{g}.$$

Then the action of R on G lifts uniquely to a smooth automorphic action of R on $\hat{G}$ such that the corresponding action of R on $\hat{\mathfrak{g}} \cong \mathfrak{g} \oplus_\omega \mathfrak{z}$ is given by

$$r.(x, z) = (r.x, z + \alpha(r, x)), \quad r \in R, x \in \mathfrak{g}, z \in \mathfrak{z}.$$

This action fixes the subgroup Z of $\hat{G}$ pointwise.

Proof. First we turn to the action on the Lie algebra $\hat{\mathfrak{g}}$. We define the a smooth map

$$\sigma_{\hat{\mathfrak{g}}} : R \times \hat{\mathfrak{g}} \to \hat{\mathfrak{g}}, \quad (r, (x, z)) \mapsto (r.x, z + \alpha(r, x)).$$

Then (2.9) implies that each map $\sigma_{\hat{\mathfrak{g}}}(r, \cdot)$ is an automorphism of $\hat{\mathfrak{g}}$:

$$r.[(x, z), (x', z')] = r.([x, x'], \omega(x, x')) = ([r.x, r.x'], \omega(x, x') + \alpha(r, [x, x']))$$
$$= ([r.x, r.x'], \omega(r.x, r.x')) = [r.(x, z), r.(x', z')],$$

and (2.10) implies that $\sigma_{\hat{\mathfrak{g}}}$ is an action of R on $\hat{\mathfrak{g}}$:

$$r_1.(r_2.(x, z)) = ((r_1 r_2).x, z + \alpha(r_2, x) + \alpha(r_1, r_2.x)) = ((r_1 r_2).x, z + \alpha(r_1 r_2, x))$$
$$= (r_1 r_2).(x, z).$$

Since every automorphism of $\hat{G}$ is uniquely determined by the corresponding automorphism of $\hat{\mathfrak{g}}$ (cf. [Mi83, Lemma 7.1]), there exists at most one automorphic action $\sigma_{\hat{G}}$ of R on $\hat{G}$ corresponding to $\sigma_{\hat{\mathfrak{g}}}$.

Now we reduce the problem to the case where $\hat{G}$ is simply connected. Suppose that the theorem holds for the special situation where $\hat{G}$ is simply connected. Then we consider the simply connected central extension $G^{\sharp}$ of G discussed in Lemma II.16. The assertion of the theorem shows that the action of R on G lifts to a smooth action of R on $G^{\sharp}$ fixing the elements of $Z^{\sharp}$ pointwise. Since G is simply connected, the exact homotopy sequence of the bundle $q: \hat{G} \to G$ implies that the natural map $\zeta: \pi_1(Z) \to \pi_1(\hat{G})$ is surjective. Identifying $\pi_1(\hat{G})$ with the kernel of the universal covering map $G^{\sharp} \to \hat{G}$, we see that ζ, viewed as a homomorphism $\pi_1(Z) \to G^{\sharp}$, is a composition of the maps

$$\pi_1(Z) \to \pi_1(Z)/\pi_1(Z^{\sharp}) \subseteq Z^{\sharp} \to G^{\sharp}.$$

Therefore the subgroup $\pi_1(\hat{G})$ of $G^{\sharp}$ is contained in $Z^{\sharp}$ and therefore fixed pointwise by R. Hence the action of R on $G^{\sharp}$ factors to a smooth action of R on $\hat{G} \cong G^{\sharp}/\pi_1(\hat{G})$. In view of the preceding argument, we may from now on assume that $\hat{G}$ is simply connected.

Next we consider the local situation in a suitable small neighborhood of the identity in $\hat{G}$. In $\hat{G}$ we have an open 1-neighborhood of the form $U \times Z \subseteq \hat{G}$, where the multiplication is given for $x, x', xx' \in U$ by

$$(x, z)(x', z') = (xx', zz' f^Z(x, x'))$$

for a local smooth cocycle $f^Z: U \times U \to Z$ ([Ne00b, Lemma IV.8]). To see how this description of the multiplication can be used to obtain an action of R, we have to recall the construction of the local cocycle f^Z from the Lie algebra cocycle ω (cf. [Ne00b, Lemma IV.8]). Let us assume that, in addition, $U \subseteq G$ is diffeomorphic to a convex subset in $\mathfrak{g}$. Using the Poincaré Lemma ([Ne00b, Lemma III.3]), we write $\Omega|_U = d\theta$ for a $\mathfrak{z}$-valued 1-form θ on U with $\theta_1 = 0$. Then, on an open symmetric 1-neighborhood $W \subseteq U$ with $W^2 \subseteq U$ and also diffeomorphic to a convex set, we determine the function

$$f: W \times W \to \mathfrak{z}$$

by

$$df(x, \cdot) = \lambda_x^* \theta|_W - \theta|_W.$$

Now let $r \in R$ and $W_1 \subseteq W$ be open and diffeomorphic to a convex set with $r.W_1 \subseteq W$. Let α_r be the left invariant $\mathfrak{z}$-valued 1-form on G with $\alpha_r(\mathbf{1}) = \alpha(r, \cdot)$. Then (2.9) implies that

$$r^*\Omega - \Omega = -d\alpha_r.$$

On W_1 we therefore have $d(r^*\theta - \theta + \alpha_r) = 0$, so that there exists a unique function $h_r \colon W_1 \to \mathfrak{z}$ with $h_r(\mathbf{1}) = 0$ and $dh_r = r^*\theta - \theta + \alpha_r$.

On $W_1 \times W_1$ we consider the function $(r^*f)(x, y) := f(r.x, r.y)$ and put $f_x := f(x, \cdot)$. Then $(r^*f)_x = r^*(f_{r.x})$, so that on W_1 we have

$$d\big((r^*f)_x\big) = r^*df_{r.x} = r^*(\lambda^*_{r.x}\theta - \theta) = \lambda^*_x r^*\theta - r^*\theta.$$

Now the left invariance of α_r leads to

$$d((r^*f)_x - f_x) = \lambda^*_x(r^*\theta - \theta) - (r^*\theta - \theta)$$
$$= \lambda^*_x(r^*\theta - \theta + \alpha_r) - (r^*\theta - \theta + \alpha_r)$$
$$= \lambda^*_x dh_r - dh_r = d(\lambda^*_x h_r - h_r).$$

In view of the normalizations $f_x(\mathbf{1}) = 0 = h_r(\mathbf{1})$, we therefore obtain

$$(2.11) \qquad (r^*f)(x, y) - f(x, y) = h_r(xy) - h_r(y) - h_r(x)$$

for x, y near to $\mathbf{1}$.

Let $q_Z \colon \mathfrak{z} \to Z$ be the quotient map, $f^Z := q_Z \circ f$ and $h^Z_r := q_Z \circ h_r$. Then we have a $\mathbf{1}$-neighborhood of the form $W_2 \times Z$ in $\hat{G}$, where $W_2 \subseteq W_1$, and the multiplication is given by

$$(g, z)(g', z') = (gg', zz' f^Z(g, g')).$$

Pick an open symmetric connected $\mathbf{1}$-neighborhood $W_3 \subseteq W_2$ with $r.W_3 \subseteq W_2$ and (2.11) holding on W_3. Then the map

$$(2.12) \qquad \sigma_{\hat{G}}(r) \colon W_3 \times Z \to W_2 \times Z \subseteq \hat{G}, \quad (g, z) \mapsto (r.g, z h^Z_r(g))$$

is a smooth homomorphism of local groups. Using Lemma II.3 in [Ne00b] and the simple connectedness of $\hat{G}$, we see that $\sigma_{\hat{G}}(r)$ extends to a smooth homomorphism $\hat{G} \to \hat{G}$. The derivative of this automorphism in $\mathbf{1} \in \hat{G}$ is given by

$$d\sigma_{\hat{G}}(r)(\mathbf{1})(x, z) = (r.x, z + dh^Z_r(\mathbf{1})(x)) = (r.x, z + dh_r(\mathbf{1})(x))$$
$$= (r.x, z + \alpha(r, x) + \theta(\mathbf{1})(r.x) - \theta(\mathbf{1})(x))$$
$$= (r.x, z + \alpha(r, x)) = \sigma_{\hat{\mathfrak{g}}}(r, x).$$

This proves that every automorphism $\sigma_{\hat{\mathfrak{g}}}(r)$, $r \in R$, integrates to a smooth endomorphism $\sigma_{\hat{G}}(r)$ of $\hat{G}$. The uniqueness of this extension implies that $\sigma_{\hat{G}}(r_1 r_2) = \sigma_{\hat{G}}(r_1)\sigma_{\hat{G}}(r_2)$ for $r_1, r_2 \in R$, hence in particular that each $\hat{\sigma}_{\hat{G}}(r)$ is an automorphism of $\hat{G}$. Let $\sigma_{\hat{G}} \colon R \times \hat{G} \to \hat{G}$ denote the corresponding action of R on $\hat{G}$.

It remains to show that this action is smooth. Since R acts by smooth automorphism on $\hat{G}$, it suffices to show that the action is smooth in a neighborhood of $(\mathbf{1}, \mathbf{1})$ and that all orbits maps $R \to \hat{G}$ are smooth in a neighborhood of $\mathbf{1}$. Since the latter property can be derived from the first one, it remains to see that the action is smooth in a neighborhood of $(1, 1)$. To this end, we slightly adjust the choices of W_1 and W_3 above. First we choose an open 1-neighborhood V in R and W_1 such that, in addition, $V.W_1 \subseteq W$. Likewise we choose $V_1 \subseteq V$ and $W_3 \subseteq W_2$ with $V_1.W_3 \subseteq W_2$. Then the function $(r, x) \mapsto h_r(x)$ is defined on $V \times W_1$, and the construction of h_r with the Poincaré Lemma implies that this function is smooth in a neighborhood of $(1, 1)$ (cf. [Ne00b, Lemma III.3]). This implies that the action map $\sigma_{\hat{G}}(r, x)$ is smooth on a neighborhood of $(1, 1)$ contained in $V_1 \times W_3$, and this completes the proof. □

Corollary II.19. *Let* $\sigma_G \colon R \times G \to G$ *be a smooth automorphic action of the Lie group* R *on the connected Lie group* G. *Assume that* G *is simply connected and that* $r^*\omega = \omega$ *for all* $r \in R$. *Then the action of* R *on* G *lifts uniquely to a smooth automorphic action of* R *on* $\hat{G}$ *such that the corresponding action of* R *on* $\hat{\mathfrak{g}} \cong \mathfrak{g} \oplus_\omega \mathfrak{z}$ *is given by*

$$r.(x, z) = (r.x, z), \quad r \in R, x \in \mathfrak{g}, z \in \mathfrak{z}.$$

This action fixes the subgroup Z *of* $\hat{G}$ *pointwise.*

Proof. We apply Theorem II.18 with $\alpha = 0$.   □

Corollary II.20. *If* K *is a compact simple simply connected Lie group and* $\tilde{L}K$ *is the central extension from* Theorem II.5, *then the rotation action of* $\mathbb{T}_r$ *on* LK *lifts to a smooth action of* $\mathbb{T}_r$ *on* $\tilde{L}K$. *The same holds for the complex groups* $LK_{\mathbb{C}}$, *resp.,* $\tilde{L}K_{\mathbb{C}}$.

Proof. Since $\pi_2(K) = \pi_1(K) = \{1\}$, the group LK is simply connected. Further the $\mathbb{T}_r$-action on $L\mathfrak{k}$ fixes the cocycle ω, so that Corollary II.19 applies.   □

Remark II.21. (a) If the commutator algebra $[\mathfrak{g}, \mathfrak{g}]$ is dense in $\mathfrak{g}$ ($\mathfrak{g}$ is topologically perfect), then in (2.9) the continuous linear map $\alpha_r := \alpha(r, \cdot) \colon \mathfrak{g} \to \mathfrak{z}$ is uniquely determined by $r^*\omega - \omega = -d\alpha_r$. Therefore

$$-d\alpha_{r_1 r_2} = (r_1 r_2)^*\omega - \omega = r_2^*(r_1^*\omega - \omega) + r_2^*\omega - \omega = -r_2^* d\alpha_{r_1} - d\alpha_{r_2}$$

implies the relation (2.10). In this sense (2.10) is only needed if $\mathfrak{g}$ is not topologically perfect.

(b) If $\hat{G}$ is a regular Lie group in the sense of [Mi83], then every automorphism of $\hat{\mathfrak{g}}$ integrates to a unique automorphism of $\hat{G}$ ([Mi83, Th. 8.1]). In our context it does not make sense to work with this additional assumption because we anyway need the more explicit information obtained in the proof of Theorem II.18 to show that the action is smooth.

(c) We consider the situation of Corollary II.19, where ω is fixed by R. Then we have a Lie group $G_R := G \rtimes R$ defined by the smooth action of R on G, and Corollary II.19 provides a central extension $\hat{G}_R \cong \hat{G} \rtimes R$. It is interesting to try to construct this group and therefore the smooth action of R on $\hat{G}$ more directly as a central Lie group extension.

We write $\hat{\mathfrak{g}} = \mathfrak{g} \oplus_\omega \mathfrak{z}$, and since ω is R-invariant, we obtain a natural action of R on $\hat{\mathfrak{g}}$ by $r.(x, z) := (r.x, z)$. The derived representation of this smooth action leads to a continuous action $\mathfrak{r} \times \hat{\mathfrak{g}} \to \hat{\mathfrak{g}}$ by derivations (cf. [Ne00a]). Let $\hat{\mathfrak{g}}_R := \hat{\mathfrak{g}} \rtimes \mathfrak{r}$ denote the corresponding Lie algebra. Then $\mathfrak{z} \subseteq \hat{\mathfrak{g}}_R$ is central with the topological complement $\mathfrak{g} \times \mathfrak{r}$, and the corresponding cocycle of the central extension $\hat{\mathfrak{g}}_R \to \mathfrak{g}_R$ is given by

$$\omega_R(x + y, x' + y') := \omega(x, x'), \quad x, x' \in \mathfrak{g}, y, y' \in \mathfrak{r}.$$

That this formula defines a cocycle can also verified more directly. It is clear that ω_R is continuous and skew symmetric. To see that it is a cocycle, we have to verify that for $a, b, c \in \mathfrak{g}_R$ the alternating expression

$$\omega_R(a, [b, c]) + \omega_R(b, [c, a]) + \omega_R(c, [a, b])$$

vanishes. For $a, b, c \in \mathfrak{g}$ this follows from the cocycle property of ω. If $a \in \mathfrak{r}$ and $b, c \in \mathfrak{g}$, then

$$\omega_R(a, [b, c]) + \omega_R(b, [c, a]) + \omega_R(c, [a, b]) = \omega(b, [c, a]) + \omega([b, a], c) = -(a.\omega)(b, c) = 0$$

because ω is R-invariant, which implies that $\mathfrak{r}.\omega = 0$ in $Z_c^2(\mathfrak{g}, \mathfrak{z})$. If two or three among a, b, c are in $\mathfrak{r}$, then each summand vanishes. Hence $\omega_R \in Z_c^2(\mathfrak{g}_R, \mathfrak{z})$.

One now might try to construct a group $\hat{G} \rtimes R$ as a central extension of G_R corresponding to ω_R. Since $\pi_2(G_R) \cong \pi_2(G) \times \pi_2(R)$ and the restriction of the left invariant 2-form Ω_R to the subgroup R of G_R vanishes, the image of the period map $\mathrm{per}_{\omega_R} : \pi_2(G_R) \to \mathfrak{z}$ coincides with the image of per_ω, hence is contained in $\pi_1(Z)$. What is not clear in this situation is how to lift the G-action on $\mathfrak{g}$ to an action on $\hat{\mathfrak{g}}_R$. If this could be done, then the criteria in [Ne00b] would imply the existence of a central Lie group extension of G_R corresponding to the Lie algebra $\hat{\mathfrak{g}}_R$. $\square$

III. Root decompositions

In this section we will consider the root decomposition of the Lie algebra $\hat{L}\mathfrak{k}_{\mathbb{C}}$. We will use the same notation as in Section I for the finite-dimensional Lie algebra $\mathfrak{k}$, we write $\mathfrak{t}_{\mathfrak{k}}$ for a Cartan subalgebra, and $T_K = \exp \mathfrak{t}_{\mathfrak{k}}$ for the corresponding maximal torus.

Let $z^k : \mathbb{S}^1 \to \mathbb{C}$ denote the function given by $z^k(\theta) := e^{ik\theta}$ and note that for $X \in \mathfrak{k}_{\mathbb{C}}$ the function $z^k X : \mathbb{S}^1 \to \mathfrak{k}_{\mathbb{C}}$ defines an element of $L\mathfrak{k}_{\mathbb{C}}$. The elements of the subalgebra

$$L_{\mathrm{pol}}\mathfrak{k}_{\mathbb{C}} := \mathbb{C}[z, \tfrac{1}{z}] \otimes \mathfrak{k}_{\mathbb{C}} = \sum_{k \in \mathbb{Z}} z^k \mathfrak{k}_{\mathbb{C}}$$

are called *polynomial loops* (with values in $\mathfrak{k}_{\mathbb{C}}$).

Lemma III.1. *$L_{\mathrm{pol}}\mathfrak{k}_{\mathbb{C}}$ is a dense subalgebra of $L\mathfrak{k}_{\mathbb{C}}$. Moreover, for $\gamma \in L\mathfrak{k}_{\mathbb{C}}$ its Fourier series*

$$\gamma = \sum_{n \in \mathbb{Z}} z^n \hat{\gamma}(n), \quad \text{where} \quad \hat{\gamma}(n) = \frac{1}{2\pi} \int_0^{2\pi} e^{-in\theta} \gamma(\theta) \, d\theta$$

converges in the Fréchet topology of $L\mathfrak{k}_{\mathbb{C}}$.

Proof. It is clear that the second assertion is much sharper than the first one, so it suffices to prove the second one.

It is easily seen that the rotation action of $\mathbb{T}_r$ on $L\mathfrak{k}_{\mathbb{C}}$ defines a smooth action, i.e., for each element $\gamma \in L\mathfrak{k}_{\mathbb{C}}$ the orbit map $\mathbb{T}_r \to L\mathfrak{k}_{\mathbb{C}}$ is smooth (cf. [Ne00a, Th. III.5]). Therefore the convergence of the Fourier series follows from Harish-Chandra's Theorem ([Wa72, Th. 4.4.2.1]) because $\gamma \mapsto z^n \hat{\gamma}(n)$ is the projection onto an isotypical $\mathbb{T}_r$-submodule. The latter fact follows directly from the observation that for $\gamma = z^m X$ we have

$$\hat{\gamma}(n) = \frac{1}{2\pi} \int_0^{2\pi} e^{im\theta} e^{-in\theta}\, d\theta X = \delta_{n,m} X. \qquad \square$$

Remark III.2. A similar assertion as in Lemma III.1 applies to the $\mathbb{T}_r$-action on the extended Lie algebra $\tilde{L}\mathfrak{k}_{\mathbb{C}}$. $\qquad \square$

The subgroup $T_r := \mathbb{T}_r \times T_K \subseteq L_r K$ acts on $L\mathfrak{k}_{\mathbb{C}}$ via the adjoint action. To identify the corresponding weights, we identify the character group $\hat{T}_r$ with

$$\hat{\mathbb{T}}_r \times \hat{T}_K \cong \mathbb{Z} \times \hat{T}_K \subseteq \mathbb{Z} \times \mathfrak{t}^*$$

as in Section I. Then the set of weights occuring in the adjoint representation is given by

$$\{(k, \alpha) \colon k \in \mathbb{Z}, \alpha \in \Delta_{\mathfrak{k}} \cup \{0\}\},$$

where

$$(L\mathfrak{k}_{\mathbb{C}})^{(k,\alpha)} = z^k \mathfrak{k}_{\mathbb{C}}^\alpha, \quad \alpha \in \Delta_{\mathfrak{k}}, \quad \text{and} \quad (L\mathfrak{k}_{\mathbb{C}})^{(k,0)} = z^k \mathfrak{t}_{\mathbb{C}}.$$

We write $\Delta_{L\mathfrak{k}} := \{(k, \alpha) \neq (0,0) \colon k \in \mathbb{Z}, \alpha \in \Delta_{\mathfrak{k}} \cup \{0\}\}$ for the set of *roots of* $L\mathfrak{k}_{\mathbb{C}}$ with respect to $\mathfrak{t}_r \oplus \mathfrak{t}$. The Lie algebra $\mathfrak{t}_{L\mathfrak{k}} := \mathfrak{t}_r \oplus \mathfrak{t}$ can be identified with the space $\mathbb{R}\frac{d}{dt} \oplus \mathfrak{t} \cong \mathbb{R} \oplus \mathfrak{t}$, where $\frac{d}{dt} = (1,0)$, and accordingly we have $R_\theta = \exp(-\theta \frac{d}{dt})$, so that

$$R_\theta . z^k = e^{-ik\theta} . z^k \quad \text{and} \quad \frac{d}{dt} z^k = ik z^k.$$

1. The Weyl group

Let $\mathcal{W} := N_K(T)/T$ denote the *Weyl group* of K. Similarly we define the *affine Weyl group*

$$\mathcal{W}_{\mathrm{aff}} := N_{L_r K}(T_r)/T_r,$$

and note that $T_r = Z_{L_r K}(T_r)$, i.e., T_r is maximal abelian in $L_r K$. In fact, a loop commuting with $\mathbb{T}_r$ must be constant, and if, in addition, it commutes with T_K, the fact that $T = Z_K(T)$ implies that its value is in T.

Since the group $\check{T}_K := \operatorname{Hom}(\mathbb{T}, T_K)$ consists in particular of smooth loops, it can be identified with a subgroup of $LT_K \subseteq L_r K$.

Proposition III.3. *The group $\mathcal{W}_{\mathrm{aff}}$ is isomorphic to the semidirect product $\check{T}_K \rtimes \mathcal{W}$, and it acts on $\mathfrak{t}_{L\mathfrak{k}}$ via*

$$(Z, w).(t, X) = (t, w.X - tZ)$$

for $Z \in \check{T}_K$ and $w \in \mathcal{W}$.

Proof. The first part follows from [PS86, p.71]. To verify the formula for the action, it clearly suffices to compute the action of elements of $\check{T}_K$. So let $\gamma \in \check{T}_K$ be given by $\gamma(\theta) = \exp(\theta Z)$, $Z \in \mathfrak{t}_{\mathfrak{k}}$.

We have

$$\gamma \exp(t, 0)\gamma^{-1} = \gamma R_{-t}\gamma^{-1} = R_{-t}(R_t\gamma R_{-t}\gamma^{-1}) = \exp(t, 0)((R_t.\gamma)\gamma^{-1})$$

with

$$((R_t.\gamma)\gamma^{-1})(\theta) = \gamma(\theta - t)\gamma^{-1}(\theta) = \gamma(\theta - t)\gamma(-\theta) = \gamma(-t) = \exp(-tZ).$$

Thus $\mathrm{Ad}(\gamma).(t, 0) = (t, -tZ)$, and the assertion follows. $\square$

It follows in particular from Proposition III.3 that the affine subspace $\{1\} \times \mathfrak{t}_{\mathfrak{k}}$ is invariant under the action of the affine Weyl group, and that the action on this hyperplane induces an affine action on $\mathfrak{t}_{\mathfrak{k}}$ given by

$$(Z, w).X = w.X - Z.$$

Note that $Z \in \check{T}_K \subseteq \mathfrak{t}_{\mathfrak{k}}$ acts by translation in the opposite direction. Then the kernel of a root $(k, \alpha) \in \Delta_{LK}$ corresponds to the affine hyperplane

$$H_{k,\alpha} := \{X \in \mathfrak{t}_{\mathfrak{k}} : \alpha(X) = -k\}.$$

The set $\bigcup_{\alpha \in \Delta_{\mathfrak{k}}, k \in \mathbb{Z}} H_{k,\alpha}$ is called the *diagram of LK*. If $\Delta_{\mathfrak{k}}^+ \subseteq \Delta_{\mathfrak{k}}$ is a positive system, then the set

$$C_0 := \{X \in \mathfrak{t}_{\mathfrak{k}} : (\forall \alpha \in \Delta_{\mathfrak{k}}^+) 0 < \alpha(X) < 1\}$$

is called a *fundamental alcove*. It is a fundamental domain for the affine action of $\mathcal{W}_{\mathrm{aff}}$ on $\mathfrak{t}_{\mathfrak{k}}$ ([PS86, Prop. 5.1.4]). A root $\underline{\alpha} = (\alpha, k)$ is said to be *positive* if $\underline{\alpha}(\{1\} \times C_0) \subseteq \mathbb{R}^+$ which is equivalent to $\alpha(C_0) \subseteq [-k, \infty[$. Therefore the set $\Delta_{L\mathfrak{k}}^+$ of positive roots is given by

$$\Delta_{L\mathfrak{k}}^+ = \{(k, \alpha) : (k > 0) \text{ or } (k = 0, \alpha \in \Delta_{\mathfrak{k}}^+)\}.$$

A root $\underline{\alpha} = (k, \alpha)$ is called *simple* if $H_{k,\alpha}$ contains a *wall* of the fundamental alcove C_0. To each root $\underline{\alpha} = (k, \alpha)$ corresponds an affine reflection on $\mathfrak{t}_{\mathfrak{k}}$ in the hyperplane $H_{k,\alpha}$ which is given by the formula

$$s_{\underline{\alpha}}(X) = s_\alpha(X) - kh_\alpha, \quad \text{i.e.,} \quad s_{\underline{\alpha}} = (kh_\alpha, s_\alpha) \in \hat{T}_K \rtimes \mathcal{W}$$

is the decomposition according to Proposition III.3.

Remark III.4. If $\mathfrak{k}$ is simple, $\alpha_1, \ldots, \alpha_r$ are the fundamental roots of $\mathfrak{k}$ with respect to $\Delta_{\mathfrak{k}}^+$, and $\alpha_0 \in \Delta_{\mathfrak{k}}^+$ is the highest root, then

$$(0, \alpha_1), \ldots, (0, \alpha_r), (1, -\alpha_0)$$

is a system of simple roots in $\Delta_{L\mathfrak{k}}^+$ (cf. [PS86, p. 73]). $\square$

2. Root decomposition of the central extension

We write

$$T_{\hat{L}K} := \mathbb{T}_r \times T_K \times \mathbb{T}_c \subseteq \hat{L}K,$$

for the maximal torus in $\hat{L}K$, where $\mathbb{T}_c$ stands for the central torus defining the central extension from Theorem II.5. We identify the character group $\hat{T}_{\hat{L}K}$ of $T_{\hat{L}K}$ with $\hat{\mathbb{T}}_r \times \hat{T}_K \times \hat{\mathbb{T}}_c \cong \mathbb{Z} \times \hat{T}_K \times \mathbb{Z} \subseteq \mathbb{Z} \times \mathfrak{t}_\mathfrak{k}^* \times \mathbb{Z}$. Since $\mathbb{T}_c$ acts trivially via the adjoint action, we can identify the root system $\Delta_{L\mathfrak{k}} \cong \Delta_{L\mathfrak{k}} \times \{0\}$ with a subset of $\mathfrak{t}_{L_r\mathfrak{k}}^*$.

Identifying $L\mathfrak{k}_{\mathbb{C}}$ with a vector subspace of $\tilde{L}\mathfrak{k}_{\mathbb{C}}$, we have

$$(\tilde{L}\mathfrak{k}_{\mathbb{C}})^{(k,\alpha)} = z^k \mathfrak{k}_{\mathbb{C}}^\alpha, \quad \alpha \in \Delta_\mathfrak{k}, \quad \text{and} \quad (\tilde{L}\mathfrak{k}_{\mathbb{C}})^{(k,0)} = z^k \mathfrak{t}_{\mathbb{C}}.$$

Via the adjoint representation the affine Weyl group $\mathcal{W}_{\text{aff}}$ can also be identified with $N(T_{\hat{L}K})/T_{\hat{L}K}$. It acts on $\hat{T}_{\hat{L}K}$ by

$$Z.(k, \lambda, h) = \left(k + \lambda(Z) + \tfrac{1}{2}h\kappa(Z, Z), \lambda + hZ^*, h\right)$$

for $Z \in \check{T}_K$, where $Z^* \in \mathfrak{t}_\mathfrak{k}$ is defined by the embedding $\check{T}_K \to \hat{T}_K, Z \mapsto Z^*$ which is well defined because $\hat{T}_K \cong \text{Hom}(\check{T}_K, \mathbb{Z})$ and $\kappa(\check{T}_K, \check{T}_K) \subseteq \mathbb{Z}$ ([PS86, Prop. 4.9.5]).

Let $\underline{\alpha} = (k, \alpha)$ denote a root of $\tilde{L}\mathfrak{k}_{\mathbb{C}}$. We put

$$e_{\underline{\alpha}} := z^k e_\alpha \quad \text{and} \quad e_{-\underline{\alpha}} := \overline{e_{\underline{\alpha}}} = z^{-k} e_{-\alpha}.$$

Then

$$[e_{\underline{\alpha}}, e_{-\underline{\alpha}}] = \left([e_\alpha, e_{-\alpha}], \omega(e_{\underline{\alpha}}, e_{-\underline{\alpha}})\right),$$

where

$$\omega(e_{\underline{\alpha}}, e_{-\underline{\alpha}}) = \frac{1}{2\pi} \int_0^{2\pi} \kappa(z^k e_\alpha, (-ik)z^{-k} e_{-\alpha}) \, d\theta$$

$$= \frac{1}{2\pi} \int_0^{2\pi} (-ik)\kappa(e_\alpha, e_{-\alpha}) \, d\theta = -i\frac{k}{2}\kappa(h_\alpha, h_\alpha)$$

(Lemma I.1). Therefore

$$[e_{\underline{\alpha}}, e_{-\underline{\alpha}}] = i\left(h_\alpha, -\tfrac{k}{2}\kappa(h_\alpha, h_\alpha)\right).$$

Putting

$$h_{\underline{\alpha}} := \left(h_\alpha, -\tfrac{k}{2}\kappa(h_\alpha, h_\alpha)\right),$$

we see that $(e_{\underline{\alpha}}, e_{-\underline{\alpha}}, h_{\underline{\alpha}})$ is a basic $\mathfrak{su}_2$-triple (cf. Section I). The corresponding reflection $s_{\underline{\alpha}}$ acts on $(\mathfrak{t}_{\hat{L}K})^*$ via

$$s_{\underline{\alpha}}(\underline{\lambda}) = \underline{\lambda} - \langle \underline{\lambda}, h_{\underline{\alpha}} \rangle \underline{\alpha}.$$

We write $\tilde{L}K(\underline{\alpha})$ for the corresponding three-dimensional subgroup of $\tilde{L}K$.

Having introduced the relevant notation, we can give a simple proof for the necessity of condition (2.4).

Lemma III.5. *The condition $\kappa(h_\alpha, h_\alpha) \in 2\mathbb{Z}$ for all $\alpha \in \Delta_\mathfrak{k}$ is necessary for the existence of the central extension $\tilde{L}K$.*

Proof. Suppose that we have a central extension

$$\{1\} \to \mathbb{R} \longrightarrow \tilde{L}K \overset{q}{\longrightarrow} LK \to \{1\}$$

corresponding to the given central extension on the level of Lie algebras. For each root $\underline{\alpha} \in \Delta_{L\mathfrak{k}}$ the corresponding three-dimensional subgroup $\tilde{L}K(\underline{\alpha})$ is compact and isomorphic to $SU(2)$ or $SO(3,\mathbb{R})$. Therefore

$$1 = \exp 2\pi h_{\underline{\alpha}} = \Big(\exp 2\pi h_{\alpha}, \exp(-k\pi\kappa(h_{\alpha}, h_{\alpha})) \Big).$$

If we apply this to a root $\underline{\alpha} = (0, \alpha)$, we see that $\exp(2\pi h_{\underline{\alpha}}) = 1$, so that $\exp(-k\pi\kappa(h_{\alpha}, h_{\alpha})) = 1$ holds for all $k \in \mathbb{N}$. This implies that $\kappa(h_{\alpha}, h_{\alpha}) \in 2\mathbb{Z}$. $\qquad\square$

IV. Representations of loop groups

Definition IV.1. Let V be a complete complex locally convex space. Then a *representation* (π, V) of a topological group G on V is a group homomorphism $\pi \colon G \to \mathrm{GL}(V)$ for which the mapping

$$G \times V \to V, \quad (g, v) \mapsto \pi(g).v$$

is continuous, i.e., the action of G on V is continuous.

   If G is a Lie group (modeled on a sequentially complete locally convex space, cf. [Ne00a, Sect. I]), then a vector $v \in V$ is called *smooth* if the orbit mapping $G \to V, g \mapsto \pi(g).v$ is smooth. We write V^{∞} for the space of smooth vectors in V. A representation (π, V) is said to be *smooth* if V^{∞} is dense in V.

   As we have seen in [Ne00a, Lemma IV.2], we obtain a representation of the Lie algebra $\mathfrak{g}$ of G on V^{∞} by putting

$$d\pi(X).v := d\phi_v(1).X,$$

where $\phi_v(g) := \pi(g).v$ denotes the orbit map of v. If G is real, then the representation $d\pi$ of $\mathfrak{g}$ on V^{∞} naturally extends to a complex linear representation of the complexified Lie algebra $\mathfrak{g}_{\mathbb{C}}$ on V^{∞}. $\qquad\square$

Definition IV.2. (a) Let (π, V) be a continuous representation of the group $\hat{L}K$. The subspace

$$V(k) := \{v \in V \colon (\forall \theta \in \mathbb{R}) \, R_{\theta}.v = e^{-ik\theta}.v\}$$

is called the *energy subspace of degree* k. We note that the Peter-Weyl Theorem applied to the representation $\pi|_{\mathbb{T}_r}$ on V implies that the *finite energy subspace*

$$\check{V} := \sum_{k \in \mathbb{Z}} V(k)$$

is dense in V and that there exist continuous projections

$$p_k \colon V \to V(k), \quad v \mapsto \frac{1}{2\pi} \int_0^{2\pi} e^{ik\theta} R_{\theta}.v \, d\theta.$$

Note that the integrals make sense because V is a complete locally convex space. Moreover, if $v \in V^\infty$ is a smooth vector, then the Fourier series $v = \sum_{k \in \mathbb{Z}} p_k(v)$ converges in V according to Harish-Chandra's Theorem (cf. [Wa72, Th. 4.4.2.1]).

(b) If V is a module of the Lie algebra $\hat{L}\mathfrak{k}_{\mathbb{C}}$, we call $V(k) := \{v \in V : X_r.v = ikv\}$, where $X_r := \frac{d}{dt}$ is the canonical basis element of $\mathfrak{t}_r$, the *energy subspace of degree* k, and $\check{V} := \sum_{k \in \mathbb{Z}} V(k)$ the *finite energy subspace*.

(c) We say that (π, V) is a *representation of positive energy* if $V(k) = 0$ for $k < 0$. ☐

Example IV.3. If $V = \tilde{L}\mathfrak{k}_{\mathbb{C}}$, where the representation of $\tilde{L}K$ is the adjoint representation, then $V(k) = z^k \mathfrak{k}_{\mathbb{C}}$ for $k \neq 0$, $V(0) = (\mathfrak{t}_r)_{\mathbb{C}} \oplus (\mathfrak{t}_c)_{\mathbb{C}} \oplus \mathfrak{k}_{\mathbb{C}}$, and therefore $\check{V} = \tilde{L}_{\text{pol}}\mathfrak{k}_{\mathbb{C}}$. ☐

Lemma IV.4. *For every representation* (π, V) *of* $\tilde{L}K$ *the space* $\check{V}^\infty$ *of smooth vectors of finite energy is invariant under* $\tilde{L}_{\text{pol}}\mathfrak{k}_{\mathbb{C}}$.

Proof. First we show that the action map $\tilde{L}\mathfrak{k}_{\mathbb{C}} \otimes V^\infty \to V^\infty$ is equivariant with respect to the action of the torus $\mathbb{T}_r$ on both sides. In fact, since $\mathbb{T}_r$ acts on $\tilde{L}K$ by conjugation, it is clear that the action map $\tilde{L}K \times V \to V$ is $\mathbb{T}_r$-equivariant. This means that for each $v \in V$ and $\theta \in \mathbb{R}$ we have

$$\phi_v \circ R_\theta = \pi(R_\theta) \circ \phi_{R_{-\theta}.v},$$

where $\phi_w(g) := \pi(g).w$ denotes the orbit map of w. Taking derivatives in **1**, this leads to

$$(R_\theta.X).v = \pi(R_\theta)\big(X.(R_{-\theta}.v)\big).$$

This proves our assertion because of the complex linearity of the action map. We conclude that

$$\tilde{L}\mathfrak{k}_{\mathbb{C}}(n).V^\infty(m) \subseteq V^\infty(n + m)$$

holds for all $n, m \in \mathbb{Z}$. Since $\tilde{L}_{\text{pol}}\mathfrak{k}_{\mathbb{C}} = \sum_{k \in \mathbb{Z}} \tilde{L}\mathfrak{k}_{\mathbb{C}}(k)$, this proves the assertion. ☐

We obtain a more refined picture by looking at the representation of the larger torus subgroup $T_{\tilde{L}K}$. We have the weight spaces $V_{(k,\lambda,h)} \subseteq V(k)$ corresponding to the weight $\underline{\lambda} = (k, \lambda, h)$. We write

$$\mathcal{P}_V := \{\underline{\lambda} \in \hat{T}_{\tilde{L}K} : V_{\underline{\lambda}} \neq \{0\}\}$$

for the *set of all weights* of V.

Definition IV.5. If the representation (π, V) is irreducible or, more generally, generated by a $\mathbb{T}_c$-eigenvector, then the invariance of the larger $\mathbb{T}_c$-eigenspaces under $\hat{L}K$ shows that there exists $h \in \mathbb{Z}$ with

$$\mathcal{P}_V \subseteq \mathbb{Z} \times \hat{T}_K \times \{h\}.$$

In this case h is called the *level of the representation* (π, V). ☐

1. Lowest weight vectors and antidominant weights

Definition IV.6. (a) Let V be a module of the Lie algebra $\hat{L}\mathfrak{k}_{\mathbb{C}}$. A non-zero weight vector $v_\lambda \in V_\lambda$ is called a *lowest weight vector* if

$$(L\mathfrak{k}_{\mathbb{C}})^{\underline{\alpha}}.v_{\underline{\lambda}} = \{0\}$$

for all $\underline{\alpha} \in \Delta_{L\mathfrak{k}}^-$. In this case λ is called the corresponding *lowest weight*. If (π, V) is a representation of $\hat{L}K$, then a lowest weight vector means a lowest weight vector for the derived representation of the Lie algebra $\hat{L}\mathfrak{k}_{\mathbb{C}}$ on V^∞.

(b) A module V of $\hat{L}\mathfrak{k}_{\mathbb{C}}$ is said to be a *lowest weight module* if it is generated by a lowest weight vector. A representation (V, π) of $\hat{L}K$ is called a *lowest weight representation* if it contains a lowest weight vector generating V.

(c) A weight $\underline{\lambda}$ satisfying $\langle \underline{\lambda}, h_{\underline{\alpha}} \rangle \in -\mathbb{N}_0$ for all $\underline{\alpha} \in \Delta_{L\mathfrak{k}}^+$ is called *antidominant* (with respect to the positive system $\Delta_{L\mathfrak{k}}^+$). $\qquad\square$

The following observation is crucial for the whole theory.

Proposition IV.7. (i) *If (π, V) is a smooth representation of $\hat{L}K$ with positive energy and $V(0) \neq \{0\}$, then $V(0)$ contains a lowest weight vector.*

(ii) *Each lowest weight $\underline{\lambda}$ is antidominant.*

Proof. (i) According to [Ne00a, Prop. IV.5], the representation of G on the complete locally convex space V^∞ is continuous. Since this applies in particular to the subgroup $\mathbb{T}_r$, it follows that the projection

$$p_0 \colon V \to V(0), \quad v \mapsto \frac{1}{2\pi}\int_0^{2\pi} R_\theta.v \, d\theta \quad \text{satisfies} \quad p_0(V^\infty) \subseteq V^\infty(0) \subseteq V^\infty.$$

In view of the smoothness assumption on (π, V), the subspace $V^\infty(0)$ which is invariant under the compact group K (the constant loops) is dense in $V(0)$.

The Peter-Weyl Theorem implies the existence of a finite-dimensional irreducible K-subspace $F \subseteq V^\infty(0)$. Let $v_\lambda \in F$ be a lowest weight vector for the representation of $\mathfrak{k}_{\mathbb{C}}$ on F. Since $\mathbb{T}_c$ commutes with K, we may w.l.o.g. assume that $v_\lambda \in V_{(0,\lambda,h)}$. Then the fact that

$$(\tilde{L}\mathfrak{k}_{\mathbb{C}})^{(k,\alpha)}.v_{\underline{\lambda}} \in V_{(k,\lambda+\alpha,h)} \subseteq V(k) = \{0\}$$

for $k < 0$ implies that $v_{\underline{\lambda}}$ is a lowest weight vector for $\tilde{L}\mathfrak{k}_{\mathbb{C}}$.

(ii) If λ is a lowest weight of a representation (π, V), then $\mathfrak{sl}_2$-theory applied to the group $\tilde{L}K(\underline{\alpha})$ shows that $\langle \underline{\lambda}, h_{\underline{\alpha}} \rangle \in -\mathbb{N}_0$ because otherwise $e_{-\underline{\alpha}}.v_\lambda \neq \{0\}$. $\qquad\square$

Remark IV.8. (a) Another possibility to prove Proposition IV.7(ii) is to note that the fact that the representation of the Lie algebra $\tilde{L}\mathfrak{k}(\underline{\alpha})$ integrates to a representation of the corresponding group implies that

$$s_{\underline{\alpha}}(\underline{\lambda}) = \underline{\lambda} - \langle \underline{\lambda}, h_{\underline{\alpha}} \rangle \underline{\alpha} \in \mathcal{P}_V$$

is also a weight in the smallest $\tilde{L}\mathfrak{k}(\underline{\alpha})$-invariant subspace containing v_λ.

If $\underline{\lambda} = (n_0, \lambda, h)$ is a weight with minimal n_0, then we see that for each root $\underline{\alpha} = (k, \alpha, 0)$ with $k > 0$ we have $\langle \underline{\lambda}, h_{\underline{\alpha}} \rangle \leq 0$. Further the W-orbit of $\underline{\lambda}$ contains an element which is antidominant for all positive roots of the type $\underline{\alpha} = (0, \alpha, 0)$ which then yields the existence of an antidominant weight.

(b) We want to make the antidominance condition more explicit. So let $\underline{\lambda} = (n, \lambda, h)$. Then the antidominance means that

$$\langle \underline{\lambda}, h_{\underline{\alpha}} \rangle = \langle \lambda, h_\alpha \rangle - \frac{hk}{2} \kappa(h_\alpha, h_\alpha) \leq 0$$

for all $\underline{\alpha} = (k, \alpha, 0) \in \Delta_{L\mathfrak{k}}^+$. This means that $\lambda(h_\alpha) \leq 0$ for all $\alpha \in \Delta_{\mathfrak{k}}^+$ and that

$$(4.1) \qquad\qquad \lambda(h_\alpha) \leq \frac{hk}{2} \kappa(h_\alpha, h_\alpha)$$

for all $k \in \mathbb{N}$ and $\alpha \in \Delta_{\mathfrak{k}}$. Using $\Delta_{\mathfrak{k}} = -\Delta_{\mathfrak{k}}$, we see that (4.1) leads to

$$h\kappa(h_\alpha, h_\alpha) \geq 2 \max\{\lambda(h_\alpha), \lambda(h_{-\alpha})\} = 2 \max\{\lambda(h_\alpha), -\lambda(h_\alpha)\} = 2|\lambda(h_\alpha)| \geq 0,$$

and hence to

$$(4.2) \qquad\qquad h\kappa(h_\alpha, h_\alpha) \geq 0.$$

For $\underline{\alpha} = (1, -\alpha)$, $\alpha \in \Delta_{\mathfrak{k}}^+$, we see that whenever (4.2) is satisfied, then $\underline{\lambda}$ is antidominant if and only if

$$(4.3) \qquad\qquad -\frac{h}{2}\kappa(h_\alpha, h_\alpha) \leq \lambda(h_\alpha) \leq 0$$

holds for all $\alpha \in \Delta_{\mathfrak{k}}^+$. $\qquad\qquad\qquad\qquad\qquad\qquad\qquad\qquad\qquad\qquad\qquad\qquad$ □

For each fixed h condition (4.3) specifies a finite set of integral linear functionals on $\mathfrak{t}_{\mathfrak{k}}$ and for $h = 0$ the only functional satisfying this condition is $\lambda = 0$. As the following proposition shows, this has serious consequences for the representation theory of the group $\hat{L}K$.

Proposition IV.9. *For a lowest weight representation (π, V) of $\hat{L}K$ of lowest weight $\underline{\lambda}$ the following assertions hold:*
  (i) *If $v_{\underline{\lambda}}$ is a lowest weight vector and $\underline{\alpha}$ is a root, then $v_{\underline{\lambda}}$ generates a finite-dimensional irreducible $\tilde{L}K(\underline{\alpha})$-representation of lowest weight $\underline{\lambda}(h_{\underline{\alpha}})$.*
  (ii) *If $\underline{\lambda} = (n, \lambda, 0)$, i.e., if (π, V) is a representation of level 0, then (π, V) is the trivial representation on the identity component of $\tilde{L}K$. If, in addition, K is simply connected, then it is given by a character of $\mathbb{T}_r$.*

Proof. (i) Since $v_{\underline{\lambda}}$ is assumed to be a smooth vector, we can write out its Fourier series with respect to the compact group $\tilde{L}K(\underline{\alpha})$:

$$v_{\underline{\lambda}} = \sum_{m \in \mathbb{N}_0} v_{\underline{\lambda}}(m),$$

where we have identified the set of equivalence classes of irreducible representations of $\tilde{L}K(\underline{\alpha})$ with a subset of $\mathbb{N}_0$ in such a way that $m \in \mathbb{N}_0$ corresponds to the representation with lowest weight μ_m satisfying $\mu_m(h_{\underline{\alpha}}) = -m$.

The facts that v_λ is a smooth vector for $\tilde{L}K(\underline{\alpha})$ and that the projections onto the isotypical components are continuous entail that the vectors $v_\lambda(m)$ are also smooth with respect to $\tilde{L}K(\underline{\alpha})$. Then these vectors have to be lowest weight vectors for $\tilde{L}K(\underline{\alpha})$ and therefore $h_{\underline{\alpha}}.v_\lambda(m) = -imv_\lambda(m)$. In view of the fact that $h_{\underline{\alpha}}.v_\lambda = i\underline{\lambda}(h_{\underline{\alpha}})v_\lambda$, the uniqueness of the Fourier expansion shows that

$$v_\lambda = v_\lambda(m) \quad \text{for} \quad m = -\underline{\lambda}(h_{\underline{\alpha}}),$$

and the assertion follows.

(ii) We have seen in Remark IV.8 that the antidominance of λ and $h = 0$ imply that $\lambda = 0$. Therefore (i) implies that for each root $\underline{\alpha} \in \Delta_{L\ell}$ the $\tilde{L}K(\underline{\alpha})$-submodule generated by v_λ is a lowest weight module with trivial lowest weight, hence a trivial module. We conclude that the three-dimensional groups $\tilde{L}K(\underline{\alpha})$ fix the lowest weight vector v_λ.

On the other hand v_λ is fixed by the torus $T_{\tilde{L}K}$. Since each element $X \in \tilde{L}_{\text{pol}}\ell$ is a finite sum of elements contained in a three-dimensional subalgebra $\tilde{L}\ell(\underline{\alpha})$, an application of the Trotter-Product-Formula (Theorem II.1) shows that

$$\exp X.v_\lambda = v_\lambda$$

holds for all $X \in \tilde{L}_{\text{pol}}\ell$, hence for all $X \in \tilde{L}\ell$ because $\tilde{L}_{\text{pol}}\ell$ is dense (cf. Lemma III.1).

We further know that the group $\hat{L}K$ is connected whenever K is simply connected (cf. [PS86, p.48]). Since the central circle $\mathbb{T}_c$ acts trivially, we may assume that we have a representation of the group $L_rK \cong \hat{L}K/\mathbb{T}_c$. Now the fact that the exponential function of LK is a local diffeomorphism (Theorem II.1) implies that the identity component of $L_rK = \mathbb{T}_r \ltimes LK$ is generated by the image of the exponential function. From that it follows that v_λ is fixed by the whole group $(\tilde{L}K)_0$, i.e., (π, V) is a one-dimensional representation whenever K is simply connected. $\qquad\square$

The following result shows that the non-trivial lowest weight representation of $\hat{L}K$ do not factor to the quotient L_rK, hence that the central extension of this group is necessary to obtain non-trivial lowest weight representations.

Corollary IV.10. *Each lowest weight representation of L_rK is one-dimensional.*

Proof. Let (π, V) be a lowest weight representation of L_rK which can also be considered as a representation of the central extension $\hat{L}K$ which is trivial on the center. Then Proposition IV.9 shows that (π, V) is a one-dimensional representation. $\qquad\square$

Remark IV.11. In the special case where ℓ is simple and κ is *the fundamental invariant form* which is normalized by $\kappa(h_{\alpha_0}, h_{\alpha_0}) = 2$ for the highest root α_0 (cf. [PS86, p.49]), then $\underline{\lambda} = (0, \lambda, h)$ is antidominant if and only if λ is antidominant and $\lambda(h_{\alpha_0}) \geq -h$. The set of all integral functionals satisfying this condition can be represented as $\sum_{k=0}^{l} n_k \varpi_k$, $n_k \in \mathbb{N}_0$, where

$$\varpi_0 = (0, 0, 1) \quad \text{and} \quad \varpi_k = \big(0, -\varpi_k, \varpi_k(h_{\alpha_0})\big), \quad k = 1, \dots, l,$$

where $\varpi_1, \dots, \varpi_l$ denote the fundamental weights of $\mathfrak{k}$. The above weights are the dual basis to

$$\underline{h_0} = (0, h_{\alpha_0}, -1) \quad \text{and} \quad \underline{h_k} = (0, h_{\alpha_k}, 0), \quad k = 1, \dots, l. \qquad \square$$

2. The Casimir operator

In this subsection we will describe how to define a Casimir operator for the Lie algebras $\tilde{L}_{\mathrm{pol}}\mathfrak{k}_{\mathbb{C}}$ which will be used later on to show that certain contravariant hermitian forms on lowest weight modules of this Lie algebra are positive definite.

We recall that the *Casimir operator* of $\mathfrak{k}$ associated to an invariant non-degenerate positive definite bilinear form κ on $\mathfrak{k}$ is given by

$$\Omega_{\mathfrak{k}} = -\tfrac{1}{2} \sum_{j=1}^{N} e_j^* e_j,$$

where $e_1, \dots, e_N$ denotes a basis of $\mathfrak{k}$ and $e_1^*, \dots, e_N^*$ denotes the dual basis of $\mathfrak{k}$ with respect to κ. Then $\Omega_{\mathfrak{k}}$ does not depend on the chosen basis and is a central element in the enveloping algebra $\mathcal{U}(\mathfrak{k})$. This implies in particular that $\Omega_{\mathfrak{k}}$ acts as a scalar multiple of the identity in every finite-dimensional irreducible representation (π_λ, V_λ) of highest weight λ. According to [PS86, Prop. 9.4.2], the corresponding scalar is given by

$$c_\lambda = \tfrac{1}{2}\big(\|\lambda - \rho\|^2 - \|\rho\|^2\big) = \tfrac{1}{2}\|\lambda\|^2 - \kappa(\lambda, \rho),$$

where $\rho = \tfrac{1}{2} \sum_{\alpha \in \Delta_{\mathfrak{k}}^+} \alpha$.

To find an operator which has similar properties for the Lie algebra $\tilde{L}_{\mathrm{pol}}\mathfrak{k}$, let $e_1, \dots, e_N$ form an orthonormal basis in $\mathfrak{k}$, write $e_j^n := z^n e_j$, and let $c \in \mathbb{R}$ denote the eigenvalue of the Casimir operator $\Omega_{\mathfrak{k}}$ in the adjoint representation acting on $\mathfrak{k}_{\mathbb{C}}$. If $\mathfrak{k}$ is simple, then we define the *Casimir operator* of $\tilde{L}_{\mathrm{pol}}\mathfrak{k}_{\mathbb{C}}$ by

$$\Omega := \Omega_0 + (I + ic)\frac{d}{dt},$$

where $I = (0, 0, 1) \in \mathfrak{k}_c$ is the generator of the center and

$$\Omega_0 = -\sum_{j=1}^{N} \sum_{n>0} e_j^n e_j^{-n} - \frac{1}{2} \sum_{j=1}^{N} e_j^2 = -\sum_{j=1}^{N} \sum_{n>0} e_j^n e_j^{-n} + \Omega_{\mathfrak{k}}.$$

If $\mathfrak{k} = \mathfrak{k}_1 \oplus \dots \oplus \mathfrak{k}_m$ is the decomposition into the center $\mathfrak{k}_1$ and the simple ideals $\mathfrak{k}_j$, $j \geq 2$, then we obtain an action of the m-dimensional torus $\mathbb{T}_r := \mathbb{T}^m$ on

$L\mathfrak{k} \cong \bigoplus_{j=1}^{m} L\mathfrak{k}_j$ by rotations in each summand. Lifting this action to the central extension $\tilde{L}\mathfrak{k}$ we obtain a larger Lie algebra

$$\tilde{\mathfrak{t}}_r \ltimes \tilde{L}\mathfrak{k} \cong (\mathbb{R}^m) \ltimes \tilde{L}\mathfrak{k}.$$

In this sense we define the Casimir operator of $\tilde{L}\mathfrak{k}$ as

$$\Omega := \Omega_0 + \sum_{j=1}^{m} (I + ic_j) \frac{d}{dt_j}, \qquad \text{where} \qquad \frac{d}{dt} := \sum_{j=1}^{m} \frac{d}{dt_j},$$

where c_j are the eigenvalues of the Casimir operators on the ideals $\mathfrak{k}_j$.

Lemma IV.12. *If $(\pi_{\underline{\lambda}}, V_{\underline{\lambda}})$ is a lowest weight module of the Lie algebra $\tilde{\mathfrak{t}}_r \ltimes \tilde{L}\mathfrak{k}_{\mathbb{C}}$ with lowest weight $\underline{\lambda} = (\underline{n}, \lambda, \underline{h})$, $\underline{h} = (h, \ldots, h)$, then the Casimir operator Ω acts on V by $c_{\underline{\lambda}}\mathbf{1}$, where*

$$c_{\underline{\lambda}} = \frac{1}{2}\|\lambda\|^2 - \kappa(\lambda, \rho) - \sum_{j=1}^{m}(h + c_j)n_j = c_\lambda - \sum_{j=1}^{m}(h + c_j)n_j.$$

If we put $\underline{\rho} = (0, \rho, -\underline{c})$, where $\underline{c} = (c_1, \ldots, c_m)$, then this can also be written as

$$c_{\underline{\lambda}} = \frac{1}{2}(\|\underline{\lambda} - \underline{\rho}\|^2 - \|\underline{\rho}\|^2),$$

where the scalar product on $\tilde{\mathfrak{t}}_r \oplus \mathfrak{t}_{\mathfrak{k}} \oplus \mathbb{R}^k$ is given by

$$\kappa_c\big((\underline{n}, \lambda, \underline{h}), (\underline{n}', \lambda', \underline{h}')\big) = \kappa(\lambda, \lambda') - \langle \underline{n}, \underline{h}' \rangle - \langle \underline{n}', \underline{h} \rangle.$$

Proof. [PS86, Prop. 9.4.9] $\qquad\qquad\qquad\qquad\qquad\qquad\qquad\qquad\qquad\qquad$ $\square$

In the following we call a hermitian form $\langle \cdot, \cdot \rangle$ on a module of a real Lie algebra $\mathfrak{g}$ *invariant* if $\langle X.v, w \rangle = -\langle v, X.w \rangle$ for all $v, w \in V$, $X \in \mathfrak{g}$. Note that this means that it is *contravariant* for $\mathfrak{g}_{\mathbb{C}}$ in the sense that

$$\langle X.v, w \rangle = -\langle v, \overline{X}.w \rangle$$

holds for all $v, w \in V$, $X \in \mathfrak{g}_{\mathbb{C}}$, with respect to the natural extension of the representation of $\mathfrak{g}$ to $\mathfrak{g}_{\mathbb{C}}$.

Theorem IV.13. (Garland's Theorem) *If $(\pi_{\underline{\lambda}}, V_{\underline{\lambda}})$ is a lowest weight module of $\tilde{\mathfrak{t}}_r \ltimes \tilde{L}_{\mathrm{pol}}\mathfrak{k}_{\mathbb{C}}$ such that*

  (i) *$\underline{\lambda}$ is antidominant, and*

  (ii) *for each $\underline{\alpha} \in \Delta_{\mathfrak{k}}$ the representation of $\tilde{L}\mathfrak{k}(\underline{\alpha})$ integrates to a representation of the associated simply connected group,*

then each contravariant hermitian form on $V_{\underline{\lambda}}$ which is positive definite on a lowest weight vector $v_{\underline{\lambda}}$ is positive definite.

Proof. By tensoring π with an appropriate one-dimensional representation of $\tilde{\mathfrak{t}}_r$, we may w.l.o.g. assume that $\underline{\lambda} = (0, \lambda, h)$.

Since $V := V_{\underline{\lambda}}$ is a lowest weight module, we conclude that $V = \sum_{m \in \mathbb{N}_0} V(m)$ and that $V(0)$ is a lowest weight module of $\mathfrak{k}_{\mathbb{C}}$ with lowest weight λ. Since the set of weights of $V(0)$ in $\mathfrak{t}$ is invariant under the Weyl group $\mathcal{W}$, the module $V(0)$ is finite-dimensional, hence an irreducible module for $\mathfrak{k}_{\mathbb{C}}$. Moreover, the Poincaré-Birkhoff-Witt Theorem shows that the $\mathfrak{k}_{\mathbb{C}}$-submodules $V(m)$, $m \in \mathbb{N}$, are finite-dimensional.

We show by induction over k that the hermitian form on V is positive definite on the submodule $V(k)$. Since the form is contravariant for $\tilde{\mathfrak{t}}_r \oplus \mathfrak{t}_{\mathfrak{k}}$, it suffices to check positivity on the weight spaces V_{μ} for $\tilde{\mathfrak{t}}_r \oplus \mathfrak{t}$, where $\mu = (\underline{m}, \mu, h)$ and $\sum_{j=1}^m m_j = k$. Since these weight spaces can be decomposed as orthogonal sums, where each piece is contained in an irreducible $\mathfrak{k}_{\mathbb{C}}$-submodule on which, according to the uniqueness of the form on irreducible submodules, the form is either positive or negative, it suffices to assume that $v_{\mu} \in V^{\mu}$ is a non-zero weight vector which is contained in an irreducible $\mathfrak{k}_{\mathbb{C}}$-submodule. We may even assume that v_{μ} is a lowest weight vector for the representation of $\mathfrak{k}_{\mathbb{C}}$ on this irreducible submodule.

If $k = 0$, then the positivity of the form on $V(0)$ follows from the positivity on v_{λ} and the irreducibility of $V(0)$ as a $\mathfrak{k}_{\mathbb{C}}$-module which implies the uniqueness of the form up to a real scalar multiple. Thus we may assume that $k > 0$. We recall that for each root $\underline{\alpha} \in \Delta_{\mathfrak{k}}$ the representation of $\tilde{L}\mathfrak{k}(\underline{\alpha})$ integrates to a representation of the associated simply connected group, so that for each reflection $s_{\underline{\alpha}} \in \mathcal{W}_{\mathrm{aff}}$ there exists an operator $\tilde{s}_{\underline{\alpha}}$ on V leaving the form invariant and which has the property that

$$\tilde{s}_{\underline{\alpha}}.V^{\underline{\mu}} = V^{s_{\underline{\alpha}} \cdot \underline{\mu}}.$$

If $\underline{\alpha} \in \Delta_{L\mathfrak{k}}^+ \setminus \Delta_{\mathfrak{k}}$, then $\underline{\alpha}$ has positive energy. If $\underline{\mu}(h_{\underline{\alpha}}) > 0$, then

$$s_{\underline{\alpha}}.\underline{\mu} = \underline{\mu} - \langle \underline{\mu}, h_{\underline{\alpha}} \rangle \underline{\alpha}$$

has lower energy. Hence our induction proves that

$$\langle v_{\mu}, v_{\mu} \rangle = \langle \tilde{s}_{\underline{\alpha}}.v_{\mu}, \tilde{s}_{\underline{\alpha}}.v_{\mu} \rangle > 0.$$

Thus we may from now on assume in addition that $\underline{\mu}(h_{\underline{\alpha}}) \leq 0$ for all $\underline{\alpha} \in \Delta_{L\mathfrak{k}}^+ \setminus \Delta_{\mathfrak{k}}$. Since our assumption that v_{μ} is a $\mathfrak{k}_{\mathbb{C}}$-lowest weight vector implies that $\mu(h_{\alpha}) \leq 0$ holds for all $\alpha \in \Delta_{\mathfrak{k}}^+$, we see that $\underline{\mu}$ is an antidominant weight.

Now we have

$$\Omega_{\mathfrak{k}}.v_{\underline{\mu}} = \tfrac{1}{2}\big(\|\mu - \rho\|^2 - \|\rho\|^2\big)v_{\underline{\mu}}$$

and so we obtain with

$$\Omega = -\sum_{j=1}^{N}\sum_{n>0} e_j^n e_j^{-n} + \Omega_{\mathfrak{k}} + \sum_{j=1}^{m}(I + ic_j)\frac{d}{dt_j},$$

and Lemma IV.12 that

$$c_{\underline{\lambda}}\langle v_{\underline{\mu}}, v_{\underline{\mu}}\rangle = \langle \Omega.v_{\underline{\mu}}, v_{\underline{\mu}}\rangle = \left(c_{\mu} - \sum_{j=1}^{m}(h+c_j)m_j\right)\langle v_{\underline{\mu}}, v_{\underline{\mu}}\rangle - \sum_{j=1}^{N}\sum_{n>0}\langle e_j^n e_j^{-n}.v_{\underline{\mu}}, v_{\underline{\mu}}\rangle$$

$$= c_{\mu}\langle v_{\underline{\mu}}, v_{\underline{\mu}}\rangle + \sum_{j=1}^{N}\sum_{n>0}\langle e_j^{-n}.v_{\underline{\mu}}, e_j^{-n}.v_{\underline{\mu}}\rangle \geq c_{\mu}\langle v_{\underline{\mu}}, v_{\underline{\mu}}\rangle$$

because $\langle e_j^{-n}.v_{\underline{\mu}}, e_j^{-n}.v_{\underline{\mu}}\rangle \geq 0$ by the induction hypothesis. To show that $\langle v_{\underline{\mu}}, v_{\underline{\mu}}\rangle$ is non-negative, it now suffices to prove that $c_{\lambda} > c_{\mu}$. In the notation from above we have

$$2(c_{\underline{\lambda}} - c_{\underline{\mu}}) = \|\underline{\lambda} - \rho\|^2 - \|\underline{\mu} - \rho\|^2 = -\langle \underline{\lambda} + \underline{\mu} - 2\rho, \underline{\mu} - \underline{\lambda}\rangle.$$

This expression is positive because $\underline{\mu} - \underline{\lambda}$ is a sum of positive roots, and

$$\langle \underline{\lambda} + \underline{\mu} - 2\rho, \underline{\alpha}\rangle < 0$$

follows for each simple root α from the antidominance of $\underline{\lambda}$ and $\underline{\mu}$ together with the fact that $\langle \rho, \underline{\alpha}\rangle = 1$.

This proves that $\langle v_{\underline{\mu}}, v_{\underline{\mu}}\rangle \geq 0$. If $\langle v_{\underline{\mu}}, v_{\underline{\mu}}\rangle = 0$, then we have equality in the above chain of inequalities and hence

$$\langle e_j^{-n}.v_{\underline{\mu}}, e_j^{-n}.v_{\underline{\mu}}\rangle = 0$$

for all $j = 1, \ldots, N$ and $n > 0$. Thus $v_{\underline{\mu}}$ is a lowest weight vector for the whole Lie algebra $\tilde{L}_{\text{pol}}\mathfrak{k}_{\mathbb{C}}$. Since Ω acts by the scalar $c_{\underline{\lambda}}$ on $V_{\underline{\lambda}}$ and by the scalar $c_{\underline{\mu}}$ on the lowest weight module generated by $v_{\underline{\mu}}$ (Lemma IV.12), we conclude that $c_{\underline{\lambda}} = c_{\underline{\mu}}$, contradicting the observation made above. This proves that $\langle \cdot, \cdot \rangle$ is positive definite on $V_{\underline{\lambda}}$. $\square$

We will see in Theorem V.6 that the antidominance of $\underline{\lambda} \in \hat{T}_{\tilde{L}K}$ implies the existence of a lowest weight module satisfying the assumptions of Theorem IV.13. In general not every lowest weight module with lowest weight $\underline{\lambda}$ has this property. For instance Verma modules do not (cf. [PS86]).

It is an interesting consequence of the assumptions (i) and (ii) in Theorem IV.13 that the lowest weight module $V_{\underline{\lambda}}$ is irreducible. Otherwise there would be a lowest weight module $V_{\underline{\mu}}$ properly contained in $V_{\underline{\lambda}}$, and this is exactly what we have shown to be impossible in the last part of the proof.

V. Representations of involutive semigroups

Before we start with the detailed analysis of the positive energy representations of the group $\hat{L}K$, we need some background from the abstract theory of representations of involutive semigroups. This background will make the constructions and results in Section VI more transparent.

Definition V.1. An *involutive semigroup* is a semigroup S endowed with an involutive antiautomorphism $s \mapsto s^*$, which means that $(s^*)^* = s$ and $(st)^* = t^*s^*$ holds for $s, t \in S$. ☐

Example V.2. (a) If G is a group and τ is an involutive automorphism of G, then $g^* := \tau(g)^{-1}$ defines the structure of an involutive group on G. A particularly important case is $\tau = \mathrm{id}_G$.

The examples that will play a central role in Section VI are the groups $\tilde{L}K_{\mathbb{C}}$ with $\gamma^* = \overline{\gamma}^{-1}$, where $g \mapsto \overline{g}$ denotes complex conjugation with respect to the real form $\tilde{L}K$. Sometimes we will also consider the extended group $\mathbb{T}_r \ltimes \tilde{L}K_{\mathbb{C}}$, with

$$(R_\theta, \gamma)^* = (R_{-\theta}, R_\theta.\gamma^*).$$

Note that on the subgroup $\hat{L}K = \mathbb{T}_r \ltimes \tilde{L}K$ this involution is the inversion.

(b) Let $\mathfrak{g}$ be a complex Lie algebra endowed with an involutive antilinear antiisomorphism $\omega \colon \mathfrak{g} \to \mathfrak{g}$. Then ω induces an involutive antilinear antiisomorphism $D \mapsto D^*$ on the enveloping algebra $\mathcal{U}(\mathfrak{g})$ satisfying $X^* = \omega(X)$ for all $X \in \mathfrak{g}$.

If $\mathfrak{g} = \mathfrak{h}_{\mathbb{C}}$ is the complexification of a real Lie algebra, then $\omega(X) = -\overline{X}$, where $\overline{X}$ denotes complex conjugation, defines an involutive antilinear antiisomorphism of $\mathfrak{g}$. Conversely, each involutive antilinear antiisomorphism ω defines the real form $\mathfrak{h} := \{X \in \mathfrak{g} \colon \omega(X) = -X\}$.

The example that will arise in Section VI is the Lie algebra $\mathfrak{g} = \hat{L}\mathfrak{k}_{\mathbb{C}}$ with the real form $\hat{L}\mathfrak{k}$.

(c) Let V be a pre-Hilbert space. We write $B_0(V) \subseteq \mathrm{End}_{\mathbb{C}}(V)$ for the set of all linear operators A on V for which there exists an operator $A^\sharp \in \mathrm{End}_{\mathbb{C}}(V)$ with $\langle A.v, w \rangle = \langle v, A^\sharp.w \rangle$ for all $v, w \in V$. Note that such an operator $A^\sharp$ is uniquely determined by this property whenever it exists. It is easy to see that $B_0(V)$ is an involutive semigroup with respect to the involution $A \mapsto A^\sharp$ and composition of operators (cf. [Ne99, Lemma II.3.2]). ☐

Definition V.3. (a) Let M be a set. A function $Q \colon M \times M \to \mathbb{C}$ is called a *positive definite kernel* if for each finite subset $\{x_1, \dots, x_n\} \subseteq M$ the matrix $\big(Q(x_i, x_j)\big)_{i,j=1,\dots,n}$ is positive semidefinite. This condition is equivalent to the following one (cf. [Ne99, Th. I.1.6]): There exists a Hilbert space $\mathcal{H} \subseteq \mathbb{C}^M$ with continuous point evaluations represented by the functions $Q_x \colon y \mapsto Q(y, x)$, i.e., $f(x) = \langle f, Q_x \rangle$ for all $f \in \mathcal{H}$. Then Q is called the *reproducing kernel* of $\mathcal{H}$ and since $\mathcal{H}$ is, as a subspace of $\mathbb{C}^M$, uniquely determined by Q, we put $\mathcal{H}_Q := \mathcal{H}$ and call it the *reproducing kernel space* associated to Q. The dense subspace of $\mathcal{H}_Q$ spanned by the functions Q_x, $x \in M$, is denoted $\mathcal{H}_Q^0$.

(b) A function $\phi \colon S \to \mathbb{C}$ on an involutive semigroup S is called *positive definite* if the kernel $Q \colon S \times S \to \mathbb{C}$ defined by $Q(s, t) := \phi(st^*)$ is positive definite.

(c) If we have a left action $S \times M \to M$ of an involutive semigroup S on the set M, then a function $Q \colon M \times M \to \mathbb{C}$ is called an *invariant kernel* if

$$Q(s.x, y) = Q(x, s^*.y)$$

for $x, y \in M$ and $s \in S$. This terminology is inspired by the group case where $g^* = g^{-1}$, so that invariance means that $Q(g.x, g.y) = Q(x, y)$ for all $g \in G$.

(d) If $(S, *)$ is an involutive semigroup, then a morphism of involutive semigroups $\pi : S \to B_0(V)$ is called a *hermitian representation* of S on the pre-Hilbert space V. This means that π is a homomorphism of semigroups and that $\pi(s)^\sharp = \pi(s^*)$ for all $s \in S$. $\square$

The following lemma relates the invariance of a positive definite kernel to the existence of certain hermitian representations ([Ne99, Prop. II.4.3]).

Proposition V.4. *Let Q be a positive definite kernel on the set M and $S \times M \to M$ a left action of the involutive semigroup S. Then Q is invariant if and only if the action of S on $\mathbb{C}^M$ given by $s.f(x) := f(s^*.x)$ leaves the space*

$$\mathcal{H}_Q^0 = \operatorname{span}\{Q_x : x \in M\}$$

invariant and defines on this spaces a hermitian representation $(\pi, \mathcal{H}_Q^0)$. In this case we have $\pi(s).Q_x = Q_{s.x}$ for $x \in X$ and $s \in S$. $\square$

Remark V.5. We note that for each hermitian representation (π, V) of the involutive semigroup S and $v \in V$ the function defined by $\phi_v(s) := \langle \pi(s).v, v \rangle$ is positive definite.

If, conversely, S has an identity element $\mathbf{1}$ and ϕ is a positive definite function on S, then the kernel defined by $Q(s, t) := \phi(st^*)$ is positive definite and invariant under the left action of S on S given by $s.x := xs^*$. Since $\phi = Q_\mathbf{1} \in \mathcal{H}_Q^0$ is contained in the corresponding pre-Hilbert space on which the action of S is given by $(s.f)(x) := f(xs)$, we obtain

$$\langle s.\phi, \phi \rangle = \langle s.\phi, Q_\mathbf{1} \rangle = (s.\phi)(\mathbf{1}) = \phi(s).$$ $\square$

So far these concepts do not refer to any topology or differentiable structure on the semigroups or the spaces involved. Now we turn to the additional properties of the representation that will be available if the kernels of the actions have additional regularity properties.

Definition V.6. Let M be a complex manifold (modeled over a sequentially complete locally convex space). We write $\overline{M}$ for the same manifold endowed with the opposite complex structure, i.e., the identity $\operatorname{id}_M : M \to \overline{M}$ is an antiholomorphic map.

A kernel $Q : M \times M \to \mathbb{C}$ is called *holomorphic* if it is holomorphic as a function $M \times \overline{M} \to \mathbb{C}$. This means that it is holomorphic in the first and antiholomorphic in the second argument. $\square$

Proposition V.7. *Let Q be a continuous positive definite kernel on the topological space M satisfying the first countability axiom. Then the following assertions hold:*

(i) *The Hilbert space $\mathcal{H}_Q$ consists of continuous functions on M and the inclusion $\mathcal{H}_Q \to C(M)$ is continuous if $C(M)$ is endowed with the topology of uniform convergence on compact subsets of M.*

(ii) *If $G \times M \to M$ is an action of the topological group G on M leaving Q invariant, then $(g.f)(x) := f(g^{-1}.x)$ defines a unitary representation of G on $\mathcal{H}_Q$ which is continuous in the sense that the map $G \times \mathcal{H}_Q \to \mathcal{H}_Q$ is continuous.*

Proof. (i) Since Q is continuous, we find for each compact subset $C \subseteq M$ a constant $c > 0$ with $Q(x, x) \leq c$ for all $x \in C$. For $f \in \mathcal{H}_Q$ we then have

$$|f(x)| = |\langle f, Q_x \rangle| \leq \|f\| \cdot \|Q_x\| = \|f\| \sqrt{\langle Q_x, Q_x \rangle} = \|f\| \sqrt{Q(x,x)} \leq \sqrt{c}\|f\|.$$

This proves that the mapping $\mathcal{H}_Q \to \mathbb{C}^M$ is continuous with respect to the topology of uniform convergence on compact subsets of M on the space $\mathbb{C}^M$. For each $x \in M$ the function $Q_x : y \mapsto Q(y, x)$ is continuous. Therefore the statement follows from the closedness of $C(M)$ in $\mathbb{C}^M$ which is the same as the completeness of $C(M)$ (cf. [Ne00a, Prop. II.12(i)]).

(ii) For each pair $x, y \in M$ the function

$$G \to \mathbb{C}, \quad g \mapsto \langle g.Q_x, Q_y \rangle = \langle Q_{g.x}, Q_y \rangle = Q(y, g.x)$$

is continuous and since G acts by isometries on $\mathcal{H}_Q$, it follows that the representation $G \to U(\mathcal{H}_Q)$ is continuous if $U(\mathcal{H}_Q)$ is endowed with the weak operator topology which on $U(\mathcal{H}_Q)$ coincides with the strong operator topology. Hence it suffices to show that with respect to this topology the action map

$$U(\mathcal{H}_Q) \times \mathcal{H}_Q \to \mathcal{H}_Q, \quad (g, v) \mapsto g.v$$

is continuous. In fact, suppose that $v_n \to v$ and $g_i \to g$. Then

$$\|g_i.v_n - g.v\| \leq \|g_i.(v_n - v)\| + \|(g_i - g).v\| = \|v_n - v\| + \|(g_i. - g).v\| \to 0.$$

This completes the proof. $\qquad\square$

Proposition V.8. *If Q is a holomorphic positive definite kernel on the complex Fréchet manifold M, then $\mathcal{H}_Q$ consists of holomorphic functions on M and the inclusion $\mathcal{H}_Q \to \mathrm{Hol}(M)$ is continuous if $\mathrm{Hol}(M)$ is endowed with the topology of uniform convergence on compact subsets of M.*

Proof. Since Q is holomorphic, it is in particular continuous, and Proposition V.7 applies and shows that $\mathcal{H}_Q \subseteq C(M)$ and that $\mathcal{H}_Q \to C(M)$ is continuous with respect to the topology of uniform convergence on compact sets. Now the assertion follows from the observation that the dense subspace $\mathcal{H}_Q^0$ consists of holomorphic functions and the closedness of $\mathrm{Hol}(M)$ in $\mathbb{C}^M$, which is the same as the completeness of $\mathrm{Hol}(M)$ (cf. [Ne00a, Th. III.9]). $\qquad\square$

VI. Borel-Weil theory

In this section we turn to the Borel-Weil theory for loop groups. This means that we study representations that can be realized in certain homogeneous complex line bundles for loop groups. One of the main points of these constructions is that once the appropriate geometric information on this homogeneous space is available, then everything works quite analogous to the finite-dimensional case.

First we explain how to construct certain complex line bundles parametrized by the characters λ of the torus $T_{\hat{L}K}$. We then study the corresponding representation of $\hat{L}K$ in the space Γ_λ of holomorphic sections. Finally we derive a criterion for the corresponding space to be non-zero (Theorem VI.8). In Section VII we will see that in some sense the representations obtained by this construction exhaust all irreducible representations with positive energy.

We recall how the fundamental homogeneous space LK/T_K of the loop group LK can be realized as a homogeneous space of the complexified loop group $LK_{\mathbb{C}}$.

Definition VI.1. Let $B_0^+ \subseteq K_{\mathbb{C}}$ be the Borel subgroup with the Lie algebra $\mathfrak{b}_0^+ := \mathfrak{t}_{\mathbb{C}} \oplus \sum_{\alpha \in \Delta_{\mathfrak{t}}^+} \mathfrak{k}_{\mathbb{C}}^\alpha$, i.e., the Borel subalgebra of $\mathfrak{k}_{\mathbb{C}}$ corresponding to the positive system $\Delta_{\mathfrak{t}}^+$. We write $N_0^\pm \subseteq K_{\mathbb{C}}$ for the nilpotent subgroup corresponding to the nilpotent Lie algebra

$$\mathfrak{n}_0^\pm = \sum_{\alpha \in \Delta_{\mathfrak{t}}^\pm} \mathfrak{k}_{\mathbb{C}}^\alpha.$$

In $LK_{\mathbb{C}}$ we consider the subgroup B^+ consisting of all smooth boundary values of holomorphic maps $\gamma \colon \{z \in \mathbb{C} \colon |z| < 1\} \to K_{\mathbb{C}}$ with $\gamma(0) \in B_0^+$. Its Lie algebra is given by

$$\mathfrak{b}^+ := \left\{ \sum_{k=0}^{\infty} z^k a_k \in L\mathfrak{k}_{\mathbb{C}} \colon a_k \in \mathfrak{k}_{\mathbb{C}}, a_0 \in \mathfrak{b}_0^+ \right\}.$$

Likewise we consider the subgroup $N^- \subseteq LK_{\mathbb{C}}$ consisting of all smooth boundary values of holomorphic maps $\gamma \colon \{z \in \mathbb{C} \colon |z| > 1\} \cup \{\infty\} \to K_{\mathbb{C}}$ with $\gamma(\infty) \in N_0^-$. Its Lie algebra is given by

$$\mathfrak{n}^- := \left\{ \sum_{k \leq 0} z^k a_k \in L\mathfrak{k}_{\mathbb{C}} \colon a_k \in \mathfrak{k}_{\mathbb{C}}, a_0 \in \mathfrak{n}_0^- \right\}.$$

Similarly one defines the subgroup $N^+ := \{\gamma \in B^+ \colon \gamma(0) \in N_0^+\}$ with the Lie algebra

$$\mathfrak{n}^+ = \left\{ \sum_{k \geq 0} z^k a_k \in L\mathfrak{k}_{\mathbb{C}} \colon a_k \in \mathfrak{k}_{\mathbb{C}}, a_0 \in \mathfrak{n}_0^+ \right\}.$$

The appendix of [GW84] contains a detailed discussion of this group.

The inclusions $N_0^\pm \hookrightarrow N^\pm$ (as constant maps) are homotopy equivalences via the map

$$H \colon N^\pm \times [0,1] \to N_0^\pm, \quad (\gamma, t) \mapsto (z \mapsto \gamma(tz)), \quad |z| \leq 1.$$

Since $N_0^\pm$ are unipotent, hence contractible, it follows that the groups $N^\pm$ are contractible and in particular simply connected.   □

Proposition VI.2. *The following assertions hold:*

(i) *The exponential functions of the subgroups B^+ and N^- are local diffeomorphisms in 0.*

(ii) *The multiplication map $N^- \times B^+ \to LK_{\mathbb{C}}$ is a diffeomorphism onto an open subset of $LK_{\mathbb{C}}$.*

(iii) $B^+ \cap LK = T_K$.

(iv) *The group $LK_{\mathbb{C}}$ acts transitively on LK/T_K, which leads to a diffeomorphism $Y := LK_{\mathbb{C}}/B^+ \cong LK/T_K$.*

Proof. [PS86, Th. 8.7.2].   □

We write $q \colon \tilde{L}K_{\mathbb{C}} \to LK_{\mathbb{C}}$ for the quotient mapping defining the central extension (2.6). Let $\tilde{K}_{\mathbb{C}} := q^{-1}(K_{\mathbb{C}})$ and $\tilde{T}_{\mathbb{C}} := q^{-1}(T_{\mathbb{C}})$. Then

$$\{1\} \to (\mathbb{T}_c)_{\mathbb{C}} \cong \mathbb{C}^\times \longrightarrow \tilde{K}_{\mathbb{C}} \overset{q}{\longrightarrow} K_{\mathbb{C}} \to \{1\}$$

is a central extension of the connected complex group $K_{\mathbb{C}}$ whose Lie algebra cocycle is trivial. If K and hence $K_{\mathbb{C}}$ is simply connected, then this extension splits, and $\tilde{K}_{\mathbb{C}} \cong K_{\mathbb{C}} \times (\mathbb{T}_c)_{\mathbb{C}}$, where $K_{\mathbb{C}} \cong (\tilde{K}_{\mathbb{C}}, \tilde{K}_{\mathbb{C}})$ is the commutator subgroup. In this case we also have

$$\tilde{T}_{\mathbb{C}} \cong T_{\mathbb{C}} \times (\mathbb{T}_c)_{\mathbb{C}} \cong (T \times \mathbb{T}_c)_{\mathbb{C}}.$$

Now let $\lambda \in \hat{\tilde{T}}_{\tilde{L}K}$ be a character which is trivial on the rotation group $\mathbb{T}_r$. Then λ extends to a holomorphic character of the complexification $\tilde{T}_{\mathbb{C}} = \tilde{T} \exp(i\tilde{t})$. Since the following diagram which is defined by the evaluation and inclusion morphisms in the bottom row and the corresponding pullbacks in the top row is commutative, we have a holomorphic homomorphism $\tilde{B}^+ \to \tilde{T}_{\mathbb{C}}$ which permits us to extend holomorphic characters from $\tilde{T}_{\mathbb{C}}$ to $\tilde{B}^+$ by pulling them back.

$$
\begin{array}{ccccccc}
\tilde{B}^+ & \longrightarrow & \tilde{B}_0^+ & \longrightarrow & \tilde{T}_{\mathbb{C}} & \hookrightarrow & \tilde{L}K_{\mathbb{C}} \\
\downarrow{\scriptstyle q} & & \downarrow{\scriptstyle q} & & \downarrow{\scriptstyle q} & & \downarrow{\scriptstyle q} \\
B^+ & \overset{eval}{\longrightarrow} & B_0^+ = N_0^+ \rtimes T_{\mathbb{C}} & \longrightarrow & T_{\mathbb{C}} & \hookrightarrow & LK_{\mathbb{C}}.
\end{array}
$$

Definition VI.3. In view of the preceding discussion, we may consider λ as a holomorphic character of the group $\tilde{B}^+$. Let

$$L_\lambda := \tilde{L}K_{\mathbb{C}} \times_{\tilde{B}^+} \mathbb{C} \to \tilde{L}K_{\mathbb{C}}/\tilde{B}^+ \cong LK_{\mathbb{C}}/B^+ = Y.$$

denote the associated holomorphic line bundle, i.e., the quotient of $\tilde{L}K_{\mathbb{C}} \times \mathbb{C}$ modulo the action of $\tilde{B}^+$ which is given by

$$b.(g, z) = (gb^{-1}, \lambda(b).z).$$

The coordinates on this bundle can be obtained by using Proposition VI.2(ii). Since the central subgroup $\mathbb{C}^\times = (\mathbb{T}_c)_{\mathbb{C}}$ of $\tilde{L}K_{\mathbb{C}}$ is contained in $\tilde{B}^+$, we obtain a natural isomorphism

$$\tilde{L}K_{\mathbb{C}}/\tilde{B}^+ \rightarrow LK_{\mathbb{C}}/B^+. \qquad\qquad \square$$

Lemma VI.4. *The natural holomorphic action of the complex group $\tilde{L}K_{\mathbb{C}}$ on the bundle L_λ can be extended to the group $\mathbb{T}_r \ltimes \tilde{L}K_{\mathbb{C}}$ by holomorphic automorphisms via*

$$R_\theta.[g, z] := [R_\theta.g, z].$$

Proof. First we observe that $\lambda(R_\theta b R_{-\theta}) = \lambda(b)$ for all $b \in \tilde{B}^+$ follows from the fact that $\mathbb{T}_r$ acts trivially on $T \times \mathbb{T}_c$. We let $\mathbb{T}_r$ act on $\tilde{L}K_{\mathbb{C}} \times \mathbb{C}$ by $R_\theta.(g, z) := (R_\theta.g, z)$, where the action of $\mathbb{T}_r$ on $\tilde{L}K_{\mathbb{C}}$ is the canonical lift of the rotation action on $LK_{\mathbb{C}}$ (cf. Section II). Then

$$R_\theta.\big(b.(g, z)\big) = R_\theta.\big(gb^{-1}, \lambda(b).z\big) = \big(R_\theta g b^{-1} R_{-\theta}, \lambda(b).z\big)$$
$$= \big(R_\theta g R_{-\theta} R_\theta b^{-1} R_{-\theta}, \lambda(R_\theta.b).z)\big) = (R_\theta.b).\big(R_\theta.(g, z)\big)$$

implies that the action of $\mathbb{T}_r$ on $\tilde{L}K_{\mathbb{C}} \times \mathbb{C}$ factors to an action on the bundle L_λ given by $R_\theta.[g, z] := [R_\theta.g, z]$. This is an action by holomorphic automorphisms. We thus obtain an action of the group $\mathbb{T}_r \ltimes \tilde{L}K_{\mathbb{C}}$ by holomorphic automorphisms of the bundle L_λ over $LK_{\mathbb{C}}/B^+$. $\qquad\qquad \square$

Lemma VI.5. *Let Γ_λ denote the space of all holomorphic sections of L_λ. We identify this space with a space of holomorphic functions on $LK_{\mathbb{C}}$ by assigning to a section $s\colon \tilde{L}K_{\mathbb{C}} \rightarrow L_\lambda$ the function $f_s \in \mathrm{Hol}(\tilde{L}K_{\mathbb{C}})$ defined by*

$$s(g\tilde{B}^+) = [g, f_s(g)].$$

Then a holomorphic function f on $\tilde{L}K_{\mathbb{C}}$ defines a section of L_λ if and only if

$$[gb^{-1}, f(gb^{-1})] = [g, f(g)]$$

for all $b \in \tilde{B}^+$ and $g \in \tilde{L}K_{\mathbb{C}}$, and this is equivalent to

$$f(gb^{-1}) = \lambda(b)f(g)$$

for $g \in \tilde{L}K_{\mathbb{C}}$, $b \in \tilde{B}^+$. $\qquad\qquad \square$

Proposition VI.6. *We endow the space $\mathrm{Hol}(\tilde{L}K_{\mathbb{C}})$ with the topology of uniform convergence on compact sets. Then the following assertions hold:*

  (i) *The space $\mathrm{Hol}(\tilde{L}K_{\mathbb{C}})$ is a complete locally convex space, the natural action of $\mathbb{T}_r \ltimes \tilde{L}K_{\mathbb{C}}$ induced by the action on $\tilde{L}K_{\mathbb{C}}$ is continuous, and the action map*

$$\tilde{L}K_{\mathbb{C}} \times \mathrm{Hol}(\tilde{L}K_{\mathbb{C}}) \rightarrow \mathrm{Hol}(\tilde{L}K_{\mathbb{C}})$$

    is holomorphic.

  (ii) *The subspace $\Gamma_\lambda \subseteq \mathrm{Hol}(\tilde{L}K_{\mathbb{C}})$ is a closed left invariant subspace, hence a complete locally convex space.*

Proof. (i) The completeness of $\mathrm{Hol}(\tilde{L}K_{\mathbb{C}})$ follows from [Ne00a, Th. III.11] because $\tilde{L}K_{\mathbb{C}}$ is modeled over the Fréchet space $\tilde{L}\mathfrak{k}_{\mathbb{C}}$. The remaining assertions follow from [Ne00a, Th. III.14].

(ii) This follows from the fact that the functions $f \in \Gamma_\lambda$ are characterized by the condition that for all $b \in \tilde{B}^+$ and $g \in \tilde{L}K_{\mathbb{C}}$ we have $f(gb^{-1}) = \underline{\lambda}(b)f(g)$. □

Remark VI.7. We note that multiplying the character $\underline{\lambda}$ with a character $\chi \in \hat{\mathbb{T}}_r$ of the group $\mathbb{T}_r$ of rotations corresponds to tensoring the corresponding representation of $\hat{L}K$ with the one-dimensional representation defined by the characer of χ of $\mathbb{T}_r \cong \hat{L}K/\tilde{L}K$. Therefore we may w.l.o.g. assume in the following that $\underline{\lambda}$ is a character of the form $(0, \lambda, h)$, i.e., trivial on $\mathbb{T}_r$. □

Theorem VI.8. *If* $\underline{\lambda} = (0, \lambda, h)$ *and* $\Gamma_{\underline{\lambda}} \neq \{0\}$, *then the following assertions hold:*
   (i) $\underline{\lambda}$ *is antidominant.*
   (ii) *The representation of* $\hat{L}K$ *on* $\Gamma_{\underline{\lambda}}$ *is a positive energy representation.*
   (iii) $\Gamma_{\underline{\lambda}}(0)$ *is an irreducible representation of* $K \times \mathbb{T}_c$ *of lowest weight* (λ, h).
   (iv) *If* μ *is a weight of* $\Gamma_{\underline{\lambda}}$, *then* $\mu - \lambda$ *is a sum of positive roots.*
   (v) *The module* $\Gamma_{\underline{\lambda}}$ *is of finite type in the sense that* $\dim \Gamma_{\underline{\lambda}}(n) < \infty$ *holds for all* $n \in \mathbb{N}$.
   (vi) $\Gamma_{\underline{\lambda}}$ *contains up to scalar multiple exactly one lowest weight vector* f. *As a function on* $\tilde{L}K_{\mathbb{C}}$, *this function is characterized by*

(6.1) $$f(n^- t n^+) = \underline{\lambda}(t)f(1)$$

   for $n^- \in N^-$, $t \in \tilde{T}_{\mathbb{C}}$ *and* $n^+ \in N^+$.

Proof. (i) Let $\underline{\alpha} = (k, \alpha) \in \Delta^+_{\tilde{L}\mathfrak{k}}$ and $K_{\mathbb{C}}(\underline{\alpha}) \subseteq \tilde{L}K_{\mathbb{C}}$ the corresponding 3-dimensional complex subgroup. Since $\Gamma_{\underline{\lambda}}$ is a left invariant subspace of $\mathrm{Hol}(\tilde{L}K_{\mathbb{C}})$, our assumption implies that it contains a function f with $f(1) \neq 0$.

The subalgebra $\mathfrak{b}(\underline{\alpha}) := \mathbb{C}h_\alpha + \mathfrak{k}^\alpha_{\mathbb{C}} z^k$ is a Borel subalgebra of the three-dimensional simple complex Lie algebra $\mathfrak{k}_{\mathbb{C}}(\underline{\alpha})$. Let $B(\underline{\alpha})$ denote the corresponding Borel subgroup of $K_{\mathbb{C}}(\underline{\alpha})$. Then the restriction of $\Gamma_{\underline{\lambda}}$ to $K_{\mathbb{C}}(\underline{\alpha})$ is non-zero and contained in the space

$$\Gamma(\underline{\alpha}) := \{f \in \mathrm{Hol}\left(K_{\mathbb{C}}(\underline{\alpha})\right): (\forall b \in B(\underline{\alpha}))f(gb^{-1}) = \underline{\lambda}(b)f(g)\}.$$

Now the finite-dimensional Borel-Weil Theory shows that the fact that this space is non-zero implies that $\underline{\lambda}(h_{\underline{\alpha}}) \in -\mathbb{N}_0$. We conclude that $\underline{\lambda}$ is antidominant.

(ii) If N^- is the subgroup defined above, then we will need the fact that $U := N^- B^+/B^+ \subseteq Y := LK_{\mathbb{C}}/B^+$ is an open dense subset and that N^- can also be identified with a subgroup of the central extension $\tilde{L}K_{\mathbb{C}}$ because the cocycle defining the central extension is trivial on its Lie algebra $\mathfrak{n}^-$, and the group N^- is simply connected, so that $\tilde{N}^- \cong \mathbb{C}^\times \times N^-$. Here we refer to [Ne00b, Th. V.4] for the fact that central extensions of simply connected Lie groups with trivial Lie algebra cocycles are trivial.

Now we consider the following sequence of maps:

(6.2) $\Gamma_{\underline{\lambda}} \xrightarrow{\ \alpha\ } \mathrm{Hol}(N^-) \xrightarrow{\ \beta\ } \mathrm{Hol}(\mathfrak{n}^-) \xrightarrow{\ \gamma\ } \prod_{p \geq 0} S^p(\mathfrak{n}^-)'$,

where $\alpha(f) = f|_{N^-}$ is the restriction map, $\beta(f) = f \circ \exp_{N^-}$, where $\exp\colon \mathfrak{n}^- \to N^-$ is the holomorphic exponential function of the group N^- which is a local diffeomorphism (Proposition VI.2) and γ is defined by the Taylor expansion of a holomorphic function on the complex Fréchet space $\mathfrak{n}^-$, where $S^p(\mathfrak{n}^-)'$ denotes the vector space of symmetric p-linear continuous maps $(\mathfrak{n}^-)^p \to \mathbb{C}$. Let $\Phi := \gamma \circ \beta \circ \alpha$. Since $U = N^- B^+/B^+$ is open in $Y = LK_{\mathbb{C}}/B^+$, the set $N^- \tilde{B}^+ \subseteq \tilde{L}K_{\mathbb{C}}$ is open. Hence the maps α and β are injective. Moreover, if $f \in \mathrm{Hol}(\mathfrak{n}^-)$, then f is uniquely determined by its Taylor expansion because this holds for the restriction of f to each one-dimensional subspace. Therefore Φ is a $\mathbb{T}_r$-equivariant injection

$$\Gamma_{\underline{\lambda}} \to \prod_{p \geq 0} S^p(\mathfrak{n}^-)'.$$

Next we explain how the action of the torus $T_{\hat{L}K}$ on $\Gamma_{\underline{\lambda}}$ can also be seen on the spaces on the right hand side of (6.2). If s is a section of $L_{\underline{\lambda}}$ and f the corresponding holomorphic function on $\tilde{L}K_{\mathbb{C}}$, then $t \in T_{\hat{L}K}$ acts on s via

$$(t.s)(gB^+) := t.s(t^{-1}gB^+) = t.\big(s(t^{-1}gtB^+)\big)$$
$$= t.[t^{-1}gt, f(t^{-1}gt)] = [gt, f(t^{-1}gt)] = [g, \underline{\lambda}(t)f(t^{-1}gt)].$$

Thus α is equivariant with respect to the action of $T_{\hat{L}K}$ on $\mathrm{Hol}(N^-)$ given by

$$(t.f)(n) := \underline{\lambda}(t)f(t^{-1}nt).$$

Similarly β is equivariant with respect to the action of $T_{\hat{L}K}$ on $\mathrm{Hol}(\mathfrak{n}^-)$ given by

$$(t.f)(X) := \underline{\lambda}(t)f\big(\mathrm{Ad}(t^{-1}).X\big).$$

Since the spaces $S^p(\mathfrak{n}^-)'$ are subspaces of $\mathrm{Hol}(\mathfrak{n}^-)$, this formula also defines an action of $T_{\hat{L}K}$ on these spaces, hence on their cartesian products, and we see that β and γ are equivariant with respect to these actions.

To see that the module $\Gamma_{\underline{\lambda}}$ is of positive energy, it now suffices to see that the module on the right hand side has this property because Φ is injective and $\mathbb{T}_r$-equivariant. In view of $\underline{\lambda} = (0, \lambda, h)$, for $t \in \mathbb{T}_r$ we have

$$(t.f)(X) := f\big(\mathrm{Ad}(t^{-1}).X\big).$$

The $\mathfrak{t}_r$-weights for the action on the space

$$\mathfrak{n}^- = \mathfrak{n}_0^- + \sum_{k<0} \mathfrak{k}_{\mathbb{C}} z^k$$

are contained in $-\mathbb{N}_0$, so that the corresponding action on the dual has weights in $\mathbb{N}_0$. So the weights on the spaces $S^p(\mathfrak{n}^-)'$ are sums of non-negative integers and therefore non-negative integers. This shows that the representation of $\mathbb{T}_r$ on the product of the space $S^p(\mathfrak{n}^-)'$ has only non-negative weights. This proves that $\Gamma_{\underline{\lambda}}$ is a positive energy module.

(iii) The $\mathbb{T}_r$-equivariance of Φ implies that it maps $\Gamma_\lambda(0)$ into

$$\left(\prod_{p \in \mathbb{N}_0} S^p(\mathfrak{n}^-)'\right)(0) = \prod_{p \in \mathbb{N}_0} S^p(\mathfrak{n}^-)'(0) \cong \prod_{p \in \mathbb{N}_0} S^p(\mathfrak{n}_0^-)'(0).$$

We conclude that each function $f \in \Gamma_\lambda(0)$ restricts to a function on N^- which factors over the evaluation morphism $N^- \to N_0^-, \gamma \mapsto \gamma(\infty)$ (cf. [Ne00a, Prop. IV.9(ii)]). Hence the $K \times \mathbb{T}_c$-representation on the space $\Gamma_\lambda(0)$ can be realized on the space $\mathrm{Hol}(N_0^-)$, where it corresponds to the action of $K \times \mathbb{T}_c$ on the space $\Gamma^0_{(\lambda,h)}$ of holomorphic sections of the bundle

$$L_{(\lambda,h)} = K_{\mathbb{C}} \times_{B_0^+} \mathbb{C} \to K_{\mathbb{C}}/B_0^+.$$

The Borel-Weil Theorem for finite-dimensional groups now implies that the representation of $K \times \mathbb{T}_c$ on $\Gamma^0_{(\lambda,h)}$ is irreducible with lowest weight (λ, h). Since $\Gamma_\lambda(0)$ embeds into this space in a $(K \times \mathbb{T}_c)$-equivariant way, we conclude that the embedding is surjective, which proves (iii).

(iv) In view of the $T_{\hat{L}K}$-equivariance of Φ, it suffices to prove the corresponding statement for the representation of $T_{\hat{L}K}$ on $\prod_{p \in \mathbb{N}} S^p(\mathfrak{n}^-)'$.

    The weights for the adjoint action of $T_{\hat{L}K}$ on the symmetric algebra $S(\mathfrak{n}^-)$ are given as sums of negative roots in $\Delta^-_{\hat{L}\mathfrak{k}}$. Hence the weights on the dual space are sums of positive roots and so the same holds for the weights on the product $\prod_{p \in \mathbb{N}} S^p(\mathfrak{n}^-)'$. This proves (iv).

(v) First we note that the subspace $\Gamma_\lambda(n)$ is $K_{\mathbb{C}}$-invariant for all $n \in \mathbb{N}$. Since the group K is compact, the Big Peter-Weyl Theorem ([HoMo98, Th. 3.51]) applies to the representation of K on this space and shows that the sum of all irreducible finite-dimensional subspaces is dense. The representations of $K_{\mathbb{C}}$ on these spaces are lowest weight representations. Thus each one contains a one-dimensional subspace of N_0^--fixed points. Therefore to show that $\Gamma_\lambda(n)$ is finite-dimensional, it suffices to show that the subspace $\Gamma_\lambda(n)^{N_0^-}$ of N_0^--fixed vectors in this space is finite-dimensional.

    To see how this space looks like, we follow the action of N_0^- through the mappings α, β and γ. The action of N^- on $\mathrm{Hol}(N^-)$ which makes α equivariant is simply the action by left translations $(n.f)(x) := f(n^{-1}x)$. Hence the N_0^--fixed points are the functions which are invariant on the N_0^--right cosets $N_0^- n, n \in N^-$. Since $N^- \cong N_1^- \rtimes N_0^-$ is a semidirect product, where $N_1^- \subseteq N^-$ is the kernel of the natural map to N_0^-, the N_0^--fixed points correspond to functions on the group N_1^- with Lie algebra

$$\mathfrak{n}_1^- = \sum_{k<0} \mathfrak{k}_{\mathbb{C}} z^k \cong \mathfrak{k}_{\mathbb{C}} \otimes z^{-1} \mathbb{C}[z^{-1}].$$

Form that we conclude that the weights of $\mathbb{T}_r$ on this space lie in $-\mathbb{N}$, hence that the weights of $\mathbb{T}_r$ on $S^p(\mathfrak{n}_1^-)'$ are contained in $p + \mathbb{N}$. Therefore

$$\Big(\prod_{p \in \mathbb{N}_0} S^p(\mathfrak{n}_1^-)' \Big)(n) = \prod_{p \in \mathbb{N}_0} S^p(\mathfrak{n}_1^-)'(n) \cong \sum_{p=0}^n S^p(\mathfrak{n}_1^-)'(n).$$

Further

$$S^p(\mathfrak{n}_1^-) \subseteq (\mathfrak{n}_1^-)^{\otimes p} = (\mathfrak{k}_{\mathbb{C}})^{\otimes p} \otimes \big(z^{-1}\mathbb{C}[z^{-1}] \big)^{\otimes p},$$

and

$$\big((\mathfrak{k}_{\mathbb{C}})^{\otimes p} \otimes \big(z^{-1}\mathbb{C}[z^{-1}] \big)^{\otimes p} \big)(-n) \cong (\mathfrak{k}_{\mathbb{C}})^{\otimes p} \otimes P_n,$$

where P_n is linearly isomorphic to the space of all polynomials in $z_1^{-1}, \ldots, z_p^{-1}$ of degree n. Since this space is finite-dimensional, the spaces $S^p(\mathfrak{n}_1^-)'(n)$, $p \le n$, are also finite-dimensional, and thus $\Gamma_\lambda(n)^{N_0^-}$ is finite-dimensional because it embeds in a finite sum of these spaces.

(vi) If $f \in \Gamma_\lambda$ is a lowest weight vector, then on the open subset $N^-\tilde{B}^+ = N^-\tilde{T}_{\mathbb{C}}N^+ \subseteq \tilde{L}K_{\mathbb{C}}$ this function satisfies (6.1) because $\mathfrak{n}^-.f = \{0\}$ implies that f is constant on the cosets N^-g of N^- (cf. [Ne00a, Prop. IV.9(ii)]). We conclude in particular that f is uniquely determined by $f(\mathbf{1})$.

To prove the existence of f, we recall from (iii) that the representation of $K \times \mathbb{T}_c$ on $\Gamma_\lambda(0)$ is an irreducible representation of lowest weight (λ, h). Using Proposition IV.6, we find a lowest weight vector $f \in \Gamma_\lambda(0)$ for $\hat{L}K$. This proves (vi). $\qquad\square$

Theorem VI.9. *If $\Gamma_\lambda \neq \{0\}$, then the following assertions hold:*
(i) *The module Γ_λ is essentially unitary in the sense that there exists a dense $\hat{L}K$-invariant reproducing kernel Hilbert space $\mathcal{H}_\lambda \subseteq \Gamma_\lambda$ on which $\hat{L}K$ acts continuously by an irreducible unitary representation.*
(ii) *The representation of $\hat{L}K$ on Γ_λ is irreducible.*
(iii) *The antidual $\Gamma_\lambda^\sharp$, i.e., the space of all continuous antilinear functionals on Γ_λ, can be identified with a subspace of $\mathcal{H}_\lambda$ in such a way that it contains the dense subspace $\mathcal{H}_\lambda^0$ spanned by the point evaluations.*

Proof. In view of Remark VI.7, we may w.l.o.g. assume that $\lambda = (0, \lambda, h)$. Let $\Gamma(0) := \Gamma_\lambda(0) \subseteq \Gamma := \Gamma_\lambda$ denote the energy subspace of degree 0. Using Theorem VI.8(vi), we choose a lowest weight function $f \in \Gamma(0)$ which is normalized by $f(\mathbf{1}) = 1$.

As a crucial piece of the proof, we will show below that the function f is a holomorphic positive definite function on the complex group $\tilde{L}K_{\mathbb{C}}$ endowed with the antiholomorphic involution given by $g^* := \bar{g}^{-1}$, where $\bar{g}$ denotes complex conjugation with respect to the real subgroup $\tilde{L}K$. The differential of this involution is an antilinear involution on the Lie algebra level which is denoted $X \mapsto X^*$. For $X \in \tilde{L}\mathfrak{k}$ we have $X^* = -X$. Since $\underline{\lambda}(t^*) = \overline{\underline{\lambda}(t)}$ for all $t \in \tilde{T}_{\mathbb{C}}$, and $(N^+)^* = N^-$, formula (6.1) implies that

(6.3) $$f(g^*) = \overline{f(g)}$$

for all $g \in \tilde{N}^{-}\tilde{B}^{+}$ and hence for all $g \in \tilde{L}K_{\mathbb{C}}$ because both sides are antiholo-morphic function which agree on an open subset of $\tilde{L}K_{\mathbb{C}}$ (cf. [Ne00a, Lemma III.13(ii)]).

For $b \in \tilde{B}^{+}$ this implies that

$$(\bar{b}.f)(g) = f(b^{*}g) = \overline{f(g^{*}b)} = \overline{\underline{\lambda}(b^{-1})f(g^{*})} = \overline{\underline{\lambda}(b^{-1})}f(g),$$

i.e.,

(6.4) $\bar{b}.f = \overline{\underline{\lambda}(b^{-1})}f.$

Now let $\Gamma^{\sharp}$ denote the antidual of the locally convex space Γ, i.e., the space of all continuous antilinear functionals $\Gamma \to \mathbb{C}$. We define a map

$$\beta \colon \Gamma^{\sharp} \to \mathrm{Hol}(\tilde{L}K_{\mathbb{C}}), \quad \beta(\alpha)(g) := \alpha(\bar{g}.f).$$

Since the orbit map $\tilde{L}K_{\mathbb{C}} \to \Gamma, g \mapsto g.f$ is holomorphic (Proposition VI.6) and $g \mapsto \bar{g}$ is antiholomorphic, the functions $\beta(\alpha)$ are indeed holomorphic functions on $\tilde{L}K_{\mathbb{C}}$. We claim that $\beta \colon \Gamma^{\sharp} \to \Gamma$ is an $\tilde{L}K_{\mathbb{C}}$-equivariant map, where $\tilde{L}K_{\mathbb{C}}$ acts on $\Gamma^{\sharp}$ via $(g.\alpha)(v) := \alpha(g^{*}.v)$. For the continuity of this action we refer to [Ne00a, Lemma IV.4(ii)].

The inclusion $\beta(\alpha) \in \Gamma$ follows from the observation that for $b \in \tilde{B}^{+}$ and $g \in \tilde{L}K_{\mathbb{C}}$ we have in view of (6.4):

$$\beta(\alpha)(gb^{-1}) = \alpha(\overline{\bar{g}b}^{-1}.f) = \alpha(\bar{g}\overline{\underline{\lambda}(b)}.f) = \underline{\lambda}(b)\alpha(\bar{g}.f) = \underline{\lambda}(b)\beta(\alpha)(g).$$

The equivariance of β follows from

$$(g.\beta(\alpha))(x) = \beta(\alpha)(g^{-1}x) = \alpha(\bar{g}^{-1}\bar{x}.f) = (g.\alpha)(\bar{x}.f) = \beta(g.\alpha)(x)$$

for $g \in \tilde{L}K_{\mathbb{C}}$. The continuity of the representation on Γ implies that for each compact subset $C \subseteq \tilde{L}K_{\mathbb{C}}$ the subset $C.f \subseteq \Gamma$ is compact (cf. Proposition IV.1), hence that β is continuous.

Now we can define a sesquilinear form on $\Gamma^{\sharp}$ by

$$\langle \alpha, \alpha' \rangle := \alpha(\beta(\alpha')).$$

Note that this form is sesquilinear because β is linear and α is antilinear. Our major goal is to show that this form turns $\Gamma^{\sharp}$ into a pre-Hilbert space such that the representation of $\tilde{L}K_{\mathbb{C}}$ is hermitian (cf. Definition IV.3(d)).

First we show that this form is hermitian. For $g \in \tilde{L}K_{\mathbb{C}}$ let $\delta_{g} \in \Gamma^{\sharp}$ be defined by $\delta(g)(f) = \overline{f(g)}$. Then

$$\langle \delta_{g}, \delta_{g'} \rangle = \delta_{g}(\beta(\delta_{g'})) = \overline{\beta(\delta_{g'})(g)} = \overline{\delta_{g'}(\bar{g}.f)}$$

$$= \overline{(\bar{g}.f)(g')} = \overline{f(g^{*}g')} = \overline{\overline{f(g'^{*}g)}} = \langle \delta_{g'}, \delta_{g} \rangle.$$

We endow the space $\Gamma^{\sharp}$ with the topology of compact convergence. Then the form $\langle \cdot, \cdot \rangle$ is separately continuous in each argument. In fact, the continuity in the first argument is trivial, whereas the continuity in the second argument follows from the continuity of β.

The subset $\{\delta_g : g \in \tilde{L}K_{\mathbb{C}}\} \subseteq \Gamma^\sharp$ spans a dense subspace because its annihilator in Γ is trivial and Γ can be identified with the antidual of $\Gamma^\sharp$ ([Ne00a, Th. II.8(ii)]). This proves that

$$\overline{\langle \delta_g, \alpha' \rangle} = \langle \alpha', \delta_g \rangle$$

holds for all $g \in \tilde{L}K_{\mathbb{C}}$, $\alpha' \in \Gamma^\sharp$, and applying the same argument a second time, that

$$\overline{\langle \alpha, \alpha' \rangle} = \langle \alpha', \alpha \rangle$$

for $\alpha, \alpha' \in \Gamma^\sharp$, i.e., that $\langle \cdot, \cdot \rangle$ is a hermitian form on $\Gamma^\sharp$.

That the representation of $\tilde{L}K_{\mathbb{C}}$ on $\Gamma^\sharp$ is hermitian with respect to this form follows from

$$\langle g.\alpha, \alpha' \rangle = (g.\alpha)\big(\beta(\alpha')\big) = \alpha\big(g^*.\beta(\alpha')\big) = \alpha\big(\beta(g^*.\alpha')\big) = \langle \alpha, g^*.\alpha' \rangle$$

for $\alpha, \alpha' \in \Gamma^\sharp$ and $g \in \tilde{L}K_{\mathbb{C}}$, where we have used that β is equivariant.

The hardest part is to show that the hermitian form that we have constructed on $\Gamma^\sharp$ is positive definite. This is the point where we have to use the essential pieces of information from Lie algebra representation theory and in particular Garland's Theorem. We write $\Gamma^\sharp(n) \subseteq \Gamma^\sharp$ for the energy subspace of degree n with respect to the action of $\mathbb{T}_r$ defined by $(t.\alpha)(v) := \alpha(t^{-1}.v)$. Then the sesquilinear pairing $\Gamma^\sharp \times \Gamma \to \mathbb{C}$ is invariant under $\mathbb{T}_r$ and so we see that $\langle \Gamma^\sharp(n), \Gamma(m) \rangle \neq \{0\}$ implies $n = m$. Since all the spaces $\Gamma(n)$ are finite-dimensional (Theorem VI.8(v)) and their sum is a dense subspace of Γ, we conclude that $\Gamma^\sharp(n) \cong \Gamma(n)^\sharp$, hence that $\Gamma^\sharp(n)$ is finite-dimensional.

The Lie algebra $\tilde{L}\mathfrak{k}_{\mathbb{C}}$ acts naturally on $\mathrm{Hol}(\tilde{L}K_{\mathbb{C}})$ by right-invariant vector fields which is compatible with the derived action of the Lie algebra on Γ (cf. Proposition VI.6). Since the elements of the Lie algebra act by continuous endomorphisms of $\Gamma = \Gamma^\infty$, where the latter space carries the subspace topology of $C^\infty(\tilde{L}K_{\mathbb{C}}, \Gamma)$ (cf. [Ne00a, Rem. IV.7]), they also act on $\Gamma^\sharp \subseteq \Gamma^{-\infty}$ by $(X.\alpha)(v) := \alpha(X^*.v)$, where $X^* = -\overline{X}$ and $\Gamma^{-\infty} := (\Gamma^\infty)^\sharp$ denotes the space of all continuous antilinear functionals on Γ^∞. Since for this action the map

$$\tilde{L}\mathfrak{k}_{\mathbb{C}} \otimes \Gamma^{-\infty} \to \Gamma^{-\infty}$$

is also $\mathbb{T}_r$-equivariant, it follows as in Lemma IV.4 that the finite energy subspace $\check{\Gamma}^{-\infty}$ is invariant under the subalgebra $\tilde{L}_{\mathrm{pol}}\mathfrak{k}_{\mathbb{C}}$. Note that the finite dimensionality of the subspaces $\Gamma^{-\infty}(n) \cong \Gamma(n)^\sharp$ implies that $\Gamma^{-\infty}(n) = \Gamma^\sharp(n)$, hence that $\check{\Gamma}^{-\infty} = \check{\Gamma}^\sharp$.

We write $F \subseteq \check{\Gamma}^\sharp$ for the $\tilde{L}_{\mathrm{pol}}\mathfrak{k}_{\mathbb{C}}$-submodule generated by δ_1. From

$$(t.\delta_1)(h) = \delta_1(t^{-1}.h) = \overline{(t^{-1}.h)(1)} = \overline{h(1)} = \delta_1(h)$$

for $t \in \mathbb{T}_r$ and $h \in \Gamma$ we conclude that $\delta_1 \in \Gamma^\sharp(0)$, hence that F is a $\mathbb{T}_r$-submodule of $\check{\Gamma}^\sharp$.

We claim that F is dense in $\Gamma^\sharp$. First we show that $\overline{F}$ is $\tilde{L}K_{\mathbb{C}}$-invariant. The annihilator $F^\perp$ of F in Γ is clearly invariant under the Lie algebra $\tilde{L}_{\mathrm{pol}}\mathfrak{k}_{\mathbb{C}}$. Since for each $v \in \Gamma$ the orbit map $\tilde{L}\mathfrak{k}_{\mathbb{C}} \to \Gamma, X \mapsto X.v = d\phi_v(1)(X)$, where $\phi_v : g \mapsto \pi(g).v$

is the orbit map, is continuous ([Ne00a, Cor. IV.6]), the density of $\tilde{L}_{\mathrm{pol}}\mathfrak{k}_{\mathbb{C}}$ in $\tilde{L}\mathfrak{k}_{\mathbb{C}}$ (Lemma III.1) implies that for each $v \in F^{\perp}$ we also have $\tilde{L}\mathfrak{k}_{\mathbb{C}}.v \subseteq F^{\perp}$, i.e., that $F^{\perp}$ is invariant under $\tilde{L}\mathfrak{k}_{\mathbb{C}}$. In view of the fact that $\tilde{L}K_{\mathbb{C}}$ has a good exponential function (Theorem II.1), and the orbit maps of elements in $\Gamma \subseteq \mathrm{Hol}(\tilde{L}K_{\mathbb{C}})$ are holomorphic, [Ne00a, Th. III.11] implies that $F^{\perp}$ is also invariant under the group $\tilde{L}K_{\mathbb{C}}$. Further duality implies that $\overline{F} = (F^{\perp})^{\perp}$ is invariant under $\tilde{L}K_{\mathbb{C}}$ (cf. [Ne00a, Th. II.8(ii)]). Since $F \subseteq \check{\Gamma}$ contains the functional $\delta_1 : f \mapsto \overline{f(1)}$ which clearly is cyclic for $\tilde{L}K_{\mathbb{C}}$ because the annihilator of its orbit is trivial, we see that $F^{\perp} = \{0\}$ and thus F is dense in $\Gamma^{\sharp}$.

If there exists an $n \in \mathbb{N}_0$ with $\Gamma^{\sharp}(n) \not\subseteq F$, then we find a non-zero element $v \in \Gamma(n) \cap F^{\perp} = \{0\}$. So we even see that $F = \check{\Gamma}^{\sharp}$, i.e., the finite energy subspace of $\Gamma^{\sharp}$ is a cyclic $\tilde{L}\mathfrak{k}_{\mathbb{C}}$-module.

We want to apply Garland's Theorem (Theorem IV.13) to see that the hermitian form on $F = \check{\Gamma}^{\sharp}$ is positive definite. For that we have to note that by a similar argument as for the rotation group $\mathbb{T}_r$ the action of the m-dimensional torus $\tilde{\mathbb{T}}_r$ on LK lifts to an action on $\tilde{L}K_{\mathbb{C}}$ leaving $\tilde{B}^+$ and the character λ invariant. Hence it induces an action on the bundle L_{λ} and also on the space Γ_{λ} of holomorphic sections. In this sense we can think of $\check{\Gamma}^{\sharp}$ as a lowest weight module of the extended Lie algebra $\tilde{\mathfrak{t}}_r \ltimes \tilde{L}_{\mathrm{pol}}\mathfrak{k}_{\mathbb{C}}$. Now Garland's Theorem applies and shows that the form on $\check{\Gamma}^{\sharp}$ is positive definite.

This has several consequences. The kernel of β is a closed $\mathbb{T}_r$-invariant subspace of $\Gamma^{\sharp}$, so that
$$\ker \beta \cap \check{\Gamma}^{\sharp} = \sum_{n \in \mathbb{N}}(\ker \beta)(n)$$
is dense in $\ker \beta$. On the other hand $\beta(\alpha) = 0$ means that $\langle \alpha, \Gamma^{\sharp} \rangle = 0$. Therefore the positive definiteness of the hermitian form on $\check{\Gamma}^{\sharp}$ entails that $\ker \beta$ does not intersect $\check{\Gamma}^{\sharp}$, hence that β is injective. The fact that β is injective on each finite energy subspace $\Gamma^{\sharp}(n)$ entails in particular that $\beta(\Gamma^{\sharp}(n)) = \Gamma(n)$ for all $n \in \mathbb{N}_0$, and thus $\beta(\check{\Gamma}^{\sharp}) = \check{\Gamma}$. Further $\beta(\delta_1)(g) = \delta_1(\overline{g}.f) = \overline{f(g^*)} = f(g)$, i.e., $\beta(\delta_1) = f$ which shows that $\check{\Gamma}$ is a cyclic $\tilde{L}_{\mathrm{pol}}\mathfrak{k}_{\mathbb{C}}$-module generated by the function f.

To see that $\langle \cdot, \cdot \rangle$ is positive definite on $\Gamma^{\sharp}$, let $\alpha \in \Gamma^{\sharp}$ and $\alpha_n \in \Gamma^{\sharp}(n)$ its restriction to the subspace $\Gamma(n)$. Then $\beta(\alpha) = \sum_{n \in \mathbb{N}_0} \beta(\alpha)_n$ converges in Γ because it is a smooth vector for the representation of $\mathbb{T}_r$ (cf. [Wa72, Th. 4.4.2.1]), and so
$$\langle \alpha, \alpha \rangle = \alpha(\beta(\alpha)) = \sum_{n \in \mathbb{N}_0} \alpha(\beta(\alpha_n)) = \sum_{n \in \mathbb{N}_0} \langle \alpha_n, \alpha_n \rangle \geq 0$$
because the restriction of $\langle \cdot, \cdot \rangle$ to $\check{\Gamma}^{\sharp}$ is positive definite. The preceding formula also shows that $\check{\Gamma}^{\sharp}$ is dense in the Hilbert space completion of $\Gamma^{\sharp}$ with respect to $\langle \cdot, \cdot \rangle$, and therefore that $\check{\Gamma}^{\sharp}$ is also dense in $\Gamma^{\sharp}$ with respect to the corresponding norm.

For $g \in \tilde{L}K_{\mathbb{C}}$ we have $(g.\delta_1)(h) = \delta_1(g^*.h) = \overline{h(\overline{g})}$, i.e., $g.\delta_1 = \delta_{\overline{g}}$. Therefore
$$\langle g.\delta_1, \delta_1 \rangle = \langle \delta_{\overline{g}}, \delta_1 \rangle = \delta_{\overline{g}}(\beta(\delta_1)) = \delta_{\overline{g}}(f) = \overline{f(\overline{g})} = f(g^{-1}).$$

Hence $f(s^*t) = \langle t^{-1}\bar{s}.\delta_1, \delta_1 \rangle = \langle \bar{s}.\delta_1, \bar{t}.\delta_1 \rangle$, and we see that f is a positive definite function on the involutive group $\tilde{L}K_{\mathbb{C}}$. Let $\mathcal{H}_\lambda \subseteq \text{Hol}(\tilde{L}K_{\mathbb{C}})$ denote the corresponding reproducing kernel Hilbert space with kernel $Q(s,t) := f(t^*s)$ (cf. Proposition IV.8). This kernel Q on $\tilde{L}K_{\mathbb{C}}$ is left invariant, i.e.,

$$Q(sx, y) = Q(x, s^*y)$$

for $s, x, y \in \tilde{L}K_{\mathbb{C}}$. Therefore the left translations $(g.f)(x) := f(g^{-1}x)$ define a natural hermitian representation of $\tilde{L}K_{\mathbb{C}}$ on $\mathcal{H}_\lambda^0$ which, according to Proposition IV.7, yields a continuous unitary representation of $\hat{L}K$ on the Hilbert space $\mathcal{H}_\lambda$. The functions Q_s, $s \in \tilde{L}K_{\mathbb{C}}$, are given by $t \mapsto Q(t,s) = f(s^*t) = (\bar{s}.f)(t)$, i.e., $Q_s = \bar{s}.f$. From the left invariance of Γ and $f \in \Gamma$ we now obtain $\mathcal{H}_\lambda^0 \subseteq \Gamma$, and since Γ is closed (Proposition VI.6(ii)) and $\mathcal{H}_\lambda \to \text{Hol}(\tilde{L}K_{\mathbb{C}})$ is continuous (Proposition IV.8), we see that $\mathcal{H}_\lambda \subseteq \Gamma$ with a continuous inclusion map.

Next we claim that $\beta(\Gamma^\sharp) \subseteq \mathcal{H}_\lambda$. Let $\gamma: \Gamma^\sharp \to \mathcal{H}_\lambda^\sharp \cong \mathcal{H}_\lambda$ denote the antiadjoint of the continuous inclusion map $\mathcal{H}_\lambda \hookrightarrow \Gamma$. For each $s \in \tilde{L}K_{\mathbb{C}}$ we then have

$$\gamma(\alpha)(s) = \langle \gamma(\alpha), \bar{s}.f \rangle = \alpha(\bar{s}.f) = \beta(\alpha)(s).$$

This shows that

$$\beta(\Gamma^\sharp) = \gamma(\Gamma^\sharp) \subseteq \mathcal{H}_\lambda.$$

Now we can show that the representation on Γ is irreducible. Let $W \subseteq \Gamma$ be a closed $\tilde{L}K_{\mathbb{C}}$-invariant subspace. If $\delta_1(W) = \{0\}$, then $W \subseteq \delta_1{}^\perp$ and the fact that $\delta_1 \in \Gamma^\sharp$ is cyclic shows that $W = \{0\}$. If $\delta_1(W) \neq \{0\}$, then $W(0) \neq \{0\}$, and since $\Gamma(0)$ is an irreducible K-module (Theorem VI.8(iii)), we see that $f \in V(0) = W(0)$. Then $\check{\Gamma} = \mathcal{U}(\tilde{L}_{\text{pol}}\mathfrak{k}_{\mathbb{C}}).f \subseteq W$, and therefore $\check{\Gamma} \subseteq W$ and the closedness of W finally show that $W = \Gamma$. The same proof shows that the representation of $\tilde{L}K$ on the Hilbert space $\mathcal{H}_\lambda$ is irreducible. $\square$

Corollary VI.10. *Every holomorphic function on $\tilde{L}K_{\mathbb{C}}/\tilde{B}^+ \cong LK_{\mathbb{C}}/B^+$ is constant.*

Proof. For $\lambda = 0$ we have

$$\Gamma_\lambda \cong \text{Hol}(\tilde{L}K_{\mathbb{C}}/\tilde{B}^+).$$

Now the subspace of constant functions is a $\hat{L}K$-invariant subspace, and, on the other hand, Theorem VI.9(ii) asserts that $\hat{L}K$ acts irreducibly on Γ_λ. This implies the assertion. $\square$

Remark VI.11. In the proof of Theorem VI.9 we have seen that the complex group acts on the spaces Γ_λ and therefore on the antidual spaces $\Gamma_\lambda^\sharp$. Moreover, the construction of the Hilbert space $\mathcal{H}_\lambda$ also gave us a hermitian representation of $\tilde{L}K_{\mathbb{C}}$ on the dense subspace of $\mathcal{H}_\lambda$ generated by the point evaluations. The whole framework is described by the triple

$$\Gamma_\lambda^\sharp \hookrightarrow \mathcal{H}_\lambda \hookrightarrow \Gamma_\lambda.$$

As the general theory of hermitian representations shows, an element $g \in \tilde{L}K_{\mathbb{C}}$ acts continuously on this pre-Hilbert space if and only if it leaves the Hilbert space $\mathcal{H}_\lambda$ invariant (cf. [Ne99, Prop. II.4.9]). For an element $\exp iX$, $X \in \mathfrak{t}_\mathfrak{k}$, this happens if and only if the set of weights is bounded from below on X.   □

Up to this point we have always studied the spaces Γ_λ under the assumptions that they are non-zero. Now we will show that the necessary condition of the antidominance of λ that we have encountered in Theorem VI.8 is sufficient for the non-triviality of the spaces Γ_λ.

Theorem VI.12. *For $\lambda \in \hat{T}_{\hat{L}K}$ the space Γ_λ is non-zero if and only if the weight λ is antidominant.*

Proof. If Γ_λ is non-zero, then Theorem VI.8(i) states that λ is antidominant.

Suppose that λ is antidominant. We have to show that there exists a non-zero function in Γ_λ. We will construct such a function f with the additional property that f is N^--invariant.

Let $w \in N_{LK}(T)$ be an element representing the corresponding element $[w]$ of the Weyl group $\mathcal{W}_{\mathrm{aff}} \cong \check{T} \rtimes \mathcal{W}_\mathfrak{k}$. Then the Bruhat decomposition of $LK_\mathbb{C}$ yields a decomposition

$$Y = LK_\mathbb{C}/B^+ = \bigcup_{[w] \in \mathcal{W}_{\mathrm{aff}}} w.U,$$

where $U = N^- B^+/B^+ \subseteq Y$ is the open subset, and $w.U = (wN^-w^{-1}).y_w$ with $y_w = wB^+ \in Y$ are other open domains in Y over which the bundle is trivial ([PS86, Th. 8.7.2]). We write

$$wN^-w^{-1} = N_w^- A_w$$

with

$$N_w^- = N^- \cap wN^-w^{-1} \quad \text{and} \quad A_w = N^+ \cap wN^-w^{-1},$$

where A_w is a finite-dimensional nilpotent group whose dimension is the length $l([w])$ of the Weyl group element $[w]$ ([PS86, Th. 8.7.2]). We are looking for a family of holomorphic functions $f_w \colon wN^-w^{-1} \to \mathbb{C}$ which define a holomorphic section of the bundle $L_\lambda \to Y$ in the sense that for each w the function f_w is obtained from trivializing the bundle L_λ over the open set $wN^-w^{-1}.y_w = wN^-.y_1 = w.U$. Since, in addition, we want the functions f_w to be N_w^--invariant, these functions will be determined by their values on the finite-dimensional group A_w.

From the Bruhat decomposition we know that

$$\dim A_w = \operatorname{codim}_{N^-} N_w^- = \operatorname{codim}_Y(N_w^-.y_w) = \operatorname{codim}_Y(N^-.y_w),$$

where $N_w^-.y_w$ is a stratum of the Bruhat decomposition ([PS86, Th. 8.7.2]). We also use this reference to see that

$$w.U \cap \left(\bigcup_{l([w']) < l([w])} w'.U \right) = w.U \setminus N^-.y_w = (wNw^{-1}).y_w \setminus N^-.y_w$$

and that the action of A_w on the domain $w.U$ yields a diffeomorphism with $A_w \times N^-.y_w \cong A_w \times N_w^-$. By induction we thus have a function on the domain

$$w.U \setminus N^-.y_w \cong (A_w \setminus \{1\}) \times N_w^-.$$

Now we construct the function f_w on the group A_w by induction over the length $l([w])$ of the Weyl group element $[w]$. We put $f_1 = 1$ on $N^- \cong U = N^- . y_1$, where $A_1 = \{1\}$.

If $l([w]) = 1$, i.e., if $w = s_{\underline{\alpha}}$ is a reflection corresponding to a simple root $\underline{\alpha}$, then we can calculate explicitly in $SL(2, \mathbb{C})$. Here $e_{\underline{\alpha}}$ corresponds to the matrix $\begin{pmatrix} 0 & 1 \\ 0 & 0 \end{pmatrix}$, so that

$$A_w = \exp(\mathbb{C} e_{\underline{\alpha}}) \cong \left\{ \begin{pmatrix} 1 & z \\ 0 & 1 \end{pmatrix} : z \in \mathbb{C} \right\}$$

and $\exp(z e_{\underline{\alpha}}).y_w \notin U$ if and only if $z = 0$. From the relation

$$\begin{pmatrix} 1 & z \\ 0 & 1 \end{pmatrix} \begin{pmatrix} 0 & 1 \\ -1 & 0 \end{pmatrix} = \begin{pmatrix} 1 & 0 \\ z^{-1} & 1 \end{pmatrix} \begin{pmatrix} -z & 0 \\ 0 & -z^{-1} \end{pmatrix} \begin{pmatrix} 1 & -z^{-1} \\ 0 & 1 \end{pmatrix}$$

in $SL(2, \mathbb{C})$ we obtain

$$\exp(z e_{\underline{\alpha}}) w = \exp(-z^{-1} e_{-\underline{\alpha}}) h_{\underline{\alpha}}(-z) \exp(-z^{-1} e_{\underline{\alpha}}),$$

where $h_{\underline{\alpha}}(z)$ denotes the image of the matrix $\begin{pmatrix} z & 0 \\ 0 & z^{-1} \end{pmatrix}$ under the natural morphism $SL(2, \mathbb{C}) \to \tilde{L}K_{\mathbb{C}}(\underline{\alpha})$. We recall from (1.1) in Section I that

$$\underline{\lambda}(h_{\underline{\alpha}}(z)) = z^{\underline{\lambda}(h_{\underline{\alpha}})}.$$

Now we observe that the point

$$[\exp(z e_{\underline{\alpha}}) w, x] \in \tilde{L}K_{\mathbb{C}} \times_{\tilde{B}^+} \mathbb{C} = L_{\underline{\lambda}}$$

is the same as

$$[\exp(z e_{\underline{\alpha}}) w, x] = [\exp(-z^{-1} e_{-\underline{\alpha}}), \underline{\lambda}(h_{\underline{\alpha}}(-z)) x] = [\exp(-z^{-1} e_{-\underline{\alpha}}), (-z)^{\underline{\lambda}(h_{\underline{\alpha}})} x].$$

In particular we obtain for $z \neq 0$:

$$[\exp(-z^{-1} e_{-\underline{\alpha}}), 1] = [\exp(-z^{-1} e_{-\underline{\alpha}}), f_1(\exp(-z^{-1} e_{-\underline{\alpha}}))]$$
$$= [\exp(z e_{\underline{\alpha}}) w, f_w(\exp(z e_{\underline{\alpha}}))] = [\exp(-z^{-1} e_{-\underline{\alpha}}), (-z)^{\underline{\lambda}(h_{\underline{\alpha}})} f_w(\exp(z e_{\underline{\alpha}}))],$$

i.e.,

$$f_w(\exp(z e_{\underline{\alpha}})) = (-z)^{-\langle \underline{\lambda}, h_{\underline{\alpha}} \rangle}.$$

In view of $\underline{\lambda}(h_{\underline{\alpha}}) \in -\mathbb{N}_0$, the function on the right hand side is a polynomial, hence extends from $\mathbb{C}^\times$ to the whole complex plane.

If $l(w) > 1$, then $\dim A_w \geq 2$ and we can use Hartog's Theorem to extend the holomorphic function given on the finite-dimensional complex manifold $A_w \setminus \{1\}$ to a holomorphic function on A_w which we then use to obtain a left N_w^--invariant function f_w on $w N w^{-1}$ which is consistent with all the functions $f_{w'}$ for $l([w']) < l([w])$. By the uniqueness of analytic extension, this also implies that this function is consistent with other functions $f_{w'}$ with $l([w']) = l([w])$ which have already been constructed. This completes the proof. $\qquad \square$

Theorem VI.12 rounds off the picture presented in this section in the sense that it is a characterization of those cases where the space Γ_λ is non-trivial. Together with the structural information on these spaces obtained in Theorem VI.8 and VI.9, we have quite a good picture of the representations in the spaces Γ_λ.

VII. Consequences for general representations

In this section we will see how Theorems VI.8, VI.9, and VI.12 from Section VI can be used to obtain information on the general positive energy representations of the groups $\tilde{L}K$ and $\hat{L}K$. More precisely, we will see that for any smooth irreducible positive energy representation a dense subspace can be embedded into some Γ_λ.

Proposition VII.1. *Let (π, V) be a smooth positive energy representation of $\hat{L}K$, $V^{-\infty}$ the antidual of V^∞, and $\varepsilon \in V^{-\infty}$ a lowest weight vector of weight λ for the representation of $\hat{L}\mathfrak{k}_{\mathbb{C}}$. Then the prescription $\Phi(v)(g) := \overline{\varepsilon(g^{-1}.v)}$ defines a continuous equivariant map*

$$\Phi: V^\infty \to \Gamma_\lambda \subseteq C^\infty(\hat{L}K).$$

Proof. We recall that $V^{-\infty}$ denotes the space of all continuous antilinear functionals on the locally convex space V^∞. The continuity of the inclusion map $V^\infty \to V$ ([Ne00a, Rem. IV.7(b)]) then yields a natural map $V^\sharp \to V^{-\infty}$. Since the representation (π, V) was assumed to be smooth, i.e., V^∞ is dense in V, this map is injective, and so we can think of $V^\sharp$ as a subspace of $V^{-\infty}$. Then we have a natural action of the Lie algebra $\hat{L}\mathfrak{k}_{\mathbb{C}}$ on $V^{-\infty}$ by $(X.\alpha)(v) := \alpha(X^*.v)$, where $X^* = -\overline{X}$ (cf. [Ne00a, Cor. IV.6]).

Next we note that $\hat{L}K$ acts continuously on $V^{-\infty}$ by $(g.\alpha)(v) := \alpha(g^{-1}.v)$ (cf. [Ne00a, Lemma IV.4(ii), Prop. IV.5]). We may w.l.o.g. assume that 0 is the minimal non-zero energy degree (Remark VI.7). Then the subspace $V^{-\infty}(0) \cong V^\infty(0)^\sharp$ is invariant under the action of the compact subgroup K. Let $M \subseteq V(0)^{-\infty}$ be an irreducible subspace for K and $\varepsilon \in M$ be a lowest weight vector of lowest weight $\lambda = (0, \lambda, h)$ for the corresponding representation of $\mathfrak{k}_{\mathbb{C}}$ with respect to $\Delta_{\mathfrak{k}}^+$. Then we define a linear map

$$\Phi: V^\infty \to C^\infty(\hat{L}K) \quad \text{by} \quad \Phi(v)(g) := \overline{\varepsilon(g^{-1}.v)}$$

which is continuous because inversion induces a continuous endomorphism of $C^\infty(\hat{L}K, V)$, the topology on V^∞ is defined by the embedding into $C^\infty(\hat{L}K, V)$, and for each continuous linear functional α on V the corresponding natural map $C^\infty(\hat{L}K, V) \to C^\infty(\hat{L}K), f \mapsto \alpha \circ f$ is continuous.

For $t \in T_{\hat{L}K}$ we further have

$$\Phi(v)(gt^{-1}) = \overline{\varepsilon(tg^{-1}.v)} = \overline{(t^{-1}.\varepsilon)(g^{-1}.v)} = \overline{\underline{\lambda}(t^{-1})\varepsilon(g^{-1}.v)} = \underline{\lambda}(t)\Phi(v)(g).$$

This shows that $\Phi(v) \in C^\infty(\hat{L}K)$ represents a smooth section of the bundle $\Gamma_\lambda \to LK/T_K \cong Y$.

Now the construction of ε and the fact that (π, V) has positive energy implies that $\varepsilon \in V^{-\infty}$ is a lowest weight vector for the representation of the Lie algebra $\hat{L}\mathfrak{k}_{\mathbb{C}}$. If $\tilde{X}$ is the left invariant vector field with $\tilde{X}(\mathbf{1}) = X$, then

$$\tilde{X}.\Phi(v)(g) = \frac{d}{dt}\bigg|_{t=0} \Phi(v)(g \exp tX) = \frac{d}{dt}\bigg|_{t=0} \overline{\varepsilon\big(\exp(-tX)g^{-1}.v\big)}$$

$$= -\overline{\varepsilon\big(X.(g^{-1}.v)\big)} = \overline{(X.\varepsilon)(g^{-1}.v)}$$

and by complex linear extension, it follows that

$$\tilde{X}.\Phi(v)(g) = \overline{(X^*.\varepsilon)(g^{-1}.v)}$$

for all $X \in \hat{L}\mathfrak{k}_{\mathbb{C}}$. For $X \in \tilde{\mathfrak{b}}^+$ this leads to

$$\tilde{X}.\Phi(v) = \overline{\lambda(X^*)}\Phi(v).$$

These differential equations characterize the holomorphic section of the bundle $\Gamma_\lambda \to L\tilde{K}_{\mathbb{C}}/\tilde{B}^+$ among the smooth sections (cf. [PS86, p.222]), whence $\Phi(V^\infty) \subseteq \Gamma_\lambda$. $\qquad\square$

Let us assume that V is an LF space ([Ne00a, Def. II.1]) and that ε is contained in the subspace $V^\sharp \subseteq V^{-\infty}$ of those functionals which extend continuously to V. If we endow $\Gamma_\lambda \subseteq \mathrm{Hol}(\tilde{L}K_{\mathbb{C}})$ with the subspace topology, i.e., the topology of uniform convergence of compact subsets of $\tilde{L}K_{\mathbb{C}}$ resp. $\tilde{L}K$, then Φ is continuous. In fact, for the element $\varepsilon \in V^\sharp$ the orbit map $\tilde{L}K \to V^\sharp$ is continuous with respect to the topology of compact convergence ([Ne00a, Lemma IV.4(ii)]). Hence the image of a compact subset of $\tilde{L}K$ is a compact subset of $V^\sharp$, therefore equicontinuous because V is an LF space and hence barreled ([Ne00a, Prop. II.2(vi), II.11(iii)]), so that convergence in V implies uniform convergence on this compact set.

Corollary VII.2. *For each smooth irreducible positive energy representation there exists a dense invariant subspace $V_0 \subseteq V$ which injects in an equivariant way in the irreducible unitary representation $(\pi_\lambda, \mathcal{H}_\lambda)$ on the Hilbert subspace $\mathcal{H}_\lambda \subseteq \Gamma_\lambda$.*

Proof. According to Proposition VII.1, we have a continuous injection $\Phi: V^\infty \to \Gamma_\lambda$. Then Φ maps the dense finite energy space $\check{V}^\infty$ into $\check{\Gamma}_\lambda = \check{\mathcal{H}}_\lambda$. This shows that $V_0 := \Phi^{-1}(\mathcal{H}_\lambda) \subseteq V^\infty$ is a dense invariant subspace. This proves the assertion. $\qquad\square$

If (π, V) is a smooth positive energy representation and $\varepsilon_\lambda \in V^\sharp$ is a lowest weight vector with respect to the action of the Lie algebra on $V^\sharp$, then the ray $[\varepsilon_\lambda] \in \mathbb{P}(V^\sharp)$ yields an equivariant map

$$\tilde{L}K/\tilde{T} \cong LK/T \to \mathbb{P}(V^\sharp), \quad gT \mapsto g.[\varepsilon_\lambda].$$

Thus one obtains a realization of the representation in the space Γ_λ. This is the geometric picture corresponding to the construction of Proposition VII.1. For a more detailed discussion of this construction we refer to Section V in [Ne00a].

References

[Alb93]   Albeverio, S., R. J. Høegh-Krohn, J. A. Marion, D. H. Testard, and B. S. Torresani, "Noncommutative Distributions – Unitary representations of Gauge Groups and Algebras," Pure and Applied Mathematics **175**, Marcel Dekker, New York, 1993

[Bo58]   Bott, R., *The space of loops on a Lie group*, Michigan Math. J. **5** (1958), 35–61

[Br93]   Bredon, G. E., "Topology and Geometry," Graduate Texts in Mathematics **139**, Springer-Verlag, Berlin, 1993

[GW84]   Goodman, R., and N. R. Wallach, *Structure and unitary cocycle representations of loop groups and the group of diffeomorphisms of the circle*, J. reine ang. Math. **347** (1984), 69–133

[GW85]   —, *Projective unitary positive energy representations of* Diff($\mathbb{S}^1$), J. Funct. Anal. **63** (1985), 299–312

[Ha82]   Hamilton, R., *The inverse function theorem of Nash and Moser*, Bull. Amer. Math. Soc. **7** (1982), 65–222

[He89]   Hervé, M., "Analyticity in Infinite-dimensional Spaces," de Gruyter, Berlin, 1989

[HoMo98]   Hofmann, K. H., and S. A. Morris, "The Structure of Compact Groups," Studies in Math., de Gruyter, Berlin, 1998

[Ka85a]   Kac, V. G., Ed., "Infinite-dimensional Groups with Applications," Mathematical Sciences Research Institute Publications **4**, Springer-Verlag, Berlin, Heidelberg, New York, 1985

[Ka85b]   —, *Constructing groups associated to infinite-dimensional Lie algebras*, in [Ka85a]

[Ka90]   —, "Infinite-dimensional Lie Algebras," Cambridge University Press, 3rd printing, 1990

[KP83]   Kac, V. G., and D. H. Peterson, *Regular functions on certain infinite-dimensional groups*, in "Arithmetic and Geometry", Vol. 2, Ed., M. Artin and J. Tate, Birkhäuser, Boston, 1983

[KP84]   —, *Unitary structure in representations of infinite-dimensional groups and a convexity theorem*, Invent. Math. **76** (1984), 1–14

[Mi83]   Milnor, J., *Remarks on infinite-dimensional Lie groups*, Proc. Summer School on Quantum Gravity, B. DeWitt ed., Les Houches, 1983

[Mi95]   Mimura, M., *Homotopy theory of Lie groups*, in "Handbook of algebraic topology," I. M. James, ed., Amsterdam, Elsevier Science B. V., 1995, 951–991

[NRW99]   Natarajan, L., E. Rodriguez-Carrington, and J. A. Wolf, *The Bott-Borel-Weil theorem for direct limit groups*, Transactions of the Amer. Math. Soc., to appear

[Ne99]   Neeb, K.-H., "Holomorphy and Convexity in Lie Theory," Expositions in Mathematics **28**, de Gruyter Verlag, Berlin, 1999

[Ne00a]   —, *Infinite-dimensional groups and their representations*, in this volume

[Ne00b]   —, *Central extensions of infinite-dimensional Lie groups*, Preprint, TU Darmstadt, 2000

[Ner83]   Neretin, Y., *Boson representation of the diffeomorphisms of the circle*, Sov. Math. Dokl. **28** (1983), 411–414

[Ner87]   —, *On spinor representations of* $O(\infty, \mathbb{C})$, Sov. Math. Dokl. **34:1** (1987), 71–74

[Pe86]   Perelomov, "Generalized Coherent States and their Applications," Springer, Berlin, 1986

[PK83]   Peterson, D. H., and V. G. Kac, *Infinite flag varieties and conjugacy theorems*, Proc. Nat. Acad. Sci. USA **80** (1983), 1778-1782

[PS86]   Pressley, A., and G. Segal, "Loop Groups," Oxford University Press, Oxford, 1986

[Su97]   Suto, K., *Borel-Weil type theorem for the flag manifold of a general Kac-Moody algebra*, J. of Algebra **193** (1997), 529–551

[Wa72]   Warner, G., "Harmonic Analysis on Semisimple Lie Groups I," Springer, Berlin, Heidelberg, New York, 1972

[Wu00]   Wurzbacher, T., *Fermionic second quantization and the geometry of the restricted Grassmannian*, in this volume

Mathematical Subject Classification (2000)
Primary: 22E67.
Secondary: 22E65, 32L10, 81R10.

Technische Universität Darmstadt
Fachbereich Mathematik
Schlossgartenstrasse 7
D-64289 Darmstadt
Germany
E-mail address: **neeb@mathematik.tu-darmstadt.de**

Coadjoint Representation of Virasoro-type Lie Algebras and Differential Operators on Tensor-densities

Valentin Yu. Ovsienko

Abstract. We discuss the geometrical nature of the coadjoint representation of the Virasoro algebra and some of its generalizations. The isomorphism of the coadjoint representation of the Virasoro group to the $\mathrm{Diff}(S^1)$-action on the space of Sturm-Liouville operators was discovered by A.A. Kirillov and G. Segal. This deep and fruitful result relates this topic to the classical problems of projective differential geometry (linear differential operators, projective structures on S^1 etc.) The purpose of this talk is to give a detailed explanation of the A.A. Kirillov method [14] for the geometric realization of the coadjoint representation in terms of linear differential operators. Kirillov's method is based on Lie superalgebras generalizing the Virasoro algebra. One obtains the Sturm-Liouville operators directly from the coadjoint representation of these Lie superalgebras. We will show that this method is universal. We will consider a few examples of infinite-dimensional Lie algebras and show that the Kirillov method can be applied to them. This talk is purely expository: all the results are known.

To my teacher Alexander Alexandrovich Kirillov

Introduction

The coadjoint representation of infinite-dimensional Lie groups and Lie algebras is one of the most interesting subjects of Kirillov's orbit method. Geometrical problems related to this subject link together such fundamental domains as: symplectic and Kähler geometry, harmonic analysis, integrable systems and many others.

The main purpose of this talk is to describe a "geometrical picture" due to Kirillov, for the coadjoint representation of the Virasoro group and the Virasoro algebra. We will also consider some of their generalizations.

1. The Virasoro group is the unique (modulo equivalence) nontrivial central extension of the group of diffeomorphisms of the circle. The corresponding Lie algebra, called the Virasoro algebra, is defined as the unique (modulo equivalence) nontrivial central extension of the Lie algebra of vector fields on S^1. The coadjoint

representation of the Virasoro group and the Virasoro algebra was studied in pio-
neering works by A.A. Kirillov [13] and G. Segal [27]. Their result is as follows.

The dual space to the Virasoro algebra can be realized as the space of Sturm-
Liouville operators:

$$L = c\frac{d^2}{dx^2} + u(x) \tag{1}$$

where $u(x + 2\pi) = u(x)$ is a periodic function, $c \in \mathbf{R}$ (or $\mathbf{C}$) is a constant. The
coadjoint representation of the Virasoro group coincides with the natural action
of the group of diffeomorphisms of S^1 on the space of operators (1).

This realization gives a geometric interpretation of the coadjoint representation of the Virasoro group (and the Virasoro algebra). It relates the coadjoint
representation of the Virasoro group to classical works on differential operators
and projective differential geometry [29],[3]. The main object in this theory which
links together its different parts, is the classical Schwarzian derivative. It appears
(in the Virasoro context) as a 1-cocycle on the group of diffeomorphisms of S^1
with values in the coadjoint representation.

2. The realization of the coadjoint representation of the Virasoro group was first
discovered as a simple coincidence. Soon after, A.A. Kirillov has suggested a sys-
tematic method using Lie superalgebras (see [14]). He considered two Lie superalgebras (called now Ramond and Neveu-Schwarz superalgebras) containing the
Virasoro algebra as the even part. Sturm-Liouville operator appears in the coadjoint action of Ramond and Neveu-Schwarz superalgebras.

To my knowledge, Kirillov's method is the only known way to obtain the
Sturm-Liouville operators (in an automatic way) directly from the coadjoint representation. This makes this method particularly useful for generalizations. However, for a long time, Kirillov's method has not been tested in any other case then
for the Virasoro algebra.

3. We consider two different generalizations of the above geometrical picture.

A. There exist series of infinite-dimensional groups, Lie algebras and Lie superalgebras generalizing the Virasoro group and the Virasoro algebra (see [21],
[25]). Geometrical realization of the coadjoint representation leads to interesting generalization of the Sturm-Liouville operator and projective structures.

B. The space of higher order linear differential operators has an interesting
structure of infinite-dimensional Poisson manifold with respect to the Adler-
Gelfand-Dickey Poisson bracket. This Poisson structure is related to so-called
W-algebras and is very popular in Mathematical Physics (see e.g. [28]). We
will discuss the relations of the Adler-Gelfand-Dickey bracket to the Diff(S^1)-
module structure on the space of linear differential operators on S^1 (studied
by classics, see [29], [3]). Following [24], we show that the Adler-Gelfand-
Dickey Poisson structure can be defined in terms of the Moyal-Weyl star-
product.

4. An important point, common for all known examples is the following "tensor sense". Arguments of differential operators are considered as tensor-densities on S^1. The well-known classical example is the Sturm-Liouville operator, (1) acting from the space of $-1/2$-densities to the space of $3/2$-densities. This defines a natural Diff(S^1)-action on the space of differential operators and intrinsically contains all the information about the related algebraic structures.

Remark here that the structure of the module over the group of diffeomorphisms on the space of linear differential operators on a manifold was studied in a series of recent papers [5], [20], [6].

Acknowledgments. I am grateful to A.A. Kirillov for his constant help and to Ch. Duval, L. Guieu, P. Marcel and C. Roger for collaboration and numerous stimulating discussions.

I. Coadjoint representation of Virasoro group and Sturm-Liouville operators; Schwarzian derivative as a 1-cocycle

This introductory section is based on the articles of A.A. Kirillov [13], [14] and G. Segal [27]. We will give the definition of the Virasoro group and the Virasoro algebra and prove the following result.

Theorem 1. *The coadjoint representation of the Virasoro group is naturally isomorphic, i.e., isomorphic as a module over the group of diffeomorphisms, to the space of Sturm-Liouville operators.*

The classical Schwarzian derivative appears as a 1-cocycle on the group of diffeomorphisms with values in the coadjoint representation.

1. Virasoro group and Virasoro algebra

Consider the Lie algebra Vect(S^1) of smooth vector fields on the circle:

$$X = X(x)\frac{d}{dx},$$

where $X(x + 2\pi) = X(x)$. The commutator in Vect(S^1) is given by the formula:

$$[X(x)\frac{d}{dx}, Y(x)\frac{d}{dx}] = (X(x)Y'(x) - X'(x)Y(x))\frac{d}{dx},$$

where $X' = dX/dx$.

Definition 1.1 The *Virasoro algebra* is the unique (up to isomorphism) non-trivial central extension of Vect(S^1). It is given by the *Gelfand-Fuchs cocycle*:

$$\omega(X(x)\frac{d}{dx}, Y(x)\frac{d}{dx}) = \frac{1}{2}\int_0^{2\pi}\begin{vmatrix} X'(x) & Y'(x) \\ X''(x) & Y''(x) \end{vmatrix}dx.$$

The Virasoro algebra is, therefore a Lie algebra on the space $\text{Vect}(S^1) \oplus \mathbf{R}$ defined by the commutator

$$[(X, \alpha), (Y, \beta)] = ([X, Y]_{\text{Vect}(S^1)}, \omega(X, Y)),$$

where α and $\beta \in \mathbf{R}$ are elements of the center.

One easily checks (using integration by part) the 2-cocycle condition:

$$\omega(X, [Y, Z]) + \omega(Y, [Z, X]) + \omega(Z, [X, Y]) = 0,$$

equivalent to the Jacobi identity for the Virasoro algebra.

Remark. The defined Lie algebra has been discovered by I.M. Gelfand and D.B. Fuchs [8] and was later rediscovered in the physics literature.

Consider the group $\text{Diff}^+(S^1)$ of diffeomorphisms of the circle preserving its orientation: $x \mapsto f(x)$, where $x \pmod{2\pi}$ is a parameter on S^1, $f(x + 2\pi) = f(x) + 2\pi$.

Definition 1.2 The *Virasoro group* is the unique (up to isomorphism) non-trivial central extension of $\text{Diff}^+(S^1)$. It is given by so-called *Thurston-Bott cocycle* (see [2]):

$$B(f, g) = \int_0^{2\pi} \log((f \circ g)') d\log(g')$$

By definition, the Virasoro group is given by the following product on $\text{Diff}^+(S^1) \times \mathbf{R}$:

$$(f, \alpha)(g, \beta) = (f \circ g, \beta + B(f, g)).$$

Associativity of this product is equivalent to the condition: $B(f, g \circ h) + B(g, h) = B(f \circ g, h)$ which means that B is a 2-cocycle.

Notation. Let us denote the Virasoro algebra by vir.

2. Regularized dual space

The dual space to the Virasoro algebra:

$$vir^* \cong \text{Vect}(S^1)^* \oplus \mathbf{R}$$

consists of pairs: (u, c) where u is a distribution on S^1 and $c \in \mathbf{R}$. Following A.A. Kirillov (cf. [13]), we will consider only the *regular part* of the dual space, vir^*_{reg} corresponding to distributions given by smooth functions. In other words, $vir^*_{\text{reg}} \cong C^\infty(S^1) \oplus \mathbf{R}$.

Geometrically, objects dual to vector fields on S^1 have the sense of quadratic differentials: $u = u(x)(dx)^2$ (cf. [13], formulae (2) and (4) below). Note, that as a vector space the space of quadratic differentials $\mathcal{F}_2 \cong C^\infty(S^1)$. One obtains the following realization:

$$vir^*_{\text{reg}} = \mathcal{F}_2(S^1) \oplus \mathbf{R}$$

with the pairing:

$$\langle (u(x)(dx)^2, c), (X(x)\frac{d}{dx}, \alpha) \rangle = \int_0^{2\pi} u(x)X(x)dx + c\alpha.$$

Let us calculate the coadjoint action of the Virasoro algebra and of the Virasoro group on space vir^*_{reg}.

3. Coadjoint representation of the Virasoro algebra

Let us recall the definition. The coadjoint representation of a Lie algebra g is the action on its dual space g^*, defined by:

$$\langle ad^*_X(\mu), Y \rangle = -\langle \mu, [X, Y] \rangle,$$

for every $X \in g$ and $\mu \in g^*$.

The coadjoint action of the Virasoro algebra preserves the regular part of the dual space.

Lemma 1.3 *The coadjoint action of the Virasoro algebra on the regular part of its dual space is given by the formula*

$$ad^*_{(X(x)\frac{d}{dx}, \alpha)}(u(x)(dx)^2, c) = (L_X(u) - c \cdot X'''(x)(dx)^2, \, 0) \qquad (2)$$

where $L_X(u)$ is the Lie derivative of a quadratic differential u:

$$L_X(u) = (X(x)u'(x) + 2X'(x)u(x))(dx)^2.$$

Proof. By definition

$$\langle ad^*_{(X(x)\frac{d}{dx}, \alpha)}(u(x), c), \, (Y(x)\frac{d}{dx}, \beta) \rangle = -\langle (u(x), c), \, \left[(X(x)\frac{d}{dx}, \alpha), (Y(x)\frac{d}{dx}, \beta) \right] \rangle$$

$$= -\int_0^{2\pi} u(XY' - X'Y)dx - \frac{c}{2}\int_0^{2\pi} (X'Y'' - X''Y')dx.$$

Integrating by parts, one obtains the expression:

$$\int_0^{2\pi} (Xu' + 2X'u - cX''')Y \, dx.$$

The lemma follows.

Note, that the coadjoint action of the Virasoro algebra is in fact, a $Vect(S^1)$-action (the center acts trivially).

Remarks.
  (a) The case $c = 0$ corresponds to the coadjoint action of $Vect(S^1)$ (without central extension). This is just the natural $Vect(S^1)$-action by the Lie derivative on the space $\mathcal{F}_2$ of quadratic differentials.
  (b) The linear map:

$$s : X(x)\frac{d}{dx} \mapsto X'''(x)(dx)^2 \qquad (3)$$

is a *1-cocycle* on $Vect(S^1)$ with values in $\mathcal{F}_2$. It satisfies the relation:

$$L_X(s(Y)) - L_Y(s(X)) = s([X, Y]).$$

4. The coadjoint action of Virasoro group and Schwarzian derivative

The coadjoint action of the Virasoro group on the regular part of vir^* is the "group version" of the $\mathrm{Vect}(S^1)$-action (2). As in the case of the Virasoro algebra, the center acts trivially and therefore, the coadjoint representation of the Virasoro group is just a $\mathrm{Diff}^+(S^1)$-representation.

It is clear that this action is of the form:

$$Ad^*_{f^{-1}}(u, c) = (u \circ f - c \cdot S(f), \ c) \tag{4}$$

where

$$u \circ f = u(f(x))(df)^2$$

is the natural $\mathrm{Diff}^+(S^1)$-action on $\mathcal{F}_2$ and S is some 1-cocycle on $\mathrm{Diff}^+(S^1)$ with values in $\mathcal{F}_2$. Indeed, this action corresponds to the coadjoint action of the Virasoro algebra (2).

The explicit formula for the 1-cocycle S was calculated in [13] and [27].

Proposition 1.4 [13] *The coadjoint action of the Virasoro group on the regular dual space* vir^*_{reg} *is defined by the 1-cocycle*

$$S(f) = \left(f'''/f' - \frac{3}{2}(f''/f')^2 \right)(dx)^2 \tag{5}$$

Notation. The cocycle (5) is called the *Schwarzian derivative*.

An elegant proof of the formulae (4), (5) directly from the definition of the coadjoint representation can be found in [13].

One can also deduce these formulae from (2). To do this, it is sufficient to check the following two properties:

(a) The formula (4) indeed defines an action of $\mathrm{Diff}^+(S^1)$:

$$Ad^*_f \circ Ad^*_g = Ad^*_{f \circ g}.$$

This follows from the well-known property of the Schwarzian derivative:

$$S(f \circ g) = S(f) \circ g + S(g).$$

Which means that the mapping $f \mapsto S(f)$ is a *1-cocycle* on $\mathrm{Diff}^+(S^1)$ with values in vir^*.

(b) The action (2) is the infinitesimal version of (4).

5. Space of Sturm-Liouville equations as a $\mathrm{Diff}^+(S^1)$-module

It turns out that the formulae (2) and (4) has already been known to classics for a long time before the discovering of the Virasoro algebra.

Consider the (affine) space of Sturm-Liouville operators (1). There exists a natural $\mathrm{Diff}^+(S^1)$-action on this space (cf. [3],[29]). It turns out that this action coincides with the coadjoint action (4).

Definition 1.5 Consider a one-parameter family of actions of $\text{Diff}^+(S^1)$ on the space of functions on S^1:

$$g_\lambda^* a = a \circ g^{-1} \left((g^{-1})' \right)^\lambda \tag{6}$$

Notation. Denote $\mathcal{F}_\lambda$ the $\text{Diff}^+(S^1)$-module structure (6) on space $C^\infty(S^1)$.

Remark. Geometrically speaking, a has the sense of tensor-density of degree λ on S^1:

$$a = a(x)(dx)^\lambda$$

and the action (6) becomes simply $g^* a = a \circ g^{-1}$.

Let us look for a $\text{Diff}^+(S^1)$-action on the space of Sturm-Liouville operators in the form: $g_{\lambda\mu}^*(L) = g_\lambda^* \circ L \circ (g_\mu^*)^{-1}$ for some λ, μ. It is easy to check that this formula preserves the space of Sturm-Liouville operators (this means, the differential operator $g_{\lambda\mu}^*(L)$ is again an operator of the form (1)) if and only if $\lambda = 3/2, \mu = -1/2$.

Definition 1.6 The action of group $\text{Diff}^+(S^1)$ on the space of differential operators (1) is defined by:

$$g^*(L) := g_{\frac{3}{2}}^* \circ L \circ (g_{-\frac{1}{2}}^*)^{-1} \tag{7}$$

(see [29],[3]).

In other words, Sturm-Liouville operators are considered as acting on tensor-densities:

$$L : \mathcal{F}_{-1/2} \to \mathcal{F}_{3/2}.$$

The following statement has already been known to classics.

Proposition 1.7 *The result of the action (7) is again a Sturm-Liouville operator:* $g^*(L) = c \cdot d^2/dx^2 + u^g$ *with the potential*

$$u^g = u \circ g^{-1} \left((g^{-1})' \right)^2 + \frac{c}{2} \cdot S(g^{-1}).$$

Proof. Straightforward.

6. The isomorphism

The last formula coincides with the coadjoint action of the Virasoro group (4) (up to the multiple $-1/2$ in the last term). This remarkable coincidence shows that the space of Sturm-Liouville operators is isomorphic as a $\text{Diff}^+(S^1)$-module to the coadjoint representation.

The isomorphism is given by the formula:

$$(u, c) \longmapsto -2c \cdot d^2/dx^2 + u(x) \tag{8}$$

Theorem 1 is proven.

7. Vect(S^1)-action on the space of Sturm-Liouville operators

The infinitesimal version of the $\mathrm{Diff}^+(S^1)$-action on the space of Sturm-Liouville operators is given by the commutator with the Lie derivative:

$$\mathrm{ad}\, L_X(L) := L_X^{3/2} \circ L - L \circ L_X^{-1/2} \tag{9}$$

where L_X^λ is the operator of Vect(S^1)-action on $\mathcal{F}_\lambda$. In other words, L_X^λ is the operator of Lie derivative on the space of tensor-densities of degree λ:

$$L_X^\lambda = X\frac{d}{dx} + \lambda \cdot X' \tag{10}$$

Proposition 1.8 *The result of the action (9) is a scalar operator of multiplication by:*

$$\mathrm{ad}\, L_X(L) = Xu' + 2X'u - c \cdot X'''$$

Proof. This formula can be proven by simple direct calculations.

The proposition follows also from the isomorphism (8). Indeed, the coadjoint action of X associates to the pair $(u(x)(dx)^2, c)$ the expression $((Xu' + 2X'u - c \cdot X''')(dx)^2, 0)$ corresponding to the scalar operator.

Remarks.

(a) The operator L (and therefore $\mathrm{ad}\, L_X(L)$) maps from $\mathcal{F}_{-1/2}$ to $\mathcal{F}_{3/2}$. This means that the scalar operator $\mathrm{ad}\, L_X(L)$ is rather an operator of multiplication by a tensor-density of degree 2 (a quadratic differential) then by a function: $\mathrm{ad}\, L_X(L) \in \mathcal{F}_2$.

(b) The formula (9) corresponds to the $\mathrm{Diff}^+(S^1)$-action (7). However, the "point of view" of Lie algebras is much more universal: it works in the case (of Lie algebras more general then the Virasoro algebra) when there is no corresponding Lie group and there is no analogue of the formula (7).

II. Projectively invariant version of the Gelfand-Fuchs cocycle and of the Schwarzian derivative

In this section we follow G. Segal (see [27]).

As it was mentioned, the Virasoro algebra is the unique nontrivial central extension of Vect(S^1). However, the 2-cocycle on Vect(S^1) defined this extension can be chosen in different ways (up to a coboundary). There exists a unique way to define the Gelfand-Fuchs cocycle and the cocycles (3),(5) such that they are projectively invariant.

Consider the subalgebra of Vect(S^1) generated by the vector fields:

$$\frac{d}{dx}, \quad \sin x\frac{d}{dx}, \quad \cos x\frac{d}{dx}. \tag{11}$$

It is isomorphic to $sl_2(\mathbf{R})$ and the corresponding Lie group is $PSL(2,\mathbf{R})$ acting on S^1 ($\cong \mathbf{RP}^1$) by *projective transformation*.

1. Modified Gelfand-Fuchs cocycle

Consider the following "modified" Gelfand-Fuchs cocycle on $\text{Vect}(S^1)$:

$$\bar\omega(X(x)\frac{d}{dx}, Y(x)\frac{d}{dx}) = \int_0^{2\pi} (X''' + X')Y\,dx \tag{12}$$

It is clear that this cocycle is cohomologous to the Gelfand-Fuchs cocycle and therefore, the corresponding central extension is isomorphic to the Virasoro algebra. Indeed, the additional term in (12) is a coboundary: the functional

$$\int_0^{2\pi} X'Y\,dx = \frac{1}{2}\int_0^{2\pi} (X'Y - XY')dx$$

depends only on the commutator of X and Y.

The cocycle (12) is sl_2-equivariant. This means,

$$\bar\omega([Z, X], Y) + \bar\omega(X, [Z, Y]) = 0$$

for every $X, Y \in \text{Vect}(S^1)$ and $Z \in sl_2$.

Proposition 2.1 The cocycle (12) is the unique (up to a constant) sl_2-equivariant 2-cocycle on $\text{Vect}(S^1)$.

Proof. Let $\tilde\omega$ be a sl_2-equivariant 2-cocycle on $\text{Vect}(S^1)$. The equivariance condition is equivalent to:

$$\tilde\omega(X, Z) \equiv 0, \quad Z \in sl_2$$

Indeed, since $\tilde\omega$ is a cocycle, one has: $\tilde\omega([Z, X], Y) + \tilde\omega([X, Y], Z) + \tilde\omega([Y, Z], X) = 0$ The sl_2-equivariance condition gives now: $\tilde\omega([X, Y], Z) = 0$ for every $X, Y \in \text{Vect}(S^1)$ and $Z \in sl_2$. But the commutant in $\text{Vect}(S^1)$ coincides with $\text{Vect}(S^1)$.

The Gelfand-Fuchs theorem (see [8]) states that $H^2(\text{Vect}(S^1)) = \mathbf{R}$, and therefore, every nontrivial cocycle is proportional to the Gelfand-Fuchs cocycle up to a coboundary. One has:

$$\tilde\omega = \kappa\omega + b,$$

where b is a coboundary: $b(X, Y) = \langle u, [X, Y]\rangle$ for some $u \in \text{Vect}(S^1)^*$.

The sl_2-equivariance condition means that $b(X, Z) = 0$ for $Z \in sl_2$ and an arbitrary $X \in \text{Vect}(S^1)$. This implies $u = 0$.

2. Modified Schwarzian derivative

It is easy to check that the modified action of $\text{Vect}(S^1)$ on vir_{reg} is as follows:

$$\overline{ad}^*_{X\frac{d}{dx}}(u(dx)^2, c) = (L_X(u) - c \cdot (X''' + X')(dx)^2,\ 0) \tag{13}$$

The modified $\text{Diff}(S^1)$-action on vir_{reg} is:

$$\overline{Ad}^*_{f^{-1}}(u, c) = (u \circ f - c \cdot \bar S(f),\ c) \tag{14}$$

where $\bar S(f)$ is the following (modified) Schwarzian derivative:

$$\bar S(f) = \left(\frac{f'''}{f'} - \frac{3}{2}\left(\frac{f''}{f'}\right)^2 + \frac{1}{2}(f'^2 - 1)\right)(dx)^2 \tag{15}$$

The 1-cocycles S and $\bar S$ on $\text{Diff}^+(S^1)$ are cohomologous.

Remarks

(a) The following amazing fact often leads to confusion. Take the affine parameter $t = \text{tg}(x/2)$. Then, the modified Schwarzian derivative $\bar{S}$ is given by the expression: $\bar{S}(f(t)) = \dddot{f} / \dot{f} - (3/2)(\ddot{f}/\dot{f})^2$, where $\dot{f} = df/dt$. This expression coincides with the formula (5) for S (but $\bar{S} \neq S$).

(b) The PSL_2-equivariant Schwarzian (15) has been considered in [27] (see also [17]).

3. Energy shift

The projectively invariant coadjoint action corresponds to another realization of the dual space to the Virasoro algebra as the space of Sturm-Liouville operators. The map

$$(u, c) \longmapsto -2c\frac{d^2}{dx^2} + u(x) + \frac{c}{2} \tag{16}$$

is an isomorphism of $\text{Diff}^+(S^1)$-module (16) and the module of Sturm-Liouville operators.

Indeed, the fact that the quantity $U = (u(x) + c/2)(dx)^2$ transforms according to the formulae (13) and (14), means that the quadratic differential $u(x)(dx)^2$ transforms under the $\text{Diff}^+(S^1)$-action via the formulae (2) and (4).

4. Projective structures

Let us recall well-known definitions.

An atlas (U_i, t_i) on S^1 is called a *projective atlas* if the coordinate transformations $t_j \circ t_i^{-1}$ are linear-fractional functions. Two atlas' are called equivalent if their union is again a projective atlas. A class of equivalent projective atlas' is called a *projective structure* on S^1.

Every projective structure on S^1 defines a (local) action of the Lie algebra $sl_2(\mathbf{R})$ generated by the vector fields

$$\frac{d}{dt}, \quad t\frac{d}{dt}, \quad (t)^2\frac{d}{dt},$$

where $t = t_i$ is a local coordinate of the projective structure. This action is invariant under the linear-fractional transformations of t_i.

Remark. This sl_2-action coincides with the action (11) for the angular parameter $x = \text{arctg}(t)$.

There exists a natural isomorphism between the space of Sturm-Liouville operators and the space of projective structures on S^1. Given a Sturm-Liouville operator (1), consider the corresponding differential equation: $c \cdot \phi'' + u(x)\phi = 0$. Local coordinates of projective structure associated to this operator are defined as functions of two independent solutions:

$$t = \frac{\phi_1}{\phi_2},$$

on an interval with $\phi_2 \neq 0$. An important remark is that for the local coordinate t, the potential of the Sturm-Liouville operators is identically zero:

$$c\frac{d^2}{dx^2} + u(x) = c\frac{d^2}{dt^2}.$$

A beautiful definition of the corresponding sl_2-action was proposed by A.A. Kirillov (see [14]). It is given by products of solutions: the generators are as follows:

$$\phi_1^1, \quad \phi_1\phi_2, \quad \phi_2^1$$

Note, that the solutions are $-1/2$-tensor-densities, therefore their product is a vector field. Use $W(\phi_1, \phi_2) = \phi_1\phi_2' - \phi_1'\phi_2 = \text{const}$ to verify that the these vector fields indeed generate a sl_2-subalgebra.

III. Kirillov's method of Lie superalgebras

The "mysterious" coincidence between the coadjoint representation of the Virasoro group and the natural $\text{Diff}^+(S^1)$-action on the space of Sturm-Liouville operators (Theorem 1) was explained by A.A. Kirillov in [14], where furthermore an algebraic explanation of the fact that Sturm-Liouville operators act on tensor-densities is given. It is an amazing fact, that the natural interpretation of these geometrical results uses Lie superalgebras.

1. Lie superalgebras

A Lie superalgebra is a $\mathbf{Z}_2$-graded algebra

$$g = g_0 \oplus g_1$$

with the multiplication called the commutator satisfying two conditions:

(1) superized skew symmetry: $[X, Y] + (-1)^{\widetilde{X}\widetilde{Y}}[Y, X] = 0$,

(2) superized Jacobi identity:

$$(-1)^{\widetilde{X}\widetilde{Z}}[X, [Y, Z]] + (-1)^{\widetilde{X}\widetilde{Y}}[Y, [Z, X]] + (-1)^{\widetilde{Y}\widetilde{Z}}[Z, [X, Y]] = 0,$$

where $\widetilde{X}$ is the degree ($\widetilde{X} = 0$ for $X \in g_0$ and $\widetilde{X} = 1$ for $X \in g_1$).

The simplest properties of Lie superalgebras are as follows:

(a) $g_0 \subset g$ is a Lie subalgebra,

(b) g_0 acts on g_1 via the commutator: if $X \in g_0$ and $\xi \in g_1$ is $[X, \xi] \in g_1$

(c) the product of $\xi, \eta \in g_1$ called an *anticommutator* is defined by a symmetric bilinear map

$$[\,,\,]_+ : g_1 \otimes g_1 \to g_0.$$

2. Ramond and Neveu-Schwarz superalgebras

Consider the space $\mathcal{F}_{-1/2}$ of $-1/2$-tensor densities on S^1: $\phi = \phi(x)(dx)^{-1/2}$. As a vector space $\mathcal{F}_{-1/2}$ is isomorphic to $C^\infty(S^1)$ and the $\text{Vect}(S^1)$-action on $\mathcal{F}_{-1/2}$ is given by the Lie derivative:

$$L_X^{-1/2}(\phi) = (X\phi' - (1/2)X'\phi)(dx)^{-1/2}.$$

There exists a natural Lie superalgebra structure on the space $\text{Vect}(S^1) \oplus \mathcal{F}_{-1/2}$, The anticommutator

$$[\,,\,]_+ : \mathcal{F}_{-1/2} \otimes \mathcal{F}_{-1/2} \longrightarrow \text{Vect}(S^1)$$

is just the product of tensor-densities:

$$[\xi(x)(dx)^{-1/2}, \eta(x)(dx)^{-1/2}]_+ := \xi(x)\eta(x)\frac{d}{dx}.$$

Thus, the commutator in the defined Lie superalgebra is given by the following formula:

$$\left[(X,\xi),(Y,\eta)\right] = ([X,Y]_{\text{Vect}(S^1)} + \xi \cdot \eta,\ L_X(\eta) - L_Y(\xi))$$

Definition 3.1 There exists a unique (modulo isomorphism) nontrivial central extension of the defined Lie superalgebra. It can be given by the following 2-cocycle:

$$\Omega\left((Xd/dx, \xi(dx)^{-1/2}), (Yd/dx, \eta(dx)^{-1/2})\right) = \int_0^{2\pi} (X''Y' + 2\xi'\eta')dx \quad (17)$$

The Lie superalgebra defined by this central extension is called the *Ramond* algebra.

Remark. The even part of the Ramond Lie superalgebra coincides with the Virasoro algebra.

Consider now the space of *anti-periodic* $-1/2$-densities on S^1:

$$\xi(x)(dx)^{-1/2}, \quad \xi(x + 2\pi) = -\xi(x).$$

This space is also a $\text{Vect}(S^1)$-module. Let us denote it: $\mathcal{F}_{-1/2}^{(-)}$. Note that the product of two anti-periodic $-1/2$-densities is a ("periodic") vector field well-defined on S^1.

Definition 3.2 The same formulae as above define a Lie superalgebra structure on the space

$$\text{Vect}(S^1) \oplus \mathbf{R} \oplus \mathcal{F}_{-1/2}^{(-)}.$$

This Lie superalgebra is called the *Neveu-Schwarz* algebra.

Remarks

(a) The Lie superalgebras on $\text{Vect}(S^1) \oplus \mathcal{F}_{-1/2}$ and $\text{Vect}(S^1) \oplus \mathcal{F}_{-1/2}^{(-)}$ can be defined as the Lie superalgebras of contact vector fields on $S^{1|1}$ and $\mathbf{RP}^{1|1}$ reciprocally (see [21]).

(b) The Ramond and Neveu-Schwarz superalgebras are particular cases of a series of so-called string superalgebras (see [21]).

3. Coadjoint representation

The (regularized) dual space to the Ramond algebra is naturally isomorphic to:

$$\mathcal{F}_2 \oplus \mathbf{R} \oplus \mathcal{F}_{3/2}.$$

Indeed, the module $\mathcal{F}_{3/2}$ is dual to $\mathcal{F}_{-1/2}$ with respect to the pairing

$$\langle \phi(x)(dx)^{3/2}, \xi(x)(dx)^{-1/2} \rangle = \int_0^{2\pi} \phi(x)\xi(x)dx.$$

Thus, the regular dual space to the Ramond algebra consists of the elements: $(u, c, \phi) = (u(x)(dx)^2, c, \phi(x)(dx)^{3/2})$.

In the same way, the regular dual space to the Neveu-Schwarz algebra is:

$$\mathcal{F}_2 \oplus \mathbf{R} \oplus \mathcal{F}_{3/2}^{(-)},$$

where $\mathcal{F}_{3/2}^{(-)}$ is the space of antiperiodic 3/2-densities.

Lemma 3.3 *The coadjoint representation of the Ramond and Neveu-Schwarz superalgebras are given by the formula:*

$$ad^*_{(X,\xi)} \begin{pmatrix} u(dx)^2 \\ c \\ \phi(dx)^{3/2} \end{pmatrix} = \begin{pmatrix} (Xu' + 2X'u - c \cdot X''' + \xi\phi'/2 + 3\xi'\phi/2)(dx)^2 \\ 0 \\ (X\phi' + 3X'/2 + u\xi - 2c \cdot \xi'')(dx)^{3/2} \end{pmatrix} \tag{18}$$

Proof. The formula (18) can be obtained directly from the definition of the coadjoint representation. The easy calculations are similar to those from the proof of Lemma 1.3.

The Sturm-Liouville operator appears as the coadjoint action of the odd part of the Ramond and Neveu-Schwarz superalgebras. Indeed,

$$ad^*_{(0,\xi)}(u, c, 0) = \left(-2c\frac{d^2}{dx^2} + u \right)\xi.$$

4. Projective equivariance and Lie superalgebra $osp(1|2)$

Consider the Lie superalgebra generated by the vector fields (11) and two more odd generators:

$$\sin\left(\frac{x}{2}\right)(dx)^{-1/2} \quad \text{and} \quad \cos\left(\frac{x}{2}\right)(dx)^{-1/2}$$

This Lie superalgebra is a subalgebra of the Neveu-Schwarz superalgebra isomorphic to the $osp(1|2)$, that has a natural interpretation as the algebra of symmetries of the projective superspace $\mathbf{P}^{1|1}$.

As in the case of the Virasoro algebra (cf. Section 2) it is possible to write the cocycle giving the central extension in a canonic ($osp(1|2)$-invariant) form.

Lemma 3.4. *The 2-cocycle*

$$\bar{\Omega}\left((X d/dx, \xi(dx)^{-1/2}), (Y d/dx, \eta(dx)^{-1/2})\right) = \int_0^{2\pi} ((X''' + X')Y + 2(\xi'' + 4\xi)\eta)dx$$

is the unique (up to a constant) nontrivial 2-cocycle on the Lie superalgebra $\mathrm{Vect}(S^1) \oplus \mathcal{F}_{-1/2}^{(-)}$ *equivariant with respect to the subalgebra* $osp(1|2)$.

Proof. Similar to those of the proof of Proposition 2.1.

IV. Invariants of coadjoint representation of the Virasoro group

It follows from Theorem 1 that the invariants of the coadjoint representation of the Virasoro group are the invariants of the $\mathrm{Diff}^+(S^1)$-action on the space of Sturm-Liouville operator (1). This is quite old and classical problem was considered in [18],[19] and in [13],[27], [30], [10] in the context of the Virasoro algebra.
    In this section we will describe the invariants of the $\mathrm{Diff}^+(S^1)$-action following [13] and [27].

1. Monodromy operator as a conjugation class of $\widetilde{SL}(2,\mathbf{R})$

Consider the Sturm-Liouville equation $2c\psi'' + u(x)\psi = 0$. Since the potential $u(x)$ is a periodic function, the translation

$$M\psi(x) = \psi(x + 2\pi)$$

defines is a linear operator on the space of solutions. This operator is called the *monodromy operator*.

We need the following two remarks.

(1) The monodromy operator defines a conjugation class of the group $SL(2,\mathbf{R})$. Indeed, the Wronsky determinant of any two solutions

$$W(\psi_1, \psi_2) = \psi_1 \psi_2' - \psi_1' \psi_2$$

is a constant function. Thus, W defines a bilinear skew-symmetric form on the space of solutions and operator M preserves W. Now, an arbitrary choice of the basis ψ_1, ψ_2 such that $W(\psi_1, \psi_2) = 1$ associates to M a matrix from $SL(2,\mathbf{R})$. The conjugation class of this matrix does not depend on the choice of the basis.

(2) Moreover, the monodromy operator defines a conjugation class of the universal covering $\widetilde{SL}(2,\mathbf{R})$. Indeed, for every value $x = x_0$, identify the space of solutions with $\mathbf{R}^2$ choosing the initial conditions: $T_{x_0} : \psi \mapsto (\psi(x_0), \psi'(x_0))$. Define a family of linear operators on the space of solutions:

$$T(x) := T_x^{-1} \circ T_0$$

The family $T(x)$ joins the monodromy operator: $M = T(2\pi)$ with the identity: $T(0) = \mathrm{Id}$. It can be lift (up to a conjugation) to $\widetilde{SL}(2,\mathbf{R})$.

We will confound the monodromy operator with the corresponding conjugation class of $\widetilde{SL}(2, \mathbf{R})$.

2. Classification theorem

The following theorem is the classification of the invariants of the $\text{Diff}^+(S^1)$-action on the space of Sturm-Liouville operators. According to Theorem 1, it classifies also the invariants of the coadjoint action of the Virasoro group. Various approaches to the classification see in [18],[19],[13] and [27].

Theorem 2. *The monodromy operator is the unique invariant of the $Diff^+(S^1)$-action.*

Proof. Let us give a simple proof (different from those of [18], [19], [13] and [27]) based on [23].

First, it is clear that the monodromy operator *is* an invariant, since the $\text{Diff}^+(S^1)$-action is just a coordinate transformation and the monodromy operator is defined intrinsically.

To prove that there is no more (independent of M) invariants, we will use the homotopy method. One should show that:

(a) Every two Sturm-Liouville operators with the same monodromy are homotopic to each other in the class of operators with the fixed monodromy. In other words, the set of operators with fixed monodromy is connected.

(b) Given a smooth family of operators with fixed monodromy:

$$L_s = 2c\frac{d^2}{dx^2} + u_s(x), \quad s \in [0, 1],$$

there exists a vector field $X \in \text{Vect}(S^1)$ such that

$$Xu' + 2X'u + cX''' = \dot{u}, \quad \text{where} \quad \dot{u} = \left.\frac{\partial}{\partial s}u_s(x)\right|_{s=0} \tag{19}$$

Statements (a) and (b) imply that every two Sturm-Liouville operators L_1 and L_1 with the same monodromy are on the same $\text{Diff}^+(S^1)$-orbit. Indeed, there exists a family L_s of operators with the fixed monodromy and a family $X_s \, \text{Vect}(S^1)$ of solutions of the homotopy equation (19) for each s. Now, there exists a flow corresponding to this family (S^1 is compact!), this is a diffeomorphism which maps L_1 to L_2.

Let us first prove (a). To each Sturm-Liouville operator, associate a family of mapping $T(x)$ (see Section 4.1). If L_1, L_2 are two operators with the same monodromy, then, the corresponding families are homotopic (as two curves in $PSL_2(\mathbf{R})$). Fix this homotopy: $T(x)^\tau$. Now, to define a family of Sturm-Liouville operators L_τ joining L_1 and L_2, one associates a Sturm-Liouville oparator to each family $T(x)^\tau \in PSL_2(\mathbf{R})$, for fixed τ (this is a standard procedure).

Let us now prove (b).

Lemma 4.1. *Given a basis ψ_1^s, ψ_2^s of solutions of the equation $L_s\psi = 0$ such that $W(\psi_1^s, \psi_2^s) \equiv 1$, the vector field*

$$X = \frac{1}{2}\begin{vmatrix} \dot{\psi}_1 & \dot{\psi}_2 \\ \psi_1 & \psi_2 \end{vmatrix} \tag{20}$$

is a solution of the equation (19).

Proof. Taking the derivative of the equality $L_s\psi^s = 0$, one gets:

$$\dot{u}\psi^{s=0} + L_{s=0}\dot{\psi} = 0.$$

To solve the equation (19), it is sufficient to find a vector fields X such that the Lie derivative $L_X(\psi) = \dot{\psi}$. Indeed, it follows from the definition (9) of the Vect(S^1)-action on the space of Sturm-Liouville operators.

Let us look for a vector field X such that $L_X(\psi_1^{s=0}) = \dot{\psi}_1$ and $L_X(\psi_2^{s=0}) = \dot{\psi}_2$. This gives a system of linear equations:

$$\begin{cases} X\psi_1' - (1/2)X'\psi_1 = \dot{\psi}_1 \\ X\psi_2' - (1/2)X'\psi_2 = \dot{\psi}_2. \end{cases}$$

Taking X and X' as independent arguments, one obtains formally:

$$X = \frac{1}{2}\begin{vmatrix} \dot{\psi}_1 & \dot{\psi}_2 \\ \psi_1 & \psi_2 \end{vmatrix}, \quad X' = \begin{vmatrix} \dot{\psi}_1 & \dot{\psi}_2 \\ \psi_1 & \psi_2 \end{vmatrix}.$$

Now, let us verify that $X' = dX/dx$. Indeed,

$$\frac{dX}{dx} = \frac{1}{2}\begin{vmatrix} \dot{\psi}_1' & \dot{\psi}_2' \\ \psi_1 & \psi_2 \end{vmatrix} + \frac{1}{2}\begin{vmatrix} \dot{\psi}_1 & \dot{\psi}_2 \\ \psi_1' & \psi_2' \end{vmatrix}.$$

The two terms in the right hand side coincide since:

$$\frac{d}{ds}\begin{vmatrix} \psi_1' & \psi_2' \\ \psi_1 & \psi_2 \end{vmatrix} = \begin{vmatrix} \dot{\psi}_1' & \dot{\psi}_2' \\ \psi_1 & \psi_2 \end{vmatrix} + \begin{vmatrix} \psi_1' & \psi_2' \\ \dot{\psi}_1 & \dot{\psi}_2 \end{vmatrix} = 0,$$

and therefore $X' = dX/dx$.

Lemma 4.1 is proven.

Let us show that Theorem 2 follows from the lemma. Indeed, since the monodromy operator does not depend on s, the basis ψ_1^s, ψ_2^s can be chosen in such a way that the corresponding monodromy matrix does not depend on s. Then, the solution (20) is periodic: $X(x + 2\pi) = \det(M)X(x) = X(x)$.

Theorem 2 is proven.

Remark. Recall that the solutions of a Sturm-Liouville equation have a sense of $-1/2$-tensor-densities. Therefore, the quadratic expression (20) is indeed a vector field.

The Kähler geometry of the coadjoint orbits of the Virasoro group has been studied in A.A. Kirillov's works [15],[16].

V. Extension of the Lie algebra of first order linear differential operators on S^1 and matrix analogue of the Sturm-Liouville operator

This section follows the recent work [22]. We will show that the Kirillov method is valid in a more general framework then the Virasoro algebra.

1. Lie algebra of first order differential operators on S^1 and its central extensions

Consider the Lie algebra of first order linear differential operators on S^1:

$$A = X(x)\frac{d}{dx} + a(x) \tag{21}$$

(This Lie algebra is in fact the semi-direct product of $Vect(S^1)$ by the module of functions $\mathcal{F}_0$).

This Lie algebra has three nonisomorphic central extensions (cf. [25]). The first one is given by the Gelfand-Fuchs cocycle and two more extensions are given by the non-trivial 2-cocycles:

$$\omega'((X\frac{d}{dx}, a), (Y\frac{d}{dx}, b)) = \int_{S^1} (X''(x)b(x) - Y''(x)a(x))dx$$

$$\omega''((X\frac{d}{dx}, a), (Y\frac{d}{dx}, b)) = 2\int_{S^1} a(x)b'(x)dx \tag{22}$$

Definition 5.1 Let us denote $\mathcal{G}$ the Lie algebra defined on the space $Vect(S^1) \oplus C^\infty(S^1) \oplus \mathbf{R}^3$ as the *universal central extension* of the Lie algebra of the operators (21). This means, $\mathcal{G}$ is the Lie algebra defined by the commutator:

$$\left[(X\frac{d}{dx}, a, \alpha), (Y\frac{d}{dx}, b, \beta)\right] = \left((XY' - X'Y)\frac{d}{dx}, Xb' - Ya', \omega\right)$$

where $\alpha = (\alpha_1, \alpha_2, \alpha_3)$, $\beta = (\beta_1, \beta_2, \beta_3) \in \mathbf{R}^3$ are in the center and

$$\omega = \left(\omega((X\frac{d}{dx}, a), (Y\frac{d}{dx}, b)), \omega'((X\frac{d}{dx}, a), (Y\frac{d}{dx}, b)), \omega''((X\frac{d}{dx}, a), (Y\frac{d}{dx}, b))\right).$$

2. Matrix Sturm-Liouville operators

The space of matrix linear differential operators on $C^\infty(S^1) \oplus C^\infty(S^1)$:

$$\mathcal{L} = \begin{pmatrix} -2c_1\dfrac{d^2}{dx^2} + u(x) & 2c_2\dfrac{d}{dx} + v(x) \\ -2c_2\dfrac{d}{dx} + v(x) & 4c_3 \end{pmatrix} \tag{23}$$

where $c_1, c_2, c_3 \in \mathbf{R}$ and $u = u(x), v = v(x)$ are 2π-periodic functions was defined in [22]. It was shown that this space gives a geometric realization for the dual space of the Lie algebra $\mathcal{G}$.

The $Vect(S^1)$-action on the space of operators (23) is defined, as in the case of Sturm-Liouville operators (1), by commutator with the Lie derivative. We consider

$\mathcal{L}$ as an operator on $Vect(S^1)$-modules:

$$\mathcal{L} : \mathcal{F}_{-\frac{1}{2}} \oplus \mathcal{F}_{\frac{1}{2}} \rightarrow \mathcal{F}_{\frac{3}{2}} \oplus \mathcal{F}_{\frac{1}{2}}.$$

Remark. The choice of degrees of tensor-densities in this formula is the *unique* choice such that the operators (23) are *selfadjoint*.

There exists a natural action of the Lie algebra of first order differential operators (21) the space of operators (23).

3. Action of Lie algebra of differential operators

There exists a nice family of modules over the Lie algebra of operators (21). Consider the space

$$\mathcal{F}_\lambda \oplus \mathcal{F}_{\lambda+1}$$

It is defined by the formula:

$$T^{(\lambda)}_{(X(x)\frac{d}{dx} + a(x))} \begin{pmatrix} \phi(x) \\ \psi(x) \end{pmatrix} = \begin{pmatrix} L^{(\lambda)}_{X\frac{d}{dx}} \phi(x) \\ L^{(\lambda+1)}_{X\frac{d}{dx}} \psi(x) - \lambda a'(x)\phi(x) \end{pmatrix} \qquad (24)$$

The action on the space of operators (23) is defined in analogous way as the action of $Vect(S^1)$ on the space of Sturm-Liouville operators given by the formula (9). Put:

$$\left[T_{(X\frac{d}{dx} + a)}, \mathcal{L} \right] := T^{(1/2)}_{(X\frac{d}{dx} + a)} \circ \mathcal{L} - \mathcal{L} \circ T^{(-1/2)}_{(X\frac{d}{dx} + a)} \qquad (25)$$

Theorem 5.2 (see [22]). *The action* (25) *coincides with the coadjoint action of the Lie algebra of first order linear differential operators.*

Proof. The explicit formula for the action (25) is:

$$\left[T_{(X\frac{d}{dx} + a)}, \mathcal{L} \right] = \begin{pmatrix} \begin{array}{c} Xu' + 2X'u - c_1 X''' \\ +va' + c_2 a'' \end{array} & \begin{array}{c} Xv' + X'v - c_2 X'' \\ +2c_3 a' \end{array} \\ \begin{array}{c} Xv' + X'v - c_2 X'' \\ +2c_3 a' \end{array} & 0 \end{pmatrix}$$

One easily verifies that this is precisely the coadjoint action of the Lie algebra of differential operators (23) (see [22] for the details).

4. Generalized Neveu-Schwarz superalgebra

The space of matrix analogues of the Sturm-Liouville operators (23) was found in [22] using a Lie superalgebra generalizing the Neveu-Schwarz algebra.

Definition 5.3 Consider the $\mathbf{Z}_2$-graded vector space : $\mathcal{S} = \mathcal{S}_0 \oplus \mathcal{S}_1$. where $\mathcal{S}_0 = \mathcal{G}$ the extension of the Lie algebra of operators (23). and $\mathcal{S}_1$ the $\mathcal{G}$-module:

$$\mathcal{S}_1 = \mathcal{F}_{-\frac{1}{2}} \oplus \mathcal{F}_{\frac{1}{2}}.$$

The even part $\mathcal{S}_0$ acts on $\mathcal{S}_1$ according to (24). Let us define the *anticommutator* $[\,,\,]_+ : \mathcal{S}_1 \otimes \mathcal{S}_1 \to \mathcal{S}_0$:

$$\left[(\phi, \alpha), (\psi, \beta)\right]_+ = (\phi\psi\frac{d}{dx}, \ \phi\beta + \alpha\psi, \ \sigma_+)$$

where $\Omega_+ = (\Omega, \Omega', \Omega'')$, where Ω is the Ramond – Neveu-Schwarz cocycle (17) and Ω', Ω'') are the continuations of the cocycles (22):

$$\Omega'((\phi, \alpha), (\psi, \beta)) = -2 \int_{S^1} (\phi'(x)\beta(x) + \alpha(x)\psi'(x))dx$$

$$\Omega''((\phi, \alpha), (\psi, \beta)) = 4 \int_{S^1} \alpha(x)\beta(x)dx$$

Theorem 5.4 (see [22]). $\mathcal{S}$ *is a Lie superalgebra.*

The differential operators (23) can be defined as a part of the coadjoint action of the superalgebra $\mathcal{G}$. Namely, one obtains:

$$\text{ad}^*{\begin{pmatrix} 0 \\ 0 \\ \phi(dx)^{-\frac{1}{2}} \\ \alpha(dx)^{\frac{1}{2}} \end{pmatrix}} \begin{pmatrix} u \\ v \\ c \\ 0 \\ 0 \end{pmatrix} = \begin{pmatrix} 0 \\ 0 \\ 0 \\ -2c_1\phi'' + u\phi + v\alpha + 2c_2\alpha' \\ -2c_2\phi' + v\phi + 4c_3\alpha \end{pmatrix}$$

Remark. The Lie algebra $\mathcal{G}$ considered in this section, is just an example from the series of seven Lie algebras generalizing the Virasoro algebra (see [25]). It turns out that Kirillov's method works for also for the other Lie algebra from this series (work in preparation of P. Marcel). It would be very interesting to apply this method to other Virasoro type Lie algebras (see [26]).

VI. Geometrical definition of the Gelfand-Dickey bracket and the relation to the Moyal-Weil star-product

In this section we follow [24]. We consider another generalization of the Virasoro algebra: the so-called *second Adler-Gelfand-Dickey* Poisson structure, which is also known as the classical W-algebras in the physics literature. The Adler-Gelfand-Dickey bracket is (an infinite-dimensional) Poisson bracket on the space of n-th order differential operators on S^1. (We will consider here only the first nontrivial case corresponding to the space of third order linear differential operators)

We will show that the Gelfand-Dickey bracket is related to the well-known *Moyal-Weyl star-product*.

The main idea is to consider arguments of differential operators as tensor-densities and use the PSL_2-equivariance of all the operations.

1. Moyal-Weyl star-product

Consider the standard symplectic plane $(\mathbf{R}^2, dp \wedge dq)$, where p, g are linear coordinates. The space of functions on $\mathbf{R}^2$ is a Lie algebra with respect to the Poisson bracket:

$$\{F, G\} = F_p G_q - F_q G_p,$$

where $F_p = \partial F / \partial p$.

The following operation:

$$F \star_\hbar G = FG + \frac{\hbar}{2}\{F, G\} + \cdots + \frac{\hbar^m}{2^m m!}\{F, G\}_m + \cdots$$

where

$$\{F, G\}_m = \sum_{i=0}^m (-1)^i \binom{m}{i} \frac{\partial^m F}{\partial p^{m-i} \partial q^i} \frac{\partial^m G}{\partial p^i \partial q^{m-i}}$$

is called the Moyal-Weyl star-product on $\mathbf{R}^2$. Here $\hbar$ is a formal parameter and the operation $\star_\hbar$ is with values in formal series in $\hbar$. (In the case polynomials, one can consider $\hbar$ as a number). The operation $\star_\hbar$ is associative.

The Moyal-Weyl star-product is a very popular object in deformation quantization.

2. Moyal-Weyl star-product on tensor-densities, the transvectants

Isomorphism 6.1 *There exists a natural isomorphism between the space $\mathcal{F}_\lambda$ (of tensor-densities of degree λ on S^1) and the space of functions on $\mathbf{R}^2 \setminus \{0\}$ homogeneous of degree -2λ. For the affine parameter on S^1: $t = tg(x)$ this isomorphism is given by the formula:*

$$\phi(t)(dt)^\lambda \longmapsto p^{-2\lambda} \phi\left(\frac{q}{p}\right) \tag{26}$$

Indeed, a function corresponding to a vector field X is: $p^2 X(q/p)$. Verify, that the Lie derivative corresponds to the Poisson bracket.

The isomorphism (26) lifts the Moyal-Weyl star-product to the space of tensor-densities.

Lemma 6.2 *The terms of this star-product are as follows:*

$$\{\phi, \psi\}_m = \frac{m!}{2^m} \sum_{i+j=m} (-1)^i m! \binom{2\lambda + m - 1}{i} \binom{2\mu + m - 1}{j} \phi^{(i)} \psi^{(j)} \tag{27}$$

where $\phi \in \mathcal{F}_\lambda, \psi \in \mathcal{F}_\mu$, $\phi^{(i)} = d^i \phi / dx^i$ and

$$\binom{k}{i} = k(k-1) \cdots (k - i + 1).$$

Proof. Straightforward.

It turns out that the operations (27) coincides (up to the constant $m!/2^m$) with so-called Gordan's *transvectants*. This operations can be defined as bilinear maps

$$\mathcal{F}_\lambda \otimes \mathcal{F}_\mu \to \mathcal{F}_{\lambda+\mu+m}$$

equivariant with respect to the action of the Lie algebra $sl_2(\mathbf{R}$ defined by (11) (projectively invariant).

Remark. The isomorphism (26) is, in fact, given by the standard projective structure on S^1. Indeed, t is the corresponding projective parameter. Given an arbitrary projective structure on S^1, one defines an isomorphism (analogue of (26)) between tensor-densities on S^1 and homogeneous functions on $\mathbf{R}^2$.

3. Space of third order linear differential operators as a $\text{Diff}^+(S^1)$-module

Consider the space of third order linear differential operators

$$A = \frac{d^3}{dx^3} + u(x)\frac{d}{dx} + v(x) \tag{28}$$

This space plays the same role that the space of Sturm-Liouville operators (1) in the case of Virasoro algebra. However, the Adler-Gelfand-Dickey bracket is not a Lie-Poisson structure. We refer [1] and [7] for the original definition and [4] for another one related to the Kac-Moody algebras.

The subject of this section was known to the classics (see [29],[3]) ... and was forgotten by the contemporary experts.

Definition 6.3 The $\text{Diff}^+(S^1)$-action on the space of operators (28) is defined by the formula:

$$g^*(A) := g_2^* \circ A \circ (g_{-1}^*)^{-1}$$

This means that the operator A is considered as acting from the space of vector fields on S^1 with values in the space of quadratic differentials:

$$A : \mathcal{F}_{-1} \to \mathcal{F}_2.$$

The corresponding action of $X(x)d/dx \in \text{Vect}(S^1)$ is:

$$\text{ad}\, L_X(A) := L_X^{(2)} \circ A - A \circ L_X^{(-1)}$$

Let us give the explicit formulae for $\text{Diff}^+(S^1)$- and $\text{Vect}(S^1)$-action. It is convenient to decompose the operator (28) as a sum of its skew-symmetric and symmetric parts:

$$A = \frac{d^3}{dx^3} + u(x)\frac{d}{dx} + \frac{u(x)}{2} + w(x),$$

where $w(x) = v(x) - u(x)/2$.

Proposition 6.4 (see [29],[3]) A diffeomorphism f transform an operator A into an operator of the form (28) with coefficients:

$$u^f = u \circ f(f')^2 + 2S(f)$$
$$w^f = w \circ f(f')^3.$$

This means, u transforms as a potential of the Sturm-Liouville operator: $4\frac{d^2}{dx^2} + u(x)$ and w has the sense of *cubic differential*: $w = w(x)(dx)^3$.

Corollary. *The projection from the space of third order operators (28) to the space of Sturm-Liouville operators:*

$$\frac{d^3}{dx^3} + u(x)\frac{d}{dx} + v(x) \longmapsto 4\frac{d^2}{dx^2} + u(x)$$

is Diff(S^1)*-equivariant (does not depend on the choice of the parameter x).*

The corresponding action of a vector field $X(x)d/dx \in \text{Vect}(S^1)$ associates to A a first order operator: $\text{ad} L_X(A) = u^X \frac{d}{dx} + \frac{u^X}{2} + w^X$, where

$$u^X = Xu' + 2X'u + 2X'''$$
$$w^X = Xw' + 3X'w.$$

Remark. The geometric interpretation of u and w is related to the projective projective differential geometry of plane curves (associated to differential operators (28)). Namely, u is interpreted as the projective curvature and w leads to the notion of projective length element: $ds = (w)^{1/3}$ (see [29],[3] and also [11]).

4. Second order Lie derivative

The notion of *second order Lie derivative* considered below was introduced in [24] (see [5] for a general definition in the multi-dimensional case). The question is as follows: given a second order contravariant tensor field $X \in \mathcal{F}_2$:

$$Z = Z(x)(dx)^{-2},$$

is it possible to define an "action" of Z on geometric quantities (like tensor-densities etc.) analogous to the Lie derivative along a vector field?

The answer is negative. There is no Diff(S^1)-equivariant bilinear differential operators

$$\mathcal{F}_2 \otimes \mathcal{F}_\lambda \to \mathcal{F}_\lambda,$$

for general values of λ (cf. [12]) and so, one can not define such an action intrinsically.

To define the second order Lie derivative, we fix a projective structure on S^1.

Definition 6.5 The second order Lie derivative over contravariant tensor field of degree 2: $Z = Z(x)(dx)^{-2}$ is a linear map

$$L_Z^2 : \mathcal{F}_\lambda \to \mathcal{F}_\lambda$$

given by:

$$L_Z^2(\phi) := \{Z, \phi\}_2$$

Remark. Note, that the operations $\{\,,\,\}_m$, $m \geq 2$ are defined if one fix a projective structure (cf. Section 6.2).

5. Adler-Gelfand-Dickey Poisson structure

A Poisson structure on a manifold is given by a linear map on each cotangent space with values in the tangent space (satisfying the Jacobi condition). Thus, to define a Poisson structure on a vector space, it is sufficient to associate a vector field to every linear functional.

Every linear functional on the space of operators (28) is a linear combination of:

$$\langle l_X^1, A \rangle = \int X(x)u(x)dx, \quad \langle l_Z^2, A \rangle = \int Z(x)w(x)dx$$

where $X = X(x)d/dx, Z = Z(x)(dx)^{-2}$.

Definition 6.6 The Adler-Gelfand-Dickey Poisson structure on the space of operators (28) associates to a linear functionals vector fields given by the commutator with the Lie derivative:

$$\dot{A}_X := [L_X, A]$$
$$\dot{A}_Z := [L_Z^2, A]$$

(see [24] for the details).

The Adler-Gelfand-Dickey Poisson structure is a very interesting and popular object in Mathematical Physics. This way to define it seems to be natural in the spirit of Section 1.

Addendum. Recently Kirillov's method has been applied for a new class of infinite-dimensional Lie algebras, see [31, 32].

References

[1] M. Adler, *On a trace functional for formal pseudo-differential operators and the symplectic structure of the Korteweg-de Vries type equation*, Invent. Math. 50:3 (1987) 219–248.

[2] R. Bott, *On the characteristic classes of groups of diffeomorphisms*, Enseign. Math. 23:3–4 (1977), 209–220.

[3] E. Cartan, *Leçons sur la théorie des espaces à connexion projective*, Gauthier-Villars, Paris, 1937.

[4] V.G. Drinfel'd & V.V. Sokolov, *Lie algebras and equations of Korteweg-De Vries type*, J. Soviet Math. 30 (1985), 1975–2036.

[5] C. Duval & V. Ovsienko, *Space of second order linear differential operators as a module over the Lie algebra of vector fields*, Advances in Math. 132 (1997), no.2, 316–333.

[6] H. Gargoubi & V. Ovsienko, *Space of linear differential operators on the real line as a module over the Lie algebra of vector fields*, IMRN (1996), N.5, 235–251.

[7] I.M. Gel'fand & L.A. Dikii, *A family of hamiltonian structures connected with integrable nonlinear differential equations* in: I.M. Gel'fand collected papers (S.G. Gindikin et al, eds), Vol. 1, Springer, 1987, 625–646.

[8] I.M. Gel'fand & D.B. Fuchs, *Cohomology of the Lie algebra of vector fields on the circle*, Funct. Anal. Appl. 2:4 (1968), 342–343.

[9] P. Gordan, Invariantentheorie, Teubner, Leipzig, 1887.

[10] L. Guieu, *Nombre de rotation, structures géométriques sur un cercle et groupe de Bott-Virasoro*, Ann. Inst. Fourier 46 (1996), no.4, 971–1009.

[11] L. Guieu & V. Ovsienko, *Structures symplectiques sur les espaces de courbes projectives et affines*, J. Geom. Phys. 16 (1995) 120–148.

[12] P. Ya. Grozman, *Classification of bilinear invariant operators over tensor fields*, Funct. Anal. Appl., 14 (1980), 58–59.

[13] A.A. Kirillov, *Infinite dimensional Lie groups : their orbits, invariants and representations. The geometry of moments*, Lect. Notes in Math., 970, Springer-Verlag (1982) 101–123.

[14] A.A. Kirillov, *Orbits of the group of diffeomorphisms of a circle and local superalgebras*, Funct. Anal. Appl., 15:2 (1980) 135–137.

[15] A.A. Kirillov, *Kähler structure on K-orbits of the group of diffeomorphisms of a circle*, Funct. Anal. Appl. 21:2 (1987) 122–125.

[16] A.A. Kirillov & D.V. Yuriev, *Kähler geometry of the infinite-dimensional homogeneous space $M = \mathrm{Diff}_+(S^1)/Rot(S^1)$*, Funct. Anal. Appl. 21:4 (1987) 248–294.

[17] B. Kostant & S.Sternberg, *The Schwarzian derivative and the conformal geometry of the Lorentz hyperboloid*, in: Quantum Theories and Geometry (M. Cahen and M. Flato eds.) Kluwer, 1988, 113–125.

[18] N.H. Kuiper, *Locally projective spaces of dimension one*, Michigan Math. J., 2 (1954) 95–97.

[19] V.F. Lazutkin & T.F. Pankratova, *Normal forms and versal deformations for Hill's equations*, Funct. Anal. Appl 9:4 (1975), 306–311.

[20] P.B.A. Lecomte, P. Mathonet & E. Tousset, *Comparison of some modules of the Lie algebra of vector fields*, Indag. Math. (N.S.) 7 (1996), no.4, 461–471.

[21] D.A. Leites, B.A. Feigin, *New Lie superalgebras of string theories*, Group-Theoretic Methods in Physics, v.1, Moscow (1983) 269–273.

[22] P. Marcel, V. Ovsienko & C.Roger, *Extension of the Virasoro and Neveu-Schwarz algebras and generalized Sturm-Liouville operators*, Lett. Math. Phys. 40 (1997), no.1, 31–39.

[23] V. Ovsienko, *Classification of third-order linear differential equations and symplectic sheets of the Gel'fand-Dikii bracket*, Math. Notes, 47:5 (1990) 465–470.

[24] O. Ovsienko & V. Ovsienko, *Lie derivative of order n on a line. Tensor meaning of the Gelfand-Dickey bracket*, Adv. in Soviet Math., 2, 1991.

[25] V. Ovsienko, C. Roger, *Extension of Virasoro group and Virasoro algebra by modules of tensor densities on S^1*, Funct. Anal. Appl. 30 (1996), no.4, 290–291.

[26] C. Roger, *Extensions centrales d'algèbres et de groupes de Lie de dimension infinie, algèbre de Virasoro et généralisations*, Rep. Math. Phys. 35 (1995), no.2–3, 225–266.

[27] G.B. Segal, *Unitary representations of some infinite dimensional groups*, Comm. Math. Phys., 80:3 (1981) 301–342.

[28] G.B. Segal, *The geometry of the KdV equation* in : Trieste Conference on topological methods in quantum field theories – W. Nahm & al, eds – World Scientific (1990), 96–106.

[29] E.J. Wilczynski, Projective differential geometry of curves and ruled surfaces, Leipzig – Teubner, 1906.

[30] E. Witten, *Coadjoint orbits of the Virasoro group*, Comm. Math. Phys., 114:1, (1988) 1–53.

[31] P. Marcel, *Extensions of the Neveu-Schwarz Lie superalgebra*, Comm. Math. Phys. 207 (1999), no. 2, 291–306.

[32] P. Marcel, *Generalizations of the Virasoro algebra and matrix Sturm-Liouville operators*, J. Geom. Phys. 36 (2000), 211–222.

Mathematical Subject Classification (2000)
Primary: 17B68.
Secondary: 17B65, 81R10.

Centre de Physique Théorique
C.N.R.S.
Luminy – Case 907
F-13288 Marseille Cedex 9
France
E-mail address: Valentin.Ovsienko@cpt.univ-mrs.fr

From Group Actions to Determinant Bundles Using (Heat-kernel) Renormalization Techniques

Sylvie Paycha

Introduction

The functional quantization of gauge field theories leads to interesting infinite dimensional geometric problems from which one expects to extract some information on the original gauge theory. The geometric framework underlying a gauge field theory is essentially that of an infinite dimensional Lie group $\mathcal{G}$, the gauge group, acting on an infinite dimensional manifold, the manifold of paths $\mathcal{P}$. Here we shall consider a setting in which this group action gives rise to a principal bundle $\pi : \mathcal{P} \to \mathcal{P}/\mathcal{G}$.

On the one hand, one can investigate the geometry of the orbits of this group action via the *second fundamental form of the orbit* which generalizes without problem to the infinite dimensional setting. Following this line of thought, when trying to define quantities that involve traces such as the *the trace of this second fundamental form* or determinants corresponding to the *volume of the orbits*, one comes across divergent expressions. One way to get rid of these divergences is to apply (heat-kernel) renormalization methods similar to the ones used in field theories.

On the other hand, group actions naturally lead to determinant bundles, the geometry of which can also give some insight into the original gauge theory. Determinant bundles are one dimensional vector bundles associated to a family of operators parametrized by some space. When trying to equip these bundles with a metric and a connection, similar divergences to the ones described above occur since one needs take determinants of operators acting on infinite dimensional manifolds as well as variations of these determinants. Again, one way to deal with these divergences is to apply (heat-kernel) renormalization methods, via which one can describe the geometry of determinant bundles, defining the *Quillen metric* [Q], the *Bismut-Freed* connection [BF] as well as other objects such as the curvature.

The leap from gauge group actions to determinant bundles can be made more understandable using a stepping stone, namely a super bundle built up from the group action. Starting from a group action $\mathcal{G} \times \mathcal{P} \to \mathcal{P}$, one builds a super-bundle

$\mathcal{E} = \mathcal{E}^+ \oplus \mathcal{E}^-$ with $\mathcal{E}^+$ given by a trivial bundle $\mathcal{P} \times Lie(\mathcal{G})$ where $Lie(\mathcal{G})$ is the Lie algebra of $\mathcal{G}$, $\mathcal{E}^- = VT\mathcal{P}$ being the vertical tangent bundle to $\mathcal{P}$. In the finite dimensional framework, one could build *the* determinant bundle associated to $\mathcal{E}$ (or rather to its complexification) namely Det $\mathcal{E} := (\Lambda^{max}\mathcal{E}^+)^{-1} \otimes \Lambda^{max}\mathcal{E}^-$. In finite dimensions the curvature on the determinant bundle Det $\mathcal{E}$ coincides with minus the first Chern form on the vector bundle $\mathcal{E}$. This finite dimensional setting is rather trivial since both $\mathcal{E}^+$ and $\mathcal{E}^-$ are trivial bundles, thus giving rise to vanishing Chern classes. But in the infinite dimensional setting we are about to consider, the "weighted first Chern form" (built similarly to the ordinary one up to the fact that we use "weighted traces" instead of ordinary traces) is not a priori closed and there is therefore no natural notion of first Chern class from this point of view. Thus we shall focus on Chern forms and as in the example of section V - an example inspired from string theory and Teichmüller theory - we should keep in mind that we might want to consider non trivial connections on trivial bundles, this leading to non vanishing weighted first Chern forms.

In infinite dimensions, the curvature on the determinant bundle gives some information on the physics of the underlying field theory and we would like to see how such a picture:

Group action → Superbundle $\mathcal{E}$ → Associated determinant bundle   Det $\mathcal{E}$

carries out to infinite dimensions.

This gives rise to at least two basic questions:

- What meaning should we give to the expression "*the* determinant bundle associated to a given superbundle" ?

- How can one define the *first Chern form* on the super vector bundle and relate the curvature (defined à la Bismut and Freed) on the determinant bundle to it ?

These two questions are of course related and we suggest here an answer within the framework of *weighted* vector bundles briefly described in these notes which offer a survey of some results scattered in [AP], [CDMP],[PR2], see also [P] for a review. The idea is to introduce a field of "elliptic operators" on the vector bundle which we shall use for various purposes. From the data $(\mathcal{E}, D)$ where $D := \begin{bmatrix} 0 & D^- \\ D^+ & 0 \end{bmatrix}$ is a field of "self-adjoint elliptic operators" on the vector bundle $\mathcal{E}$, on the one hand we can build what we shall call a *weighted vector bundle* $(\mathcal{E}, Q := D^2)$, Q now being a field of positive elliptic operators and on the other hand we can build *the associated determinant bundle* $Det(\mathcal{E}, D)$ defined following Quillen's construction as the determinant bundle associated to the family of operators given by the field D. To make this more concrete, one can think of $\mathcal{E}$ as an infinite rank vector bundle with fibres given by spaces of sections of some spinor bundle and of D as a field of Dirac operators parametrized by the base space of the bundle acting on these sections.

The weight Q provides us with a way of regularizing the otherwise divergent traces like for example the trace of the curvature involved in the definition of the first Chern form and hence renormalize them using again heat-kernel techniques. As we shall see in these notes, starting from a group action, the field D arises naturally as the tangent operator (or rather its complexification) to the map that takes an element of the group into the orbit of a given point. In this framework of "weighted geometric structures" we can generalize the above picture to infinite dimensions:

$$\text{Group action} \to (\mathcal{E}, D) \to Det(\mathcal{E}, D).$$

Unlike the finite dimensional case, for which the curvature on the determinant bundle coincides up to a sign with the first Chern form on the original vector bundle, in the infinite dimensional case, obstructions arise from the presence of a weight. This obstruction preventing an identification between the (Bismut-Freed) curvature on $Det(\mathcal{E}, D)$ and minus the weighted first Chern form on $(\mathcal{E}, Q = D^2)$ can be expressed in terms of Wodzicki residues so that

$$\Omega^{Det(\mathcal{E}, D)} = -r_1^{\mathcal{E}, Q} + \text{Wodzicki residue terms}$$

where $r_1^{\mathcal{E}, Q}$ is the Q-weighted first Chern form on $\mathcal{E}$ and $\Omega^{Det(\mathcal{E}, D)}$ the curvature of the Bismut-Freed connection on the determinant bundle $Det(\mathcal{E}, D)$.

Let us describe the structure of this contribution. We start in the first part with the description of what we understand by renormalization procedure, a procedure on which the remaining constructions in this paper rely. The second part gives a brief presentation of the concept of weighted vector bundle which we need to define a notion of first Chern form on a vector bundle. The third part is devoted to the description of the geometry of gauge orbits and the fourth part to that of determinant bundles. We shall see how renormalized determinants of operators acting on an infinite dimensional space arise in a fundamental way, as volumes of orbits in the orbit picture, as norms of sections in the determinant bundle picture and how their infinitesimal variations play a fundamental role, since these arise as traces of the second fundamental form of the orbits (up to a constant factor) in the first picture and as the real part of the connection (up to a constant factor) on the determinant bundle in the second one. In the fifth part of this paper, we shall illustrate the above constructions with the example of diffeomorphisms acting on complex structures. In this particular case there is a natural complexification and the first Chern form on the hermitian vector bundle induced by the group action can be expressed in terms of (weighted) traces of some multiplication operators.

Let us start with the general prescription for "renormalization".

I. Renormalization techniques

Warning

In this section, we set up the basic techniques which we shall be using afterwards to make sense of divergent traces.

Strictly speaking, from a quantum field theoretic point of view, the techniques we are about to present are "regularization techniques", but in the context we are considering here, they are commonly called "renormalization techniques".

1. Renormalized limits

This presentation is close in spirit to that of [BGV].

We shall call *renormalizable around zero* a function $f : \mathbb{R}^+/\{0\} \to \mathbb{R}$ with the property that there are positive integers m, n, $m \neq 0$ and real numbers α, $a_j, b_j, c_j, j \in \mathbb{N}$ such that f behaves asymptotically as follows around zero:

$$f(\varepsilon) \sim_0 \sum_{j=0}^{\infty} a_j \varepsilon^{\lambda_j} + \sum_{j=0, \lambda_j \in \mathbb{Z}}^{\infty} b_j \varepsilon^{\lambda_j} \log \varepsilon + \sum_{j=0}^{\infty} c_j \varepsilon^j \qquad (R)$$

where $\lambda_j := \frac{j - \alpha - n}{m}$. The symbol $\sim$ means that for any $J \in \mathbb{N}$, $\exists K_J := [\alpha] + mJ + n \in \mathbb{N}$ such that

$$f(\varepsilon) = \sum_{j=0}^{K_J} a_j \varepsilon^{\lambda_j} + \sum_{j=0, \lambda_j \in \mathbb{Z}}^{K_J} b_j \varepsilon^{\lambda_j} \log \varepsilon + \sum_{j=0}^{J} c_j \varepsilon^j + o(\varepsilon^J).$$

Note that if $\alpha \in \mathbb{Z}$, there is a redundancy in this expression since there can be constant terms arising in the first and last sum. For a given real parameter μ, its $\mu-$ *renormalized limit* at zero is defined by

$$Lim_{\varepsilon \to 0}^{\mu} f(\varepsilon) := \text{ constant term} - \mu \cdot \text{ coefficient of } \quad \log \varepsilon$$
$$= a_{\alpha + n} + c_0 - \mu b_{\alpha + n}$$

where we set $a_{\alpha+n} = 0$ and $b_{\alpha+n} = 0$ if $\alpha + n \notin \mathbb{N}$. We shall also write for short $Lim^{\mu} f$. When $b_{\alpha+n} = 0$, we shall simply write $Lim f$ which is the finite part of the otherwise divergent limit of $f(\varepsilon)$ when $\varepsilon \to 0$.

Of course any renormalizable function which converges to l at point 0 has renormalized limit l.

Here is an example of a renormalizable function around zero which will play an important role in what follows.

Example. *For $\lambda > 0$, let us define $f_\lambda : \mathbb{R}^+/\{0\} \to \mathbb{R}$ by:*

$$f_\lambda(\varepsilon) \equiv -\int_\varepsilon^\infty \frac{e^{-t\lambda}}{t} dt.$$

Using the fact that $\lambda^{-1} = -\int_0^\infty e^{-t\lambda} dt$ and that the Euler constant reads $\gamma = \int_0^1 \frac{1-e^{-t}}{t} dt - \int_1^\infty \frac{e^{-1}}{t} dt$, an easy computation gives:

$$lim_{\varepsilon \to 0}(f_\lambda(\varepsilon) - log\varepsilon) = log\lambda + \gamma.$$

Hence f_λ is renormalizable around zero and

$$Lim^\mu_{\varepsilon \to 0} f_\lambda(\varepsilon) = log\,\lambda + \gamma - \mu.$$

In particular, for $\mu = \gamma$ we have:

$$Lim^\gamma_{\varepsilon \to 0} f_\lambda(\varepsilon) = log\,\lambda.$$

2. Renormalization procedures

Here we introduce the notion of renormalization procedure, adopting the terminology introduced in [Z].

We shall call a *renormalization procedure* (or *R.P.* for short) on a separable Hilbert space H a family of Hilbert Schmidt operators $\mathcal{K} = \{K_\varepsilon, \varepsilon > 0\}$ on H such that for all $u \in H$,

$$\lim_{\varepsilon \to 0} \|K_\varepsilon u - u\| = 0$$

Remark. *The parameter space over which ε runs is $\mathbb{R}^+$ and this will be enough for our needs in this paper. However one needs to extend the parameter space to open subsets of the complex plane in order to introduce "zeta renormalization procedures".*

For an R. P. $\mathcal{K} = \{K_\varepsilon, \varepsilon > 0\}$, we define a function $Z^\mathcal{K} : \mathbb{R}^+\{0\} \to \mathbb{R}$ by

$$Z^\mathcal{K}(\varepsilon) := \|K_\varepsilon\|^2_{HS} = tr(K_\varepsilon^* K_\varepsilon)$$

where $\| \cdot \|_{HS}$ denotes the Hilbert-Schmidt norm of K_ε. The asymptotic behavior of $Z^\mathcal{K}$ both around zero and at infinity will play an important role in what follows. Were the space H to be of finite dimension, the limit $\lim_{\varepsilon \to 0} Z^\mathcal{K}(\varepsilon)$ would coincide with the dimension of H. In the infinite dimensional case, $Z^\mathcal{K}(\varepsilon)$ does not converge in general as ε goes to zero and one of our tasks here is to circumvent this problem and try to "renormalize" the dimension of the Hilbert space H to a finite dimension. In order to use the tools at hand, we shall choose the renormalization procedure $\mathcal{K}$ in such a way that $Z^\mathcal{K}$ is renormalizable around zero. It is essential for our needs that the renormalized dimension of H one obtains does not depend on the parameter μ. This is why we shall restrict ourselves to a subclass of renormalization procedures, namely those that verify the following

Assumption (R.P.1). $Z^\mathcal{K}$ *satisfies (R) (i.e it is renormalizable) with $\alpha = 0$ and $b_j = 0 \; \forall j \in \mathbb{N}$.*

In particular, for an R.P. $\mathcal{K}$ satisfying (R.P.1), the function $\varepsilon \to Z^{\mathcal{K}}(\varepsilon)$ is renormalizable around zero and we can define the *effective dimension of H* by

$$dim_{\mathcal{K}}(H) = Lim^{\mu} Z^{\mathcal{K}} = Lim_{\varepsilon \to 0} tr(K_{\varepsilon}^* K_{\varepsilon}) = Lim_{\varepsilon \to 0} \|K_{\varepsilon}\|^2_{HS}$$

which is independent of the choice of μ since there is no logarithmic divergence.

For a densely defined operator B on H such that $K_{\varepsilon}^* B K_{\varepsilon}$ is trace-class for all $\varepsilon > 0$ we set

$$tr_{K_{\varepsilon}}(B) := tr(K_{\varepsilon}^* B K_{\varepsilon}).$$

In particular $tr_{K_{\varepsilon}}(B)$ is well defined for any bounded operator on H.

In a similar way, for a bounded symmetric bilinear form $\mathcal{B}$ on H, we set

$$tr_{K_{\varepsilon}} \mathcal{B} := \sum_n \mathcal{B}(K_{\varepsilon} e_n, K_{\varepsilon} e_n) = \sum_n \mathcal{B}(K_{\varepsilon} e_n, K_{\varepsilon} e_n)$$

which is a finite expression independent of the choice of the C.O.N.S $(e_n)_{n \in \mathbb{N}}$ of H.

If $\varepsilon \to tr_{K_{\varepsilon}}(B)$ (resp. $\varepsilon \to tr_{K_{\varepsilon}}(\mathcal{B})$) is a renormalizable function around zero, we shall say that B (resp. $\mathcal{B}$) has a *renormalizable trace* with respect to the renormalization procedure $\mathcal{K} = \{K_{\varepsilon}, \varepsilon > 0\}$ and call $tr_{\mathcal{K}}^{\mu}(B) = Lim_{\varepsilon \to 0}^{\mu} tr_{K_{\varepsilon}}(B)$ (resp. $tr_{\mathcal{K}}^{\mu}(\mathcal{B}) = Lim_{\varepsilon \to 0}^{\mu} tr_{K_{\varepsilon}}(\mathcal{B})$) the *renormalized trace* of B (resp. of $\mathcal{B}$) with respect to the renormalization procedure $\mathcal{K}$.

Remark. *If B is trace-class, the renormalized trace of B coincides with the ordinary trace of B.*

Under assumption (R.P.1), the identity operator has a renormalizable trace w.r.to the R.P. $\mathcal{K}$ and its renormalized trace is the *effective dimension* of H w.r.to this renormalization procedure.

3. Heat-kernel renormalization procedures

From now on, we shall focus on renormalization procedures that arise as heat operators $\{e^{-\varepsilon Q}, \varepsilon > 0\}$, where Q is a densely defined positive (possibly with non trivial kernel) operator on H.

More precisely, let Q be a positive self-adjoint elliptic differential operator with strictly positive order acting on sections of some hermitian vector bundle E of finite rank on a compact boundaryless Riemannian manifold M. Let us denote by $Ell^+_{ord>0}(M, E)$ the class of such operators.

With the notations of paragraph 1.2, the Hilbert space H is the L^2-closure of the space of smooth sections of E for the scalar product induced by the Riemannian structure on H and the hermitian structure on V. We shall call the couple (H, Q) a *weighted Hilbert space*, the *weight* being Q.

Examples. *The Laplace operator Δ on a compact Riemannian boundaryless manifold M lies in $Ell^+_{ord>0}(M, E)$ where $E = M \times \mathbb{R}$. Here $H = L^2(M, \mathbb{R})$, the closure of the space of smooth $\mathbb{R}$ valued functions on M for the scalar product induced by the Riemannian structure on the manifold M.*

A generalized Laplacian acting on forms on a compact boundaryless manifold yields an example of weight on $Ell^+_{ord>0} (L^2(M, \Lambda T^* M))$ [G], [BGV].

Well known results concerning elliptic differential operators on compact manifolds (see e.g [G]) tell us that for any $Q \in Ell^+_{ord>0}(M, V)$, the operator $e^{-\varepsilon Q}, \varepsilon > 0$ is a positive self-adjoint Hilbert Schmidt (in fact it is infinitely smoothing) operator on H so that it yields an R.P.

$$\mathcal{K}^Q := \{K^Q_\varepsilon := e^{-\frac{\varepsilon}{2}Q}, \varepsilon > 0\}.$$

Moreover, the asymptotic behaviour of traces of heat-operators associated to an operator Q in $Ell^+_{ord>0}(M, V)$ tells us that the renormalization $\mathcal{K}^Q$ fulfills assumption $(R.P.1)$ [G]. The coefficient n in the expansion (R) corresponds to the dimension of the manifold M and m to the order of the elliptic operator Q, which is two for generalized Laplacians, the case one mostly comes across in the context of gauge theory.

Let us now apply the construction of section 2 to the renormalization procedure $\mathcal{K}^Q$. Let $CL(M, E)$ denote the algebra of classical pseudo-differential operators (P.D.Os) acting on smooth sections of E. For any $P \in CL(M, E)$, since K^Q_ε is a smoothing operator, the map $K^Q_\varepsilon P K^Q_\varepsilon$ is trace class and for any $\varepsilon > 0$

$$tr_{K^Q_\varepsilon}(P) = tr(K^Q_\varepsilon P K^Q_\varepsilon) = tr(Pe^{-\varepsilon Q})$$

is well defined. Classical results on asymptotic expansions of this type (see e.g. [L] and references therein) furthermore tell us that the map $\varepsilon \to tr_{K^Q_\varepsilon}(P)$ is a renormalizable function and hence that any operator $P \in CL(M, E)$ has a renormalizable trace w.r.to the R.P. $\mathcal{K}^Q$. We define

$$tr^{Q,\mu}(P) := Lim^\mu_{\varepsilon \to 0} tr(Pe^{-\varepsilon Q}).$$

Here we use the word "trace" in an improper way since weighted traces are not tracial [CDMP], [P]. However this terminology is convenient for our purpose since weighted traces play here a role similar to that of ordinary traces in finite dimensions.

It is useful to recall at this point that the coefficient in front of the parameter μ keeps track of the coefficient in the asymptotic expansion in front of the essential singularity $log\varepsilon$. In fact the expression:

$$res(P) := ordQ \cdot \frac{\partial}{\partial \mu} tr_{Q,\mu}(P)$$

defines a trace on the algebra of classical P.D.Os called the *Wodzicki residue* [W] (see also [K] for a review on the subject).

Since for $Q \in Ell^+_{ord>0}(M, E)$, $\mathcal{K}^Q$ satisfies assumption R.P.1, we can define the *effective dimension* of $H = L^2(M, V)$ (where E is the bundle on which sections the operator Q is acting) relative to the weight Q by:

$$dim^Q H := tr^{Q,\mu}(Id) = Lim_{\varepsilon \to 0} tr_{K^Q_\varepsilon}(Id)$$

$$= Lim_{\varepsilon \to 0} Z^{\mathcal{K}^Q}(\varepsilon) = Lim_{\varepsilon \to 0} tr(e^{-\varepsilon Q}).$$

Notice that it is independent of μ. Let us give another interpretation of this effective dimension of H.

As an operator in $Ell^{+}_{ord>0}(M, E)$ (the underlying manifold M being closed as before), Q has purely discrete spectrum. Let us denote by $\{\lambda_n, n \in \mathbb{N}\}, \lambda_n \in \mathbb{R}^+$ its spectrum. Again, well-known results concerning elliptic operators on compact manifolds tell us that the asymptotic behaviour of λ_n is given by [G]:

$$\lambda_n \simeq Cn^{\alpha} \quad \text{when n tends to} \quad \infty$$

for some $\alpha > 0$ and $C > 0$. For $s \in \mathbb{C}$ with real part large enough, we can therefore define :

$$\zeta_Q(s) \equiv \sum_{\lambda_n \neq 0} \lambda_n^{-s}.$$

This function can be extended to a meromorphic function on $\mathbb{C}$ which is holomorphic at zero so that $\zeta_Q(0)$ is well defined (see e.g. [G] and references therein):

Using the Mellin transform:

$$\lambda^{-s} = \int_0^{\infty} t^{s-1} e^{-t\lambda} dt,$$

one can rewrite the effective dimension of $H = L^2(M, E)$ showing that for any $\mu \in \mathbb{R}$,

$$dim^Q(L^2(M, E)) = \zeta_Q(0) + dim KerQ$$

where E is as before the hermitian vector bundle on which sections the operator Q is acting and $KerQ$ is the kernel of Q. This procedure using the Mellin transform is described in great details in [BGV] par. 9.

4. Renormalized determinants

Given an operator $Q \in Ell^{+}_{ord>0}(M, E)$, not only do we have information on the asymptotic behaviour of the function K^Q at zero but also at infinity.

Indeed, the function $t \to tr(K^Q(t)) = tr(e^{-tQ})$ satisfies the following [G]:

Assumption (R.P. 2).

$$\exists \lambda > 0, C > 0, K > 0, \text{ such that for } t > K, \quad tr(e^{-tQ}) < Ce^{-t\lambda}$$

which is an assumption on the asymptotic behavior at infinity of Z^{K^Q}.

With this information at hand we can define the notion of renormalized determinant. Let $Q \in Ell^{+}_{ord>0}(M, E)$, and define $f_Q : \mathbb{R}^+/\{0\} \to \mathbb{R}$ by:

$$f_Q(\varepsilon) := -\int_{\varepsilon}^{\infty} \frac{tr(e^{-tQ})}{t} dt = -\int_{\varepsilon}^{\infty} \frac{Z^{K^Q}(t)}{t} dt.$$

This function f_Q is well defined under assumptions (R.P.1) and (R.P.2).

Integrating the asymptotic expansion of $Z^{K^Q}(\varepsilon) = tr(e^{-\varepsilon Q})$, one shows that f_Q is renormalizable and

$$Lim^{\mu}_{\varepsilon \to 0} f_Q(\varepsilon) = \text{const. term in the asymptotic expansion of} \quad f_Q$$

$$+ (\gamma - \mu) \cdot \text{const. term in the asymptotic expansion of} \quad Z^{K^Q}.$$

A motivation for this choice of function f_Q lies in the fact that when brought down to the finite dimensional case- for a special value of μ, namely when μ is the Euler constant γ- it gives the logarithm of the determinant of Q.

Let H be finite dimensional of dimension d, and let $\{\lambda_1, \cdots, \lambda_d\}$ be the eigenvalues of Q, we have $f_Q(\varepsilon) = \sum_{k=1}^{d} f_{\lambda_k}(\varepsilon)$, and hence, repeatedly using the results of the example concerning f_λ in section 1 applied to each λ_k, we find

$$Lim^{\mu} f_Q = \sum_{k=1}^{d} (log\lambda_k + \gamma - \mu) = log(detQ) + d(\gamma - \mu),$$

so that if $\mu = \gamma$,

$$exp(Lim^{\mu} f_Q) = detQ.$$

At this stage, one could be tempted to set $\mu = \gamma$ because of this analogy, but as was pointed out above, the parameter μ keeps track of the infinite dimensional essence of the problem.

On the grounds of this finite dimensional analogy, we interpret the exponential of the renormalized limit of f_Q as the μ- *renormalized determinant of Q* and set:

$$det^{\mu}(Q) \equiv exp(Lim^{\mu} f_Q).$$

From the above considerations, it follows that for $\mu = \gamma$ where γ is the Euler constant, we have $det^{\mu}(Q) = det(Q)$ when Q is a symmetric invertible operator acting on a finite dimensional space.

The various μ-renormalized determinants are proportional to each other for different values of μ and we have

$$Lim^{\mu}_{\varepsilon \to 0} f_Q(\varepsilon) = Lim^{\mu'}_{\varepsilon \to 0} f_Q(\varepsilon) + (\mu' - \mu)\zeta_Q(0)$$

and hence

$$det^{\mu}(Q) = e^{(\mu' - \mu)\zeta_Q(0)} \cdot det^{\mu'}(Q).$$

On the other hand, an easy computation shows that for $\alpha > 0$, we have $f_{\alpha\lambda}(\varepsilon) = f_\lambda(\alpha\varepsilon)$ and hence $f^{\alpha Q}(\varepsilon) = f^Q(\alpha\varepsilon)$. From this it follows that:

$$det^{\mu}(\alpha Q) = \alpha^{dim^Q(L^2(M,E))} det^{\mu}(Q)$$

where $dim^Q L^2(M, E)$ is the effective dimension relative to teh weight Q of the L^2-closure of the space of sections of the vector bundle E.

There is another approach to these renormalized determinants, the *zeta function* approach (described e.g in [BGV], and [R] in the case of Laplacians). For an invertible operator $Q \in Ell^+_{ord>0}(M, E)$ we saw we could define a function ζ_Q which generalizes the ordinary zeta function $\zeta(s) = \sum_n n^{-s}$ (hence the name of the corresponding renormalization). Notice that if Q is a diagonalizable invertible operator acting on a finite dimensional space with a finite number of eigenvalues $(\lambda_n, n = 1, \cdots, d)$, then $\zeta'_Q(0) = -\sum_{n=1}^{d} \log\lambda_n = -\log\det(Q)$. The function ζ_Q is also holomorphic at point 0 and we set:

$$\det_\zeta(Q) \equiv e^{-\zeta'_Q(0)}$$

which we shall refer to as the ζ-*renormalized determinant* of Q.

A natural question is how this determinant relates to the μ-renormalized determinants introduced above. This can be seen with the help of the Mellin transform:

$$\lambda^{-s} = \Gamma(s)^{-1} \int_0^\infty t^{s-1} e^{-t\lambda} dt$$

and one finds that the ζ-renormalized determinant is exactly the γ-renormalized determinant obtained when setting $\mu = \gamma$. Combining the fact that $a_n = \zeta_Q(0)$ (since Q is invertible) with the above properties, we can write:

$$\det^\mu(Q) = e^{a_n(\gamma-\mu)} \det_\zeta(Q)$$
$$= \det_\zeta(e^{(\gamma-\mu)}Q)$$
$$= e^{\zeta_Q(0)(\gamma-\mu)} \det_\zeta(Q).$$

II. The first Chern form on a class of hermitian vector bundles

1. Renormalization procedures on vector bundles

In this section we apply the techniques of section I to give a meaning to the first Chern form on a hermitian vector bundle and to investigate some of its properties. We first extend the notion of renormalization procedure introduced in paragraph 2 of section I on Hilbert spaces to families of Hilbert spaces.

Let $\mathcal{E} \to X$ be a smooth hermitian vector bundle based on some smooth manifold X with fibres modeled on some separable Hilbert space H. Let $HS(\mathcal{E})$ denote the bundle of Hilbert-Schmidt morphisms of $\mathcal{E}$ so that a local section T of $HS(\mathcal{E})$ is pointwise a Hilbert-Schmidt operator for the hermitian product induced on the fibres by the hermitian structure on $\mathcal{E}$.

We shall call a *renormalization procedure* on $\mathcal{E}$ (or R.P. for short) a family of smooth sections $\mathcal{K} := \{K_\varepsilon, \varepsilon > 0\}$ of the bundle $HS(\mathcal{E})$ such that for any $x \in X$, for any section u of $\mathcal{E}$,

$$\lim_{\varepsilon \to 0} \| K_\varepsilon(x)u(x) - u(x) \|_x = 0$$

where $\| \cdot \|_x$ is the norm on the fibre above x induced by the hermitian product on the fibre.

In order to generalize heat-kernel renormalization procedures which are defined from an elliptic operator to the vector bundle setting, we need to understand what families of elliptic operators on such a bundle can be. For this let us restrict ourselves to a class $\mathcal{CE}$ of Riemannian or hermitian vector bundles, namely those that are modelled on some Hilbert space of sections of some finite rank hermitian vector bundle E based on a compact Riemannian manifold M with transition maps given by operators in $CL(M, E)$. In fact the latter being bounded and their inverses too, they lie in the group of invertible elliptic operators of zero order. We shall also consider the class $C\mathcal{X}$ of Riemannian or hermitian Hilbert manifolds with tangent bundles in $C\mathcal{E}$.

Examples.

1) *Finite rank vector bundles lie in $C\mathcal{E}$; let us take the manifold M reduced to a point $\{*\}$ and let us set $E := \{*\} \times \mathbb{R}^d$ where d is the dimension of the manifold. Then the space of sections of E reduces to $\mathbb{R}^d$ which is indeed the model space for M. The algebra $Ell(M, E)$ boils down to $Gl_d(\mathbb{R})$ so that the transition maps indeed lie in $Ell(\{*\}, E)$.*

2) *Hilbert manifolds $H^s(M, N), s > \frac{dim M}{2}$ of Sobolev H^s maps from a compact manifold M to a Riemannian manifold N lie in the class $C\mathcal{X}$. Indeed there are local exponential charts defined pointwise via the exponential charts on N (see [M]). They give rise to transition maps locally described in terms of multiplication operators which patch up to a P.D.Os of order 0.*

For a bundle in $C\mathcal{E}$ we can define the bundle $CL(\mathcal{E})$ of classical P.D.Os on $\mathcal{E}$ by requiring that in a local chart the operator be a classical P.D.O. The transition maps being P.D.Os, the definition of pseudo-differential operator on $\mathcal{E}$ is independent of the chosen local chart. Since ellipticity boils down to invertibility of the principal symbol which is multiplicative on the algebra of P.D.Os, the notion of ellipticity (in some local trivialization) is also independent of the chosen local trivialization. The notion of order of an operator also makes sense locally since it is also invariant under a change of chart. Provided we restrict ourselves to orthonormal transformations using the Riemannian structure on the bundle, positivity of an operator also makes sense locally.

Let us call *weighted vector bundle* a couple $(\mathcal{E}, Q)$ where $\mathcal{E} \in C\mathcal{E}$ and Q is a section of $CL(\mathcal{E})$ which is locally an elliptic operator of (constant) strictly positive order inducing fibrewise a strictly positive operator. Let us call *weighted manifold* a couple (X, Q) where $X \in C\mathcal{X}$ and (TX, Q) is a weighted vector bundle. (The notion of weight can be weakened replacing positivity by admissibility, in which case one does not need to restrict oneself to Riemannian or hermitian bundles and Riemannian or hermitian manifolds [P]).

Example. *Going back to the previous example $H^s(M, N)$ let us take $N = G$ a compact Lie group and M Riemannian. For $s > \frac{dim M}{2}$ this is a Hilbert Lie group with Lie algebra $H^s(M, Lie(G))$ where $Lie(G)$ is the Lie algebra of G. Let $Q_0 := \Delta \otimes 1_{LieG}$ where 1_{LieG} is the identity on the Lie algebra of G and Δ is the ordinary Laplacian on the functions on M. For $\gamma \in H^s(M, G)$ with s large enough we set*

$$Q_\gamma := L_\gamma Q_0 L_\gamma^{-1}.$$

Then the couple $(H^s(M,G),Q)$ *yields a weighted manifold. These are investigated in [CDMP] and [Ma].*

Given a weighted vector bundle $(\mathcal{E}, Q)$ based on a manifold X and following the construction of weighted traces on weighted Hilbert spaces, for any section P of $PDO(\mathcal{E})$ we can define the Q-weighted trace of P by:

$$tr^{Q,\mu}(P)(x) := tr^{Q_x,\mu}(P_x), \quad \forall x \in X$$

where Q_x, P_x are the operators induced on the fibre above x. This expression is to be understood in a local chart where it becomes a weighted trace on the model space H. One easily checks it is independent of the choice of local chart [P].

2. Weighted first Chern forms on infinite dimensional vector bundles

Let us first recall some geometric facts about the finite dimensional setting. Let $\mathcal{E}$ be a finite rank hermitian vector bundle based on X equipped with a hermitian connection $\nabla^{\mathcal{E}}$. The curvature tensor $\Omega^{\mathcal{E}}$ is a section of the bundle $\Gamma(T^*X \otimes T^*X \otimes Hom(\mathcal{E}))$ and the *first Chern form* is the tensor given by:

$$r_1^{\mathcal{E}}(u,v) := tr(\Omega^{\mathcal{E}}(u,v)) = tr((\nabla^{\mathcal{E}})^2(u,v)) \quad \forall u, v \in TX$$

where the trace is taken along the fibres of $\mathcal{E}$. It is a closed two form, for writing $\nabla = d + \theta$ locally then:

$$dr_1^{\mathcal{E}} = tr(d\Omega^{\mathcal{E}})$$
$$= tr(d\Omega^{\mathcal{E}}) + tr([\theta, \Omega^{\mathcal{E}}])$$

where we use the tracial property

$$= tr([\nabla, \Omega^{\mathcal{E}}])$$
$$= 0$$

where we use Bianchi identity.

It defines a cohomology class, the first Chern class. When X is a Kähler manifold and $\mathcal{E} = TX$, then the first Chern form on $\mathcal{E}$ coincides with the Ricci curvature on X:

$$r_1^{TX} = Ricc^X.$$

From the connection $\nabla^{\mathcal{E}}$ one can also build a connection on the associated determinant bundle. The determinant bundle (see e.g. [BGV] (chap. 9.7)) associated to a graded vector bundle $\mathcal{E} := \mathcal{E}^+ \oplus \mathcal{E}^-$ with $rank(\mathcal{E}^+) = rank(\mathcal{E}^-)$ is defined in terms of the maximal exterior power of $\mathcal{E}^+$ and $\mathcal{E}^-$ by $Det\,\mathcal{E} := (\Lambda^{max}\mathcal{E}^+)^{-1} \otimes (\Lambda^{max}\mathcal{E}^-)$ where the subscript -1 denotes the inverse bundle. If D^+ is a section of $Hom(\mathcal{E}^+, \mathcal{E}^-)$ then $DetD^+ := \Lambda^{max}D^+$ is a section of $Det\,\mathcal{E}$. It is sometimes convenient to work with the self-adjoint operator:

$$D := \begin{bmatrix} 0 & D^+ \\ D^- & 0 \end{bmatrix}$$

where D^- is the adjoint of D^+.

The bundle Det $\mathcal{E}$ can be equipped with a metric induced by the one on $\mathcal{E}$ defined by $|DetD^+| = \sqrt{det(D^- D^+)}$. A connection $\nabla^{\mathcal{E}}$ on $\mathcal{E}$ induces a connection on $Hom(\mathcal{E})$ defined on the section D of $Hom(\mathcal{E})$ by $[\nabla^{\mathcal{E}}, D]$. This in turn induces a connection on Det $\mathcal{E}$ defined by:

$$\nabla^{\text{Det } \mathcal{E}} DetD^+ = DetD^+ tr(D^{+-1}[\nabla^{\mathcal{E}}, D^+])$$

at any point for which the section D^+ of $Hom(\mathcal{E}^+, \mathcal{E}^-)$ is invertible. This connection is compatible with the metric since we have:

$$dlog|DetD^+| = \frac{1}{2}dlogdet(D^- D^+) = Re(tr(D^{+-1}[\nabla^{\mathcal{E}}, D^+])).$$

The curvature $\Omega^{\text{Det}\mathcal{E}}$ of this connection coincides up to a sign with the first Chern form on the vector bundle $\mathcal{E}$ (see e.g. [BGV] par. 10.6. up to sign conventions):

$$\Omega^{\text{Det } \mathcal{E}} = -r_1^{\mathcal{E}} = -tr(\nabla^{\mathcal{E}})^2.$$

The proof uses the fact that $[\nabla^{\mathcal{E}}, tr] = 0$ in an essential way.

Let us now extend the concept of first Chern form to the framework of weighted vector bundles.

The *Q-weighted first Chern form* on a weighted (hermitian) bundle $(\mathcal{E}, Q)$ with Levi-Civita connection $\nabla^{\mathcal{E}}$ and corresponding curvature $\Omega^{\mathcal{E}}$ is the 2-form given by:

$$r_1^{\mathcal{E}, Q, \mu} := tr^{Q, \mu}(\Omega^{\mathcal{E}}), \quad \mu \in \mathbb{R}$$

which is well-defined whenever $\Omega^{\mathcal{E}}(u, v)$ is a section of $CL(\mathcal{E})$ for any vector fields u, v on the base manifold X.

When the rank of the bundle $\mathcal{E}$ is finite, it coincides with the usual first Chern form because weighted traces coincide with usual traces on trace-class operators and hence on finite rank operators.

As in the finite dimensional case, when $(\mathcal{E}, Q)$ is the tangent bundle to a Kähler weighted manifold (X, Q), the Q-weighted Ricci curvature on X

$$Ricc^{X, Q, \mu}(u, v) := tr^{Q, \mu}\left(\Omega^X(u, \cdot)v\right) \quad \forall u, v \in TX$$

(with $\Omega^X := \Omega^{TX}$) coincides with the Q-weighted first Chern form on $(\mathcal{E}, Q)$. This expression makes sense provided $\Omega^X(u, \cdot)v$ defines a section of $CL(\mathcal{E})$.

Example. *In the geometric framework described by Freed for current groups [F], one can show [CDMP] that the weighted first Chern form on the current group $H^{\frac{1}{2}}(S^1, G)$ equipped with the above weight Q coincides with Freed's "two step trace" first Chern form obtained from first taking the trace on the finite dimensional Lie algebra and then the trace of the possibly trace-class operator thus obtained. For this choice of weight on the current group, the weighted first Chern form turns out to be closed.*

Unlike the first Chern form on a vector bundle of finite rank, because of the Q dependence of the weighted trace tr^Q which does not in general commute with the connection ($[\nabla^\mathcal{E}, tr^Q] \neq 0$), one does not expect the weighted first Chern form to be closed. Its exterior differential- which can be seen as an obstruction of purely infinite dimensional nature- can be expressed in terms of a Wodzicki residue [CDMP]:

$$dr_1^{\mathcal{E},Q,\mu} = -\frac{1}{ordQ}res\left([\nabla^\mathcal{E}, logQ]\Omega^\mathcal{E}\right)$$

where res is the Wodzicki residue defined on classical P.D.Os as in paragraph 1.3. Although $logQ$ is not a classical P.D.O, provided $\nabla^\mathcal{E}$ written $d + \theta$ in a local chart is such that θ can be interpreted as a classical P.D.O valued one form, $[\nabla^\mathcal{E}, logQ]$ is indeed a classical P.D.O and one can make sense of the obove residue.

III. The geometry of gauge orbits

Infinite dimensional determinants often arise in gauge theories as jacobian determinants. Typically, an infinite dimensional Lie group $\mathcal{G}$ (the gauge group) acts on an infinite dimensional space $\mathcal{P}$ (the path space) so that integrating $\mathcal{G}$-invariant functionals on the path space- when seen as integrals on the quotient space $\mathcal{P}/\mathcal{G}$- can give rise to jacobian determinants. These jacobian determinants, or rather their variation, have a geometric meaning in terms of the geometry of gauge orbits. In order to make this geometric interpretation clearer, let us first describe the finite dimensional geometric framework. This section is essentially based on joint work with M. Arnaudon [AP].

1. The finite dimensional setting

The group G is a finite dimensional compact connected Lie group acting smoothly on a Riemannian manifold P:

$$\Theta : G \times P \to P$$

$$(g, p) \to p \cdot g$$

The action is assumed to be isometric and free (the action is free means that $(p \cdot g = p \ \ \forall p \in P) \Leftrightarrow g = 1_G$). In particular this action induces a manifold structure on the quotient space $X \equiv P/G$ and the canonical projection $\pi : P \to P/G$ is a smooth principal bundle. Moreover, for any $x \in X$, the orbit O_x is a smooth submanifold of P (these are the only properties we need to remember for our purpose).

Let us consider one specific orbit. In general, a geodesic of P starting at a point $p \in P$ in a direction tangent to the orbit O_p at point p, will leave the orbit O_p from which it started.

> The orbit is called *totally geodesic* if any geodesic starting in the orbit tangentially to it stays in this orbit.

There is another way of formulating this condition.

In general, if U, V are tangent vector fields to the orbit $\nabla_U V$ is not tangent to this orbit. Let us make this precise; let O denote a generic orbit and let us consider the projection $\nabla^O = P^O\nabla$ of the connection ∇ to the tangent bundle TO to the orbit , P^O denoting here the orthogonal projection on TO. If U, V are tangent vector fields to O, the normal part $II^O(U, V) := (\nabla^O_U V)^\perp$ which in general does not vanish, is called *the second fundamental form* of the orbit O. For fixed p we write $II_p = II^{O_p}$. For any $p \in P$, II_p is a symmetric bilinear form [C]. For a horizontal tangent vector X at point p_0, one can define the shape operator $\mathcal{H}_X : T_p O_p \to T_p O_p$ by

$$\langle \mathcal{H}_X U, V \rangle = \langle II_p(U, V), X \rangle$$

which is a symmetric operator.

If the second fundamental form vanishes the connection ∇ induces a connection on the tangent bundle to the orbit, so that geodesics of O are indeed geodesics of P. In fact a totally geodesic orbit is one for which the second fundamental form vanishes.

Without being totally geodesic, an orbit can have an intermediate property, that of minimality.

> An orbit O is *minimal* whenever the trace of its second funda-
> mental form vanishes i.e $tr(II^O) := \sum_n II^O(U_n, U_n) = 0$ where
> $(U_n)_{n\in\mathbb{N}}$ is any orthonormal basis of TO.

"Minimality of the orbits" is the property we will focus on here. We shall relate minimality to extremality of some determinants.

For this, let us consider an operator which will play an important part in what follows. The action Θ induces for every $p \in P$ a map:

$$\theta_p : G \to O_p$$

$$g \to p \cdot g.$$

The tangent operator $T_g\theta_p$ obeys the following property:

$$T_g\theta_p U(g) = \frac{d}{dt}_{/t=0}\left(p \cdot e^{tU(g)}\right) = \frac{d}{dt}_{/t=0}\left((p \cdot g)e^{tU(e)}\right) = T_e\theta_{p\cdot g}U(e)$$

where $U(g) = gU(e)$ is a left invariant vector field on G. It induces an operator on the Lie algebra $\mathcal{G}$ when the latter is identified to the set of left invariant vector fields:

$$\tau_p : \mathcal{G} \to TO_{\pi(p)}$$

$$U \to \frac{d}{dt}_{/t=0}(p \cdot e^{tU(e)})$$

so that with these notations we have $T_g\theta_p U(g) = \tau_{p\cdot g}U$.

Let us now relate the volume of an orbit with a jacobian determinant. A change of variable using θ_p (combined with the technical assumption $det(\tau_p^*\tau_p)=det(\tau_{p\cdot g}^*\tau_{p\cdot g})$) yields the following formula for the volume $Vol(O_{\pi(p)})$ of the orbit of a point p:

$$Vol(O_{\pi(p)}) = \int_G \sqrt{det[(T_g\theta_p)^*(T_g\theta_p)]}dvol(g) = \sqrt{det[\tau_p^*\tau_p]}Vol(G)$$

where $Vol(G)$ is the volume of the group G equipped with the Haar measure $dvol$, and where "det" denotes the finite dimensional determinant.

Let us differentiate along a horizontal tangent vector X at point p_0 the expression:

$$\sqrt{det(\tau_p^* \tau_p)} = \left(\sum_{\sigma \in \Sigma_k} (-1)^{\varepsilon(\sigma)} \prod_i \langle \tau_p U_i(e), \tau_p U_{\sigma(i)}(e) \rangle_p \right)^{\frac{1}{2}}$$

where Σ_k is the permutation group of k elments, $\varepsilon(\sigma)$ the signature of the permutation σ, $(U_i, i \in \mathbb{N})$ a complete orthonormal system (C.O.N.S) of the Lie algebra $\mathcal{G}$ of the group G. We find that

$$\frac{1}{2} \nabla_X logdet^\mu (\tau_p^* \tau_p) = -\langle X, \sum_i \frac{II_{p_0}(\tau_{p_0} U_i(e), \tau_{p_0} U_i(e))}{\langle \tau_p U_i(e), \tau_p U_i(e) \rangle} \rangle$$

which yields a relationship between the infinitesimal variation of this jacobian determinant (and hence of the volume of the orbit) and the trace of the second fundamental form. For any vector field X we have:

$$\frac{1}{2} \nabla_X log(det(\tau_p^* \tau_p)) = \nabla_X Vol(O_p) = - < trII_p, X >= -tr\mathcal{H}_X$$

This sets up links between the various notions we introduced above (see [H]):

the orbit of p is minimal $\leftrightarrow$ the jacobian determinant $det(\tau_p^* \tau_p)$ is extremal

$\leftrightarrow$ the volume of the orbit of p is extremal.

2. The infinite dimensional setting

Our aim here is to generalize to the infinite dimensional setting the notion of minimality and volume of the orbit as well as the correspondance between them.

The group $\mathcal{G}$ is now an infinite dimensional connected Lie group, *the gauge group* (it is actually not a Lie group in practice but is close enough to a Lie group, see [T] for a detailed discussion on this point) acting smoothly on an infinite dimensional Riemannian manifold $\mathcal{P}$ (the path space):

$$\Theta : \mathcal{G} \times \mathcal{P} \to \mathcal{P}, \qquad (\gamma, p) \to p \cdot \gamma.$$

We assume the action induces a manifold structure on the quotient space $X :=$ $\mathcal{P}/\mathcal{G}$ (the moduli space) and that the canonical projection $\pi : \mathcal{P} \to \mathcal{P}/\mathcal{G}$ is a smooth principal bundle. Moreover, for any $p \in \mathcal{P}$, the orbit O_p is a smooth submanifold of $\mathcal{P}$. Let us denote by $VT\mathcal{P}$ the vertical tangent bundle. We keep the same notations as before.

Difficulties arise when trying to generalize to this infinite dimensional setting the notions of

- "volume of an orbit" which involves taking a determinant
- "minimality" which involves the trace over an infinite dimensional space of the second fundamental form, of an operator acting on an infinite dimensional space.

Let us go back to the first part of these notes in order to *renormalize* these quantities.

Let us make more precise the geometric setting so as to fit it into the framework of renormalization procedures of elliptic type and renormalized determinants.

Let M be a compact boundaryless Riemannian manifold. Let E and F be two vector bundles on M and let $\mathcal{P}$, resp. $\mathcal{G}$ be Fréchet manifolds modelled on the space of smooth sections of E, resp. on the space of smooth sections of F.

For any point p, the operator $\tau_p^* \tau_p$ which is an operator densely defined on $\mathcal{G}$ acts on smooth sections of E. We shall assume that it lies in $Ell^+_{ord>0}(M, E)$, the underlying manifold being M and the vector bundle E. The field $Q(p) := \tau_p^* \tau_p$ defines a weight on the trivial bundle $\mathcal{P} \times \mathcal{G}$.

Remark. *Here the weight Q is given by the geometry which will also be the case for the geometry of determinant bundles. This is why we will not always make the Q-dependence explicit in the notations.*

We can now consider the renormalized determinant and set for $\mu \in \mathbb{R}$:

$$Vol^\mu(O_p) = \sqrt{det^\mu(\tau_p^* \tau_p)}$$

which we shall call the μ-*renormalized volume* of the orbit.

For $p \in \mathcal{P}$, let $\tilde{Q}(p) := \tau_p \tau_p^*$ which acts on $VT_p\mathcal{P}$. Then

$$\mathcal{K}^{\tilde{Q}_p} \equiv \{e^{-\frac{\varepsilon}{2}\tilde{Q}(p)}, \varepsilon > 0\}$$

is an R.P. satisfying condition $(R.P.1)$.

Indeed, since the non zero eigenvalues of $\tau_p^* \tau_p$ coincide with that of $\tau_p \tau_p^*$ and since both $Ker(\tau_p)$ and $Ker(\tau_p^*)$ are finite dimensional, $\varepsilon \to tr(e^{-\varepsilon \tau_p^* \tau_p})$ has the same asymptotic behaviour around zero as $\varepsilon \to tr(e^{-\varepsilon \tau_p^* \tau_p})$. The operator $\tau_p^* \tau_p$ being elliptic, assumption $(R.P.1)$ is fullfilled for $\varepsilon \to tr(e^{-\varepsilon \tau_p^* \tau_p})$ and hence for $\varepsilon \to tr(e^{-\varepsilon \tau_p \tau_p^*})$.

Given a Riemannian metric on $\mathcal{P}$, and given a connection on the bundle $\mathcal{P} \to \mathcal{P}/\mathcal{G}$ induced by the Riemannian metric (we skip the details here about the different topologies involved), we can define the second fundamental form II_p [C], [J] for the embedding of the group $\mathcal{G}$ into the Riemannian manifold $\mathcal{P}$ via the map $g \to p \cdot g$ induced by the action Θ. Provided II_p can be interepreted via Riesz theorem as a pseudo differential operator on $C^\infty(M, F)$, for any $\mu \in \mathbb{R}$ we can define the μ-renormalized trace of S

$$tr^\mu II_p := Lim^\mu_{\varepsilon \to 0} tr^{\tilde{Q}_{p,\mu}} II_p.$$

Let us call an orbit O_p μ- *minimal* whenever $tr^\mu II_p = 0$.

Remark. *Notice that for two values μ and μ', we have with the notations of (R) in section 1.1 applied to $tr(II_p e^{-\varepsilon \tilde{Q}_p})$:*

$$\langle X, tr^\mu II_p \rangle = \langle X, tr^{\mu'} II_p \rangle + (\mu - \mu')da_n(X)$$

where $a_n = \zeta_{Q_p}(0) + dimKerQ_p$ is computable as the integral of a local expression in the symbol of the operator $\tau_p^* \tau_p$. An orbit might therefore be μ-minimal but not μ'-minimal except in some special cases when da_n vanishes. In this sense, the non canonical feature of the renormalization procedure which gives different notions of minimality according to the choice of parameter μ, is tempered by the fact that they differ via a local expression $da_n = d\zeta_{Q_p}(0)$.

The correspondance set up in the finite dimensional case between the infinitesimal variation of the jacobian determinant and the trace of the second fundamental form goes through to these renormalized quantities. Indeed, provided the group $\mathcal{G}$ is equipped with a fixed metric independent of p, one can show that for any horizontal vector field X we have [AP] Proposition 3.3:

$$\frac{1}{2}\nabla_X log(det^\mu(\tau_p^* \tau_p)) = \nabla_X logVol^\mu(O_p) = -tr^\mu_{\mathcal{K}_{A_p}} \mathcal{H}_X.$$

Remark. *The fact that the group $\mathcal{G}$ be equipped with a fixed metric goes without saying in the finite dimensional case. But as we shall see in section V, some gauge theories give rise to gauge groups equipped with a varying metric.*

This formula yields a generalization of Hsiang's theorem mentioned above to infinite dimensions.

The relation between minimality and extremality of the volume of the orbits therefore generalizes to this infinite dimensional setting and we have:

      the orbit of p is μ-minimal

$\leftrightarrow$ the μ-renormalized jacobian determinant $det^\mu(\tau_p^* \tau_p)$ is extremal

$\leftrightarrow$ the μ-volume of the orbit $Vol^\mu(O_{\pi(p)})$ is extremal.

This correspondence was first proved using zeta-function regularization methods (which corresponds to the in the case $\mu = \gamma$) in [KT] in the context of loop groups acting on loop algebras and later generalized in [MRT] to a specific class of gauge theories, namely Yang-Mills theories.

Remarks. *It is very difficult in practice to single out minimal orbits among other orbits and only in the two examples stated above (loop groups and Yang-Mills actions) has it been done systematically.*

Strictly speaking, the above relationships between the vanishing of the logarithmic variation of the volumes and minimality of the orbits were proved in [AP] when the group $\mathcal{G}$ is equipped with a fixed metric. However, as we show in the example of diffeomorphisms acting on complex structures of a Riemann surface, this can be extended to some cases when $\mathcal{G}$ is equipped with a family of metrics indexed by $p \in P$.

IV. The geometry of determinant bundles

Determinant bundles associated to a family of elliptic operators and their geometric properties play an important part in gauge field theories since some informations on gauge theories such as anomalies can be read off the geometry of corresponding determinant bundles. We will only give an idea here of the Quillen [Q] construction of determinant bundles, which is rather technical when described in detail (see e.g [BF], [BGV]), and we will focus on how to equip a determinant bundle with a metric and a connection, both of which will require some "renormalization" procedure.

1. Determinant bundles

We want to extend the construction of determinant bundles from a finite rank vector bundle to the infinite dimensional case; $\mathcal{E}^+$ and $\mathcal{E}^-$ are now vector bundles with fibres of infinite dimension, $D_x^+ : \mathcal{E}_x^+ \to \mathcal{E}_x^-, x \in X$ a smooth family of operators acting on infinite dimensional spaces. The first problem one comes across is how to define the maximal exterior power of an infinite dimensional fibre $\mathcal{E}_x^+$ or $\mathcal{E}_x^-$.

Before we deal with this problem, let us go back to the finite dimensional setting for a while. The exact sequence of vector bundles:

$$0 \to Ker D^+ \to \mathcal{E}^+ \to \mathcal{E}^- \to Coker D^+ \to 0$$

gives an isomorphism of vector bundles

$$(\text{Det } \mathcal{E}^+)^* \otimes \text{Det } \mathcal{E}^- \simeq (Det(Ker D))^* \otimes (Det(Coker D)).$$

In the infinite dimensional setting, the operators $D_x^+, x \in X$ that come into play are *Fredholm* so their kernel and cokernel are finite dimensional. Instead of trying to give a meaning to $(Det\mathcal{E}^+)^{-1} \otimes \text{Det } \mathcal{E}^-$, we shall consider:

$$Det D^+ := (Det(Ker D^+))^{-1} \otimes Det(Coker D^+),$$

which defines a line bundle provided the kernel and cokernel of D^+ have the same dimension, or in other words that the index of D^+ is zero. We can always restrict ourselves to this case, since for a smooth family of elliptic operators, the index is locally constant and we can modify the family of operators D^+ so as to make is locally of index 0.

We have now replaced the original problem of defining $(Det\mathcal{E}^+)^* \otimes Det\mathcal{E}^-$ to the more tractable problem of studying a family of vector spaces $(Det(Ker D_x^+))^{-1} \otimes Det(Coker D_x^+), x \in X$ built up from finite dimensional spaces. However, the fact that the dimension of the kernel of D^+ might vary makes it difficult to equip this family with a bundle structure. The actual construction of the determinant bundle relies on the spectral properties of the operators $D_x^+, x \in X$.

The spectrum of Fredholm operators is purely discrete; one covers the base manifold X with open subsets U_λ over which the operator D_x does not have λ as an eigenvalue and over each of these open subsets, one builds a subvector bundle $\mathcal{E}_\lambda^+$

of $\mathcal{E}^+$ (resp. $\mathcal{E}_\lambda^-$ of $\mathcal{E}^-$) of finite rank, the fibres of which are spanned by eigenvectors of $D^- D^+$ (resp. $D^+ D^-$ -both these operators are Fredholm) corresponding to eigenvalues smaller than λ. Over U_λ, we can build the line bundle $\mathcal{L}_\lambda \equiv (Det\mathcal{E}_\lambda^+)^* \otimes Det\mathcal{E}_\lambda^-$ and show it is isomorphic to $DetD \equiv (DetKerD^+)^{-1} \otimes Det(CokerD^+)$. The last step is to show these line bundles patch up in a consistent way into a line bundle on X called *the determinant bundle* $Det(\mathcal{E}, D)$ *associated to the family* $\{D_x, x \in X\}$ [BGV], [BF], [Q].

2. A metric on the determinant bundle

Over each of the open subsets U_λ, we had a line bundle built up from vector bundles of finite rank $\mathcal{E}_\lambda^+$ and $\mathcal{E}_\lambda^-$. A hermitian metric on these bundles will naturally yield a metric $|\cdot|_\lambda$ on $\mathcal{L}_\lambda \equiv (Det\mathcal{E}_\lambda^+)^{-1} \otimes Det\mathcal{E}_\lambda^-$. But these metrics have no reason to patch up in a consistent way! For an invertible section D of $Hom(\mathcal{E})$, $|DetD^+|_\lambda = \prod_{\lambda_n \leq \lambda} \lambda_n$ where $(\lambda_n)_{n\in\mathbb{N}}$ are the eigenvalues of $D^- D^+$ so that $|\cdot|_\lambda$ takes care of a finite number of eigenvalues of $D^- D^+$. What about the infinitely many remaining eigenvalues ? This is where renormalization comes in.

Given $\mu \in \mathbb{R}$, we shall set for $x \in U_\lambda$ such that D_x is invertible,

$$|DetD^+|_\lambda^\mu := |DetD^+|_\lambda \sqrt{det^\mu (D^- D^+)_{>\lambda}}$$

where $(D^- D^+)_{>\lambda}$ is the restriction of $D^- D^+$ to the space spanned by eigenvectors corresponding to eigenvalues larger than λ,

$$|DetD^+|_\lambda^\mu := |DetD^+|_\lambda e^{(\gamma-\mu)card\{\lambda_n<\lambda\}}.$$

Here det^μ denotes the $\mu-$renormalized determinant we defined in the first section. One can show that these metrics defined on $\mathcal{L}_\lambda$ patch up in a consistent way to a metric $|\cdot|_Q^\mu$ on $\mathcal{L}_D$ called the *Quillen metric* (when $\mu = \gamma$) first introduced by Quillen in the case of Cauchy-Riemann operators on Riemann surfaces [Q].

3. A connection on the determinant bundle

We are looking for a connection on the determinant bundle compatible with the Quillen norm. The curvature of such a connection on a determinant bundle associated to some gauge theory contains some important physical information since it expresses a local geometric anomaly of the theory.

In the infinite dimensional setting, we use weighted traces to define a connection first introduced in [BF] compatible with the Quillen metric. For this let us look at the vector bundle $\mathcal{E}^+$ equiped with a weight given by the family of operators $Q^+ := D^- D^+$ parametrized by X. The bundle map $P := (D^+)^{-1}[\nabla^{\mathcal{E}}, D^+]$ has renormalizable trace w.r.to the heat-kernel R.P. induced by the field Q^+ of elliptic operators. Its renormalized version which in the language of section II corresponds to its $D^- D^+$ weighted trace, defines the *Bismut-Freed connection*:

$$(DetD^+)^{-1}\nabla^{Det,\mu}(DetD^+)) := tr^{D^- D^+,\mu}((D^+)^{-1}[\nabla^{\mathcal{E}}, D^+])$$

whenever D^+ is invertible.

Let $DetD^+$ be an invertible section of $Det(\mathcal{E}, D)$, then for any $\mu \in \mathbb{R}$, we have $|DetD^+|_Q^\mu = \sqrt{det^\mu(D^-D^+)}$ and

$$Re((DetD^+)^{-1}\nabla^{\text{Det } \mathcal{E},\mu}(DetD^+)) = dlog|DetD^+|_Q^\mu$$

which expresses the compatibility of the Bismut -Freed connection on the determinant bundle with the Quillen metric. Let us denote by $\Omega^{\text{Det } \mathcal{E}}$ the curvature of $\nabla^{\text{Det } \mathcal{E}}$.

4. Curvature on the determinant bundle

To study the curvature, it is convenient to work with the super vector bundle $\mathcal{E} := \mathcal{E}^+ \oplus \mathcal{E}^-$. The connection induced by $\nabla^{\mathcal{E}}$ on the morphism bundle $Hom(\mathcal{E})$ can be viewed as a super connection on the superbundle $\mathcal{E}$ (see e.g. [BGV] def. 1.37). This super bundle can moreover be equipped with a weight $Q := Q^+ \oplus Q^-$ with $Q^+ = D^-D^+$ as before and setting $Q^- := D^+D^-$. Generalizing the prescription described in section III to super vector bundles, the renormalized first Chern form on the weighted bundle $(\mathcal{E}, Q = D^-D^+ \oplus D^+D^-)$ is defined as the Q-weighted supertrace of the curvature $\Omega^{\mathcal{E}}$ of $\nabla^{\mathcal{E}}$:

$$r_1^{\mathcal{E},Q,\mu} := str^{Q,\mu}(\Omega^{\mathcal{E}})$$
$$= tr^{Q^+,\mu}(\Omega^{\mathcal{E}^+}) - tr^{Q^-,\mu}(\Omega^{\mathcal{E}^-}).$$

Here again, when trying to extend the relation that holds in finite dimensions between the first Chern form and the curvature on determinant bundles, one does not expect it to hold in this infinite dimensional case because of the weight coming into play. Indeed the finite dimensional proof strongly relies on the fact that the connection commutes with the trace $[\nabla^{\mathcal{E}}, tr] = 0$, which is not any more the case when replacing ordinary traces by weighted traces tr^Q. There are additional Wodzicki residue terms arising [PR2]:

$$\Omega^{\text{Det } \mathcal{E},\mu} = -r_1^{\mathcal{E},Q,\mu} + \text{ Wodzicki residue terms.}$$

One way to compute the obstruction is to apply techniques used in [BF], [BGV] (see lemma 10.34) to compute the curvature on the determinant bundle. Following Bismut [B] we first modify the superconnection $\nabla^{\mathcal{E}}$ into a one parameter family $\{\nabla_\varepsilon^{\mathcal{E},D}, \varepsilon > 0\}$ of superconnections:

$$[\nabla_\varepsilon^{\mathcal{E},D}]_{[0]} := \sqrt{\varepsilon}D \quad \text{and} \quad [\nabla_\varepsilon^{\mathcal{E},D}]_{[1]} := \nabla^{\mathcal{E}}.$$

We then apply the transgression formula

$$\frac{d}{dt}str(f(A_t^2)) = d \, str\left(\frac{\partial A_t}{\partial t}f'(A_t^2)\right)$$

where $A_t, t \in I$ is a differentiable one parameter family of superconnections and f a polynomial function to $A_t := \nabla_t^{\mathcal{E},D}$. This way we express the curvature on the determinant bundle as the homogeneous part of degree two of the form $str(exp(-(\nabla_\varepsilon^{\mathcal{E},D})^2))$ which yields the result.

We find [PR2]:

$$\Omega^{\mathrm{Det}\ \varepsilon,\mu} = -r_1^{\varepsilon,Q,\mu} + Lim^\mu \left(\varepsilon \langle I, [\nabla^\varepsilon, D], [\nabla^\varepsilon, D] \rangle_{\varepsilon,2,Q} \right)$$

where the trace form $\langle I, [\nabla^\varepsilon, D], [\nabla^\varepsilon, D] \rangle_{\varepsilon,2,Q}$ is defined by the general formula:

$$\langle A_0, A_1, \cdots, A_k \rangle_{\varepsilon,k,Q}$$

$$:= \int_{\sigma_i \geq 0, \sum_{i=0}^k \sigma_i = 1} str \left(A_0 e^{-\varepsilon \sigma_0 Q} A_1 e^{-\varepsilon \sigma_1 Q} \cdots A_k e^{-\varepsilon \sigma_k Q} \right) d\sigma_0 \cdots d\sigma_k.$$

One then shows that renormalized limits of such trace forms can be described in terms of Wodzicki residues [PR2].

V. An example: the action of diffeomorphisms on complex structures

We illustrate here by an example (investigated in [PR1,2]) the links between the two pictures, the gauge orbit picture and the determinant bundle picture.

When quantizing strings (which describe Riemann surfaces moving in space time) one is naturally lead to the study of the action of the group of diffeomorphisms of a Riemann surface on the manifold of complex structures on this surface [AJPS]. In what follows, Z is a compact boundaryless Riemann surface of genus strictly larger than 1.

1. The orbit picture

The objects we are about to introduce here are described in the context of Teichmüller theory in [T]. Let us describe the space of paths.

The *"space of paths"* described by the moving string can be seen as the manifold $\mathcal{C}$ (resp. $\mathcal{C}^s$) of smooth (resp. Sobolev) complex structures on Z (resp. of class H^s).

Remark. *The actual space of paths of string theory is the manifold of all Riemannian metrics on Z whereas, as we shall see, taking the manifold of complex structures as a space of paths, we restrict ourselves to the submanifold of metrics with curvature -1.*

$\mathcal{C}$ (resp. $\mathcal{C}^s$) is a Fréchet (resp. a Hilbert) manifold which is diffeomorphic to the manifold $\mathcal{A}$ (resp. $\mathcal{A}^s$) of smooth (resp. Sobolev) almost complex structures (of class H^s) on Z and is modelled on a space $C^\infty(T_1^1 Z)$ (resp. $H^s(T_1^1(Z)$) of $(1,1)$ tensors on Z.

There is a one-to-one diffeomorphism

$$\Psi : \mathcal{A}^s \to \mathcal{M}^s_{-1}$$

$$J \to \Psi(J) \equiv g(J)$$

which induces a diffeomorphism

$$\Psi : \mathcal{A} \to \mathcal{M}_{-1}$$

$$J \to \Psi(J) \equiv g(J)$$

where $\mathcal{M}_{-1}$ (resp. $\mathcal{M}_{-1}^s$) is the Fréchet (resp. Hilbert) manifold of smooth (resp. Sobolev) Riemannian (resp. of class H^s) metrics with curvature equal to -1.

Finally we have the following identifications:

$$\mathcal{C} = \mathcal{A} \simeq \mathcal{M}_{-1} \quad \text{and} \quad \mathcal{C}^s = \mathcal{A}^s \simeq \mathcal{M}_{-1}^s.$$

With the notations of section 3.2, we shall set

$$\mathcal{P} := \mathcal{M}_{-1}.$$

We now describe the "gauge group" $\mathcal{G}$.

Remark. *This is in fact only a subgroup of the whole gauge group relative to string theory which corresponds to a semi-direct product of the the Weyl group and the group we are about to describe.*

Let $\mathcal{D}$ (resp. $\mathcal{D}^s$) be the group of all orientation preserving smooth (resp. Sobolev) diffeomorphisms (resp. of class H^s) of Z. We shall mainly concentrate on the subgroup $\mathcal{D}_0$ (resp. $\mathcal{D}_0^s$) of all diffeomorphisms homotopic to identity.

Both $\mathcal{D}$ and $\mathcal{D}_0$ (resp. $\mathcal{D}^s$ and $\mathcal{D}_0^s$) are Fréchet (resp. Hilbert) manifolds modelled on a space $C^\infty(TZ)$ (resp. $H^s(TZ)$) of vector fields. Both $\mathcal{D}$ and $\mathcal{D}_0$ are Fréchet Lie groups but the groups $\mathcal{D}^s$ and $\mathcal{D}_0^s$ are not Lie groups in the traditional sense since the map $f \to f \circ g$ is smooth but not $g \to f \circ g$ unless f is smooth. Moreover, the map $f \to f^{-1}$ is not differentiable.

Let us now describe the "action" of the "gauge group" $\mathcal{G}$ on the "path space" $\mathcal{P}$.

The group $\mathcal{D}_0$ (resp. $\mathcal{D}_0^{s+1}$) acts on $\mathcal{A}$, $\mathcal{C}$ and $\mathcal{M}_{-1}$ (resp. on $\mathcal{A}^s$, $\mathcal{C}^s$ and $\mathcal{M}_{-1}^s$) on the right by pull back. The action is smooth (resp. continuous), free and proper and and one can show that $\mathcal{M}_{-1} \to \mathcal{M}_{-1}/\mathcal{D}_0$ is a smooth principal bundle [T]. The quotient space is the *Teichmüller space* $\mathcal{T}$ of Z. One can furthermore show that $\mathcal{A} \to \mathcal{T}$ and $\mathcal{C} \to \mathcal{T}$ are also principal fibre bundles.

2. Riemannian structures

In order to define the second fundamental form, we need to define a (possibly weak) *Riemannian structure* on the manifold $\mathcal{P}$.

Let us equip the manifold $\mathcal{A}$ (resp. $\mathcal{A}^s$) with the L^2-metric structure (resp. which induces a weak topology) defined for h and k in $T_J\mathcal{A}$ by:

$$\langle h, k \rangle_J^- \equiv \int_Z tr_{g(J)}(hk)d\mu_J.$$

Here $d\mu_J$ is the volume measure on Z for the metric $g(J)$, tr_g the trace on tensors using contractions of tensors induced by the metric g.

In a similar way, let us equip the manifold $\mathcal{M}_{-1}$ (resp. $\mathcal{M}_{-1}^s$) with the L^2-metric structure (resp. which induces a weak topology) defined for $h, k \in T_g\mathcal{M}_{-1}$ (resp. $h, k \in T_g\mathcal{M}_{-1}^s$) by:

$$\langle h, k \rangle_g^- \equiv \int_Z tr_g(h_g^0 k_g^0)d\mu_g$$

where for a covariant two tensor h, $h_g^0 \equiv h - tr_g(h)g$ is the traceless part of h with respect to the metric g. Here $d\mu_g$ is the volume measure on Z for the metric g.

The map $\Psi : \mathcal{A} \to \mathcal{M}_{-1}$ is an isometry for these Riemannian structures ([T] par. 2.5).

We also need a Riemannian metric on the gauge group $\mathcal{G}$. In fact we shall introduce a family of left invariant metrics parametrized by $\mathcal{M}_{-1}$ on the Lie algebra $Lie(\mathcal{G})$ of $\mathcal{G}$ defined for u and v in $Lie(\mathcal{G})$ by:

$$\langle u, v \rangle_g^+ := \int_Z tr_g(uv)d\mu_g.$$

Equivalently, using the one-to-one map $\mathcal{A} \leftrightarrow \mathcal{M}_{-1}$, $J \to g(J)$ and looking at $\mathcal{E}^+$ as a bundle over $\mathcal{A}$, we can write:

$$\langle u, v \rangle_J^+ := \int_Z tr_{g(J)}(uv)d\mu_{g(J)}.$$

Notice that the fact that the Riemannian structure varies with J takes us outside the framework of section 3.2 and we can therefore not apply directly the results proven there.

3. A super vector bundle arising from the group action

We set:

$$\mathcal{V}^+ := \mathcal{A} \times H^{s+1}(TZ) \qquad and \qquad \mathcal{V}^- := T\mathcal{A}_{|\mathcal{A}}^s$$

i.e the tangent bundle to $\mathcal{A}^s$ restricted to $\mathcal{A}$. There are natural almost complex structures on $\mathcal{V}^+$ and $\mathcal{V}^-$ repectively defined by:

$$\mathcal{J}_J^+(u) := Ju, \quad \mathcal{J}_J^-(h) := hJ, \quad \forall u \in H^{s+1}(TZ), h \in T\mathcal{A}^s.$$

In order to define connections on these bundles it is useful to introduce a horizontal distribution on the principal bundle $\mathcal{A} \to \mathcal{T}$.

The action of $\mathcal{D}_0$ on $\mathcal{M}_{-1}$ induces for a given $g \in \mathcal{M}_{-1}$ a map:

$$\theta_g : \mathcal{D}_0 \to \mathcal{M}_{-1}$$
$$f \to f^*g.$$

Since the Lie algebra $Lie(\mathcal{G})$ corresponds to the space of tangent vector fields $\Gamma(TZ)$, the tangent map at identity reads:

$$T_g : \Gamma(TZ) \to T_g^0\mathcal{M}_{-1}$$
$$V \to (L_V g)_g^0$$

where $(L_V g) = \frac{d}{dt}_{|t=0} \left(g(e^{tV}) \right)$ is the Lie derivative of g in direction V and

$$T_g^0\mathcal{M}_{-1} \equiv \{h \in T_g\mathcal{M}_{-1}, tr_g(h) = 0\}.$$

One can check that the map $T_g^*T_g$ is a self-adjoint elliptic operator of order two (see [AJPS] or [T] par. 2.5). This leads to an orthogonal splitting:

$$T_g\mathcal{M}_{-1} = \mathcal{R}(T_g) \oplus T_g^0\mathcal{M}_{-1}$$

where $\mathcal{R}(\tau_g)$ is the range of τ_g which corresponds to the vertical tangent space at g. This yields a horizontal disribution on $\mathcal{M}_{-1} \to \mathcal{T}$ given by $T_g^0 \mathcal{M}_{-1}, g \in \mathcal{M}_{-1}$.

This horizontal distribution induces one on $\mathcal{A} \to \mathcal{T}$ (see [T] Theorem 2.5.6) via the isomorphism $\tilde{\Psi} := \Xi \circ \Psi^{-1}$ where Ξ is a map that sends a metric to the unique metric with curvature -1 in its conformal class (see [T] Sect. 2.5) and where Ψ is the map defined above.

The "algebraic connexion" introduced in [T] formula (5.6) on $\mathcal{A}$ defined for $k \in T_J \mathcal{A}$, $h \in \Gamma(T\mathcal{A})$ by:

$$\nabla_k h := Dh(k) - \frac{1}{2} J(h \cdot k + k \cdot h)$$

induces a connexion ∇^- on $\mathcal{V}^-$. In this formula we have used the dot to denote the matrix product between two $(1,1)$ tensors locally seen as 2×2 matrices.

In a similar way, the trivial vector bundle $\mathcal{V}^+$ can be equipped with a connexion defined for $k \in T_J \mathcal{A}$, $u \in \Gamma(\mathcal{V}^+)$ by:

$$\nabla_k^+ u := Dh(u) + \frac{1}{2} k J u.$$

Let $\mu(g)$ denote as before the Riemannian volume measure induced by the metric g. A simple computation shows that the derivative of the map $J \to \mu_{g(J)}$ vanishes in horizontal directions (see [T] Lemma 5.1.7), from which follow two (related) facts:

(i) (see [T], Theorem 5.1.10) The Weil-Petersson metric on $\mathcal{T}$ induced by the weak L^2-metric on $\mathcal{A}^s$ defined above is Kähler,

(ii) (see Appendix B of [PR2]) The connexions ∇^+ and ∇^- are L^2-Kählerian in horizontal directions, meaning by this that they are compatible in horizontal directions with the L^2-metrics and the almost complex structures defined above.

4. The determinant bundle picture

Let us now build a *determinant bundle on $\mathcal{A}$*.

A given almost complex structure $J \in \mathcal{A}$ on Z induces a complex structure on Z for which the metric $g(J) = \Psi(J)$ reads $g(J) = \rho(z) dz d\bar{z}$. Via the identification

$$T^{*(0,1)} Z \to T^{1,0} Z$$

$$\frac{\partial}{\partial z} \to \rho^2 d\bar{z}$$

we can write [AJPS]:

$$\tau_{g(J)} = \bar{\partial} \oplus \bar{\partial}^*$$

where the operator $\bar{\partial} \equiv \sqrt{2}(d\bar{z} \wedge \frac{d}{dz})$ is the $\bar{\partial}$ operator for the complex structure induced by J. We have thus associated to any almost complex structure J an operator

$$\tau_J \equiv D_J^+ \oplus D_J^-$$

where $D_J^+ = \bar{\partial}$ is an elliptic operator. By definition of the almost complex structure on $\mathcal{V}^+$, $\mathcal{E}^+ := \mathcal{V}^{+^{1,0}}$ is the bundle on $\mathcal{A}$ with fibre given by the holomorphic tangent bundle $T^{1,0}Z_J$ of the complex manifold Z_J obtained from Z equipped with the complex structure J. Similarly, from $\mathcal{V}^-$ we can build the complex bundle $\mathcal{E}^-$ on $\mathcal{A}$ with fibre $T^{1,1}Z_J$. Then

$$\mathcal{D}_J \equiv \begin{bmatrix} 0 & D_J^- \\ D_J^+ & 0 \end{bmatrix}.$$

Following the procedure described above we build the determinant bundle

$$Det(\mathcal{E} := \mathcal{E}^+ \oplus \mathcal{E}^-, D)$$

based on $\mathcal{A}$. Because of the invariance of the construction under diffeomorphisms in $\mathcal{D}_0$ this determinant bundle on $\mathcal{A}$ induces a determinant bundle $\mathcal{L}_D$ on $\mathcal{T}$. From the results of section 4.4, it follows that the curvature on this determinant bundle coincides with minus the first Chern form on the super bundle $\mathcal{E} := \mathcal{E}^+ \oplus \mathcal{E}^-$. In the final section, we compute the first Chern forms.

5. First Chern form on the vector bundle

Let us compute the curvature $\Omega^{\mathcal{E}} := \Omega^{\mathcal{E}^+} \oplus \Omega^{\mathcal{E}^-}$ on the vector bundle $\mathcal{E}$ where $\Omega^{\mathcal{E}^+} := \left(\nabla^{\mathcal{E}^+}\right)^2$ is the curvature of the connections $\nabla^{\mathcal{E}^+}$ and $\Omega^{\mathcal{E}^-} := \left(\nabla^{\mathcal{E}^-}\right)^2$ of the connection $\nabla^{\mathcal{E}^-}$. These are zero order pseudo-differential operators given for $\forall M, N \in T_J\mathcal{A}(\Lambda)$ by:

$$\Omega^{\mathcal{E}^+}(M,N) = -\frac{1}{4}\left(Q^+\right)^{-\frac{s}{2}}\left([M,N]_{op}\cdot\right)\left(Q^+\right)^{\frac{s}{2}}$$

$$\Omega^{\mathcal{E}^-}(M,N) = \left(Q^-\right)^{-\frac{s}{2}}\left(-\frac{1}{2}[M,N]_{op} + \frac{1}{2}(N(\cdot)M - M(\cdot)N)\right.$$
$$\left. - \frac{1}{4}[\{M,\cdot\},\{N,\cdot\}]\right)\left(Q^-\right)^{\frac{s}{2}}$$

where by $[M,N]_{op}$ we mean the multiplication operator by the bracket of the matrices and where $\{H,K\} := HK + KH$. Here $Q^+ \equiv D^-D^+$ and $Q^- \equiv D^+D^-$.

The first Chern form $r_1^{\mathcal{E},Q,\mu}$ on the half-weighted super vector bundle $(\mathcal{E}, D)$ equipped with the superconnection $\nabla^{\mathcal{E}} := \left(\nabla^{\mathcal{E}^+}\right)^* \otimes 1 + 1 \otimes \nabla^{\mathcal{E}^-}$ and with $Q = D^2$ is independent of the parameter μ used in the renormalization procedure. For any $M, N \in T_J\mathcal{A}(\Lambda)$ we have [PR2] for any $\mu \in \mathbb{R}$:

$$-ir_1^{\mathcal{E},Q,\mu}(M,N) = \frac{1}{2}tr^{Q^-}\left(J[M,N]_{op}\right) + \frac{1}{4}tr^{Q^-}\left(J[\{M,\cdot\},\{N,\cdot\}]\right)$$
$$- \frac{1}{2}tr^{Q^-}\left(N(\cdot)M - M(\cdot)N\right) + \frac{1}{4}tr^{Q^+}\left(J[M,N]_{op}\right).$$

Let us make a remark concerning the first Chern form. A matrix $H \in T_J\mathcal{A}^s(\Lambda)$ satisfies the property $HJ = -JH$ so that a product of two such matrices H, K

has the property $HKJ = JHK$. Hence writing

$$J = \begin{bmatrix} 0 & -1 \\ 1 & 0 \end{bmatrix}$$

in isothermal coordinates, we see that HK is of the type

$$\begin{bmatrix} \alpha & \beta \\ \beta & -\alpha \end{bmatrix}$$

and the same holds for any even product of matrices in $T_J \mathcal{A}^s(\Lambda)$. Hence $J[M, N]_{op}$ is of the type

$$\begin{bmatrix} \gamma & \delta \\ -\delta & \gamma \end{bmatrix}$$

so that its trace which arises in the expression $tr^{Q^+}(J[M, N]_{op})$ has no reason to vanish contradicting some wrong statement we had made a little too hastily in a previous article [PR1].

References

[AJPS] S. Albeverio, J. Jost, S. Paycha and S. Scarlatti, *A mathematical introduction to string theory. Variational problems, geometric and probabilistic methods*, London Mathematical Society Lecture Note Series 225, Cambridge University Press, Cambridge, 1997.

[AP] M. Arnaudon and S. Paycha, *Regularisable and minimal orbits for group actions in infinite dimensions*, Comm. Math. Phys. 191 (1998), 641–662.

[B] J.-M. Bismut, *Localization formulas, superconnections and the index theorem for families*, Comm. Math. Phys. 103 (1986), 127–166.

[BF] J.M. Bismut and D.S. Freed, *The analysis of elliptic families I*, Comm. Math. Phys. 106 (1986), 159–176.

[BGV] N. Berline, E. Getzler and M. Vergne, *Heat kernels and Dirac operators* (second edition), Grundlehren der mathematischen Wissenschaften 298, Springer, Berlin, 1996.

[C] B.Y. Chen, *Geometry of submanifolds*, Marcel Dekker, New York, 1973.

[CDMP] A. Cardona, C. Ducourtioux, J.P. Magnot and S. Paycha, *Weighted traces on algebras of pseudo-differential operators and geometry on loop groups*, Preprint 2000.

[FLPTT] N. Fagella, A. Lesne, S. Paycha, L. Tedeschini-Lalli and S.T. Tsou, *Renormalizations, Proceedings of a Workshop*, Publication of: European Women in Mathematics and *femmes et mathématiques*, Paris, 1996.

[F] D. Freed, *The geometry of loop groups*, J. Diff. Geom. 28 (1988), 223–276.

[G] P. Gilkey, *Invariance theory, the heat equation and the Atiyah-Singer index theorem* (second edition), Studies in advanced mathematics, CRC Press, Boca Raton, FL, 1995.

[GHL]   S. Gallot, D. Hulin and J. Lafontaine, *Riemannian Geometry*, Springer, Berlin-New York, 1987.

[H]   W.Y. Hsiang, *On compact homogeneous minimal submanifolds*, Proc. Nat. Acad. Sci. USA 56 (1966), 5–6.

[J]   J. Jost, *Riemannian geometry and geometric analysis* (second edition), Springer, Berlin, 1998.

[K]   C. Kassel, *Le résidu non commutatif (d'après M. Wodzicki)*, Séminaire Bourbaki, Vol. 1988/89, Astérisque No. 177–178 (1989), Exp. No. 708, 199–229.

[KT]   C. King and C.L. Terng, *Minimal submanifolds in path space*, in: "Global analysis in modern mathematics" (Orono, ME, 1991; Waltham, MA, 1992), 253–281, Publish or Perish, Houston, TX, 1993.

[KV]   M. Kontsevich, S. Vishik, *Determinants of elliptic pseudodifferential operators*, Preprint of the Max-Planck-Institut für Mathematik, 1994;

[KV2]   M. Kontsevich and S. Vishik, *Geometry of determinants of elliptic operators*, in: "Functional analysis on the eve of the 21st century", Vol. 1 (New Brunswick, NJ, 1993), 173–197, Progr. Math. 131, Birkhäuser Boston, Boston, MA, 1995.

[L]   M. Lesch, *On the noncommutative residue for pseudo-differential operators with log-polyhomogeneous symbols*, Ann. Global Anal. Geom. 17 (1999), no. 2, 151–187.

[Ma]   J.-P. Magnot, Ph.D. thesis, in preparation.

[M]   J. Marsden, *Applications of global analysis in mathematical physics*, Mathematical Lecture Series No. 2, Publish or Perish, Boston, MA, 1974.

[MRT]   Y. Maeda, S. Rosenberg and P. Tondeur, *The mean curvature of gauge orbits*, in: "Global analysis in modern mathematics" (Orono, ME, 1991; Waltham, MA, 1992), 171–217, Publish or Perish, Houston, TX, 1993;

[MRT2]   Y. Maeda, S. Rosenberg and P. Tondeur, *Minimal orbits of metrics*, J. Geom. Phys. 23 (1997), 319–349.

[P]   S. Paycha, *Renormalized traces as a looking glass into infinite dimensional geometry*, Infinite Dim. Analysis, Quantum Probability and Related Topics, to appear.

[PR1]   S. Paycha and S. Rosenberg, *About infinite dimensional group actions and determinant bundles*, in: "Analysis on infinite-dimensional Lie groups and algebras" (Marseille, 1997), 355–367, World Sci. Publ., River Edge, NJ, 1998.

[PR2]   S. Paycha and S. Rosenberg, *Curvature on determinant bundles and first Chern forms*, J. Geom. Phys., to appear.

[Q]   D. Quillen, *Determinants of Cauchy Riemann operators on Riemann surfaces*, Functional Anal. Appl. 19 (1985), no. 1, 31–34.

[R]   S. Rosenberg, *The Laplacian on a Riemannian manifold. An introduction to analysis on manifolds*, London Mathematical Society Student Texts 31, Cambridge University Press, Cambridge, 1997.

[T]   A. Tromba, *Teichmüller theory in Riemannian geometry*, Lecture notes pre-
        pared by Jochen Denzler, Lectures in Mathematics ETH Zürich, Birkhäuser,
        Basel, 1992.

[W]   M. Wodzicki, *Non-commutative residue. I. Fundamentals*, Lecture Notes in
        Math., 1289, Springer, Berlin-New York, 1987.

[Z]   Y.N. Zhang, *Lévy Laplacian and Brownian particles in Hilbert spaces*, J. Funct.
        Anal. 133 (1995), no. 2, 425–441.

Mathematical Subject Classification (2000)
Primary: 53Cxx.
Secondary: 53C10, 53C15, 58B25, 58D17, 58D19, 81T30.

Laboratoire de Mathématiques Appliquées
Complexe Universitaire des Cézeaux
F-63177 Aubière Cedex
France
E-mail address: paycha@ucfma.univ-bpclermont.fr

Fermionic Second Quantization and the Geometry of the Restricted Grassmannian

Tilmann Wurzbacher

Abstract. We explain how fermionic second quantization leads to G_{res}, the restricted Grassmannian of a polarized Hilbert space, and its homogeneous Kähler geometry, and how vice-versa G_{res} encodes – via its holomorphic determinant bundle – the basic ingredients of fermionic second quantization, as, e.g., the fermionic Fock space and the "Schwinger term". Using this approach we derive a new construction of the universal central extension of U_{res}, the restricted unitary group. Furthermore, we develop the general theory of symplectic manifolds and symplectic actions in infinite dimensions and apply it notably to the U_{res}-action on G_{res}. Finally we construct the determinant bundle on G_{res} functorially using "C^*-algebro-geometric methods" naturally arising from the use of CAR-algebras in fermionic second quantization.

Introduction

Second quantization can – from a simplifying mathematical point of view – be seen as the study of certain operators on the tensor space $\otimes K$ over a Hilbert space K. Notably one associates to an element of K the tensor multiplication and contraction operations on $\otimes K$ and, to a linear operator A on K the "second quantized" operator $\otimes A$ on $\otimes K$. Furthermore one is interested in "physical" operators such as the Hamiltonian or the scattering matrix and of course in the symmetries present. The physical aim of second quantization is to construct so-called "quantum field theories" as opposed to quantum mechanics, where the physical states are described by a Hilbert space of wave-functions, the (first) quantization of classical mechanics of point particles. (Compare the classical references [Ber], [Coo] and [Fo].) In fermionic second quantization as we consider it in this article one restricts to the antisymmetric part of $\otimes K$ and supposes that K is polarized, i.e. that there is given an orthogonal decomposition $K = K_+ \oplus K_-$. This decomposition, corresponding to the physical idea of particles and anti-particles, means we can use the space $\Lambda K_+ \otimes \Lambda \overline{K_-}$ as the "fermionic Fock space".

The restricted Grassmannian G_{res} of the polarized Hilbert space K is the set of linear subspaces of K that are – in a precise sense – "close" to K_+. It constitutes an interesting example of an infinite-dimensional Kähler manifold which is homogeneous under its holomorphic isometries, and contains several other such manifolds as equivariantly embedded submanifolds.

The goal of this paper is to explain how fermionic second quantization naturally defines G_{res} and its geometry, and how vice-versa G_{res} encodes the basic ingredients of fermionic second quantization. The connecting link is the "Schwinger term". On the one hand it is viewed as the cocycle of a central Lie algebra extension coming from a projective representation of the symmetry group U_{res}, the "restricted unitary group", arising in fermionic second quantization, and on the other hand as the Chern class of a homogeneous line bundle on the U_{res}-homogeneous space G_{res}.

The relation between projective representations and line bundles is of course well-known, especially to aficionados of geometric quantization. Nevertheless a thorough study of this single example seems to be justified by its preeminent rôle in the theory of infinite-dimensional Lie algebras and Lie groups, the numerous interesting problems arising from its infinite-dimensional nature, and its relation to many other fields in mathematics and theoretical physics. Furthermore it supports our point of view that – at least parts of – second quantization (as well as of noncommutative geometry) can be neatly expressed in terms of the differential geometry and the geometric quantization of (commutative) infinite-dimensional manifolds.

This paper of course owes much to the two outstanding books by J. Mickelsson ([Mic]) and by A. Pressley and G. Segal ([PrSe]) which cover large parts of the subject. The main difference between their approaches and ours is that we stress that G_{res} might be more easily studied by C^*-algebra theory, and that we give more space to considerations of infinite-dimensional geometry and analysis. This leads to a whole section developing from scratch the theory of symplectic actions in infinite dimensions in order to have a rigorous framework for the study of the U_{res}-moment map on G_{res}, whose description in [PrSe] has a gap which was later filled by Grosse and Maderner ([GM]). Furthermore we give a "C^*-algebraic" construction of the central extension of U_{res} as well as of the determinant bundle on G_{res}.

Let us briefly describe the contents of the sections of this article.

Section I motivates the fermionic multiparticle formalism by a "genetic" approach starting – à la Dirac – with the Klein-Gordon equation. We develop the useful "hands-on" approach of physicists to the Fock space and the CAR-algebra, the algebra of the "canonical anticommutation relations", as well as the mathematical language of C^*-algebras, states and associated representations.

Section II introduces and studies the projective representation of U_{res} on the Fock space via the theorem of Powers and Størmer, and the explicit construction of the second-quantized operators acting on it. We then discuss the basic properties of the resulting Lie algebra cohomology cocycle, also known as the "Schwinger term", and construct central Banach Lie group extensions $\widetilde{U_{\text{res}}}$ and $\widetilde{Gl_{\text{res}}}$ via deprojectivization.

Section III defines G_{res}, the restricted Grassmannian of a polarized Hilbert space as a moduli space of vacua. We show the equivalence to the usual definition as a certain set of linear subspaces of the given polarized Hilbert space and study some of its basic properties as a homogeneous Kähler manifold whose Kähler form is given by the Schwinger term.

Section IV starts with a rather detailed discussion of differential k-forms on manifolds modelled on locally convex topological vector spaces. Though Banach manifolds are in most aspects very similar to finite-dimensional manifolds, the fact that a genuine Fréchet space never has a metrizable dual causes some subtleties in this more general setting. We give some foundations of a theory of symplectic actions and (co-)moment maps, at least for Fréchet manifolds. After sketching a couple of typical examples of moment maps we study the construction of the U_{res}-moment on G_{res} in some detail, and show – following [GM] – that the Schwinger term manifests itself by the non-equivariance of this moment map, i.e., on the level of "infinite-dimensional classical mechanics". Though relatively simple, this example illustrates already the usefulness of the general framework.

Section V is again devoted to the C^*-algebraic approach to the Grassmannian. Using a simple functorial approach, we construct "a" determinant bundle and prove that it is isomorphic to the determinant bundle constructed in [PrSe]. We conclude by realizing the Fock space as holomorphic sections of the dual of the determinant bundle, which is in fact almost tautological in our approach.

This text is an exposition of the subject with many remarks pointing to related issues, but we have nevertheless tried to give complete proofs. We deviated from this only in the remarks and in a few other places, where details seemed to lead us too far afield. In particular we stated without proof the theorem of Powers and Størmer (Theorem II.1) and we were rather sketchy about the proof of the related implementation theorem of Carey and Ruijsenaars (Theorem II.4). Furthermore we have not even given the complete statement of the theorem of Pickrell (Remark (2) after Proposition V.22).

Due to lack of place and/or competence we have said very little about the relation of our subject to noncommutative geometry, to determinant bundles associated to families of elliptic operators, and to scattering theory and the external field problem. We also decided not to dwell into details of the embeddings of infinite-dimensional homogeneous spaces into G_{res} and the algebraic theory of infinite-dimensional Lie algebras. Nevertheless we give a few remarks and references concerning these domains. Unfortunately we did not find the time to include the recently proposed language of bundle gerbes that seems to be quite adapted to the mathematical treatment of the Schwinger term and other "anomalies" arising in quantum field theory. (See the review [CMM] and references therein.) Last but not least we have completely omitted the relation to the KdV and KP hierarchies of integrable equations (cf., e.g., [Sa]).

We have not even tried to fulfil the formidable task of giving a complete bibliography but have rather stuck to a highly subjective assembly of sources which we ourselves used to understand second quantization and the geometry of the Grassmannian. We apologize to all authors whose contributions to the field we have failed to mention.

This paper began as a part of rough lecture notes for an informal course in the summer of 1995 in Bochum and was principally shaped by my talks in the DMV-Seminar "Infinite dimensional Kähler manifolds" in November 1995 at the Mathematisches Forschungsinstitut Oberwolfach. The present text benefitted from the opportunity of giving several talks on the subject, notably in Clausthal, Leipzig, Lyon, Marseille, Nancy, Padova and Strasbourg.

The very nature of the subject of the "geometry of quantum field theory" lying between (and sometimes outside of) many traditional areas of mathematics necessitates the use of a variety of techniques. On the other hand this diversity may attract the interest of non-specialists. We hope that this text appeals to them as well as to advanced students, in the spirit of the DMV-Seminars. The more initiated reader might be intrigued by our attempt to advocate the importance of the relation between C^*-algebras arising from quantum field theory and Kähler geometry, and some of our foundational observations on infinite-dimensional symplectic actions.

Acknowledgements. Without having collaborated with M. Spera I could of course never have written this paper. I would like to thank him for numerous conversations on related and unrelated topics. It is also a great pleasure to thank N. Bopp, A. Carey, P. Gosselin, P. Iglesias, P. Julg, J. Mickelsson, A. Reznikov, H. Rubenthaler and M. Slupinski for discussing different mathematical questions touched upon during the course of this work. Their advice was very helpful to me. Notably I would like to mention many useful discussions with K.-H. Neeb. They allowed me to improve this text in several places. Last but not least, I would like to express my thanks to the participants of the DMV-Seminar and especially to A.T. Huckleberry for their interest in my lectures.

I. Fermionic second quantization

In this section we will first develop the Hamiltonian approach to field quantization in the concrete example of electrons in an external electromagnetic field. In this section we will make ample use of arguments and notations used in the book of Thaller ([Tha]) on the Dirac equation, which we highly recommend for a wealth of informations on this equation in the flat case. In the second subsection, we will employ the more abstract language of C^*-algebras, that appears to be more suitable for our further developments.

Though many physicists seem to prefer the functional integral point of view for quantizing fields, since "gauge-invariance renders canonical (or second) quantization an awkward subject" ([Pe], p. 79), the days of second quantization are not yet counted. In fact, many notions of Connes' non-commutative geometry (compare [Con]) stem directly from the C^*-algebraic, second-quantization approach to

quantum field theory, and even string fields and quantum gravity might possibly
be formulated in the language of non-commutative geometry (cf. e.g. [FG], [FGR]).

1. The Dirac equation and the negative energy problem

Let us start with the simplest relativistically invariant field equation on Minkowski
space $\mathbb{R}^{1,3}$, the Klein-Gordon equation for a real- (or complex-)valued function φ :

$$(\Box + m^2)\varphi = 0. \tag{1.1}$$

Here $\Box$ is the D'Alembertian $\sum\limits_{\mu,\nu} g^{\mu\nu}\partial_\mu\partial_\nu$ with $\partial_\mu = \frac{\partial}{\partial x^\mu}(\mu = 0, 1, 2, 3)$ and $g^{\mu\nu} = g_{\mu\nu} = g(\partial_\mu, \partial_\nu)$ are the coefficients of the metric tensor used in physics

$$g = \begin{pmatrix} 1 & & & \\ & -1 & & \\ & & -1 & \\ & & & -1 \end{pmatrix}.$$

Furthermore, the physical constants h and c are set to 1 and m is a non-
negative parameter, called the mass.

Singling out time as $t = x^0$, we find the equivalent equation

$$\left(i\frac{\partial}{\partial t}\right)^2 \varphi(t, x) = ((-i\nabla)^2 + m^2)\varphi(t, x), \tag{1.2}$$

where $x = (x^1, x^2, x^3)$ and ∇ is the spatial gradient.

Since the Klein-Gordon equation is of second order in the time-variable, it is
not an evolution equation of Schrödinger type. Furthermore, physicists wishing to
describe the behaviour of an electron need to incorporate the internal spin as well
as the electromagnetic field into this equation. This leads to the following natural
Ansatz:

$$\Box + m^2 = \left(i\sum_\mu \gamma^\mu\partial_\mu - m\right)\left(-i\sum_\mu \gamma^\mu\partial_\mu - m\right). \tag{1.3}$$

(Compare to $\Delta_{\mathbb{R}^2} = \left(\frac{\partial}{\partial x^1}\right)^2 + \left(\frac{\partial}{\partial x^2}\right)^2 = 4\frac{\partial}{\partial z}\frac{\partial}{\partial \bar{z}}$.)

Equivalently, the "coefficients" γ^μ have to obey:

$$\gamma^\mu \cdot \gamma^\nu + \gamma^\nu \cdot \gamma^\mu = 2g^{\mu\nu} \cdot \mathbf{1}. \tag{1.4}$$

This forces us to consider functions $\psi : \mathbb{R}^{1,3} \to \mathbb{C}^N$, where $\mathbb{C}^N$ is an appropriate
vector space of "spinors" allowing a $\mathbb{R}$-linear map $\gamma : \mathbb{R}^{1,3} \to \text{End}_{\mathbb{C}}(\mathbb{C}^N)$ which
fulfils

$$\gamma(u) \cdot \gamma(v) + \gamma(v) \cdot \gamma(u) = 2g(u, v) \cdot \mathbf{1} \tag{1.5}$$

for all u, v in $\mathbb{R}^{1,3}$.

(This construction leads of course to the general theory of Clifford algebras
and spinors.)

For definiteness in the sequel, we explicit a minimal choice in the case of $\mathbb{R}^{1,3}$:

$$N = 4 \text{ and } \gamma^0 = \begin{pmatrix} 1 & 0 \\ 0 & -1 \end{pmatrix}, \gamma^k = \begin{pmatrix} 0 & \sigma^k \\ -\sigma^k & 0 \end{pmatrix} \text{ with}$$

$$\sigma^1 = \begin{pmatrix} 0 & 1 \\ 1 & 0 \end{pmatrix}, \sigma^2 = \begin{pmatrix} 0 & -i \\ i & 0 \end{pmatrix}, \sigma^3 = \begin{pmatrix} 1 & 0 \\ 0 & -1 \end{pmatrix}.$$

Of course, the "Pauli-matrices" σ^k have size (2×2), whereas the "Dirac- or γ-matrices" have size (4×4).

We thus arrive at the famous "(free) Dirac equation" for a $\mathbb{C}^4$-valued function ψ on $\mathbb{R}^{1,3}$:

$$\left(\sum_\mu \gamma^\mu (i\partial_\mu) - m \right)\psi = 0. \tag{1.6}$$

In passing we note that enlarging the target space of the function to the space of "spinors", yields a representation of a "spin group" on these functions. In the case at hand, physicists call ψ a "Dirac spinor" of spin $1/2$ (in fact $\mathbb{C}^4$ can be identified to a sum $(1/2, 0) \oplus (0, 1/2)$ with respect to $SL(2, \mathbb{C})$, the spin group of this situation).

We can now incorporate an electromagnetic field by the "principle of minimal coupling". Let $(A^\mu)_{\mu=0,1,2,3} = (\phi_{el}, \vec{A})$ be the four-potential and $A_\mu = \sum_\nu g_{\mu\nu} A^\nu$ ($\mu = 0, \ldots, 4$) the associated four-vector.

Replacing $i\partial_\mu$ by $i(\partial_\mu + ieA_\mu) = \pi_\mu$, where e is the charge of an electron, we find the Dirac-equation of a *single* relativistic electron in an electromagnetic field:

$$\left(\sum_\mu \gamma^\mu \pi_\mu - m \right)\psi = 0. \tag{1.7}$$

Remark. This is of course the very origin of gauge theory in physics. Mathematically speaking we simply replaced all ordinary derivatives by covariant derivatives with respect to the connection coming from A.

Let us now write the free Dirac equation (1.6) in Schrödinger form:

$$i\frac{\partial f}{\partial t} = H_D f, \tag{1.8}$$

where f is in $L^2(\mathbb{R}^3, \mathbb{C}^4)$, $H_D = \vec{\alpha} \cdot \vec{p} + m\beta$ with $\alpha^k = \gamma^0 \gamma^k = \begin{pmatrix} 0 & \sigma^k \\ \sigma^k & 0 \end{pmatrix}, \beta = \gamma^0$ and $p_k = -i\partial_k$ (for $k = 1, 2, 3$). The "free Dirac-Hamiltonian" H_D is a densely-defined self-adjoint operator which can be explicitly diagonalized:

Proposition I.1 *The free one-particle Dirac-Hamiltonian H_D is unitarily conjugated to a diagonal operator by means of the "Foldy-Wouthuysen transformation" U_{FW} on $L^2(\mathbb{R}^3, \mathbb{C}^4)$:*

$$U_{FW} \circ H_D \circ U_{FW}^{-1} = \begin{pmatrix} |H_D| & 0 \\ 0 & -|H_D| \end{pmatrix}, \tag{1.9}$$

where

$$\Delta = \sum_{k=1}^{3} (\partial_k)^2 \quad and \quad |H_D| = \sqrt{-\Delta + m^2}$$

is defined as a pseudo-differential operator.

Remark. The proof (see e.g. [Tha], pp. 9-11) follows from diagonalizing the Fourier transform of H_D with parameter p in $\mathbb{R}^3$. Thus we are confronted with the physically unreasonable absence of a reference energy level, since the spectrum of H_D is bilaterally unbounded. We will sketch now Dirac's ingenious solution of this "negative energy problem".

Let us consider the one-particle Dirac-Hamiltonian of a particle with charge e in an electromagnetic field:

$$H_D(e, A) = \vec{\alpha} \cdot (\vec{p} - e\vec{A}) + m\beta + e\phi_{\text{el}}, \tag{1.10}$$

recalling that $(\phi_{\text{el}}, \vec{A}) = (A^0, A^1, A^2, A^3) = (A_0, -A_1, -A_2, -A_3)$. Denoting the unitary map $i\beta\alpha^2 = i\gamma^2 : \mathbb{C}^4 \to \mathbb{C}^4$ by U_C and setting $C \cdot f = U_C \cdot \bar{f}$ for $f \in L^2(\mathbb{R}^3, \mathbb{C}^4)$, one finds $C^2 = 1$ and

$$C \circ H_D(e, A) \circ C^{-1} = -H_D(-e, A). \tag{1.11}$$

Thus the anti-unitary transformation C maps solutions of the Dirac equation

$$i\frac{\partial f}{\partial t} = H_D(e, A)f \tag{1.12}$$

to solutions of

$$i\frac{\partial g}{\partial t} = -H_D(-e, A)g, \tag{1.13}$$

and it maps the negative energy space of $H_D(e, A)$ (defined by spectral projection) to the **positive** energy space of $H_D(-e, A)$, preserving norms: $\|Cf\|^2 = \|f\|^2$.

Considering C as the "charge conjugation operator", we arrive at Dirac's conclusion that the undesired electrons with negative energy should be considered as "anti-electrons" or "positrons", having mass m, spin 1/2, charge $(-e)$ and positive energy!

Since many elementary proccesses, as particle-antiparticle creations and scattering proccesses, are still not well-explained by this one-particle approach, physicists created the multi-particle formalism, which we will review in the next subsection.

2. Fermionic multiparticle formalism: Fock space and the CAR-algebra

Looking at the elements f of a Hilbert space K of the type $L^2(\mathbb{R}^3, \mathbb{C}^4)$ as the wavefunctions obtained by quantizing a (fictitious?) underlying mechanical system, the below described process of associating operators to its elements was coined "second quantization". Allowing phase spaces to be infinite-dimensional manifolds, it seems to be more judicious to take this term not too literally.

Let us first recall the usual scalar product on the full tensor product of a complex Hilbert space H (which can always be assumed to be separable for our purposes).

Denoting the scalar product on H by $\langle , \rangle$ (which we will always assume to be $\mathbb{C}$-linear in the second argument in these notes), we have the following scalar product on $\otimes H$

$$\langle h_1 \otimes \ldots \otimes h_n, h_1' \otimes \ldots \otimes h_m' \rangle = \delta_{n,m} \cdot \prod_{j=1}^{n} \langle h_j, h_j' \rangle. \tag{1.14}$$

Setting $h_1 \wedge \ldots \wedge h_n = \frac{1}{\sqrt{n!}} \sum_{\sigma \in S_n} \text{sign}(\sigma) \cdot h_{\sigma(1)} \otimes \ldots \otimes h_{\sigma(n)}$, we find

$$\langle h_1 \wedge \ldots \wedge h_n, h_1' \wedge \ldots \wedge h_n' \rangle = \det\left(\langle h_j, h_k' \rangle_{1 \leq j, k \leq n} \right). \tag{1.15}$$

Here and in the sequel we will always complete sums, tensor products and wedge products of Hilbert spaces without changing the symbols.

In the last subsection we split the Hilbert space $K = L^2(\mathbb{R}^3, \mathbb{C}^4)$ into the direct orthogonal sum of the non-negative and negative spectral subspace of the Dirac-Hamiltonian, i.e. $K = K_+ \oplus K_-$, and introduced an anti-unitary map $C : K \rightarrow K$, the charge conjugation. This leads to a "hands-on" construction of the Fock space used in Quantum Electrodynamics (QED) (see e.g. [Tha] or [CR]). First we define

$$\mathcal{F}_\pm^{(0)} = \mathbb{C}, \mathcal{F}_+^{(1)} = K_+ \text{ and } \mathcal{F}_-^{(1)} = CK_-$$

and set $\mathcal{F}^{(n,m)} = \Lambda^n \mathcal{F}_+^{(1)} \otimes \Lambda^m \mathcal{F}_-^{(1)}$ and $\mathcal{F} = \oplus_{n,m \geq 0} \mathcal{F}^{(n,m)}$. The constructed space $\mathcal{F}$ is called the "Fock space", the vector $\Omega = 1$ in $\mathcal{F}^{(0,0)}$ the "vacuum vector" and elements of $\mathcal{F}^{(n,m)}$ will be considered as describing a physical systems containing n particles and m anti-particles. Taking now a decomposable element $\psi^{(n,m)} = f_1 \wedge \ldots \wedge f_n \otimes C g_1 \wedge \ldots \wedge C g_m$ in $\mathcal{F}^{(n,m)}$, we can associate two operators to f in K_+:

$$a^*(f)\psi^{(n,m)} = f \wedge f_1 \wedge \ldots \wedge f_n \otimes C g_1 \wedge \ldots \wedge C g_m, \tag{1.16}$$

and

$$a(f)\psi^{(n,m)} = \sum_{j=1}^{n} (-1)^{j+1} \langle f, f_j \rangle f_1 \wedge \ldots \wedge \hat{f}_j \wedge \ldots \wedge f_n \otimes C g_1 \wedge \ldots \wedge C g_m, \tag{1.17}$$

where a hat means that the corresponding element is to be omitted. The first operator, called "particle creator", is $\mathbb{C}$-linear in f in K_+, whereas the second, the "particle annihilator", is $\mathbb{C}$-antilinear in f.

It is easy to verify that both operators extend to the whole Fock space, fulfilling

$$(a(f))^* = a^*(f) \text{ and } \|a^*(f)\| = \|a(f)\| = \|f\|.$$

Furthermore they obey the following "canonical anticommutation relations" (CAR):

$$\{a(f), a^*(f')\} = a(f)a^*(f') + a^*(f')a(f) = \langle f, f' \rangle_K \cdot 1_\mathcal{F}, \tag{1.18}$$

$$\{a^*(f), a^*(f')\} = 0 = \{a(f), a(f')\} \tag{1.19}$$

for all f, f' in K_+. The unital C^*-algebra generated by these operators in $\mathcal{B}(F)$ (the unital Banach algebra of bounded linear operators on F) yields a concrete construction of the so-called CAR-algebra $A(K_+)$ of K_+.

We define a second set of operators, this time for g in K_- :

$$b^*(g)\psi^{(n,m)} = (-1)^n f_1 \wedge \ldots \wedge f_n \otimes Cg \wedge Cg_1 \wedge \ldots \wedge Cg_m \tag{1.20}$$

$$b(g)\psi^{(n,m)} = (-1)^n f_1 \wedge \ldots \wedge f_n$$

$$\otimes \left(\sum_{j=1}^m (-1)^{j+1} \langle Cg, Cg_j \rangle Cg_1 \wedge \ldots \wedge \widehat{Cg_j} \wedge \ldots \wedge Cg_m \right). \tag{1.21}$$

These operators, called "anti-particle creator" and "anti-particle annihilator" fulfil

$$(b(g))^* = b^*(g), \|b^*(g)\| = \|b(g)\| = \|g\|, \text{ and}$$

$$\{b(g), b^*(g')\} = \langle Cg, Cg' \rangle_K \cdot 1_F = \langle g', g \rangle_K \cdot 1_F \tag{1.22}$$

$$\{b^*(g), b^*(g')\} = 0 = \{b(g), b(g')\}. \tag{1.23}$$

Observe that $b^*(g)$ depends $\mathbb{C}$-antilinearly on g in K_-. They generate the "complex conjugate" of the CAR-algebra of K_- in $\mathcal{B}(F)$ (or more intrinsically $A(\overline{K_-})$. It is readily verified that these two sets of operators anticommute:

$$\{a^\times(f), b^\times(g)\} = 0 \text{ for "}\times\text{" being "}*\text{" or "non-}*\text{".}$$

Furthermore, the vacuum vector Ω is (up to scalar multiples) the only vector in F fulfilling

$$a(f)\Omega = 0 \text{ for all } f \text{ in } K_+ \text{ and} \tag{1.24}$$

$$b(g)\Omega = 0 \text{ for all } g \text{ in } K_-. \tag{1.25}$$

We can now define an action of $A(K)$, the CAR-algebra of the whole 1-particle space by means of the so-called "field operators" (or "smeared fermionic time zero fields" in the physics literature):

$$\Psi^*(f) = a^*(f_+) + b(f_-) \text{ and} \tag{1.26}$$

$$\Psi(f) = a(f_+) + b^*(f_-) \tag{1.27}$$

for $f = f_+ + f_-$ in $K = K_+ \oplus K_-$. Observe that $\Psi^*(f)$ depends linearly on f in K and that one indeed has

$$\{\Psi(f), \Psi^*(f')\} = \langle f, f' \rangle_K \cdot 1_F \text{ and} \tag{1.28}$$

$$\{\Psi^*(f), \Psi^*(f')\} = 0 = \{\Psi(f), \Psi(f')\}. \tag{1.29}$$

Furthermore, Ω is characterized by

$$\Psi^*(f_-)\Omega = 0 = \Psi(f_+)\Omega \tag{1.30}$$

for all f_- in K_- and f_+ in K_+.

The equation (1.30) explains the term "Dirac sea" for Ω : the physical system described by Ω is filled up with negative energy particles and contains no positive energy particles. Furthermore, annihilating a negative energy state by means of

$\Psi(f_-) = b^*(f_-)$ tantamounts to creating a positive energy **anti-particle**! Precursing the general discussion of how to "second-quantize" one-particle operators in Subsection 2.1, we assure us that the multi-particle Hamiltonian has non-negative spectrum.

Let us assume for simplicity that the self-adjoint Hamiltonian $H = H_D(e, A)$ figurating in the Schrödinger equation (1.12): $i\frac{\partial f}{\partial t} = H_D(e, A)f$ has pure point spectrum, i.e. there are orthonormal bases $\{e_1, e_2, \ldots\}$ and $\{e_{-1}, e_{-2}, \ldots\}$ of K_+ and K_- fulfilling

$$He_j = \lambda_j e_j \text{ and } He_{-k} = \mu_k e_{-k}$$

for $j, k > 0$ with $\lambda_j \geq 0$ and $\mu_k < 0$. The one-particle time evolution $U_t = \exp(-itH)$ induces a C^*-algebra automorphism of the CAR-algebra $A(K)$ in $B(\mathcal{F})$ by setting $\beta_t(\Psi^\times(f)) = \Psi^\times(U_t f)$, for f in K and "$\times$" meaning "$*$" or "non-$*$", and extending this definition in the obvious way to all α in $A(K)$. Looking for a unitary map $\tilde{U}_t$ of $\mathcal{F}$ fulfilling $\beta_t(\alpha) = \tilde{U}_t \circ \alpha \circ \tilde{U}_t^{-1}$ for all α in $A(K)$, we find $\tilde{U}_t \Omega = \Omega$ and

$$\tilde{U}_t(f_1 \wedge \ldots \wedge f_n \otimes Cg_1 \wedge \ldots \wedge Cg_m) = U_t f_1 \wedge \ldots \wedge U_t f_n \otimes CU_t g_1 \wedge \ldots \wedge CU_t g_m.$$

For $\psi^{(n,m)} = e_{j_1} \wedge \ldots \wedge e_{j_n} \otimes Ce_{-k_1} \wedge \ldots \wedge Ce_{-k_m}$, it follows that

$$\tilde{U}_t \psi^{(n,m)} = \exp(-it[\lambda_{j_1} + \ldots + \lambda_{j_n} - \mu_{k_1} - \ldots - \mu_{k_m}])\psi^{(n,m)}.$$

Thus the multi-particle Hamiltonian

$$\tilde{H} = i\frac{d}{dt}\Big|_0 \tilde{U}_t$$

has non-negative spectrum.

Aiming to rephrase the above construction of the Fock space in terms of the representation theory of C^*-algebras, let us recall the basic notion of a state.

Given a complex unital C^*-algebra A, a "positive linear form" is a $\mathbb{C}$-linear functional ω such that $\omega(a^*a) \geq 0$ for all a in A. It follows that ω is bounded with $\|\omega\| = \omega(1)$. A "state on A" is defined as a positive linear form of norm 1.

(We recommend [Mur] as a general reference on C^*-algebras, and [BR] and [dlHJ] for more details on CAR-algebras).

Given a state ω the Gelfand-Naimark-Segal (GNS) construction yields a C^*-representation $\pi_\omega^A : A \longrightarrow B(\mathcal{H}_\omega^A)$ of A as bounded linear operators on a Hilbert space $\mathcal{H}_\omega^A$ and a "vacuum vector" ξ_ω^A of norm one in $\mathcal{H}_\omega^A$ such that

$$\omega(a) = \langle \xi_\omega^A, \pi_\omega^A(a)\xi_\omega^A \rangle_{\mathcal{H}_\omega^A} \tag{1.31}$$

and $\pi_\omega^A(A) \cdot \xi_\omega^A$ is dense in $\mathcal{H}_\omega^A$. Equation (1.31) is paraphrased by saying that ω equals the "vector state" associated to the representation π_ω^A and the vector ξ_ω^A in the representation space, whereas the latter property is preferred to by saying that ξ_ω^A is a "cyclic vector".

Specializing to the CAR-algebra $A(K)$ of a complex Hilbert space K with unit denoted by 1, we can associate to a self-adjoint operator q on K fulfilling

$0 \le q \le 1$ a "quasi-free state" ω_q by setting

$$\omega_q(1) = 1 \qquad \text{and}$$

$$\omega_q(a^*(k_1)\ldots a^*(k_n)a(\ell_1)\ldots a(\ell_m)) = \delta_{n,m} \det((\langle \ell_i, q(k_j)\rangle_K). \tag{1.32}$$

(We changed here from the ψ^*, ψ notation to the labels a^*, a for creators respectively annihilators associated to **arbitrary** k more customarily in texts on abstract CAR-algebras. This slightly unfortunate clash is inevitable given the notational traditions in quantum field theory respectively in the theory of operator algebras.)

We observe that ω_q is already fixed by the "two-point functions"

$$\omega_q(a^*(k)a(\ell)) = \langle \ell, q(k)\rangle_K. \tag{1.33}$$

The simplest case arises when $q = p$ is an orthogonal projector. The corresponding state ω_p is then also known as a "gauge-invariant quasi-free state" on $A(K)$. A simple calculation, using

$$\langle \ell, p(k)\rangle = \omega_p(a^*(k)a(\ell)) = \langle \xi_p, \pi_p(a^*(k)a(\ell))\xi_p\rangle_{\mathcal{H}_p} \tag{1.34}$$

where ξ_p is the vacuum vector in the GNS-space $\mathcal{H}_p = \mathcal{H}_{\omega_p}^{A(K)}$ and π_p the corresponding GNS-representation, now shows that

$$\pi_p(a(k))\xi_p = 0 \quad \text{for all } k \text{ in Ker } p, \text{ and}$$

$$\pi_p(a^*(\ell))\xi_p = 0 \quad \text{for all } \ell \text{ in Im } p.$$

Comparing this result to the behaviour of the vacuum Ω in $\mathcal{F}$, we are led to

Proposition I.2 (folklore) *The representation of $A(K)$ on $\mathcal{F}$ generated by the field operators Ψ^* and Ψ (see (1.26)–(1.30) above) is unitarily equivalent to the GNS-representation associated to the gauge-invariant quasi-free state ω_{p_-} coming from the projection to K_- on K.*

Proof. Let us calculate the two-point functions of the vector state $\omega_{\mathcal{F}}$ associated to the cyclic vector Ω in $\mathcal{F}$:

$$\omega_{\mathcal{F}}(a^*(k)a(\ell)) = \langle \Omega, \Psi^*(k)\Psi(\ell)\Omega\rangle_{\mathcal{F}}$$

$$= \langle \Psi(k)\Omega, \Psi(\ell)\Omega\rangle_{\mathcal{F}} = \langle \Psi(k_-)\Omega, \Psi(\ell_-)\Omega\rangle_{\mathcal{F}}$$

$$= \langle Ck_-, C\ell_-\rangle_{\mathcal{F}(0,1)} = \langle \bar{k}_-, \bar{\ell}_-\rangle_K = \langle \ell_-, k_-\rangle_K$$

$$= \langle \ell, p_-(k)\rangle_K = \omega_{p_-}(a^*(k)a(\ell)).$$

Standard arguments from the general theory of CAR-algebras or concrete use of the cyclicity of Ω respectively ξ_{p_-} imply now the existence of a unitary isomorphism $\mathcal{H}_{p_-}^{A(K)} \xrightarrow{\simeq} \mathcal{F}$, sending of course ξ_{p_-} to Ω and intertwinning the two representations of $A(K)$.   □

Remark. There is another conceptual way to "unterstand" the formulas of the representation of $A(K)$ on $\mathcal{F}$: Observe first that the tensor product of the GNS-representations on $\mathcal{H}_0^{A(K_+)}$ and $\mathcal{H}_0^{A(K_-)}$ associated to the zero-projectors is – due to the insertion of the signs $(-1)^n$ in the definitions (1.20) and (1.21) of the b's – identified with $\mathcal{F}$ as a representation of the **graded** tensor product

$A(K_+) \hat{\otimes} A(\overline{K_-})$. The Fock- anti-Fock correspondence (see e.g. [SW]) yields now an isomorphism with the representation $A(K_+) \hat{\otimes} A((\overline{K_-})^*)$ on $\mathcal{H}_0^{A(K_+)} \otimes \mathcal{H}_I^{A((\overline{K_-})^*)}$. A final step identifies $A(K)$ with $A(K_+) \hat{\otimes} A(K_-) \cong A(K_+) \hat{\otimes} A((\overline{K_-})^*)$ and $\mathcal{H}_{p_-}^{A(K)}$ with $\mathcal{H}_0^{A(K_+)} \otimes \mathcal{H}_I^{A(K_-)}$.

II. Bogoliubov transformations and the Schwinger term

The first goal of this section is to derive the "Schwinger term" as the Lie algebra cohomology 2-cocycle arising from the **projective** representation of the the natural symmetry group U_{res} (to be defined below) of the CAR-algebra on the Fock space. We then study some of its basic properties; notably we prove that it is non-trivial as well as that it induces well-known cocycles on various infinite dimensional Lie algebras. Finally we give the construction of a Banach Lie group structure on the "deprojectivization" $\tilde{U}_{res}$ of U_{res} associated to the above mentioned projective representation. We also recall in detail the more "hands-on" construction of a central $U(1)$-extension U_{res} given in [PrSe] and prove that the two extensions are isomorphic in the category of Banach Lie groups.

1. Implementation of operators on the Fock space

In this subsection we will analyze the question of how to associate to a one-particle operator, acting on K, an operator on the Fock space. It will turn out that the class of "implementable" symmetries is characterized by having Hilbert-Schmidt "off-diagonal entries". The corresponding unitary group $U_{res}(K, K_+)$ (see below for a precise definition) is only projectively represented on the Fock space and the resulting Lie algebra cohomology 2-cocycle is the "Schwinger term" of QED in two dimensions.

Given an evolution equation of the type

$$i\frac{\partial f}{\partial t} = (H_0 + V(t, x))f, \tag{2.1}$$

one possible approach to extract information from it is to consider the "scattering operator" S. Let us assume that the potential $V(t, x)$ is sufficiently well-behaved to assure the existence of the time evolution operators $U(t, s)$ fulfilling

$$i\frac{\partial}{\partial t}U(t, s)f = (H_0 + V(t, x))U(t, s)f$$

and

$$U(t_3, t_2)U(t_2, t_1) = U(t_3, t_1), \text{ as well as } U(t, t) = Id.$$

Denoting the free time evolution by $U_0(t) = e^{-itH_0}$, S is defined by the strong limit

$$\lim_{\substack{t \to +\infty \\ s \to -\infty}} e^{itH_0}U(t, s)e^{-isH_0}$$

and the "transition amplitudes" $|\langle f, Sg \rangle|^2$ for f, g in an appropriate Hilbert space K have the following interpretation :

$$|\langle f, Sg \rangle|^2 \approx |\langle U_0(t_+)f, U(t_+, s_-)U_0(s_-)g \rangle|^2 \qquad \text{(for } t_+, -s_- >> 0)$$

describes the probability that an incoming free physical state $U_0(s_-)g$ will be found in the outgoing free state $U_0(t_+)f$ after interacting with the field described by $H_0 + V$. Thus asymptotic properties of the interaction Hamiltonian $H_0 + V$, as cross sections and decay rates, are accessible though details of the processes involved are ignored. (See e.g. [Tha], Ch. 8 or [RS3] for a thorough discussion of mathematical scattering theory.)

In the case that H_0 is the free Dirac Hamiltonian, it is of course very important to know if there exists a multi-particle scattering operator $\mathbb{R}$ associated to S acting on the Fock space $\mathcal{F}$. (In fact, relativistic scattering in four dimensions led to the so-called "S-matrix" approach to quantum field theory, compare e.g. [I].) Let us thus look at the following general problem: Given a unitary operator $U : K \to K$, we have a natural "Bogoliubov transformation" $\beta_U : A(K) \to A(K)$ induced by

$$\beta_U(a^*(f)) = a^*(Uf). \tag{2.2}$$

(It is obvious that this definition preserves the canonical anti-commutation rules (1.28)–(1.29), which we agreed to note with Ψ replaced by a in case we are dealing with abstract CAR-algebras.)

Does there then exist a unitary transformation $\mathbb{U}$ of $\mathcal{F} \cong \mathcal{H}_{p_-}$ such that

$$\mathbb{U} \circ \pi_-(\alpha) \circ \mathbb{U}^{-1} = \pi_-(\beta_U(\alpha)) \tag{2.3}$$

for all α in $A(K)$?

(We denote the representation induced by the state ω_{p_-} by π_- here.)

The answer is given by one of the fundamental results in the theory of CAR-algebras:

Theorem II.1 ("Theorem of Powers and Størmer" or "Shale-Stinespring criterion") *The Bogoliubov transformation β_U associated to a unitary operator U on a separable complex Hilbert space K, is implemented on the GNS-space $\mathcal{H}_{p_-}$ ("the Fock space") if and only if the commutator $[p_+ - p_-, U]$ is a Hilbert-Schmidt operator.*

Remarks.

(1) A bounded linear operator B on a separable Hilbert space H is called a "trace class operator" if for one (and then all) orthonormal basis $\{e_k\}_{k \in \mathbb{Z}}$ of H one has

$$\|B\|_{\mathcal{L}^1(H)} = \operatorname{tr}|B| = \sum_{k \in \mathbb{Z}} \langle e_k, |B|e_k \rangle_H < \infty, \tag{2.4}$$

where $|B|$ denotes the non-negative self-adjoint operator $\sqrt{B^*B}$.

More generally, one can define for $p \geq 1$ the "p-th Schatten class" of operators A from H_1 to H_2 by the condition that $(A^*A)^{p/2}$ is of trace class. These spaces

$\mathcal{L}^p(H_1, H_2)$ are Banach spaces with respect to the norms defined by

$$\|A\|_{\mathcal{L}^p}^p = \sum_{k \in \mathbb{Z}} \langle e_k, (A^*A)^{p/2} e_k \rangle_{H_1}. \tag{2.5}$$

Notably one has for $p = 2$ and $H_1 = H_2 = H$ the Hilbert space $\mathcal{L}^2(H)$ of "Hilbert-Schmidt operators" fulfilling

$$\|A\|_{\mathcal{L}^2(H)}^2 = \langle A, A \rangle_{\mathcal{L}^2(H)} = \sum_{k \in \mathbb{Z}} \|Ae_k\|^2 < \infty. \tag{2.6}$$

(See e.g. [Si] for more details on the Schatten classes.)

(2) In fact, the theorem of Powers and Størmer ([PoSt]) is more general than stated above. A rather "hands-on" proof of the announced result can be found in [Tha] along the lines of [CR]; compare also the preceding references [Lu] and [Fr]. The proof of the sufficiency of the Hilbert-Schmidt condition follows from Theorem II.4 below and is also proved in [dlHJ]. The approach via infinite-dimensional Clifford algebras goes back to [ShSt] and is reviewed in detail in the textbook [PR].

(3) Though we will not reproduce the proof of this result, we find it useful and instructive to apply the explicit machinery of the "second quantization of one particle operators", trying to stick to the notations employed in [Tha] and [CR].

(4) The theorem of Powers and Størmer is especially useful in the "external field approach" to quantum field theory. Its idea is to prove the existence of the scattering operator S on the one-particle Hilbert space and then to implement S on the Fock space. Thus the "particles" are quantized, whereas the "external field" enters only as a perturbation of the free Hamiltonian, i.e. it is not to be quantized. (See e.g. [Tha], Ch. 10, [RS3], XI.15 and e.g. [LM2] as an example for a more recent treatment.)

The set of implementable unitary operators

$$U_{\text{res}} = U_{\text{res}}(K, K_+) = \{U \in U(K) \mid [p_+ - p_-, U] \in \mathcal{L}^2(K)\} \tag{2.7}$$

forms a Banach Lie group, called the "restricted unitary group of the polarized Hilbert space $K = K_+ \oplus K_-$". Its topology is induced by the embedding

$$U_{\text{res}}(K, K_+) \hookrightarrow U(K) \times \mathcal{L}^2(K), \ U \mapsto (U, [p_+ - p_-, U]).$$

Let us recall the standard (2×2)-block-matrix description of a linear operator A from $K \to K$:

$$A = \begin{pmatrix} A_+ & A_{+-} \\ A_{-+} & A_- \end{pmatrix}, \tag{2.8}$$

where $A_\pm$ go from $K_\pm$ to $K_\pm$ and A_{+-} respectively A_{-+} map K_- to K_+ respectively K_+ to K_-.

The Lie algebra $\mathfrak{u}_{\text{res}}(K, K_+)$ is now described as

$$\{A \in \mathcal{B}(K) \mid A^* = -A \quad \text{and} \quad A_{+-}, A_{-+} \text{ are Hilbert Schmidt operators}\}, \tag{2.9}$$

where $\mathcal{B}(K)$ is the space of bounded linear operators on K, and it is normed by

$$\|A\|_{\mathrm{ures}} = \|A_+\| + \|A_-\| + \|A_{+-}\|_2 + \|A_{-+}\|_2, \tag{2.10}$$

where $\|\ \|$ denotes the operator norm and $\|\ \|_2$ the Hilbert-Schmidt norm (see formula (2.6)).

Proposition II.2 *Let $K = K_+ \oplus K_-$ be a polarized Hilbert space. Then*

(i) *the exponential map $\mathfrak{u}_{\mathrm{res}} \to U_{\mathrm{res}}$, $\exp(A) = \sum_{n\geq 0} \frac{1}{n!} A^n$, gives real-analytic coordinates on U_{res} and*

(ii) *the connected components of U_{res} are the sets*

$$U_{\mathrm{res}}^k = \{U \in U_{\mathrm{res}} \mid \mathrm{index}\,(U_+) = k\},$$

where $\mathrm{index}(U_+) = \dim(\mathrm{Ker}\,U_+) - \dim(\mathrm{Coker}\,U_+)$ and k is in $\mathbb{Z}$.

Proof. Recalling that the exponential series gives the exponential map of the Banach Lie group $U(K)$ and observing that $\exp(A) = \sum_{n\geq 0} \frac{1}{n!} A^n$ is in U_{res} for A in $\mathfrak{u}_{\mathrm{res}}$ it follows easily that this series is the exponential map for U_{res} and provides real-analytic coordinates.

The second assertion contains the fact that $U_+ : K_+ \to K_+$ (and analogously U_-) is a Fredholm operator, i.e. $\dim \mathrm{Ker}\,U_+$ and $\dim \mathrm{Coker}\,U_+ = (\dim K_+/\mathrm{Im}\,U_+)$ are finite. This can easily be seen as follows

$$1 = UU^* \text{ and } 1 = U^*U \text{ imply}$$
$$1_+ = \mathrm{Id}_{K_+} = U_+(U_+)^* + U_{+-}(U_{+-})^* \text{ and}$$
$$1_+ = (U_+)^*U_+ + (U_{-+})^*U_{-+}.$$

Thus U_+ is invertible up to compact operators, i.e. U_+ is Fredholm.

Since the projection $U \mapsto U_+$ from U_{res} to $\mathrm{Fred}(K_+)$, the space of Fredholm operators on K_+, is continuous as well as the map $T \mapsto \mathrm{ind}(T)$ on the set of Fredholm operators it follows that each connected component of U_{res} lies inside one of the U_{res}^k.

The connectedness of U_{res}^k follows now from the slightly more general result Theorem 1.4 in [CHO'B]. □

Remarks.

(1) It is easily checked that $\mathrm{ind}((U \cdot V)_+) = \mathrm{ind}(U_+) + \mathrm{ind}(V_+)$, i.e. $U \mapsto \mathrm{ind}(U_+)$ is a homomorphism onto $\mathbb{Z}$ with kernel U_{res}^0.

(2) For the basics of the theory of Fredholm operators the reader might consult [Dou] or [Mur].

Lemma II.3 *The map $U_{\mathrm{res}}(K, K_+) \to U(\mathcal{H}_{p_-}), U \mapsto \mathbb{U}$ given by the theorem of Powers and Størmer yields a projective representation $U_{\mathrm{res}}(K, K_+) \to \mathbb{P}U(\mathcal{H}_{p_-})$.*

Proof. The representation of $A(K)$ on $\mathcal{H}_{p_-}$ is irreducible as can be seen from the general theory of "pure states" (compare e.g. [Mur] or [dlHJ]) or by uniqueness and cyclicity of the vacuum line $\mathbb{C} \cdot \xi_{p_-}$ (compare (1.30) and Proposition I.2). Since the product $V = (\mathbb{U}_1)^{-1}\mathbb{U}_2$ of two implementers of the same U in U_{res}

necessarily commutes with all $\{\pi_{p_-}(\alpha)|\alpha \in A(K)\}$, Schur's lemma shows that V is the multiplication by a complex number of modulus one.   □

As a corollary we can define the following central extension of U_{res} by $U(1)$:

$$U_{\mathrm{res}}^{\sim} = U_{\mathrm{res}}^{\sim}(K, K_+) = \{(U, \mathbb{V}) \in U_{\mathrm{res}}(K, K_+) \times U(\mathcal{H}_{p_-}) \mid [\mathbb{U}] = [\mathbb{V}] \text{ in } \mathbb{P}U(\mathcal{H}_{p_-})\}.$$

Unfortunately this definition of $U_{\mathrm{res}}^{\sim}$ does not furnish the structure of a Banach Lie group on it, since the homomorphism $\Theta : U_{\mathrm{res}} \to \mathbb{P}U(\mathcal{H}_+)$ is not continuous with respect to the topology on the latter group coming from the norm topology on operators. But $\mathbb{P}U(\mathcal{H}_+)$ is a metrizable topological group in the topology induced from the strong operator topology (see [DD]). Furthermore the homomorphism Θ is continuous with respect to this topology by the continuity of the action $U_{\mathrm{res}} \times CAR(K) \to CAR(K)$ and a closer inspection of the GNS-construction leading to $\mathcal{H}_+$ (See [Wa], Lemma on p.481). This does of course nevertheless not provide a Lie group structure on $U_{\mathrm{res}}^{\sim}$ since there is no Lie group structure known on the unitary group of a Hilbert space with underlying topology given by the strong operator topology.

Our strategy will now be the following: in Lemma II.5 below we will find for each A in $\mathfrak{u}_{\mathrm{res}}$ a (generally unbounded!) implementer $\mathbb{A}$ defined on and preserving a dense domain $\mathcal{D}$ in $\mathcal{F} = \mathcal{H}_+$, which is independent of A. We can then calculate the commutator $[\mathbb{A}, \mathbb{B}]$ and compare it to the operator $\mathbb{C}$ associated to the commutator $C = [A, B]$. It will turn out that $[\mathbb{A}, \mathbb{B}] - \mathbb{C}$ is a scalar multiple $s(A, B) \cdot 1_{\mathcal{F}}$ of the identity of $\mathcal{F}$ and that the cocycle s is continuous. This observation allows us for the time being to define a Banach Lie algebra structure on $\mathfrak{u}_{\mathrm{res}}^{\sim} = \mathfrak{u}_{\mathrm{res}} \oplus i\mathbb{R}$.

Ignoring the subtleties coming from the unboundedness of the implementers this procedure is the standard way of calculating the class $[s]$ in $H^2(\mathfrak{u}_{\mathrm{res}}(K, K_+), i\mathbb{R})$ associated to this central extension: a representative s of it is given by a linear section σ of the following short exact sequence of Lie algebras:

$$0 \longrightarrow i\mathbb{R} = \mathrm{Lie}\, U(1) \longrightarrow \mathfrak{u}_{\mathrm{res}}^{\sim} \xrightarrow{\sigma} \mathfrak{u}_{\mathrm{res}}(K) \longrightarrow 0 \qquad (2.11)$$

and the formula

$$s(A, B) = [\sigma(A), \sigma(B)] - \sigma([A, B]) \text{ for } A, B \text{ in } \mathfrak{u}_{\mathrm{res}}.$$

In Subsection II.3 below we will carefully analyze the construction of the corresponding Banach Lie group.

Let us fix again orthonormal bases $\{e_j | j \geq 0\}$ of K_+ and $\{e_j | j < 0\}$ of K_-. The one-particle matrix description of an A in $\mathcal{B}(K)$:

$$Ae_j = \sum_{k \in \mathbb{Z}} \langle e_k, Ae_j \rangle e_k$$

can be viewed as the annihilation of e_j followed by the creation of all possible e_k's weighted by the matrix elements $\langle e_k, Ae_j \rangle$.

Thus one arrives – in the "concrete Fock space" notations of Section I – at the following rather suggestive description of A :

$$A\Psi^*\Psi = \sum_{k\in\mathbb{Z}}\sum_{j\in\mathbb{Z}}\langle e_k, Ae_j\rangle\Psi^*(e_k)\Psi(e_j) = \sum_{j\in\mathbb{Z}}\Psi^*(Ae_j)\Psi(e_j).$$

This expression can now – at least formally – be interpreted as an operator on the Fock space $\mathcal{F}$. Decomposing $A\Psi^*\Psi$ with respect to K_+ and K_- (recall formulas (1.26) and (1.27)) yields:

$$A\Psi^*\Psi = \sum_{i,j\geq 0}\langle e_i, Ae_j\rangle a^*(e_i)a(e_j)$$

$$+ \sum_{i\geq 0, k<0}\langle e_i, Ae_k\rangle a^*(e_i)b^*(e_k)$$

$$+ \sum_{k<0, j\geq 0}\langle e_k, Ae_j\rangle b(e_k)a(e_j)$$

$$+ \sum_{k<0, \ell<0}\langle e_k, Ae_\ell\rangle b(e_k)b^*(e_\ell)$$

$$= Aa^*a + Aa^*b^* + Aba + Abb^*$$

$$= A_+a^*a + A_{+-}a^*b^* + A_{-+}ba + A_-bb^*. \tag{2.12}$$

We immediately encounter a major problem upon calculating the "vacuum expectation value" of Abb^* :

$$\langle\Omega, Abb^*\Omega\rangle_\mathcal{F} = \sum_{k,\ell<0}\langle e_k, Ae_\ell\rangle_K\langle\Omega, b(e_k)b^*(e_\ell)\Omega\rangle_\mathcal{F}$$

$$= \sum_{k,\ell<0}\langle e_k, Ae_k\rangle_K\langle b^*(e_k)\Omega, b^*(e_\ell)\Omega\rangle_\mathcal{F}$$

$$= \sum_{k,\ell<0}\langle e_k, Ae_\ell\rangle_K\langle e_k, e_\ell\rangle_K = \text{trace } A_-.$$

In other words, unless A_- fulfils an undesirable trace class condition – which is neither fulfilled by the Hamiltonian nor by the grading nor by the identity of K – the vacuum vector is not in the domain of the definition of $A\Psi^*\Psi$. Thus we boldly subtract the possibly infinite trace and redefine the operators by the so-called "normal-ordering" prescription, i.e. each time a creator **precedes** an annihilator the order is changed and a factor of (-1) is inserted. We arrive at

$${}^\circ_\circ Abb^*{}^\circ_\circ = - \sum_{k,\ell<0}\langle e_k, Ae_\ell\rangle b^*(e_\ell)b(e_k).$$

In case A_- is trace class, one has of course

$${}^\circ_\circ Abb^*{}^\circ_\circ = Ab^*b - (\text{trace } A_-).$$

All the other summands in $A\Psi^*\Psi$ already being in this order, we have

$${}^\circ_\circ A\Psi^*\Psi{}^\circ_\circ = Aa^*a + Aa^*b^* + Aba + ({}^\circ_\circ Abb^*{}^\circ_\circ) \ . \tag{2.13}$$

Let us remark that the importance of the Hilbert-Schmidt condition now already shows up in the following easy observation: Ω is in the domain of Aa^*b^* if and only if A_{+-} is Hilbert-Schmidt. Going back to the Powers-Størmer theorem one has the following beautifully concrete prescription for the implementers:

Theorem II.4 *Let A be in $\mathfrak{u}_{res}(K, K_+)$. Then ${}_\circ^\circ A\Psi^*\Psi_\circ^\circ$ is well-defined on a dense domain in $\mathcal{F}$ (independent of A) and has a unique anti-selfadjoint extension $\mathbb{A}$. Furthermore $\exp(t\mathbb{A})$ is unitary for all t in $\mathbb{R}$ and implements $U_t = \exp(tA)$, i.e. $\exp(t\mathbb{A})$ is a representative of $[U_t]$.*

Remark. Though the result seems to make part of conventional wisdom in quantum field theory, the proof in [Lu] appears to have a gap and to be only completed in [CR] (see their Proposition 2.1 and the subsequent remarks.) Similar considerations in an equivalent set-up can be found in the more recent textbook [Ot].

Sketch of a proof of Theorem II.4. Let us define projection operators P_L on the Fock space $\mathcal{F}$ by

$$P_L\left(\sum_{n,m\geq 0} \Psi^{(n,m)}\right) = \sum_{n+m\leq L} \Psi^{(n,m)}$$

and an obviously dense subspace of $\mathcal{F}$ by $\mathcal{D} = \{\xi = \sum_{n,m\geq 0} \Psi^{(n,m)} \in \mathcal{F} \mid \Psi^{(n,m)} \in \mathcal{F}^{(n,m)}$ and $\exists L$ such that $P_L(\xi) = \xi\}$. Physicists call $\mathcal{D}$ the space of "finite particle vectors" because it consists of physical states described as finite superpositions of states with a bounded number of particles and anti-particles.

The explicit form of the second-quantized operators ${}_\circ^\circ A\Psi^*\Psi_\circ^\circ$ implies that

$$P_{L+2}({}_\circ^\circ A\Psi^*\Psi_\circ^\circ)P_L = ({}_\circ^\circ A\Psi^*\Psi_\circ^\circ)P_L$$

and thus that $\mathcal{D}$ is an invariant dense domain for all ${}_\circ^\circ A\Psi^*\Psi_\circ^\circ$ with A in $\mathfrak{u}_{res}$. The crucial ingredient of the proof is the estimate

$$\|({}_\circ^\circ A\Psi^*\Psi_\circ^\circ)P_L(\xi)\|_{\mathcal{F}} \leq (L+2) \cdot \|A\|_{\mathfrak{u}_{res}} \cdot \|P_L(\xi)\|_{\mathcal{F}},$$

which can be directly derived from our formulas. An easy application of this estimate shows then that all elements of $\mathcal{D}$ are analytic vectors for all ${}_\circ^\circ A\Psi^*\Psi_\circ^\circ$. Since the ${}_\circ^\circ A\Psi^*\Psi_\circ^\circ$ are anti-symmetric we conclude by Nelson's analytic vector theorem (see e.g. [RS2]) that there is a unique anti-selfadjoint extension $\mathbb{A}$ of $({}_\circ^\circ A\Psi^*\Psi_\circ^\circ)$. The theorem of Stone-von Neumann (see e.g. [RS1]) assures now the existence of the unitary one-parameter group $\exp(t\mathbb{A})$. Direct calculations show then that $\exp(t\mathbb{A})$ implements $\exp(tA)$.   □

Remark. The above theorem implies the sufficiency of the Hilbert-Schmidt condition in Theorem II.1.

Let us proceed by calculating some useful "second-quantized operators":

Taking the identity $\mathbf{1}$ of K we find

$${}^{\circ}_{\circ}\mathbf{1}\Psi^*\Psi{}^{\circ}_{\circ} = \mathbf{1}a^*a + ({}^{\circ}_{\circ}\mathbf{1}bb^*{}^{\circ}_{\circ})$$

$$= \sum_{i,j\geq 0} \langle e_i, e_j\rangle a^*(e_i)a(e_j) - \sum_{k,\ell<0} \langle e_k, e_\ell\rangle b^*(e_k)b(e_\ell)$$

$$= \sum_{j\geq 0} a^*(e_j)a(e_j) - \sum_{k<0} b^*(e_k)b(e_k).$$

Applying this to a vector $\psi^{(n,m)} \in \mathcal{F}^{(n,m)} = \Lambda^n K_+ \otimes \Lambda^m(CK_-)$ yields

$$({}^{\circ}_{\circ}\mathbf{1}\Psi^*\Psi{}^{\circ}_{\circ})(\psi^{(n,m)}) = (n-m)\cdot\psi^{(n,m)}.$$

Thus ${}^{\circ}_{\circ}\mathbf{1}\Psi^*\Psi{}^{\circ}_{\circ}$ is the "charge operator" Q on $\mathcal{F}$.

In the same vein, one sees that

$$({}^{\circ}_{\circ}(p_+ - p_-)\Psi^*\Psi{}^{\circ}_{\circ})(\psi^{(n,m)}) = (n+m)\cdot\psi^{(n,m)},$$

i.e. ${}^{\circ}_{\circ}(p_+ - p_-)\Psi^*\Psi{}^{\circ}_{\circ}$ equals the "number operator" N.

A similar calculation shows that $\mathbb{U}_t = \exp(-it({}^{\circ}_{\circ}H\Psi^*\Psi{}^{\circ}_{\circ}))$ equals the ad-hoc construction of an implementer $\tilde{U}_t$ of the one-particle evolution operator $U_t = \exp(-iH)$ given in Section I.2 and thus we confirm the semi-boundedness from below of the spectrum of the second-quantized Hamiltonian $\mathbb{H} = ({}^{\circ}_{\circ}H\Psi^*\Psi{}^{\circ}_{\circ})$.

Theorem II.4 provides us of course with a direct way to calculate the Lie algebra cocycle defining $\mathfrak{u}^{\sim}_{\mathrm{res}}(K, K_+)$:

Lemma II.5 *On the common dense domain $\mathcal{D}$ in $\mathcal{F}$, one has for elements A, B of $\mathfrak{u}_{\mathrm{res}}(K, K_+)$ with commutator $C = [A, B]$ the following identity:*

$$[\mathbb{A}, \mathbb{B}] = [{}^{\circ}_{\circ}A\Psi^*\Psi{}^{\circ}_{\circ}, {}^{\circ}_{\circ}B\Psi^*\Psi{}^{\circ}_{\circ}] = ({}^{\circ}_{\circ}C\Psi^*\Psi{}^{\circ}_{\circ}) + s(A, B)\cdot\mathbf{1}_{\mathcal{F}}$$

with $s(A, B) = \mathrm{trace}(A_{-+}B_{+-} - B_{-+}A_{+-}).$

Proof. Let us first observe that the section $\sigma : \mathfrak{u}_{\mathrm{res}}(K, K_+) \to \{$ anti-self adjoint operators on $\mathcal{F}$ defined on and preserving $\mathcal{D}\}$ of the sequence (2.11) given by $A \mapsto \mathbb{A}$ has been in fact fixed by demanding the vanishing of vacuum expectation values:

$$\langle \Omega, \mathbb{A}\Omega\rangle_{\mathcal{F}} = 0.$$

Since the difference $[{}^{\circ}_{\circ}A\Psi^*\Psi{}^{\circ}_{\circ}, {}^{\circ}_{\circ}B\Psi^*\Psi{}^{\circ}_{\circ}] - ({}^{\circ}_{\circ}C\Psi^*\Psi{}^{\circ}_{\circ})$ is necessarily a multiple of the identity of $\mathcal{F}$ we find

$$s(A, B) = \langle\Omega, s(A, B)\cdot\Omega\rangle = \langle\Omega, [{}^{\circ}_{\circ}A\Psi^*\Psi{}^{\circ}_{\circ}, {}^{\circ}_{\circ}B\Psi^*\Psi{}^{\circ}_{\circ}]\Omega\rangle.$$

A direct calculation using (1.24), (1.25) and (1.12), (1.13) yields now:

$$\langle\Omega, ({}^{\circ}_{\circ}A\Psi^*\Psi{}^{\circ}_{\circ})({}^{\circ}_{\circ}B\Psi^*\Psi{}^{\circ}_{\circ})\Omega\rangle = \mathrm{trace}(A_{-+}B_{+-})$$

and thus the formula for s follows. $\qquad\square$

Remarks.

(1) Since the product of two Hilbert-Schmidt operators has a trace, the "commutator anomaly" s defines a continuous cocycle giving rise to an element in the Lie algebra cohomology $H^2(\mathfrak{u}_{res}, i\mathbb{R})$.

(2) In case A, B are in $\mathcal{L}^1(K)$, one has a much more direct method to compute $s(A, B)$. Observe first that then $A\Psi^*\Psi$ and $B\Psi^*\Psi$ are well-defined on $\mathcal{D}$ and indeed

$$[A\Psi^*\Psi, B\Psi^*\Psi] = ([A, B])\Psi^*\Psi.$$

Using that ${}^\circ_\circ A\Psi^*\Psi{}^\circ_\circ = A\Psi^*\Psi - (\text{trace } A_-)\mathbf{1}$ for trace class operators we find

$$[{}^\circ_\circ A\Psi^*\Psi{}^\circ_\circ, {}^\circ_\circ B\Psi^*\Psi{}^\circ_\circ] - ({}^\circ_\circ C\Psi^*\Psi{}^\circ_\circ) = (\text{trace } C_-) \cdot \mathbf{1}$$

$$= \text{trace}(A_{-+}B_{+-} - B_{-+}A_{+-}) + \text{trace}([A_-, B_-]).$$

Since trace $A_-B_- = $ trace B_-A_- for trace class operators, the last term vanishes.

Before turning to the analysis of the properties of the commutator anomaly s we will collect some information on the action of $U_{res}(K, K_+)$ on the Fock space $\mathcal{F}$. Let us first recall that the second quantization of the identity of the one-particle Hilbert space K is the charge operator Q characterized by

$$Q|_{\mathcal{F}(n,m)} = (n - m) \cdot \mathbf{1}_{\mathcal{F}(n,m)}.$$

It follows that $\mathcal{F} = \underset{k \in \mathbb{Z}}{\oplus} \mathcal{F}_k$, where the "charge-$k$ sector" is given as

$$\mathcal{F}_k := \{\psi \in \mathcal{F} | Q\psi = k \cdot \psi\} = \underset{\substack{n,m \geq 0 \\ n-m=k}}{\bigoplus} \mathcal{F}^{(n,m)}.$$

Lemma II.6

(i) All implementers $\mathbb{U}$ of a given U in $U_{res}(K, K_+)$ map $\mathcal{F}_k$ to $\mathcal{F}_{k+q(U)}$ for all k in $\mathbb{Z}$ and $q(U) = index(U_+)$.

(ii) The connected component of the neutral element of $\tilde{U}_{res}(K, K_+)$ acts irreducibly on $\mathcal{F}_k$ for all k.

(iii) The representation of $\tilde{U}_{res}(K, K_+)$ on $\mathcal{F}$ is irreducible.

Proof. Since $e^{it}U = Ue^{it}$ and

$$({}^\circ_\circ e^{it}\Psi^*\Psi{}^\circ_\circ) = ({}^\circ_\circ e^{it\mathbf{1}_K}\Psi^*\Psi{}^\circ_\circ) = e^{it({}^\circ_\circ \mathbf{1}_K\Psi^*\Psi{}^\circ_\circ)} = e^{itQ}$$

by Lemma II.4, we know that $e^{itQ}\mathbb{U} = e^{i\varphi(t)}\mathbb{U}e^{itQ}$ with a real-valued function φ on $\mathbb{R}$.

Since $e^{itQ} = e^{i\varphi(t)}\mathbb{U}e^{itQ}\mathbb{U}^{-1}$ is a 1-parameter group with period 2π it follows that $\varphi(t) = q(\mathbb{U})t$ with an integer $q(\mathbb{U})$. Observing furthermore that $\mathbb{U}e^{itQ}\mathbb{U}^{-1}$ depends only on U, one has in fact $q(\mathbb{U}) = q(U)$ in $\mathbb{Z}$. The equality $e^{itQ} = e^{itq(U)}\mathbb{U}e^{itQ}\mathbb{U}^{-1}$ implies now $Q\mathbb{U} = q(U)\mathbb{U} + \mathbb{U}Q$ on $\mathcal{D}$ and thus $\mathbb{U}(\mathcal{F}_k) = \mathcal{F}_{k+q(U)}$, which also assures us that $q(U_1 \cdot U_2) = q(U_1) + q(U_2)$. Since a vacuum vector Ω in $\mathcal{F}$ has no charge, we find (after normalizing $\|\Omega\| = 1$) that $q(U) = \langle\Omega, (Q + \mathbb{U}Q\mathbb{U}^{-1})\Omega\rangle = \langle\Omega, (\mathbb{U}Q\mathbb{U}^{-1})\Omega\rangle$ is a continuous function on U_{res}. Thus

$q(\exp(At)) = 0$ for all t in $\mathbb{R}$ and A in $\mathfrak{u}_{\text{res}}$ and q is constant on the connected components of U_{res}. Thus it follows from Proposition II.2 that q factorizes over the homomorphism $U_{\text{res}} \rightarrow \mathbb{Z}, U \mapsto \text{index}(U_+)$. An easy direct calculation with e.g. a shift operator T fulfilling $\text{index}(T_+) = 1$, yields $q(T_+) = 1$. (Compare [Tha], pp. 295-296). Thus the first assertion is proven.

Let us observe that if H is a separable complex Hilbert space and k is in $\mathbb{N}_0$ then $U(H)$ acts irreducibly on $\Lambda^k H$ by the representation $\Lambda^k(U)(v_1 \wedge \cdots \wedge v_k) = (Uv_1) \wedge \cdots \wedge (Uv_k)$. We recall furthermore that if $\pi_\alpha : G_\alpha \rightarrow U(H_\alpha)$ are irreducible unitary representations of topological groups ($\alpha = 1, 2$) then the representation $\pi_1 \otimes \pi_2 : G \times G_2 \rightarrow U(H_1 \otimes H_2)$ is irreducible as well (See e.g. [Ga], Thm. IV.3.15).

It follows that the representation of $U(H_1) \times U(H_2)$ on $\Lambda^n H_1 \otimes \Lambda^m H_2$ is irreducible for all $n, m \geq 0$.

We proceed to analyze the action of the second quantization $\mathbb{A}$ on

$$\Lambda^n K_+ \otimes \Lambda^m (CK_-) \text{ for } A = \begin{pmatrix} A_+ & 0 \\ 0 & A_- \end{pmatrix}:$$

$$({}^\circ_\circ A_+ \Psi \Psi^* {}^\circ_\circ)(e_{l_1} \wedge \cdots \wedge e_{l_n}) = \left(\sum_{i,j \geq 0} \langle e_i, Ae_j \rangle a^*(e_i)a(e_j) \right) (e_{l_1} \wedge \cdots \wedge e_{l_n})$$

$$= \sum_{r=1}^{n} e_{l_1} \wedge \cdots \wedge e_{l_{r-1}} \wedge Ae_{l_r} \wedge e_{l_{r+1}} \wedge \cdots \wedge e_{l_n} = \Lambda^p(A_+)(e_{l_1} \wedge \cdots \wedge e_{l_n}) .$$

The action of $({}^\circ_\circ A_- \Psi \Psi^* {}^\circ_\circ) = -\sum_{k,l<0}\langle e_k, A_- e_l \rangle b^*(e_l)b(e_k)$ is conveniently described by identifying the space CK_- (inside K) with the abstract conjugate Hilbert space $\overline{K_-}$ via $\Theta : CK_- \rightarrow \overline{K_-}, Cf_- \mapsto \overline{f_-}$. The map Θ is obviously a complex-linear unitary isomorphism, since the "charge conjugation" C was chosen to be complex anti-linear and norm-preserving. The real Lie algebra $\mathfrak{u}(K_-)$ is obviously represented on $\overline{K_-}$ via $\vartheta : \mathfrak{u}(K_-) \rightarrow \mathfrak{u}(\overline{K_-}), \vartheta(A_-)\overline{f_-} = \overline{(A_- f_-)}$.

A direct calculation as above for A_+ shows now that the action of $({}^\circ_\circ A_- \Psi^* \Psi {}^\circ_\circ)$ on $\Lambda^m(CK_-)$ is intertwined by $\Lambda^m(\Theta)$ to $\Lambda^m \vartheta(A_-)$ on $\Lambda^m \overline{K_-}$. Thus we deduce that $\mathfrak{u}(K_+) \times \mathfrak{u}(K_-)$ acts irreducibly on $\mathcal{F}^{(n,m)} = \Lambda^n K_+ \otimes \Lambda^m(CK_-) \cong \Lambda^n K_+ \otimes \Lambda^m(\overline{K_-})$.

We conclude thus that the Fockspace representation of the subalgebra $\mathfrak{u}(K_+) \times \mathfrak{u}(K_-)$ on $\mathcal{F}^{(n,m)}$ is simply the derivative of the (non-projective) action of $U(K_+) \times U(K_-)$ described above. Furthermore all the modules $\mathcal{F}^{(n,m)}$ are non-isomorphic irreducible $(U(K_+) \times U(K_-))$-representations.

Any bounded $U(K_+) \times U(K_-)$-invariant operator T on $\mathcal{F}_k = \bigoplus_{n-m=k} \mathcal{F}^{(n,m)}$ is therefore diagonal with $T|_{\mathcal{F}^{(n,m)}} = \lambda_{n,m} \cdot 1_{\mathcal{F}^{(n,m)}}$. Assuming that T is not a multiple of the identity of $\mathcal{F}_k$ there exist n and m such that $\lambda_{n+1,m+1} \neq \lambda_{n,m}$.

It is now straightforward to construct a vector $\psi^{(n,m)}$ in $\mathcal{F}^{(n,m)}$ and A_{+-} in $\mathcal{L}^2(K_+, K_-)$ such that $({}^\circ_\circ A_{+-} \Psi^* \Psi {}^\circ_\circ) = A_{+-} a^* b^*$ maps $\psi^{(n,m)}$ to a non-vanishing vector in $\mathcal{F}^{(n+1,m+1)}$ ("pair production" in the physics jargon). This contradicts the above assumption on T and shows that T is a multiple of $1_{\mathcal{F}_k}$. By Schur's

lemma the representation of $(U_{res}^0)^\sim = (U_{res}^\sim)^0$, the central extension of the connected component U_{res}^0 of U_{res}, on $\mathcal{F}_k$ is irreducible. This proves the second assertion. (Let us remark that A_{+-} itself is not in u_{res} but in its complexification,
i.e. it exists A, B in u_{res} such that $A + iB = A_{+-}$. Since the second quantization
map $A \mapsto \mathbb{A}$ obviously extends to this complexification our usage of the operator
A_{+-} is permitted. See also below after Corollary II.12 for more details on this
complexification.)

We write the Fock space now as the completed infinite sum over the charge
sectors: $\mathcal{F} = \bigoplus_{k \in \mathbb{Z}} \mathcal{F}_k$ and recall that $\mathcal{F}_k$ is the eigenspace associated to the eigenvalue k of the charge operator $Q = (\overset{\circ}{\circ}1_K \Psi^* \Psi \overset{\circ}{\circ})$. Since 1_K is in U_{res}^0 the $\mathcal{F}_k$ are
non-isomorphic irreducible representations of this group.

Using any U in U_{res} such that $\mathrm{ind}(U_+) = 1$ assertion (i) implies that any
implementer $\mathbb{U}$ fulfils $\mathbb{U} \cdot \mathcal{F}_k = \mathcal{F}_{k+1}$. A similar argument as used above in the proof
of the second assertion now shows that any bounded operator on $\mathcal{F}$ commuting
with the $U_{res}^\sim(K, K_+)$-action is necessarily a multiple of the identity and thus the
$U_{res}^\sim(K, K_+)$-representation on the Fock space $\mathcal{F}$ is irreducible. $\square$

Remark. The usual proof of the irreducibility of the representation of U_{res} on $\mathcal{F}$
uses an identification of K with $L^2(S^1, \mathbb{C})$ and an embedding $C^\infty(S^1, U(1))$ into
U_{res} (as in Lemma II.9 below). It turns out that $\mathcal{F}$ is already irreducible as a
$C^\infty(S^1, U(1))$ -representation. On the level of groups (as opposed to Lie algebras)
this approach seems to be independently found in [PrSe] and [CH] and makes use
of the so-called "boson-fermion correspondence"

2. The Schwinger term

The aim of this subsection is to point out several remarkable properties of the
commutator anomaly found in Lemma II.5:

$$s(A, B) = \mathrm{trace}(A_{-+} \cdot B_{+-} - B_{-+} \cdot A_{+-})$$
$$= 2i \, \mathrm{Im} \, \{\mathrm{trace}(A_{-+} B_{+-})\} \qquad (2.14)$$

for A, B in $u_{res}(K, K_+)$.

Notably we will show the non-triviality of this 2-cocycle, as well as the fact that it
induces the well-known extensions of the vector fields on the circle to the Virasoro
algebra respectively of the loop algebras to the affine Kač-Moody algebras. Furthermore we will relate it to the "japanese cocycle" on generalized Jacobi matrices.

We close this subsection by proving that current groups, the analog of loop
groups for higher dimensional sources, embed into "restricted unitary groups"
defined by Schatten ideals with $p > 2$. We sketch in this last part a beautiful idea
of Mickelsson and Rajeev of how to use pseudo-differential operators in order to
prove this result. This touches upon the relation between second quantization and
non-commutative geometry à la Connes, a subject we will nevertheless not further
pursuit here.

The commutator anomaly s was defined as the 2-cocycle measuring the de-
viation of the projective representation $U_{res}(K, K_+) \to \mathbb{P}U(\mathcal{H}_{p_-}^{A(K)})$ from being

liftable to a non-projective representation. We calculated it by identifying $\mathcal{H}_{p_-}^{A(K)}$ with the so-called Fock space $\mathcal{F}$ and implementing elements of $U_{\mathrm{res}}(K, K_+)$ via the normal ordering of second-quantized operators.

It is also known as the "Schwinger term" because of its close relation to the "current anomaly" appearing in quantization of $(1+1)$-dimensional QED. (Compare e.g. [CHO'B], pp. 365-366, [Mic], pp. 105-106 and [Rui] for the mathematical point of view and the original article of Schwinger [Sc] and [Pe], p. 651 for the physicist's approach.) The fundamental observation is that in "all" quantization schemes physicists apply to classical gauge field theory part of the classical symmetries do not pertain to the associcated Quantum field theory; we will discuss mathematical versions of this theme in the Sections IV and V.

Using the notation $\varepsilon = p_+ - p_-$ for the operator defining the polarization $K = K_+ \oplus K_-$, we can rewrite (2.9) as

$$u_{\mathrm{res}}(K, K_+) = \{A \in u(K) \mid [\varepsilon, A] \in \mathcal{L}^2(K)\} \tag{2.15}$$

and (2.14) as

$$s(A, B) = \frac{1}{4} \operatorname{trace} (\varepsilon \cdot [\varepsilon, A] \cdot [\varepsilon, B]). \tag{2.16}$$

We consider first the case of trace class operators, which includes of course the case that $\dim K < \infty$. Defining

$$t : \mathcal{L}^1(K) \to \mathbb{C}, \quad t(A) := \frac{1}{2} \operatorname{trace} (\varepsilon \cdot A),$$

we have

Lemma II.7 *The Lie algebra cohomology 2-cocycle s is the coboundary of the 1-cocycle t on* $u_{\mathrm{res}}(K, K_+) \cap \mathcal{L}^1(K)$.

Proof. We have for A, B in $u_{\mathrm{res}}(K, K_+) \cap \mathcal{L}^1(K)$

$$(\delta t)(A, B) = -t([A, B]) = -\frac{1}{2} \operatorname{trace} (\varepsilon \cdot [A, B])$$

$$= -\frac{1}{2} \operatorname{trace} (\varepsilon AB - \varepsilon BA) = \frac{1}{4} \operatorname{trace} (A\varepsilon B - AB\varepsilon - \varepsilon AB + A\varepsilon B)$$

$$= \frac{1}{4} \operatorname{trace} (\varepsilon \cdot [\varepsilon, A] \cdot [\varepsilon, B]) = s(A, B). \qquad \square$$

Corollary II.8 *On* $u_{\mathrm{res}}(K, K_+) \cap \mathcal{L}^2(K)$ *one has*

$$s(A, B) = -\frac{1}{2} \operatorname{trace} (\varepsilon \cdot [A, B]).$$

Though it will follow from Theorem II.16 below we find it useful to give a direct proof of the fact that loop groups embed into $U_{\mathrm{res}}(F, F_+)$ for appropriate polarized Hilbert spaces $F = F_+ \oplus F_-$. (To avoid a clash of notation we use in this subsection II.2 the symbol F for the one-particle Hilbert space.)

Lemma II.9 *Let K be a compact Lie group and $K \overset{\rho}{\hookrightarrow} U(\mathbb{C}^d)$ a faithful unitary representation of K. Then the loop group $LK = C^\infty(S^1, K)$ acts by unitary multiplication oprators on $F = L^2(S^1, \mathbb{C}^d)$. Polarising F as $F_+ \oplus F_-$, where F_+ respectively F_- is the subspace of F having only non-negative respectively negative Fourier components, LK is mapped into $U_{\mathrm{res}}(F, F_+)$.*

Proof. Let us abbreviate $\exp(2\pi i k t)$ by $e_k(t)$ and apply Fourier development to f in F as well as to $\rho(\varphi)$ for φ in LK :

$$f = \sum_{k \in \mathbb{Z}} f_k e_k \text{ with } f_k \text{ in } \mathbb{C}^d \text{ and}$$

$$\rho(\varphi) = \sum_{k \in \mathbb{Z}} \varphi_k e_k \text{ with } \varphi_k \text{ in } \mathrm{Mat}(d \times d, \mathbb{C}).$$

Defining M_φ as the multiplication by $\rho(\varphi)$ on F :

$$(M_\varphi(f))(t) := \rho(\varphi(t)) \cdot f(t),$$

we get

$$M_\varphi(e_\ell) = \sum_{k \in \mathbb{Z}} (M_\varphi)_{k,\ell} e_k$$

with $(M_\varphi)_{k,\ell} = \varphi_{k-\ell}$.

Furthermore M_φ is a unitary operator on F and

$$
\begin{aligned}
\|[\varepsilon, M_\varphi]\|^2_{\mathcal{L}^2(F)} &= \mathrm{tr}([\varepsilon, M_\varphi] \cdot ([\varepsilon, M_\varphi])^*) \\
&= 4 \cdot \mathrm{tr}\{((M_\varphi)_{+-}) \cdot ((M_\varphi)_{+-})^* + ((M_\varphi)_{-+}) \cdot ((M_\varphi)_{-+})^*\} \\
&= 4 \cdot \left\{ \sum_{\substack{k \geq 0 \\ \ell < 0}} (M_\varphi)_{k,\ell} \overline{((M_\varphi)_{k,\ell})} + \sum_{\substack{k < 0 \\ \ell \geq 0}} (M_\varphi)_{k,\ell} \overline{((M_\varphi)_{k,\ell})} \right\} \\
&= 4 \cdot \left\{ \sum_{r > 0} r|\varphi_r|^2 + \sum_{s < 0} (-s)|\varphi_s|^2 \right\} \\
&= 4 \cdot \left(\sum_{n \in \mathbb{Z}} |\varphi_n|^2 \cdot |n| \right).
\end{aligned}
$$

The last right hand side is of course finite for smooth φ and thus M_φ is in $U_{\mathrm{res}}(F, F_+)$. $\qquad\square$

Remark. Working with the definition of a smooth map between manifolds modelled on locally convex vector spaces à la Milnor (see, e.g., [Mil2], [N1] or Subsection IV.1 below), which generalizes the usual definition in the case of Banach manifolds, one easily shows that the exponential map exists for both groups and that they are local diffeomorphisms as well as that the map $LK \to U_{\mathrm{res}}$ is smooth.

A similar calculation yields the pull-back of the Schwinger term to $L\mathfrak{k} = C^\infty(S^1, \mathfrak{k})$, the Lie algebra of LK :

Lemma II.10 *Denoting the differential of the embedding* $LK \hookrightarrow U_{\mathrm{res}}(F, F_+)$ *of Lemma II.9 by* $j : L\mathfrak{k} \hookrightarrow \mathfrak{u}_{\mathrm{res}}(F, F_+)$, *we have*

$$(j^*s)(\xi, \eta) = \frac{1}{2\pi i} \int_0^1 tr_d\{\rho(\dot{\xi}(t)) \cdot \rho(\eta(t))\} dt,$$

where ξ and η are in $L\mathfrak{k}$, tr_d is the trace over the complex vector space $\mathbb{C}^d$ and S^1 is parametrized by $[0, 1]$.

Proof. We use again Fourier developements:

$$\rho(\xi) = \sum_{k \in \mathbb{Z}} \xi_k e_k \text{ and } \rho(\eta) = \sum_{k \in \mathbb{Z}} \eta_k e_k$$

with ξ_k, η_k in $\mathrm{Mat}(d \times d, \mathbb{C})$ and ρ denoting here the Lie algebra representation $\mathfrak{k} \hookrightarrow \mathfrak{u}(d)$. A direct calculation shows

$$(j^*s)(\xi, \eta) = s(j(\xi), j(\eta)) = s(M_\xi, M_\eta)$$
$$= \mathrm{tr}_F\{((M_\xi)_{-+}) \cdot ((M_\eta)_{+-}) - ((M_\eta)_{-+}) \cdot ((M_\xi)_{+-})\} \quad \text{by (2.14)}$$
$$= \mathrm{tr}_d\{\sum_{r<0}(r\xi_r\eta_{-r} - r\eta_r\xi_{-r})\},$$

since the trace over the tensor product $F \cong \mathbb{C}^d \otimes L^2(S^1, \mathbb{C})$ of an operator $a \otimes A$ with a in $\mathrm{Mat}(d \times d, \mathbb{C})$ and A an operator on $L^2(S^1, \mathbb{C})$ is given by $\mathrm{tr}_F(a \otimes A) = \mathrm{tr}_d(a) \cdot \mathrm{tr}_{L^2(S^1,\mathbb{C})}(A)$.

Finally

$$(j^*s)(\xi, \eta) = \mathrm{tr}_d\left(\sum_{r \in \mathbb{Z}} r\xi_r\eta_{-r}\right) = \frac{1}{2\pi i} \int_0^1 \mathrm{tr}_d(\rho(\dot{\xi}(t)) \cdot \rho(\eta(t))) dt. \qquad \square$$

We deduce the important

Corollary II.11 *The class $[j^*(s)]$ in $H^2(L\mathfrak{k}, i\mathbb{R})$ is non-trivial.*

Proof. Let us fix any one-dimensional torus $U(1)$ in K. The Lie algebra $\rho(\mathfrak{u}(1))$ in $\mathfrak{u}(d)$ consists then of the real multiples of a fixed matrix $a = \rho(\zeta)$ with ζ in $\mathfrak{u}(1) \subset \mathfrak{k}$. We define $\xi(t) = \cos(2\pi t) \cdot \zeta$ and $\eta(t) = \sin(2\pi t) \cdot \zeta$ in $L\mathfrak{k}$. Since the commutator in $L\mathfrak{k}$ is point-wise taken we have

$$[\xi, \eta](t) = [\xi(t), \eta(t)] = 0.$$

Thus ξ and η generate a two-dimensional abelian Lie algebra $\mathfrak{a}$ in $L\mathfrak{k}$:

$$\mathfrak{a} = ((\xi, \eta))_{\mathbb{R}} \overset{i_\mathfrak{a}}{\hookrightarrow} L\mathfrak{k}.$$

Pulling-back j^*s to $\mathfrak{a}$ we have

$$i_\mathfrak{a}^* \circ j^*(s)(\xi, \eta) = \frac{1}{2\pi i} \int_0^1 \mathrm{tr}_d\{\rho(\dot\xi(t)) \cdot \rho(\eta(t))\} dt$$

$$= \left(\frac{2\pi}{2\pi i}\right) \int_0^1 \sin^2(2\pi t) \cdot \mathrm{tr}_d((-a) \cdot a) dt$$

$$= \frac{\mathrm{tr}_d(a^*a)}{i} \int_0^1 \sin^2(2\pi t) dt \neq 0.$$

Thus $i_\mathfrak{a}^* \circ j^*(s)$ gives rise to a non-trivial central extension of $\mathfrak{a}$, the Heisenberg Lie algebra, and therefore $[j^*s] \neq 0$ for all $L\mathfrak{k}$. □

Obviously we then have

Corollary II.12 *The class $[s]$ in $H^2(\mathfrak{u}_{\mathrm{res}}, i\mathbb{R})$ is non-zero.*

Proof. Since all separable complex Hilbert spaces are unitarily isomorphic and all polarizations $F = F_+ \oplus F_- = F'_+ \oplus F'_-$ with $F_\pm$ and $F'_\pm$ infinite-dimensional are related by a unitary transformation, i.e. there always exists a U in $U(F)$ such that $U(F_\pm) = F'_\pm$, we can choose any model for $F = F_+ \oplus F_-$. Using $F = L^2(S^1, \mathbb{C}^d)$ together with a faithful unitary representation of a compact Lie group K on $\mathbb{C}^d$, Corollary II.11 implies the assertion, since $0 \neq [j^*s] = j^*[s]$. □

Let us observe that the involved groups U_{res} and LK allow for obvious complexifications, notably we have

$$GL_{\mathrm{res}}(F, F_+) = \{g \in GL(F) \mid [\varepsilon, g] \in \mathcal{L}^2\}, \tag{2.17}$$

where $GL(F)$ denotes the invertible bounded linear operators on F, and its Lie algebra equals

$$\mathfrak{gl}_{\mathrm{res}}(F, F_+) = \{A \in \mathfrak{gl}(F) \mid [\varepsilon, A] \in \mathcal{L}^2\},$$

where $\mathfrak{gl}(F)$ denotes the space $\mathcal{B}(F)$ of bounded linear operators on F.
Observing further that

$$LK^\mathbb{C} = C^\infty(S^1, K^\mathbb{C})$$

is a naturally defined complexification of LK (see [Gl] for universality properties of the complexification $C^0(S^1, K^\mathbb{C})$ of $C^0(S^1, K)$ in the category of Banach Lie groups), we can describe how the Schwinger term pertains to the complexifications:

Lemma II.13

(i) *The Schwinger term is well-defined for A, B in $\mathfrak{gl}_{\mathrm{res}}(F, F_+)$:*

$$s(A, B) = \mathrm{trace}(A_{-+} \cdot B_{+-} - B_{-+} \cdot A_{+-}). \tag{2.18}$$

(ii) *Given a faithful representation ρ of $K^\mathbb{C}$ on $\mathbb{C}^d$, we get an injective homomorphism $j : LK^\mathbb{C} \hookrightarrow GL_{\mathrm{res}}(F, F_+)$ (with F and F_+ as in Lemma II.9) by associating multiplication operators to elements of $LK^\mathbb{C}$.*

(iii) *The pull-back of s to $L\mathfrak{k}^\mathbb{C} = C^\infty(S^1, \mathfrak{k}^\mathbb{C})$, the Lie algebra of $LK^\mathbb{C}$, is non-trivial and thus the class $[s]$ itself is non-zero in $H^2(\mathfrak{gl}_{\mathrm{res}}, \mathbb{C})$.*

We omit the proofs which are word-by-word reproductions of the corresponding arguments in the "compact" case.

Remarks.

(1) The central extensions of $L\mathfrak{k}^{\mathbb{C}}$ for K compact and semi-simple are closely related to certain classes of Kač-Moody Lie algebras. Their rôle in the representation theory of LK can of course be found in [PrSe] and is also developed in some detail in [N2].

(2) In its extended form (2.18) on $\mathfrak{gl}_{\mathrm{res}}(F, F_+)$, the Schwinger term can easily be related to the so-called "japanese cocycle": Let $\mathfrak{gl}(\infty, \mathbb{C})$ be the Lie algebra of doubly-infinite matrices $A = (a_{ij})_{i,j \in \mathbb{Z}}$ having only finitely many non-zero entries and $\mathfrak{gl}_J(\mathbb{C})$ the Lie algebra of "generalized Jacobian matrices" having only finitely many non-zero diagonales. Otherwise stated $A = (a_{ij})_{i,j \in \mathbb{Z}}$ is in $\mathfrak{gl}_J(\mathbb{C})$ if there is a N in $\mathbb{N}$ such that $a_{ij} = 0$ for all i, j fulfilling $|i - j| > N$. Defining $\pi(a)$ as the "upper left corner" of a doubly-infinite matrix $A = (a_{ij})_{i,j \in \mathbb{Z}}$, i.e. $\pi(A)_{i,j} = a_{i,j}$ for $i, j \leq 0$ and all other entries of $\pi(a)$ are zero, we can define a two-cocycle τ on $\mathfrak{gl}_J(\mathbb{C})$ by

$$\tau(A, B) := \mathrm{tr}(\pi([A, B]) - [\pi(A), \pi(B)])$$

for A, B in $\mathfrak{gl}_J(\mathbb{C})$. (Observe that we write here a doubly-infinite matrix in a different order relative to our convertion on block matrices of operators on a polarized Hilbert space $F = F_+ \oplus F_-$.)

The fundamental result on the two-cocycle τ is the following:

Proposition II.14

(i) *The two-cocycle τ is a coboundary on $\mathfrak{gl}(\infty, \mathbb{C})$.*

(ii) *The class of the two-cocycle τ is non-trivial in $H^2(\mathfrak{gl}_J(\mathbb{C}), \mathbb{C})$.*

(iii) *The class $[\tau]$ is a generator of $H^2(\mathfrak{gl}_J(\mathbb{C}), \mathbb{C})$.*

Remark. The easy proofs of (i) and (ii) as well as references for the proof of the third assertion can be found in [Fuk].

Fixing a Hilbert basis $\{e_k | k \in \mathbb{Z}\}$ of $F = F_+ \oplus F_-$ such that $\{e_k | k \leq 0\}$ respectively $\{e_k | k > 0\}$ is a Hilbert basis for F_- respectively F_+, we can associate doubly-infinite matrices to elements of $\mathfrak{gl}_{\mathrm{res}}(F, F_+)$. (Note that we here and only here put e_0 to F_- in order to be in line with the definition of π in [Fuk].) We find that for A, B in $\mathfrak{gl}_{\mathrm{res}} \cap \mathfrak{gl}_J$

$$\begin{aligned}
\tau(A, B) &= \mathrm{tr}(\pi([A, B]) - [\pi(A), \pi(B)]) \\
&= \mathrm{tr}(A_{-+} \cdot B_{+-} - B_{-+} \cdot A_{+-}) \\
&= s(A, B).
\end{aligned}$$

Of course $\mathfrak{gl}_{\mathrm{res}} \cap \mathfrak{gl}_J$ is smaller than $\mathfrak{gl}_{\mathrm{res}}$ or $\mathfrak{gl}_J$, but the essential feature for the construction of τ and s is the same, namely the fact that the off-diagonal entries are "small", whereas the diagonal entries are "big".

Let us also observe that the Virasoro algebra can be induced from the Schwinger term. In order to be brief we will not give the complete proof here and recommend, e.g., [PrSe] for more details.

Proposition II.15

(i) *The group $Diff^+(S^1)$ of orientation-preserving smooth diffeomorphisms of the circle is embedded into $U_{res}(F, F_+)$ for $F = L^2(S^1, \mathbb{C})$.*

(ii) *The central extension of the Lie algebra $diff(S^1)$, the smooth vector fields on the circle, induced by pulling back the Schwinger term is the Virasoro algebra.*

Starting point of the proof. Let F be $L^2(S^1, \mathbb{C})$ and φ in $Diff^+(S^1)$, then we can define an unitary action of φ on F by setting:

$$(\varphi \cdot f)(t) := f(\varphi^{-1}(t)) \cdot ((\varphi^{-1})'(t))^{1/2},$$

where for an orientation-preserving diffeomorphism ψ of S^1 ψ' is the $\mathbb{R}^{>0}$-valued function defined by the following equation of 1-forms on S^1 :

$$\psi'(t)dt = d\psi = \psi^*(dt).$$

The above action is of course intrinsically described as the action on half-densities or half-forms and relates in the case of S^1 to spinors. For the completion of the proof we propose p. 91 of [PrSe] to the reader. □

Remark. For a wealth of information on the Virasoro Lie algebra and Lie group and related geometric structures the reader may consult [Ov].

In order to relate to the general problem of fermionic second quantization in $(d+1)$ dimensions and to the concept of "current anomalies", we introduce several natural objects, which are closely related to non-commutative geometry. For much more on the intriguing relation between second quantization, Schwinger terms etc and non-commutative geometry see the fundamental contribution [Ar], and [LM1] as one example of many recent articles.

We consider a smooth, compact d-dimensional spin manifold X with a fixed Riemannian metric g and a fixed spin structure. We thus have an associated bundle S of complex spinors and the self-adjoint Dirac operator D^S acting on its sections. Assuming furthermore the existence of a hermitian vector bundle E with a metric connection and compact structure group K, we have the "twisted Dirac operator" $D = D^{S \otimes E}$ on the bundle $S \otimes E$ (see e.g. [Ta], Ch. 10). The operator D has a self-adjoint extension to $F = \Gamma_{L^2}(X, S \otimes E)$ and has the Fredholm property due to its ellipticity as a first order partial differential operator. This allows a decomposition of F into $F_+ \oplus F_-$ into its subspaces $F_\pm$ corresponding to the non-negative respectively negative spectral subspaces. Ignoring the finite-dimensional kernel of D, we can describe the grading operator $\varepsilon = p_+ - p_-$ by $\frac{D}{|D|}$, "the phase of the Dirac operator".

The principal K-bundle P over X which is associated to the auxiliary vector bundle E, gives rise to two bundles $P \times_K K$ and $P \times_K \mathfrak{k}$, where K acts by conjugation on itself and by the adjoint action on its Lie algebra $\mathfrak{k}$.

Smooth sections of the former constitute the "gauge group" $\text{Aut}(E)$, whereas the sections of the latter form the "gauge algebra" $\mathfrak{aut}(E)$, the Lie algebra of $\text{Aut}(E)$. Since K and $\mathfrak{k}$ act by assumption on the typical fibre of E, $\text{Aut}(E)$ and $\mathfrak{aut}(E)$ act (pointwise) on sections of E. Both actions extend trivially to actions on sections of $S \otimes E$.

Remarks.

(1) In the case that $E \cong X \times \mathbb{C}^n$ is trivial, $\mathfrak{aut}(E)$ is isomorphic to $C^\infty(X, \mathfrak{k})$, the "current algebra" with source X and target $\mathfrak{k}$.

(2) Physically, sections of S might be interpreted as particles ("matter") whereas connections on E viewed as gauge potential give rise – via their curvature – to gauge fields as the electromagnetic or the Yang-Mills fields ("interaction carrier"). Physical quantities as Lagrangians are chosen to be gauge-invariant at least **before** either Hamiltonian second quantization, the so-called "operator formalism", or path integral quantization are applied. Violation of these symmetries after quantization leads then to "anomalous current identities" or more general "anomalies" in physics. This corresponds in the Hamiltonian approach to the appearance of Schwinger terms. (Compare the classical reference [J], and e.g. [La] for an attempt to cast this in the language of non-commutative geometry.)

The crucial property of the action of the gauge transformations is for us the following

Theorem II.16 *Let (X, S, E, D) be as above. Let M_ξ be the action of ξ in $\mathfrak{aut}(E)$ on $F = \Gamma_{L^2}(X, S \otimes E) = F_+ \oplus F_-$ and $\varepsilon = p_+ - p_-$. Then $[\varepsilon, M_\xi]$ is in the Schatten class $\mathcal{L}^p(F)$ if $p \underset{\neq}{>} d = \dim X$.*

Remarks.

(1) The statement of the theorem holds verbatim for g in $\text{Aut}(E)$.

(2) The case $X = S^1$ boils down to the case of loop groups considered in Lemma II.9. On the other hand, if $\dim X \geq 2$, the gauge transformations are only in generalizations of the restricted unitary group, namely in

$$\{U \in U(F) | [\varepsilon, U] \in \mathcal{L}^p(F)\} \quad \text{for } p \underset{\neq}{>} \dim X \geq 2.$$

Thus the theorem of Powers-Størmer does **not** apply and consequently the representation of the gauge group on the Fock space over F demands more refined constructions, which are not yet fully understood (see [Mic] for the concept of "generalized Schwinger terms"). The appearance of the condition $[\varepsilon, A] \in \mathcal{L}^p(F)$ for an operator A on F ties also closely in with the idea of a "p-summable Fredholm module" in Connes' non-commutative differential geometry (cf. [Con]). Though the sketch of a proof found in [PrSe], p. 100 for $X = (S^1)^d$ is not too difficult to elaborate, the general result demands an ingenious use of the calculus of pseudo-differential operators and is due to Mickelsson and Rajeev ([MR]). The basic idea is nevertheless simple to explain. The classical pseudo-differential operator $A = D^2 = D^*D$ has leading symbol $g_x(\xi, \xi) \cdot \mathbf{1}_{S \otimes E}$, where x is in X, g_x the metric on $T_x^* X$ and ξ is in this latter space. Ignoring the finite-dimensional kernel of D and the details of "smoothing near $\xi = 0$", the operator $A^{-1/2} = |D|^{-1}$ is again a classical pseudo-differential operator having leading symbol

$$\frac{1}{\sqrt{g_x(\xi, \xi)}} \cdot \mathbf{1}_{S \otimes E} = \frac{1}{|\xi|_x} \cdot \mathbf{1}_{S \otimes E}.$$

It follows that the symbol of $\varepsilon = D \circ |D|^{-1}$ is given in local coordinates $x = (x_1, \dots, x_d)$ by

$$a_0(x, \xi) = \sum_{\mu=1}^{d} (\gamma_\mu(x) \otimes \mathbf{1}_E) \frac{\xi_\mu}{|\xi|_x}$$

plus terms of order (-1). Here $\gamma_\mu(x)$ is the matrix of Clifford multiplication by $dx_\mu|_x$ and an element ξ in $T_x^* X$ is written as $\sum_{\mu=1}^{d} \xi_\mu \cdot (dx_\mu|_x)$ with ξ_μ in $\mathbb{R}$. Since a gauge transformation g in $\mathfrak{aut}(E)$ or $\mathrm{Aut}(E)$ is locally written as $\mathbf{1}_S \otimes M(x)$ with $M(x)$ a matrix, g is a classical pseudodifferential operator of order 0 and commutes with the operator given by the principal symbol of ε. Thus $[\varepsilon, g]$ has only order (-1) and therefore is in $\mathcal{L}^p(F)$ if p is strictly greater than the dimension of X.

3. The central extensions $\widetilde{\mathrm{U}}_{\mathrm{res}}$ and $\widetilde{\mathrm{GL}}_{\mathrm{res}}$

In this subsection we will give the explicit construction of the Banach Lie groups $\widehat{\mathrm{GL}}_{\mathrm{res}}$ and $\hat{U}_{\mathrm{res}}$ à la Pressley and Segal and show that the corresponding Lie algebra extensions of $\mathfrak{gl}_{\mathrm{res}}$ respectively $\mathfrak{u}_{\mathrm{res}}$ are given by the Schwinger term. We will then proceed to show that the topological group $\widetilde{U}_{\mathrm{res}}$, given by the "deprojectivization" of the representation of U_{res} on the Fock space, does in fact carry the structure of a Banach Lie group and that $\widetilde{U}_{\mathrm{res}}$ and $\hat{U}_{\mathrm{res}}$ are isomorphic as such.

Starting from this subsection we will go back to the convention that $K = K_+ \oplus K_-$ denotes a fixed separable polarized Hilbert space and notations as $\mathrm{GL}_{\mathrm{res}}$ will always mean $\mathrm{GL}_{\mathrm{res}}(K, K_+)$. Denoting the set of Fredholm operators on a separable Hilbert space H by $\mathrm{Fred}(H)$ we recall that $\mathrm{Fred}(H)$ is open in the norm topology on $\mathcal{B}(H)$, the bounded linear operators on H.

Proposition II.17 *The map*

$$GL_{\mathrm{res}}(K, K_+) \to Fred(K_+), A = \begin{pmatrix} a & b \\ c & d \end{pmatrix} \mapsto a$$

is a homotopy equivalence.

Sketch of the proof. Let us first remark that it is easy to see that a and d are Fredholm operators if A is in $\mathrm{GL}_{\mathrm{res}}$ (Compare Proposition II.2).

The map $\mathrm{GL}_{\mathrm{res}} \xrightarrow{\pi} \mathcal{F} := \pi(\mathrm{GL}_{\mathrm{res}}) \subset \mathrm{Fred}(K_+) \times \mathcal{L}^2(K_+, K_-)$ given by $\begin{pmatrix} a & b \\ c & d \end{pmatrix} \mapsto \begin{pmatrix} a \\ c \end{pmatrix}$ realizes the fibre bundle $\mathrm{GL}_{\mathrm{res}} \to \mathrm{GL}_{\mathrm{res}}/B$, where

$$B = \left\{ \begin{pmatrix} 1 & b \\ 0 & d \end{pmatrix} \middle| d \in \mathrm{GL}(K_-) \text{ and } b \in \mathcal{L}^2(K_+, K_-) \right\}$$

since $\mathrm{GL}_{\mathrm{res}}$ acts transitively on $\mathcal{F}$ and the isotropy of $\begin{pmatrix} 1 \\ 0 \end{pmatrix}$ is given by the group B. Since $\mathrm{GL}(K_-)$ is contractible ([Ku]) B is contractible as well and π is thus a homotopy equivalence.

Projecting $\mathcal{F}$ to $\mathrm{Fred}(K_+)$ by $p\begin{pmatrix} a \\ c \end{pmatrix} = a$ one finds by direct verification that p is surjective, $\mathcal{F}$ is open in the Banach space $\mathcal{B}(K_+) \times \mathcal{L}^2(K_+, K_-)$ and the fibre

$$p^{-1}(a) = \left\{ \begin{pmatrix} a \\ c \end{pmatrix} \middle| c \in \mathcal{L}^2(K_+, K_-) \text{ and } c|_{\ker a} \text{ is injective} \right\}$$

is contractible for all a.

A more refined inspection shows then that the map p is a homotopy equivalence and thus the same holds true for the map $A \mapsto a$.   □

Remark. The usual reference for a proof of the preceding proposition is in fact the proof of Proposition 6.2.4 in [PrSe] (pp. 81–82). Unfortunately the proof that the map p is a homotopy equivalence is not given there. The missing details can be found in [Wu2].

Following Pressley and Segal we define the set

$$\mathcal{E} = \left\{ (A, q) \in \mathrm{GL}^0_{\mathrm{res}} \times \mathrm{GL}(K_+) \,\middle|\, a - q \in \mathcal{L}^1(K_+) \right\} \tag{2.19}$$

and topologize it via the set-theoretic embedding

$$\varphi : \mathcal{E} \to \mathrm{GL}^0_{\mathrm{res}} \times \mathcal{L}^1(K_+), \varphi(A, q) = (A, a - q).$$

Lemma II.18
(i) $\mathcal{E}$ is a subgroup of $\mathrm{GL}^0_{\mathrm{res}} \times \mathrm{GL}(K_+)$.
(ii) $\varphi(\mathcal{E})$ is open in $\mathrm{GL}^0_{\mathrm{res}} \times \mathcal{L}^1(K_+)$.
(iii) $\mathcal{E}$ is a Banach Lie group with the subspace topology of $\varphi(\mathcal{E})$.
(iv) The sequence

$$\{1\} \to GL^1(K_+) \xrightarrow{\alpha} \mathcal{E} \xrightarrow{\beta} GL^0_{\mathrm{res}} \to \{1\},$$

where $GL^1(K_+) = \{q \in GL(K_+) | q - 1_+ \in \mathcal{L}^1(K_+)\}, \alpha(q) = (1, q)$ and $\beta(A, q) = A$, is a short exact sequence of Banach Lie groups.

Remark. The Banach Lie group structure on $GL^1(K_+)$ is of course induced by the embedding

$$GL^1(K_+) \hookrightarrow \mathcal{L}^1(K_+), q \mapsto q - 1_+.$$

Proof of Lemma II.18. The first assertion follows by directly verifying that the product of two elements of $\mathcal{E}$ as well as the inverse of an element of $\mathcal{E}$ are in $\mathcal{E}$. Openness of $GL(K_+)$ in $\mathcal{B}(K_+)$ implies that $\varphi(\mathcal{E})$ is open in $GL^0_{\mathrm{res}} \times \mathcal{L}^1(K_+)$.

Let us define a multiplication "$*$" on $\mathcal{A} := \mathfrak{gl}_{\text{res}} \oplus \mathcal{L}^1(K_+)$ by setting

$$(A, t) * (A', t') = (AA', bc' + at' + ta' - tt').$$

It is straightforward to check that $(\mathcal{A}, +, *)$ is a complex Banach algebra with unit $(1, 0)$ and that for elements (A, t) and (A', t') in $\varphi(\mathcal{E})$ one has:

$$(A, t) * (A', t') = \varphi([\varphi^{-1}(A, t)] \cdot [\varphi^{-1}(A', t')]).$$

Let us now observe that the set $\mathcal{A}^*$ of invertible elements of a unital Banach algebra $\mathcal{A}$ is always open in $\mathcal{A}$ and forms a Banach Lie group (modelled on $\mathcal{A}$). The set $\varphi(\mathcal{E})$ is obviously an open subgroup of $\mathcal{A}^*$, but in fact it is equal to it: let (A, t) be in $\mathcal{A}^*$ and (A', t') be its inverse. We then have $aa' = 1_+ - bc'$ and $bc' + at' + ta' - tt' = 0$ and thus $(a - t)(a' - t') = 1_+$, i.e. $a - t = q$ is invertible and $\text{ind}(a) = \text{ind}(a - t) = 0$, showing that $(A, t) = \varphi(A, q)$.

The projection β is surjective since for A in GL_{res}^0, the Fredholm index of a is zero and thus there exists a finite rank operator t such that $a + t = q$ is invertible. The kernel of β is given by $\{(1, q)|1_+ - q \in \mathcal{L}^1(K_+)\}$, the image of α. $\square$

Let us now consider the following diagram:

$$
\begin{array}{ccc}
\mathcal{E} & \longrightarrow & \text{GL}(K_+) \times \mathcal{L}^1(K_+) \\
\beta \downarrow & & \downarrow \gamma \\
\text{GL}_{\text{res}}^0 & \longrightarrow & \text{Fred}^0(K_+)
\end{array}
,
$$

where the upper horizontal map is given by $(A, q) \mapsto (q, aq^{-1} - 1_+)$, the lower horizontal map by $\begin{pmatrix} a & b \\ c & d \end{pmatrix} \mapsto a$ and γ is defined as $\gamma(q, t) = (1 + t)q$. Obviously the diagram commutes and γ is surjective with typical fibre $\text{GL}^1(K_+)$ acting as follows:

$$(1 + s) \cdot (q, t) = ((1 + s)q, (1 + t)(1 + s)^{-1} - 1).$$

The (surjective) upper horizontal map being the pullback of the lower homotopy equivalence (compare Proposition II.17) is a homotopy equivalence as well.

Corollary II.19

(i) *The Banach Lie group $\mathcal{E}$ is contractible.*

(ii) *The homotopy groups of GL_{res}^0 fulfil*

$$\pi_0(\text{GL}_{\text{res}}^0) = \{0\},$$
$$\pi_{2k+1}(\text{GL}_{\text{res}}^0) = \{0\} \quad \text{and}$$
$$\pi_{2k+2}(\text{GL}_{\text{res}}^0) \cong \mathbb{Z} \quad \text{for } k \geq 0.$$

Proof. Since the map $\mathcal{E} \to \text{GL}(K_+) \times \mathcal{L}^1(K_+)$ is a surjective homotopy equivalence onto a contractible space, it follows that $\mathcal{E}$ is contractible itself.

The exact sequence in (iv) of Lemma II.18 implies now that for $j \geq 0$

$$\pi_{j+1}(\text{GL}_{\text{res}}^0) \cong \pi_j(\text{GL}^1(K_+)).$$

By a result of Palais ([Pa]) the latter groups are given by $\pi_j(\mathrm{GL}(\infty, \mathbb{C}))$, where $\mathrm{GL}(\infty, \mathbb{C}) = \bigcup_{n \geq 1} \mathrm{GL}(n, \mathbb{C})$ is the ascending union of the finite-dimensional complex general linear groups with the inductive limit topology. Observing that $\pi_j(\mathrm{GL}(\infty, \mathbb{C})) = \pi_j(\mathrm{GL}(n, \mathbb{C}))$ for $j < 2n$ allows the application of Bott periodicity (see e.g. [Mil1]) which yields the second assertion. $\square$

In order to define the desired central extension of $\mathrm{GL}^0_{\mathrm{res}} \cong \mathcal{E}/\alpha(\mathrm{GL}^1(K_+))$ by $\mathbb{C}^*$ we first recall some basic facts on determinants.

Let H be a separable Hilbert space on $\mathbb{C}$ and $\bigotimes^n H$ respectively $\Lambda^n H$ the Hilbert space completion of its n-th tensor product respectively its n-th exterior product in the algebraic sense. Then we can associate to every bounded operator A on H an operator $\bigotimes^n(A)$ uniquely defined by

$$\bigotimes^n(A)(v_1 \otimes \ldots \otimes v_n) = Av_1 \otimes \ldots \otimes Av_n.$$

Since $\bigotimes^n(A)$ preserves the subspace $\Lambda^n H$ of $\bigotimes^n H$ we have an induced operator $\Lambda^n(A)$ which fulfils and is uniquely fixed by

$$\Lambda^n(A)(v_1 \wedge \ldots \wedge v_n) = Av_1 \wedge \ldots \wedge Av_n.$$

Obviously we have for A, B in $\mathcal{B}(H)$:

$$\bigotimes^n(A \cdot B) = \bigotimes^n(A) \cdot \bigotimes^n(B) \quad \text{and} \quad \Lambda^n(A \cdot B) = \Lambda^n(A) \cdot \Lambda^n(B).$$

Let us furthermore recall that in finite dimensions the determinant can be expressed in terms of traces: let $A = 1 + (A - 1) = 1 + B$ be a linear operator on a n-dimensional Hilbert H, then

$$\Lambda^n(A) = (\det A) \cdot 1_{\Lambda^n H} = \left[\sum_{j=0}^{n} \mathrm{tr}(\Lambda^j(B)) \right] \cdot 1_{\Lambda^n H}. \tag{2.20}$$

For $B = 1 - A$ of trace class we can mimick the above formula by setting

$$\det(1 + B) := \sum_{j=0}^{\infty} \mathrm{tr}(\Lambda^k(B)). \tag{2.21}$$

Let us resume the properties of this "(first) Fredholm determinant" (compare e.g. [RS4] for the proofs):

Proposition II.20 *For B and C in $\mathcal{L}^1(H)$ one has:*

(i) $\|\Lambda^k(B)\|_{\mathcal{L}^1(\Lambda^k H)} \leq \left(\|B\|_{\mathcal{L}^1(H)} \right)^k / (k!)$

(ii) $|\det(1 + B)| \leq \exp(\|B\|_{\mathcal{L}^1(H)})$

(iii) $1 + B$ *is invertible if and only if* $\det(1 + B) \neq 0$

(iv) $\det((1 + B)(1 + C)) = \det(1 + B) \cdot \det(1 + C)$

(v) $|\det(1 + B) - \det(1 + C)| \leq \|B - C\|_{\mathcal{L}^1(H)} \cdot \exp(\|B\|_{\mathcal{L}^1(H)} + \|C\|_{\mathcal{L}^1(H)} + 1)$.

It follows immediately that the Fredholm determinant is well-defined and yields a continuous homomorphism $\det : \mathrm{GL}^1(H) \to \mathbb{C}^*$ which allows to make

Definition II.21 Let H be a separable Hilbert space. Then the *special linear group of H with respect to the first Fredholm determinant* is defined as

$$\mathrm{SL}^1(H) = \Big\{ 1 + B \in \mathrm{GL}^1(H) \,|\, \det(1 + B) = 1 \Big\}.$$

Remark. Obviously $\mathrm{SL}^1(H)$ is a closed subgroup of $\mathrm{GL}^1(H)$. Furthermore, taking any q_0 in $\mathcal{L}^1(H)$ such that $\mathrm{tr}(q_0) \neq 0$ the space $\mathbb{C} \cdot q_0$ is a closed topological complement of the Lie algebra $\mathfrak{sl}^1(H) = T_e(\mathrm{SL}^1(H)) = \{T \in \mathcal{L}^1(H) \,|\, \mathrm{tr}(T) = 0\}$. Thus $\mathrm{SL}^1(H)$ is a (Banach) Lie subgroup of $\mathrm{GL}^1(H)$ in the sense of [Bo].

Going back to the group $\mathcal{E}$ defined by (2.19) it follows that $\alpha(\mathrm{SL}^1(K_+))$ is a closed Banach Lie subgroup of it since taking again a q_0 in $\mathcal{L}^1(K_+)$ such that $\mathrm{tr}(q_0) \neq 0$ it follows easily that $\mathfrak{gl}_{\mathrm{res}} \oplus \mathbb{C} \cdot q_0$ is a closed topological complement of $\alpha(\mathfrak{sl}^1(K_+)) = \{(0, t) \,|\, t \in \mathcal{L}^1(K_+)\}$ and $\mathrm{tr}(t) = 0\}$. Furthermore $\alpha(\mathrm{SL}^1(K_+))$ is easily seen to be normal since $\mathrm{tr}(qTq^{-1}) = \mathrm{tr}(T)$ for T in $\mathcal{L}^1(K_+)$ and q an invertible bounded operator on K_+. We define the quotient group by

$$\widehat{\mathrm{GL}}_{\mathrm{res}}^{0} := \mathcal{E}/\alpha(\mathrm{SL}^1(K_+)). \qquad (2.22)$$

Proposition 11 in Paragraph 1.6 of [Bo] implies that $\widehat{\mathrm{GL}}_{\mathrm{res}}^{0}$ is a Banach Lie group. The homomorphism $\beta : \mathcal{E} \to \mathrm{GL}_{\mathrm{res}}^{0}$ induces a short exact sequence

$$\{1\} \to \ker \pi \to \widehat{\mathrm{GL}}_{\mathrm{res}}^{0} \xrightarrow{\pi} \mathrm{GL}_{\mathrm{res}}^{0} \to \{1\}, \qquad (2.23)$$

where $\pi([A, q]) = A$ and $(A, q) \sim (A, q')$ if and only if it exists $1 + T$ in $\mathrm{SL}^1(K_+)$ such that $q' = q(1 + T)$.

A direct calculation yields that

$$\ker \pi = \{[1, r] \in \widehat{\mathrm{GL}}_{\mathrm{res}}^{0} \,|\, r \in \mathrm{GL}^1(K_+)\} \to \mathbb{C}^*, [1, r] \mapsto \det(r)$$

is a complex-analytic isomorphism and that, furthermore, $\ker \pi$ is central in $\widehat{\mathrm{GL}}_{\mathrm{res}}^{0}$, since

$$[A, q][1, r]([A, q])^{-1} = [1, qrq^{-1}] \mapsto \det(qrq^{-1}) = \det(r).$$

Thus (2.23) defines a central extension of $\mathrm{GL}_{\mathrm{res}}^{0}$ by $\mathbb{C}^*$ in the category of Banach Lie groups.

The associated Lie algebra cocycle on $\mathfrak{gl}_{\mathrm{res}}$ can either be calculated as in ([PrSe], pp. 88-89) or directly as follows. Let us recall that $\mathcal{E} \cong \varphi(\mathcal{E})$, the set of invertible elements in a certain Banach algebra $\mathcal{A}$ with multiplication "$*$". It follows that the Lie bracket on the vector space $\mathrm{Lie}\,\mathcal{E} \cong \mathcal{A} \cong \mathfrak{gl}_{\mathrm{res}} \oplus \mathcal{L}^1(K_+)$ is given by the commutator on $\mathcal{A}$:

$$[(A, t), (A', t')]_{\mathcal{E}} = (A, t) * (A', t') - (A', t') * (A, t)$$
$$= (AA' - A'A, bc' + at' + ta' - tt' - b'c - a't - t'a + t't).$$

Recalling that the Lie algebras of $\alpha(\mathrm{GL}^1(K_+))$ respectively $\alpha(\mathrm{SL}^1(K_+))$ in $\mathcal{A} \cong$ (Lie $\mathcal{E}$) are given by $\{(0, t) \,|\, t \in \mathcal{L}^1(K_+)\}$ respectively by $\{(0, t) \,|\, t \in \mathcal{L}^1(K_+)$ and ,

$\operatorname{tr}(t) = 0\}$, it follows that $\operatorname{Lie} \widehat{GL}_{\text{res}}^0 \cong (\operatorname{Lie} \mathcal{E})/\alpha(\mathfrak{sl}^1(K_+)) \cong \mathfrak{gl}_{\text{res}} \oplus \mathbb{C}$ upon identifying $[(A,t)]$ with $(A, \operatorname{tr} t)$.

The Lie bracket on $\widehat{\mathfrak{gl}}_{\text{res}}$ is given by

$$\left[\left[A,t\right], \left[A',t'\right]\right]_{\widehat{\mathfrak{gl}}_{\text{res}}} =$$
$$\left[\left([A,A']_{\mathfrak{gl}_{\text{res}}}, \operatorname{tr}(bc' - b'c + at' - t'a + ta' - a't - tt' + t't)\right)\right]$$
$$= \left[[A,A']_{\mathfrak{gl}_{\text{res}}}, \operatorname{tr}(bc' - b'c)\right].$$

(Brackets without subscripts denote equivalence classes in this calculation.) Identifying $\widehat{\mathfrak{gl}}_{\text{res}}$ as above with $\mathfrak{gl}_{\text{res}} \oplus \mathbb{C}$ as a vector space, we find

$$\left[(A,z), (A',z')\right] = \left([A,A'], \operatorname{tr}(bc' - b'c)\right).$$

The section $\sigma(A) = (A,0)$ of the projection $\widehat{\mathfrak{gl}}_{\text{res}} \cong \mathfrak{gl}_{\text{res}} \oplus \mathbb{C} \to \mathfrak{gl}_{\text{res}}$ yields now obviously the cocycle associated to the central extension:

$$[\sigma(A), \sigma(B)] - \sigma([A,B])$$
$$= (0, \operatorname{tr}(A_{-+}B_{+-} - B_{-+}A_{+-}))$$
$$= (0, s(A,B)),$$

i.e. the (complexified) Schwinger term on $\mathfrak{gl}_{\text{res}}$:

Proposition II.22 *The Lie algebra cocycle associated to the central $\mathbb{C}^*$-extension $\widehat{GL}_{\text{res}}^0 \longrightarrow GL_{\text{res}}^0$ is given by the Schwinger term on $\mathfrak{gl}_{\text{res}}$.*

Let us recall that the connected components of GL_{res} are the sets

$$\{A \in GL_{\text{res}} \mid \operatorname{ind}(A_{++}) = k\} \quad \text{with} \quad k \text{ in } \mathbb{Z}$$

(compare Proposition II.2 and Lemma II.17). Furthermore the map $GL_{\text{res}} \to \mathbb{Z}, A \mapsto \operatorname{ind}(A_{++})$ is a surjective group homomorphism. Taking an operator γ with $\operatorname{ind}(\gamma_{++}) = \pm 1$ we have a subgroup $\Gamma = \{\gamma^n \mid n \in \mathbb{Z}\}$ isomorphic to $\mathbb{Z}$. For definiteness, let us take a Hilbert basis $\{e_j \mid j \geq 0\}$ for K_+ and $\{e_j \mid j < 0\}$ for K_-, then $\gamma(e_j) = e_{j+1}$ defines an element of U_{res} with $\operatorname{ind}(\gamma_{++}) = -1$. It follows that γ acts on GL_{res}^0 by the group automorphism $\sigma(A) := \gamma A \gamma^{-1}$ and that we have a Banach Lie group isomorphism

$$GL_{\text{res}}^0 \rtimes \mathbb{Z} \longrightarrow GL_{\text{res}}, \ (A,n) \mapsto A \cdot \gamma^n. \tag{2.24}$$

One can easily lift σ to an injective group homomorphism $\tilde{\sigma} : \mathcal{E} \to \mathcal{E}$ by setting $\tilde{\sigma}(A,q) = (\sigma(A), q_\sigma)$, where $q_\sigma := \gamma|_{K_+} \circ q \circ \left(\gamma|_{K_+}\right)^{-1}$ on $\gamma(K_+)$ and $q_\sigma := \operatorname{Id}$ on the one-dimensional space $\left(\gamma|_{K_+}(K_+)\right)^\perp = ((e_0))_\mathbb{C} \subset K_+$. Of course $\tilde{\sigma}$ is not surjective, but since $\det(q_\sigma) = \det(q)$ for q in $GL^1(K_+)$ it is not difficult to check that the map $\hat{\sigma} : \widehat{GL}_{\mathrm{res}}^{0} \to \widehat{GL}_{\mathrm{res}}^{0}, \hat{\sigma}([A,q]) = \left[\tilde{\sigma}(A,q)\right] = \left[\sigma(A),q_{\sigma}\right]$ is an

analytic group automorphism of $\widehat{GL}_{\mathrm{res}}^{0}$.

Definition and Proposition II.23 The group $\widehat{GL}_{\mathrm{res}}^{0} \rtimes \mathbb{Z}$, where n in $\mathbb{Z}$ acts as $(\hat{\sigma})^{n}$

on $\widehat{GL}_{\mathrm{res}}^{0}$, is called $\widehat{GL}_{\mathrm{res}}$ and gives a central $\mathbb{C}^{*}$-extension of GL_{res} which restricts

to $\widehat{GL}_{\mathrm{res}}^{0}$ over GL_{res}^{0}.

Remark. Obviously the whole construction has a unitary version: let

$$\mathcal{E}_U = \{(A,q) \in \mathcal{E} | A \in U_{\mathrm{res}} \text{ and } q \in U(K_+)\},$$
$$U^1(K_+) = GL^1(K_+) \cap U(K_+),$$
$$SU^1(K_+) = SL^1(K_+) \cap U(K_+).$$

Then $\widehat{U}_{\mathrm{res}}^{0} = \mathcal{E}/\alpha(SU^1(K_+))$ defines a central $U(1)$-extension of U_{res}^{0}. Recalling that the shift operator γ is in U_{res}, it follows that there is also a central $U(1)$-extension $\widehat{U}_{\mathrm{res}} = \widehat{U}_{\mathrm{res}}^{0} \rtimes \mathbb{Z}$ of U_{res}.

Our next goal will be to compare the extensions $\widehat{U}_{\mathrm{res}}$ and $U_{\mathrm{res}}^{\sim}$. The result of Proposition II.22 strongly indicates that these should in fact be isomorphic. Since $U_{\mathrm{res}}^{\sim} = \left\{(U,V) \in U_{\mathrm{res}} \times U(\mathcal{F}) | \rho(U) = [V]\right\}$, where $\rho : U_{\mathrm{res}} \to \mathbb{PU}(\mathcal{F})$ is the projective representation of U_{res} on the Fock space stemming from the theorem of Powers and Størmer and $[V]$ the image of V in $\mathbb{PU}(\mathcal{F})$, carries a priori only the structure of a topological group, we will use a general approach to prove smoothness of a deprojectivization which we learned from [To] and which goes back to Bargmann ([Ba]).

Let ξ be in $\mathcal{F} \backslash \{0\}$ and $\mathcal{V}_{\xi} = \left\{[V] \in \mathbb{PU}(\mathcal{F}) | \langle \xi, V\xi \rangle \neq 0\right\}$. Obviously $\mathcal{V}_{\xi}$ is open in the topology induced by the strong topology on $U(\mathcal{F})$ (and thus a fortiori in the topology induced by the norm topology on $U(\mathcal{F})$). Denoting the projection $U(\mathcal{F}) \to \mathbb{PU}(\mathcal{F})$ by p the set $p^{-1}(\mathcal{V}_{\xi})$ is open as well. Let us define a map

$$\alpha_{\xi} : p^{-1}(\mathcal{V}_{\xi}) \to U(1), \alpha_{\xi}(V) = \frac{\langle \xi, V\xi \rangle}{|\langle \xi, V\xi \rangle|}. \tag{2.25}$$

We observe that $\alpha_{\xi}(t \cdot V) = t \cdot \alpha_{\xi}(V)$ for all t in $U(1)$. Furthermore the map

$$\phi : p^{-1}(\mathcal{V}_{\xi}) \to \mathcal{V}_{\xi} \times U(1), \phi(V) = ([V], \alpha_{\xi}(V))$$

is a homeomorphism with inverse

$$\phi^{-1}([V], s) = V \cdot (\alpha_{\xi}(V))^{-1} \cdot s,$$

where V is any element of $p^{-1}(\mathcal{V}_{\xi})$ such that $p(V) = [V]$. Notably we can in this local $U(1)$-equivariant trivialization of the $U(1)$-bundle $U(\mathcal{F}) \to \mathbb{PU}(\mathcal{F})$ realize the multiplication in $U(\mathcal{F})$ as $x * y = \phi(\phi^{-1}(x) \cdot \phi^{-1}(y))$ and the inverse as $I(x) = \phi((\phi^{-1}(x))^{-1})$ (for x, y in $\mathcal{V}_{\xi} \times U(1)$ sufficiently close to $([1_{\mathcal{F}}], 1)$).

Explicitely one has

$$([V], s) * ([V'], s') = \left([V] \cdot [V'], ss' \cdot \frac{\alpha_\xi(V \cdot V')}{\alpha_\xi(V) \cdot \alpha_\xi(V')} \right), \qquad (2.26)$$

where V, V' are in $p^{-1}(\mathcal{V}_\xi)$ fulfilling $p(V) = [V], p(V') = [V']$ but the quotient $\frac{\alpha_\xi(V \cdot V')}{\alpha_\xi(V) \cdot \alpha_\xi(V')}$ depends only on $[V]$ and $[V']$.

Observing that $\alpha_\xi(V^{-1}) = (\alpha_\xi(V))^{-1}$ we have furthermore

$$I([V], s) = ([V^{-1}], s^{-1}) = ([V]^{-1}, s^{-1}).$$

Since ρ is continuous if we use the strong topology on $\mathbb{P}U(\mathcal{F})$, the set $\mathcal{O}_\xi = \{U \in U_{\text{res}} | \langle \xi, \rho(U)\xi \rangle \neq 0\}$ is an open neighborhood of the identity in U_{res}.
It follows that

$$\mathcal{U}_\xi = \{(U, V) \in U_{\text{res}}^\sim | U \in \mathcal{O}_\xi\}$$

is an open neighborhood of the neutral element of the topological group $U_{\text{res}}^\sim$.
Furthermore the condition $\rho(U) = [V]$ implies that V is in $p^{-1}(\mathcal{V}_\xi)$ for all couples (U, V) in $\mathcal{U}_\xi$. Thus the composition

$$\mathcal{U}_\xi \longrightarrow \mathcal{O}_\xi \times p^{-1}(\mathcal{V}_\xi) \longrightarrow \mathcal{O}_\xi \times \mathcal{V}_\xi \times U(1)$$

yields a homeomorphism

$$\psi : \mathcal{U}_\xi \longrightarrow \mathcal{O}_\xi \times U(1), \psi(U, V) = (U, \alpha_\xi(V))$$

with inverse $\psi^{-1}(U, s) = (U, V \cdot \alpha_\xi(V)^{-1} \cdot s)$, where V is any element in $U(\mathcal{F})$ such that $[V] = \rho(U)$.
Abusing slightly the notation the local multiplication on $\mathcal{O}_\xi \times U(1)$ will now again be noted by "$*$" and reads as

$$(U, s) * (U', s') = \left(UU', ss' \cdot \frac{\alpha_\xi(V \cdot V')}{\alpha_\xi(V) \cdot \alpha_\xi(V')} \right), \qquad (2.27)$$

for a choice of V respectively V' such that $[V] = \rho(V)$ respectively $[V'] = \rho(V')$. Of course the quotient appearing in (2.27) is independent of these choices. The local formula for the inverse is easily checked to be $I(U, s) = (U^{-1}, s^{-1})$.
Obviously I is analytic on $\mathcal{O}_\xi \times U(1)$ and it remains to prove smoothness of the multiplication. Let us first consider an $\epsilon > 0$ such that the exponential map of U_{res} is a diffeomorphism on $B_\epsilon(0) = \{X \in \mathfrak{u}_{\text{res}} \mid \|X\|_{\mathfrak{u}_{\text{res}}} < \epsilon\}$ and such that $\mathcal{O}_{\xi, \epsilon} = \exp(B_\epsilon(0))$ lies in $\mathcal{O}_\xi$. By Theorem II.4 we have a local covering of $\rho :$ $\mathcal{O}_{\xi, \epsilon} \to \mathbb{P}U(\mathcal{F})$ by $\tilde{\rho} : \mathcal{O}_{\xi, \epsilon} \to U(\mathcal{F}), \tilde{\rho}(U) := \exp_{U(\mathcal{F})}(\mathbb{A}(U))$, where $\mathbb{A}(U)$ is the second quantization of

$$A(U) = \left(\exp_{U_{\text{res}}} |_{B_\epsilon(0)} \right)^{-1} (U)$$

in $\mathfrak{u}_{\text{res}}$.

It remains to prove that there are vectors ξ in $\mathcal{F}$ such that the map

$$\mathcal{O}_{\xi,\epsilon} \times \mathcal{O}_{\xi,\epsilon} \to U(1), (U, U') \mapsto \frac{\alpha_\xi(\tilde{\rho}(U) \cdot \tilde{\rho}(U'))}{\alpha_\xi(\tilde{\rho}(U)) \cdot \alpha_\xi(\tilde{\rho}(U'))} \qquad (2.28)$$

is smooth.

This can be derived from more general results in [To] (compare Corollary 4.2.2 and Proposition 5.3.1 there). The starting point for applying these results is the definition of a "Laplacian" $\Delta = N+1$, where $N = (\frac{8}{}(p_+ - p_-)\Psi^*\Psi\frac{8}{})$ is the number operator (one might equally well employ the "Hamiltonian") on $\mathcal{F}$ (compare the text before Lemma II.5) and of the (possibly infinite) functions $p_n(\xi) = \|\Delta^n\xi\|_{\mathcal{F}}$ for $n \in \mathbb{N}$ and $\xi \in \mathcal{F}$. The space

$$\mathcal{D}_\infty = \left\{ \xi = \sum_{p,q\geq 0} \psi^{(p,q)} \in \mathcal{F} \,\middle|\, \forall n \quad p_n(\xi) = \left(\sum_{p,q\geq 0} (p+q+1)^{2n} \|\psi^{(p,q)}\|_{\mathcal{F}}^2 \right)^{1/2} < \infty \right\}$$

of "smooth vectors" is obviously a Fréchet space whose topology is defined by the increasing family of semi-norms $\{p_n | n \geq 0\}$ and contains the finite-particle subspace $\mathcal{D}$ defined in the first remark after Theorem II.4. One might now prove the estimates necessary to apply the machinery of [To].

In the given situation we can also proceed in a more direct way. Let P_L again denote the projector on the "space of at most L particles":

$$P_L\left(\sum_{p,q\geq 0} \Psi^{(p,q)} \right) = \sum_{p+q\leq L} \Psi^{(p,q)}$$

and $\mathcal{D}$ the finite particle space. We already mentioned in the (sketch of a) proof of Theorem II.4 that for $P_L(\xi) = \xi$ one has $P_{L+2}(\mathbb{A}\,\xi) = \mathbb{A}\,\xi$ and

$$\|\mathbb{A}\,\xi\| \leq (L+2) \cdot \|A\|_{\text{ures}} \cdot \|\xi\|. \qquad (2.29)$$

Thus $\sum_{k\geq 0} \frac{t^k}{k!}\mathbb{A}^k\xi$ is absolutely convergent for $|t| < (2 \cdot \|A\|_{\text{ures}})^{-1}$. Assuming now that ϵ is smaller than $\frac{1}{2}$ it follows that $\sum_{k\geq 0} \frac{1}{k!}\mathbb{A}^k\xi$ is absolut convergent for all A in $B_\epsilon(0)$. Hence the map

$$B_\epsilon(0) \to \mathcal{F}, \quad A \mapsto \sum_{k\geq 0} \frac{1}{k!}\mathbb{A}^k\xi = \exp(\mathbb{A})\cdot\xi$$

is real-analytic in A for fixed ξ in $\mathcal{D}$. (It is in fact enough to show that this map is smooth. Compare Subsection IV.1 for the definition of a smooth map in infinite dimensional situations or, e.g., [Muj] for the notions of smooth and analytic maps in the here sufficing Banach space setting.)

Thus the maps

$$A \mapsto \langle \xi, \exp(\mathbb{A})\cdot\xi \rangle \quad \text{and}$$

$$(A, A') \mapsto \langle \xi, \exp(\mathbb{A}) \cdot \exp(\mathbb{A}')\cdot\xi \rangle = \langle \exp(-\mathbb{A})\cdot\xi, \exp(\mathbb{A}')\cdot\xi \rangle$$

are analytic as well. Therefore the map defined by (2.28) and the local multiplication in $\mathcal{U}_{\xi,\epsilon} = \{(U, V) \in \mathcal{U}_\xi \mid U \in \mathcal{O}_{\xi,\epsilon}\} \cong \mathcal{O}_{\xi,\epsilon} \times U(1)$ are smooth. It is

now easy to show that a topological group having a neighborhood $\mathcal{U}$ of its neutral element which is homeomorphic to an open set of a Banach space such that the "transported local multiplication" and the "transported local inversion" are smooth carries itself the structure of a smooth Banach Lie group.
Summarizing we proved

Proposition II.24 *The topological group*

$$\widetilde{U}_{\text{res}} = \{(U, V) \in U_{\text{res}} \times U(\mathcal{F}) \mid \rho(U) = [V]\}$$

carries a compatible structure of a smooth Banach Lie group such that the projection $(U, V) \mapsto U$ realizes $\widetilde{U}_{\text{res}}$ as a smooth central $U(1)$-extension of U_{res}.

Remark. By Theorem II.4 and the very construction of the map $\tilde{\rho}$ above it follows that the Lie algebra cocycle of the extension $\widetilde{U}_{\text{res}}$ of U_{res} is given by the Schwinger term.

In order to show the isomorphy of $\widehat{U}_{\text{res}}$ and $\widetilde{U}_{\text{res}}$, we formulate

Proposition II.25 *Let G be a connected and simply connected Lie group and E, E', E'' be central $U(1)$-extension of G. Then*

(i) *$E \rightarrow G$ is trivial if and only if the associated Lie algebra extension is trivial and*

(ii) *E' and E'' are isomorphic extensions if and only if the associated Lie algebra cocycles are cohomologous.*

Remarks. Both assertions follow in the case that G is smoothly paracompact from the Sections (4.4) and (4.5) in [PrSe] since – as is pointed out in [To] – the arguments used there for loop groups extend without any change. In the general case the result follows from Corollary V.10 in [N3]. □

Corollary II.26 *The central $U(1)$-extensions $(\widetilde{U}_{\text{res}})^0$ and $\widehat{U}_{\text{res}}^0$ are isomorphic (in the category of Banach Lie group extensions).*

We conclude now for the restricted unitary group in showing

Proposition II.27 *The central $U(1)$-extensions $\widetilde{U}_{\text{res}}$ and $\widehat{U}_{\text{res}}$ are isomorphic in the category of smooth Banach-Lie group extensions.*

Proof. Let γ be the shift operator considered in II.23 and the succeeding remark defining $\widehat{U}_{\text{res}}$ by $\widehat{U}_{\text{res}}^0 \rtimes \mathbb{Z}$. Let V_γ be an element of $U(\mathcal{F})$ such that $[V_\gamma] = \rho(\gamma)$. Denoting the projection $\widetilde{U}_{\text{res}} \rightarrow U_{\text{res}}$ by π we observe that $\pi^{-1}(U_{\text{res}}^0) = (\widetilde{U}_{\text{res}})^0$, the connected component containing the neutral element and that the other components are given by $\pi^{-1}\{U_{\text{res}}^0 \cdot \gamma^n\}$.

We define a $\mathbb{Z}$-action by group automorphisms on $(\widetilde{U}_{\text{res}})^0$ by

$$n \cdot (U, V) = (\gamma^n U \gamma^{-n}, (V_\gamma)^n V (V_\gamma)^{-n}).$$

This action covers the $\mathbb{Z}$-action on U_{res}^0 and yields a Banach-Lie group isomorphism

$$(\widetilde{U}_{\text{res}})^0 \rtimes \mathbb{Z} \rightarrow \widetilde{U}_{\text{res}}, \quad ((U, V), n) \mapsto (U\gamma^n, V(V_\gamma)^n)$$

covering the isomorphism

$$U_{\text{res}}^0 \rtimes \mathbb{Z} \to U_{\text{res}}, \quad (U, n) \mapsto U\gamma^n.$$

Thus the map

$$(U_{\text{res}}^{\sim})^0 \rtimes \mathbb{Z} \to \widehat{U}_{\text{res}}^0 \rtimes \mathbb{Z}, \quad ((U, V), n) \mapsto (\Theta(U, V), n),$$

where Θ is an isomorphism between the Banach Lie group extensions $(U_{\text{res}}^{\sim})^0$ and $\widehat{U}_{\text{res}}^0$, gives us the desired isomorphism. $\qquad\qquad\qquad\qquad\qquad\square$

Remarks.

(1) Obviously there is an analogously defined (complex-analytic) Banach Lie group $GL_{\text{res}}^{\sim}$, and $GL_{\text{res}}^{\sim}$ and $\widehat{GL}_{\text{res}}$ are isomorphic in the category of complex-analytic Banach-Lie group $\mathbb{C}^*$-extensions.

(2) The embedding $j : LK \to U_{\text{res}}$ being smooth (compare the remark before Lemma II.10) one has an induced smooth $U(1)$-extension $\tilde{L}K := j^*(U_{\text{res}}^{\sim})$. Using Proposition II.25, Lemma II.10 and the results in [N2] it follows that the above defined $\tilde{L}K$ and the extension of LK given there are isomorphic (if the involved invariant scalar products on the Lie algebra of the compact group K coincide). Mutatis mutandis, the same holds true for the complexified group $LK^{\mathbb{C}}$ and the embedding $j^{\mathbb{C}} : LK^{\mathbb{C}} \to GL_{\text{res}}$.

III. The restricted Grassmannian of a polarized Hilbert space

In this section we will introduce the restricted Grassmannian as a natural moduli space associated to the problem of implementation of Bogoliubov transformations on representation spaces of the CAR-algebra and explain its basic properties. We will here especially stress the fact that the restricted Grassmannian is a homogeneous Hilbert manifold, whose geometry as a hermitian symmetric space is derived from the interpretation of the Schwinger term as its Kähler form.

Since our "genetic approach" to second quantization followed in Section I the idea of Dirac to consider the vacuum as filled with the undesirable negative energy states, we arrived at the conclusion that the physical Fock space corresponds to the state ω_{p_-} on $\text{CAR}(K)$ induced from the projection onto K_-. Unfortunately the standard reference [PrSe] chooses the notation K_+ for the reference space to whom the other elements of the Grassmannian should be "close". We will thus in the sequel change the rôles of K_+ and K_-, which will not affect the definition of U_{res} but will introduce a new Schwinger term, which is in fact (-1) times the old Schwinger term given by (2.14) (see below in Subsection III.2).

1. The restricted Grassmannian as a homogeneous complex manifold

In Subsection II.1 we considered the problem of how to implement a Bogoliubov transformation β_U associated to a U in $U(K)$ on the representation space $\mathcal{F} \cong \mathcal{H}_{K_-}$ given by the GNS-construction and learned from the theorem of Powers and Størmer that this is possible if and only if U is in $U_{\text{res}}(K, K_+)$. As announced above we will now replace the state ω_{K_-} by the state ω_{K_+} and restate the implementation problem (2.3) as follows. Does there exists a unitary $\mathbb{U}$ on $\mathcal{H}_{K_+}$ such that

$$\mathbb{U} \circ \pi_{K_+}(\alpha) \circ \mathbb{U}^{-1} = \pi_{K_+}(\beta_U(\alpha)) \tag{3.1}$$

for all α in $A(K)$.

Since π_{K_+} is given by the state ω_{K_+} (compare (1.32)–(1.33)) it is easy to calculate the state $\omega_{K_+} \circ \beta_{U^{-1}}$:

$$\omega_{K_+} \circ \beta_{U^{-1}}(a^*(k)a(\ell))$$
$$= \omega_{K_+}(a^*(U^{-1}k)a(U^{-1}\ell)$$
$$= \langle U^{-1}\ell, p_{K_+}(U^{-1}k)\rangle_K$$
$$= \langle \ell, U p_{K_+} U^{-1}(k)\rangle_K$$
$$= \langle \ell, p_{UK_+}(k)\rangle_K,$$

where p_{UK_+} denotes the projector on the subspace $UK_+ = \{Uk | k \in K_+\}$ of K. Thus $\omega_{K_+} \circ \beta_{U^{-1}}$ equals ω_{UK_+}. We can now reformulate the implementation problem (3.1) as an equivalence problem of GNS-representations associated to quasi-free states on the CAR-algebra $A(K)$.

Lemma III.1 *For U in $U(K)$ and K polarized as $K = K_+ \oplus K_-$ the following are equivalent:*

(i) *There exists a $\mathbb{U}$ in $U(\mathcal{H}_{K_+})$ such that*

$$\pi_{K_+}(\beta_U(\alpha)) = \mathbb{U} \circ \pi_{K_+}(\alpha) \circ \mathbb{U}^{-1}$$

for all α in $A(K)$.

(ii) *There exists a unitary isomorphism $\tilde{U}$ from the GNS-space $\mathcal{H}_{K_+}$ to the GNS-space $\mathcal{H}_{UK_+}$ such that the conjugation by $\tilde{U}$ intertwines the GNS-representations π_{K_+} and π_{UK_+}, i.e.*

$$\pi_{UK_+}(\alpha) = \tilde{U} \circ \pi_{K_+}(\alpha) \circ \tilde{U}^{-1}.$$

Proof. Let $\mathcal{H}_{K_+}$ be a fixed realization of the GNS-space associated to ω_{K_+}. It is easily seen that the representation $\pi_{K_+} \circ \beta_{U^{-1}}$ of $A(K)$ on $\mathcal{H}_{K_+}$ is a GNS-representation of $A(K)$ associated to the state ω_{UK_+}, since the vacuum ξ_+ of $\mathcal{H}_{K_+}$ is cyclic and gives rise to the state ω_{UK_+} :

$$\langle \xi_+, (\pi_{K_+} \circ \beta_{U^{-1}}(\alpha))\xi_+\rangle$$
$$= \omega_{K_+}(\beta_{U^{-1}}(\alpha)) = (\omega_{K_+} \circ \beta_{U^{-1}})(\alpha)$$
$$= \omega_{UK_+}(\alpha) \quad \text{for all } \alpha \text{ in } A(K).$$

By the uniqueness of the GNS-representation (up to unitary equivalence, compare e.g. [Mur]) the representation $\pi_{K_+} \circ \beta_{U^{-1}}$ is thus a realization of π_{UK_+}. The equivalence of (i) and (ii) is now obvious. $\qquad\square$

Remark. The relation between the implementation problem and the equivalence problem is also stressed in [PR], where these two problems are studied for Clifford algebras and orthogonal groups. Since we were led to consider the polarization of a one-particle space K by the analysis of a given Dirac-Hamiltonian and to the implementation problem by the question if the obtained Fock space is still useful for perturbations of this operator by potentials, it is natural to consider the set of "second quantizations that are equivalent to a given one":

Definition III.2 Given a separable complex Hilbert space K together with an orthogonal decomposition $K = K_+ \oplus K_-$ such that K_+ and K_- are both infinite-dimensional, *the restricted Grassmannian (of the polarized Hilbert space $K = K_+ \oplus K_-$)* is the set of gauge-invariant, quasi-free states on the CAR-algebra $A(K)$ whose associated GNS-representations are unitarily equivalent to the GNS-representation induced from the state ω_{K_+}. We will denote this set in the sequel by $G_{\text{res}}(K, K_+)$ or simply by G_{res}.

Remarks.

(1) The set G_{res} is by the preceding lemma and the theorem of Powers and Størmer of course the orbit of ω_{K_+} under the action of $U_{\text{res}}(K, K_+)$ on the space of states on $A(K)$.

(2) Since for each state in G_{res} there is a "vacuum line", the multiples of the GNS-vector in the GNS-representation, the restricted Grassmannian can also be considered as a moduli space of vacua of fermionic second quantization. This point of view leads to a natural construction of the determinant bundle (see [SpWu1]) which we will review in Section 5 below.

Let us first show that this definition coincides with the more standard one found e.g. in [PrSe] (p. 101).

Lemma III.3 *The restricted Grassmannian $G_{\text{res}}(K, K_+)$ is the set of all closed subspaces W of K such that*

(i) *the orthogonal projection $p_+ : W \to K_+$ is a Fredholm operator and*

(ii) *the orthogonal projection $p_- : W \to K_-$ is a Hilbert-Schmidt operator.*

Proof. An element of the Grassmannian is given by a quasi-free state, which in turn is given by a closed subspace of K.

We have thus only to show that a subspace W fulfils (i) and (ii) if and only if it is the image of K_+ under an element

$$U = \begin{pmatrix} U_+ & U_{+-} \\ U_{-+} & U_- \end{pmatrix} \text{ of } U_{\text{res}}(K, K_+).$$

Assuming $W = UK_+$ with U in U_{res} we find that (i) is fulfilled since U_+ is Fredholm and (ii) since U_{-+} is Hilbert-Schmidt.

Given now W with the properties (i) and (ii), there exists of course an element

$$U = \begin{pmatrix} U_+ & U_{+-} \\ U_{-+} & U_- \end{pmatrix} \text{ in } U(K) \text{ such that } UK_+ = W.$$

The assertions (i) and (ii) imply now that U_+ is Fredholm and U_{-+} Hilbert-Schmidt. By the unitarity of U it follows that U_- is a Fredholm operator as well. Thus it exists V_- such that $U_- V_- = 1_- + F_-$ with F_- of finite rank. We conclude – again by the unitarity of U – that

$$U_{+-} = -U_+ (U_{-+})^* V_- - U_{+-} F_-$$

is Hilbert-Schmidt, since $L^2(F)$ is a two-sided ideal in the bounded operators. □

Corollary III.4

(i) *The restricted Grassmannian* $G_{res}(K, K_+)$ *is a homogeneous space under* $U_{res}(K, K_+)$ *and* $GL_{res}(K, K_+)$. *The corresponding isotropy groups of the state* ω_{K_+} *(or equivalently of the subspace* K_+ *of* K) *are*

$$P = \left\{ \begin{pmatrix} a & b \\ c & d \end{pmatrix} \in GL_{res}(K, K_+) \, \middle| \, c = 0 \right\}$$

respectively

$$H = \left\{ \begin{pmatrix} a & b \\ c & d \end{pmatrix} \in U_{res}(K, K_+) \, \middle| \, b = 0 \text{ and } c = 0 \right\}.$$

(ii) *Let* W *be a closed complex subspace of* K *and* p_W *the orthoprojector onto* W. *Then* W *is in* $G_{res}(K, K_+)$ *iff the difference* $p_W - p_+$ *is a Hilbert-Schmidt operator.*

Proof. The restricted Grassmannian is homogeneous under $U_{res}(K, K_+)$ by Remark (1) after its definition and also acted upon by $GL_{res}(K, K_+)$ since this latter group preserves the conditions (i) and (ii) in Lemma III.3. The determination of the isotropies is trivial.

Since both conditions in the second assertion obviously imply that W is of infinite dimension and infinite codimension there exists a U in $U(K)$ such that $W = U \cdot K_+$. Since then $p_W = U p_+ U^{-1}$ it follows by a direct calculation that $p_W - p_+$ is Hilbert-Schmidt iff $[U, p_+]$ is Hilbert-Schmidt and the latter condition is of course equivalent to U in $U_{res}(K, K_+)$. □

Remark. The very natural second assertion seems to be first noted in [SV].

Since the book of Pressley and Segal ([PrSe]) gives a very thorough account of the fundamental properties of G_{res} as a manifold, we will present here only the most important ones.

Proposition III.5

(i) *The restricted Grassmannian* $G_{res}(K, K_+)$ *is a complex-analytic manifold modelled on the separable Hilbert space* $L^2(K_+, K_-)$.

(ii) *The actions of* $GL_{res}(K, K_+)$ *and* $U_{res}(K, K_+)$ *are complex-analytic respectively real-analytic.*

(iii) *The linear isotropy representation in the point* K_+ *is given by the map*

$$Ad : P \to GL(L^2(K_+, K_-)), \quad Ad\begin{pmatrix} a & b \\ 0 & d \end{pmatrix} \cdot \gamma = a \circ \gamma \circ d^{-1}$$

for γ in $L^2(K_+, K_-) \cong \mathfrak{gl}_{\text{res}}/\mathfrak{p} = T_{K_+} G_{\text{res}}$. $(GL(L^2(K_+, K_-))$ denotes the bounded, invertible operators on the Hilbert space $L^2(K_+, K_-).)$

(iv) *The connected components of G_{res} are given by the sets of subspaces having "virtual dimension k" (k in $\mathbb{Z}$) :*

$$G_{\text{res}}^k = \{W \in G_{\text{res}}| \text{index}\,(p_+ : W \to K_+) = k\}.$$

(v) *There are holomorphic embeddings of $G(\mathbb{C}^{2N}) = \bigcup_{n=0}^{2N} G_n(\mathbb{C}^{2N})$, the total Grassmannian of $\mathbb{C}^{2N}$, into G_{res} such that their images $G_{\text{res}(N)}$ are increasing and the union $\bigcup_{N \geq 1} G_{\text{res}(N)} = G_{\text{res}(\infty)}$ is dense in G_{res}.*

Furthermore the intersection $G_{\text{res}(N)} \cap G_{\text{res}}^k$ is biholomorphic to $G_{N+k}(\mathbb{C}^{2N})$.

Proof. (Compare also Chapter 7 in [PrSe]).
As in the case of finite-dimensional Grassmannians there are natural coordinate charts near a fixed W in G_{res} defined for those subspaces that are graphs over W. Consider the set $\mathcal{U}_W = \{W' \subset K | W'$ a closed complex subspace of K and $P_W : W' \to W$ is an isomorphism $\}$.

The elements of $\mathcal{U}_W$ are then in bijection with the Hilbert space $\mathcal{L}^2(W, W^\perp)$ since they are graphs of such maps. The complex-analyticity of the coordinate changes on the intersection of two graph coordinate patches follows now as in finite dimensions. Let W be in $\mathcal{U}_{01} = \mathcal{U}_{W_0} \cap \mathcal{U}_{W_1}$ for W_0 and W_1 in G_{res}. The "graph coordinates" are the maps $\phi_j : \mathcal{L}^2(W_j, W_j^\perp) \to \mathcal{U}_{W_j}, \phi_j(T_j) = \{w_j + Tw_j | w_j \in W_j\}$ for $j = 0, 1$.

Writing the identity of K as a block matrix $\begin{pmatrix} a & b \\ c & d \end{pmatrix}$ with a a map from W_0 to W_1 etc, it follows from Corollary III.4 (ii) that $b = p_{W_1} : (W_0)^\perp \to W_1$ and $c = p_{(W_1)^\perp} : W_0 \to (W_1)^\perp$ are Hilbert-Schmidt (and thus a and d are Fredholm).

Let us assume that $W = \phi_0(T_0) = \phi_1(T_1)$ for appropriate T_j in $\mathcal{L}^2(W_j, W_j^\perp)$. Since $p_{W_1} : W \to W_1$ is an isomorphism the map

$$a + bT_0 = p_{W_1} \circ Id_K \circ \begin{pmatrix} 1 \\ T_0 \end{pmatrix} : W_0 \to W_0 \oplus (W_0)^\perp \to W_1 \oplus (W_1)^\perp \to W_1$$

is an isomorphism as well. It follows that the maps $\begin{pmatrix} a & b \\ c & d \end{pmatrix} \circ \begin{pmatrix} 1 \\ T_0 \end{pmatrix}$ and $\begin{pmatrix} 1 \\ T_1 \end{pmatrix} \circ$ $(a + bT_0)$ from W_0 to $W_1 \oplus (W_1)^\perp$ are equal. Thus $T_1 = (c + dT_0)(a + bT_0)^{-1} = \phi_{10}(T_0)$ is a holomorphic map from $\phi_0^{-1}(\mathcal{U}_{01}) = \{T_0 \in \mathcal{L}^2(W_0, W_0^\perp) | a + bT_0$ is invertible$\}$ to $\phi_1^{-1}(\mathcal{U}_{01})$.

Let us also remark that this yields a useful countable atlas on G_{res}. Let $\mathcal{S}$ denote the set $\{S \subset \mathbb{Z}| \text{card}(S - \mathbb{N}_0)$ and $\text{card}(\mathbb{N}_0 - S)$ are finite $\}$ and fix an orthonormal basis (ONB in the sequel) $\{e_j | j \in \mathbb{Z}\}$ of K such that $\{e_j | j < 0\}$ respectively $\{e_j | j \geq 0\}$ generate K_- respectively K_+. We associate to S in $\mathcal{S}$ the closed complex subspace K_S generated by $\{e_j | j \in S\}$, which is easily seen to be in G_{res}. The important observation is now that for each W in G_{res} there exists a

S in $\mathcal{S}$ such that W is in U_{K_S} (see [PrSe], pp. 102-104). Thus $\{U_{K_S}|S \in \mathcal{S}\}$ forms a countable open cover of G_{res} by graph coordinate sets.

To prove the second assertion recall that GL_{res} is a complex-analytic Banach Lie group modelled on its Lie algebra

$$\mathfrak{gl}_{\text{res}} = \{A \in \mathcal{B}(K)|[A, \varepsilon] \in L^2(K)\}$$

by means of its exponential map, $A \mapsto e^A$, and that the isotropy group P of K_+ is a closed complex subgroup of it. It follows that the complex-analytic action

$$\text{GL}_{\text{res}} \times \text{GL}_{\text{res}}/P \to \text{GL}_{\text{res}}/P$$

yields – a possibly different – complex-analytic structure on G_{res} as the coset GL_{res}/P. In order to compare their structures one has to observe that the orbit of the subgroup

$$N_- = \left\{ \begin{pmatrix} 1_+ & 0 \\ \gamma & 1_- \end{pmatrix} \middle| \gamma \in L^2(K_+, K_-) \right\}$$

through K_+ equals the graph coordinate set U_{K_+} by means of the set-theoretic bijection $\text{GL}_{\text{res}}/P \cong G_{\text{res}}$, since

$$\begin{pmatrix} 1_+ & 0 \\ \gamma & 1_- \end{pmatrix} \cdot K_+ = \{(k, \gamma k) \in K_+ \oplus K_- = K|k \in K_+\}.$$

Since the complex-analytic coordinates coming from the coset-point of view are on $N_- \cdot (eP)$ given by

$$\mathfrak{n}_- = \left\{ \begin{pmatrix} 0 & 0 \\ \gamma & 0 \end{pmatrix} \middle| \gamma \in L^2(K_+, K_-) \right\} \to N_- \cdot (eP),$$

$$\begin{pmatrix} 0 & 0 \\ \gamma & 0 \end{pmatrix} \mapsto \exp \begin{pmatrix} 0 & 0 \\ \gamma & 0 \end{pmatrix} \cdot (eP) = \begin{pmatrix} 1_+ & 0 \\ \gamma & 1_- \end{pmatrix} \cdot (eP),$$

we conclude that these coordinates are exactly the graph coordinates on U_{K_+}.

The formula (iii) for the linear isotropy representation follows now from general principles for homogeneous spaces.

Since the isotropy of the U_{res}-action is the contractible Banach Lie group $U(K_+) \times U(K_-)$ (see [Ku]) the homotopy type of G_{res} is determined by U_{res}. It follows now from Proposition II.2 that the connected components of G_{res} are given by the open sets

$$G_{\text{res}}^k = \{W = U \cdot K_+|U \in U_{\text{res}} \text{ and index}\,(U_+) = k\}$$
$$= \{W \in G_{\text{res}}| \text{ index}\,(p_+ : W \to K_+) = k\},$$

the subspaces having virtual dimension k (k in $\mathbb{Z}$). A direct verification shows that if S is in $\mathcal{S}$ and the "virtual cardinality"

$$\text{virt. card. } S = (\text{cardinality of } (S - \mathbb{N}_0)) - (\text{cardinality of } (\mathbb{N}_0 - S))$$

equals k, then the virtual dimension of K_S is k.

In order to construct a subset $G_{\text{res}(\infty)}$ as in assertion (v), we fix again an ONB $\{e_j | j \in \mathbb{Z}\}$ as above in the proof. Defining $K_N = ((e_N, e_{N+1}, \dots ,))_{\mathbb{C}}$ and $K_{-N} = ((e_{-N}, e_{-N+1}, \dots ,))_{\mathbb{C}}$ the set

$$G_{\text{res}(N)} = \{W \in G_{\text{res}} | K_N \subset W \subset K_{-N}\}$$

for a $N \geq 1$ is seen to be isomorphic to $G(\mathbb{C}^{2N}) = \bigcup_{n=0}^{2N} G_n(\mathbb{C}^{2N})$, the "full Grassmannian" of $\mathbb{C}^{2N}$, by mapping W to $W/K_N \subset K_{-N}/K_N \cong \mathbb{C}^{2N}$. The inclusion $G_{\text{res}(N)} \subset G_{\text{res}(N+1)}$ corresponds then to sending V in $G(\mathbb{C}^{2N})$ to $\mathbb{C} \oplus V \oplus \{0\}$ in the Grassmannian $G(\mathbb{C}^{2(N+1)})$, where $\mathbb{C}^{2(N+1)} = \mathbb{C} \oplus \mathbb{C}^{2N} \oplus \mathbb{C}$.

Let us for k and ℓ in $\mathbb{Z}$ define an operator $E_{k,\ell} : K \to K$ as follows: $E_{k,\ell}(e_{\ell'}) = \delta_{\ell,\ell'} \cdot e_k$ for ℓ' in $\mathbb{Z}$ and linear extension to K. Then the set $\{E_{k,\ell} | k < 0$ and $\ell \geq 0\}$ forms an ONB of $L^2(K_+, K_-)$. It is now easy to check that $G_{\text{res}(\infty)} \cap \mathcal{U}_{K_+}$ corresponds in the graph coordinates of $\mathcal{U}_{K_+}$ to operators that are finite linear combinations in the above ONB of $L^2(K_+, K_-)$. These operators are of course dense in $L^2(K_+, K_-)$. Similar considerations in the graph coordinates centered in the subspaces K_S with $S \in \mathcal{S}$ show that $G_{\text{res}(\infty)}$ is everywhere dense.

Let us observe that this also shows that $G(\mathbb{C}^{2N})$ is holomorphically embedded into G_{res} for all N since the graph coordinates of G_{res} restrict to those of the finite-dimensional Grassmannians.

The last part of assertion (v) follows by a direct calculation. Let $\tilde{p}_+$ denote the projection $W/K_N \to K_+/K_N$. Then

$$\begin{aligned}
k &= \text{index}\,(p_+ : W \to K_+) \\
&= \text{index}\,(\tilde{p}_+) = \dim(\ker \tilde{p}_+) - \dim(\text{coker}\,\tilde{p}_+) \\
&= \dim(W/K_N) - N. \qquad \square
\end{aligned}$$

Corollary III.6 *All holomorphic functions on G_{res} are locally constant.*

Proof. Since $\mathcal{U}_{\text{res}}$ acts by holomorphic transformations it is enough to consider a function f that is holomorphic on G_{res}^0, the connected component of G_{res} containing K_+. Given now two points W_1 and W_2 in $G_{\text{res}}^0 \cap G_{\text{res}(\infty)}$ there is a N such that W_1 and W_2 are in $G_{\text{res}}^0 \cap G_{\text{res}(N)}$. Since the latter is biholomorphic to the connected compact complex manifold $G_N(\mathbb{C}^{2N})$ the function f assumes the same value in W_1 and W_2. Thus f is constant on $G_{\text{res}}^0 \cap G_{\text{res}(\infty)}$ and by the density of this set in G_{res}^0 the assertion follows. $\qquad \square$

2. The basic differential geometry of the restricted Grassmannian

According to the remarks in the beginning of Section III, we exchange the rôles of K_+ and K_- in the formula (2.14) of the Schwinger term s and arrive at the "new Schwinger term" $\tilde{s}$ given by

$$\tilde{s}(A, B) = -s(A, B) = \text{tr}(A_{+-}B_{-+} - B_{+-}A_{-+}) \qquad (3.2)$$

for A, B in $\mathfrak{u}_{\text{res}}(K, K_+)$.

By a sequence of easy considerations we will now show that $\tilde{s}$ gives rise to a U_{res}-invariant Kählerian structure on the restricted Grassmannian.

First we define a real-valued antisymmetric bilinear form on $\mathfrak{u}_{res}(K, K_+)$ by setting:

$$\hat{\Omega}_{K_+}(A, B) := (-i)\tilde{s}(A, B). \tag{3.3}$$

Remark. Unfortunately the symbol Ω was – following traditional customs – also used for the vacuum vector of the Fock space $\mathcal{F}$. It should be nevertheless easy for the reader to distinguish the two usages from the context.

Lemma III.7 *The bilinear form* $\hat{\Omega}_{K_+}$ *on* $\mathfrak{u}_{res}(K, K_+)$ *vanishes on the isotropy sub-algebra* $\mathfrak{u}(K_+) \times \mathfrak{u}(K_-)$ *and is invariant under the linear isotropy representation of* $U(K_+) \times U(K_-)$.

Corollary III.8 *The bilinear form* $\hat{\Omega}_{K_+}$ *on* $\mathfrak{u}_{res}(K, K_+)$ *descends to a form* Ω_{K_+} *on*

$$\mathfrak{u}_{res}(K, K_+)/\mathfrak{u}(K_+) \times \mathfrak{u}(K_-) \cong L^2(K_+, K_-) \cong T_{K_+}G_{res},$$

which is invariant by $U(K_+) \times U(K_-)$:

$$\Omega_{K_+}(\gamma, \delta) = 2 \operatorname{Im} \operatorname{tr}(\gamma^*\delta) \quad \text{for } \gamma, \delta \in L^2(K_+, K_-). \tag{3.4}$$

Remark. The lemma and its corollary follow from obvious calculations using the assertion (iii) of Proposition III.5 and the fact that $A_{-+} = -(A_{+-})^*$ for A in $\mathfrak{u}_{res}(K, K_+)$.

Having at our disposal the following complex structure J_{K_+} on $T_{K_+}G_{res} \cong L^2(K_+, K_-)$ which is of also $U(K_+) \times U(K_-)$-invariant:

$$J_{K_+}\gamma = i\gamma \quad \text{for all } \gamma \in L^2(K_+, K_-) \tag{3.5}$$

it is easy to supply $L^2(K_+, K_-)$ with a $U(K_+) \times U(K_-)$-invariant structure of a "Kählerian vector space":

$$g_{K_+}(\gamma, \delta) := \Omega_{K_+}(\gamma, J_{K_+}\delta) = 2 \operatorname{Re} \operatorname{tr}(\gamma^*\delta) \tag{3.6}$$

and

$$h_{K_+}(\gamma, \delta) := g_{K_+}(\gamma, \delta) + i\Omega_{K_+}(\gamma, \delta) = 2 \operatorname{tr}(\gamma^*\delta). \tag{3.7}$$

Since h_{K_+} equals two times the standard scalar product on the complex Hilbert space $L^2(K_+, K_-)$, it follows that h_{K_+} is strongly non-degenerate in the sense that the map

$$L^2(K_+, K_-) \to (L^2(K_+, K_-))^*, \gamma \mapsto h_{K_+}(\gamma, \cdot)$$

is a complex anti-linear isomorphism. Considering now $L^2(K_+, K_-)$ as a real vector space $L^2(K_+, K_-)_{\mathbb{R}}$, its standard scalar product is given by $\operatorname{Re} \operatorname{tr}(\gamma^*\delta)$ and thus g_{K_+} is of course also strongly non-degenerate. Finally we observe that the map

$$L^2(K_+, K_-)_{\mathbb{R}} \to (L^2(K_+, K_-)_{\mathbb{R}})^*, \gamma \mapsto \Omega_{K_+}(\gamma, \cdot),$$

is an isomorphism as well and we conclude that Ω_{K_+} is a strongly non-degenerate symplectic form.

Remark. It is crucial at this point that we work with a Hilbert space. In a Banach or Fréchet space setting one considers often more generally "weakly non-degenerate symplectic forms" fulfilling the condition that the map $\gamma \mapsto \Omega(\gamma, \cdot)$ is a continuous injection of the given space into its dual (Compare Subsection IV.2).

Corollary III.9 *The restricted Grassmannian is a Kähler manifold which is homogeneous under its Kähler isometries.*

Proof. Since $U(K_+) \times U(K_-)$-invariant tensors on $L^2(K_+, K_-) \cong T_{K_+} G_{\mathrm{res}}$ correspond to U_{res}-invariant tensors on G_{res} we obviously have a U_{res}-invariant almost-Kähler structure (Ω, J, h, g) on G_{res}.

Recalling that the almost-complex structure J comes in fact from holomorphic coordinates on G_{res} (compare Proposition III.5 and its proof), completes now the assertion of the corollary. □

Remark. Let us stress again that the Kähler form on G_{res} is equivalent to the Schwinger term and thus the anomaly of fermionic second quantization is encoded in the geometry of G_{res}. In Section V we will see that the Fock space itself can be deduced from the holomorphic sections of a line bundle on G_{res}, which has the Kähler form as its Chern class.

In order to relate the symplectic form Ω to the cohomology of G_{res}, we note

Lemma III.10 *Let* $\mathbb{P}_1(\mathbb{C}) = \left\{ [z_{-1}, z_0] \,\big|\, \binom{z_{-1}}{z_0} \in \mathbb{C}^2 \backslash \{0\} \right\}$ *be identified with* $G_1(\mathbb{C}^2)$. *The holomorphic embedding* $j : G_1(\mathbb{C}^2) \hookrightarrow G_{\mathrm{res}}$ *with image* $G_{\mathrm{res}}^0 \cap G_{\mathrm{res}(1)}$ *pulls the Kähler form* Ω *back to the Fubini-Study form on* $\mathbb{P}_1(\mathbb{C})$, *given by*

$$\Omega_{FS} = i\partial\bar{\partial} \log(|z_{-1}|^2 + |z_0|^2).$$

Notably we have $\int_{\mathbb{P}_1(\mathbb{C})} j^* \Omega = 2\pi$.

Proof. We observe first that the coordinates $[z_{-1}, z_0] \mapsto \frac{z_{-1}}{z_0}$ on $\{z_0 \neq 0\}$ in $\mathbb{P}_1(\mathbb{C})$ are exactly the graph coordinates on $(G_{\mathrm{res}}^0 \cap G_{\mathrm{res}(1)}) \cap \mathcal{U}_{K_+}$ in G_{res}. A direct calculation in the point mapped to K_+ under j, yields then the equality of $j^*\Omega$ and Ω_{FS} in this point. Furthermore both forms as well as the associated hermitian metrics are $U(2)$-invariant on $\mathbb{P}_1(\mathbb{C})$. It remains only to recall that $\mathbb{P}_1(\mathbb{C}) = U(2)/(U(1) \times U(1))$ is an isotropy-irreducible homogeneous manifold and that thus all $U(2)$-invariant metrics on $\mathbb{P}_1(\mathbb{C})$ are proportional (compare [GHL], p. 67). □

Remark. As mentioned in the proof of Proposition III.5 the homotopy type of G_{res} is given by that of GL_{res} or, isomorphically, by that of U_{res}. It follows by Corollary II.19 and the Hurewicz isomorphism that $H_2(G_{\mathrm{res}}^0, \mathbb{Z}) \cong \pi_2(G_{\mathrm{res}}^0) \cong \mathbb{Z}$. The universal coefficient theorem implies now that $H^2(G_{\mathrm{res}}^0, \mathbb{Z}) \cong (H_2(G_{\mathrm{res}}^0, \mathbb{Z}))^* \cong \mathbb{Z}$ and $H^2(G_{\mathrm{res}}^0, \mathbb{R}) \cong \mathbb{R}$. Since G_{res} is a second-countable Hilbert manifold it is smoothly paracompact and the theorem of de Rham holds (see [E] or more recently, e.g., Thm. 34.7 in [KM]). Thus we know that $H_{dR}^2(G_{\mathrm{res}}^0, \mathbb{R}) \cong H^2(G_{\mathrm{res}}^0, \mathbb{R}) \cong (H_2(G_{\mathrm{res}}^0, \mathbb{R}))^*$ and that this isomorphism specializes to integration on smooth 2-cycles. Hence it follows from the preceding lemma that $j : G_1(\mathbb{C}^2) \cong S^2 \hookrightarrow G_{\mathrm{res}}$

defines a nontrivial element of $\pi_2(G_{\text{res}}^0)$ which generates $H_2(G_{\text{res}}^0, \mathbb{R})$, and that $[\Omega]$ generates $H_{dR}^2(G_{\text{res}}^0, \mathbb{R})$.

More refined homotopy-theoretic arguments show that j generates already the second homotopy group and thus that $[(2\pi)^{-1}\Omega]$ generates the second cohomology group with integral coefficients.

In closing this section we would like to mention some further differential-geometric properties of the restricted Grassmannian. Decomposing $\mathfrak{u}_{\text{res}}$ as $\mathfrak{u}_{\text{res}} = \mathfrak{h} \oplus \mathfrak{m}$, where

$$\mathfrak{h} = \mathfrak{u}(K_+) \oplus \mathfrak{u}(K_-) \text{ and } \mathfrak{m} = \left\{ \begin{pmatrix} 0 & -\gamma^* \\ \gamma & 0 \end{pmatrix} \Big| \gamma \in L^2(K_+, K_-) \right\} \cong L^2(K_+, K_-)$$

we observe that

$$[\mathfrak{h}, \mathfrak{m}] \subset \mathfrak{m} \text{ and } [\mathfrak{m}, \mathfrak{m}] \subset \mathfrak{h}.$$

This leads to

Proposition III.11 ([SpWu2]) *The restricted Grassmannian is a hermitian symmetric space. Thus it is notably geodesically complete and the geodesic exponential map* Exp_0 *in the point* K_+ *is given by* $\pi \circ \exp : \mathfrak{m} \cong T_{K_+}G_{\text{res}} \to G_{\text{res}}$, *where* $\exp$ *is the exponential map of* $G = U_{\text{res}}$ *and* π *the projection* $G \to G/H \cong G_{\text{res}}$. *The Riemann curvature tensor of* G_{res} *is completely fixed by its value in the point* K_+, *where it is given by the formula*

$$R^{\mathfrak{m}}(\gamma, \delta) \cdot \varepsilon = [[\gamma, \delta], \varepsilon],$$

where $\gamma, \delta, \varepsilon$ *are in* $\mathfrak{m} \cong T_{K_+}G_{\text{res}}(K, K_+)$ *and* $[[\gamma, \delta], \varepsilon]$ *denotes here the lower-left corner of*

$$\left[\left[\begin{pmatrix} 0 & -\gamma^* \\ \gamma & 0 \end{pmatrix}, \begin{pmatrix} 0 & -\delta^* \\ \delta & 0 \end{pmatrix} \right], \begin{pmatrix} 0 & -\varepsilon^* \\ \varepsilon & 0 \end{pmatrix} \right],$$

and $R^{\mathfrak{m}}$ *is the Riemann curvature tensor viewed as map from* $\mathfrak{m}^{\otimes 3}$ *to* $\mathfrak{m}$. *Furthermore the trace corresponding to the Ricci curvature of* G_{res} *is "linearly divergent".*

A more widely known feature of the restricted Grassmannian is the fact that many important infinite-dimensional manifolds are embedded into it:

Proposition III.12

(i) *The Fréchet manifold* $\text{Diff}^+(S^1)/S^1$ *of the orientation-preserving diffeomorphisms of the circle modulo the rotations of the circle is smoothly embedded into the restricted Grassmannian.*

(ii) *The complex Fréchet manifold of based loops in a compact Lie group* K *is holomorphically embedded into the restricted Grassmannian.*

Idea of a proof. (Compare [PrSe]) In both cases one uses an embedding of the transitively acting groups $\text{Diff}^+(S^1)$ respectively $LK = C^\infty(S^1, K)$ into $U_{\text{res}}(F, F_+)$ for an appropriate choice of a polarized complex separable Hilbert space $F = F_+ \oplus F_-$ and calculates then the stabilizers of the point F_+ under the action of these groups on the Grassmannian.

A careful analysis of the involved manifold structures yields then the result. $\square$

Remarks.

(1) Since the Schwinger term induces the universal central extensions of u_{res} as well as of $L\ell$ and Lie $\text{Diff}^+(S^1)$ (compare [PrSe]), and descends to the Kähler form of the Grassmannian, these embeddings give rise to the interpretation of the central extensions of $L\ell$ and Lie $\text{Diff}^+(S^1)$ in terms of infinite-dimensional Kählerian and symplectic geometry. (Compare [N2] and [Ov] in this volume.)

(2) Finer aspects of the embedding of based loop groups into the restricted Grassmannian can be found in [SpWu2].

IV. The non-equivariant moment map of the restricted Grassmannian

In this section we will first develop the basics of symplectic and Hamiltonian actions in an infinite-dimensional set-up. Notably we prove that the continuity of the co-moment map is equivalent to the smoothness of the associated moment map in the case of Fréchet manifolds (see Subsection IV.3). After a short inspection of some examples we will concentrate on the case of the U_{res}-action on the restricted Grassmannian and show – following Grosse and Maderner ([GM]) – that the Schwinger term is the obstruction preventing the U_{res}-moment map to be equivariant.

1. Differential k-forms in infinite dimensions

The aim of this subsection is to recall the notion of a smooth map in infinite dimensions and notably to define smooth differential k-forms on infinite-dimensional manifolds.

We adopt the definition of a continuously differentiable map between locally convex Hausdorff topological vector spaces (l.c. spaces in the sequel) used in [Mil2] and thoroughly studied in [Ha] in the Fréchet case. The extension of the fundamental properties found in the latter reference to the case of sequentially complete l.c. spaces (s.c.l.c. spaces in the sequel) is given in [N1].

Definition IV.1 Let E and F be l.c. spaces and U open in E.

(i) A map f from U to F is called C^1 or *continuously differentiable* if the partial derivatives

$$d^1 f(x)(h) = df(x)(h) = \lim_{\varepsilon \to 0} \frac{1}{\varepsilon} \{ f(x + \varepsilon h) - f(x) \}$$

exist for all x in U and h in E, and the map

$$df : U \times E \to F, (x, h) \mapsto df(x)(h)$$

is continuous.

(ii) A map f is called C^2 or *twice continuously differentiable* if $df : U \times E \to F$ is C^1. We will employ the notation

$$d^2 f(x)(h_1, h_2) = \lim_{\varepsilon \to 0} \frac{1}{\varepsilon} \{ df(x + \varepsilon h_2)(h_1) - df(x)(h_1) \}.$$

(iii) A map f is called C^∞ or *smooth* if the iterated differentials $d^n f = d(d^{n-1} f)$ exist and are continuous for all $n \geq 0$ $(d^o f = f)$.

Remarks.

(1) The above definition was also studied using "convergence structures" on vector spaces (compare e.g. [Ke]).

(2) In the case of Fréchet spaces smoothness in the above sense coincides with the definition given in [KM].

Given two l.c. spaces E and F the space $L^k(E, F)$ of k-linear jointly continuous maps from $E^k = E \times \ldots \times E$ (k factors) to F can be given the structure of a l.c. space as follows. We call a family $\mathcal{B}$ of bounded subsets of E a *bornology* if the following conditions are fulfilled

$$(\mathcal{B}_0) \qquad \bigcup_{B \in \mathcal{B}} B = E$$

$$(\mathcal{B}_1) \qquad \forall\, A, B \in \mathcal{B} \;\; \exists\, C \in \mathcal{B} \text{ such that } A \cup B \subset C$$

$$(\mathcal{B}_2) \qquad \forall\, A \in \mathcal{B} \;\; \forall\, \lambda \in \mathbb{R} \;\; \exists\, B \in \mathcal{B} \text{ such that } \lambda \cdot A \subset B.$$

(This is not the most general definition possible, but sufficient for our purposes.)

We define for B in $\mathcal{B}$ and q a continuous seminorm on F the following seminorm on $L^k(E, F)$:

$$q_B(T) = \sup\{q(T(h^{(1)}, \ldots, h^{(k)})) \mid h^{(j)} \in B \text{ for } j = 1, \ldots, k\}. \qquad (4.1)$$

The vector space $L^k(E, F)$ together with the topology defined by all seminorms obtained in this way (for a fixed $\mathcal{B}$) will be noted by $L_{\mathcal{B}}^k(E, F)$ and is a l.c. space. The coarsest topology of this type is of course given by the topology of uniform convergence on all finite sets, the weak topology, and the finest by the topology of uniform convergence on all bounded sets, the strong topology. We will denote $L^k(E, F)$ with the latter by $L_b^k(E, F)$ and with the topology of uniform convergence on all compact subsets by $L_c^k(E, F)$.

If E is a Fréchet space and $\mathcal{B}$ contains all compact sets then $L_{\mathcal{B}}^m(E, L_{\mathcal{B}}^n(E, F))$ is canonically isomorphic to $L_{\mathcal{B}}^{m+n}(E, F)$ as a topological vector space. If furthermore F is complete then $L_{\mathcal{B}}^k(E, F)$ is complete. (See [Ke] and [Tr].) Denoting $L^1(E, F)$ by $L(E, F)$ and $L^k(E, \mathbb{R})$ by $L^k(E)$ we then have – under the above condition on $\mathcal{B}$ – that $L_{\mathcal{B}}^1(E, \mathbb{R}) = E_{\mathcal{B}}'$, the space of continuous linear functionals on E with the topology of uniform convergence on all elements of $\mathcal{B}$, and $L_{\mathcal{B}}^2(E, \mathbb{R}) = L_{\mathcal{B}}^2(E) \cong L_{\mathcal{B}}(E, E_{\mathcal{B}}')$, and both spaces are complete. One can now mimick the usual definition of differentiability in a Banach space (and in finite dimensions!) and call a map $C_{\mathcal{B}}^1$ if the partial derivatives exist and the induced map $df : U \to L_{\mathcal{B}}(E, F), (df(x))(h) = df(x)(h)$ is continuous. Similarly one defines $C_{\mathcal{B}}^p$ and $C_{\mathcal{B}}^\infty$.

Lemma IV.2 *Let E be a Fréchet space, F a l.c.s.c. space, and $\mathcal{B}$ a bornology on E containing all compact sets. If U is open in E and $f : U \to F$ a map, then f is $C_{\mathcal{B}}^\infty$ if and only if f is C^∞ in the sense of Definition IV.1.*

Remarks.

(1) The basic ingredient of the proof is the (one-dimensional) integral formula $g(x + h) - g(x) = \int_0^1 d^1 g(x + th)(h) dt$ which is still valid for a C^1-map (See Thm. 3.2.2 in [Ha] and Lemma I.1 in [N1]). Details of the proof as well as more on the relations between various notions of differentiability in locally convex spaces can be found in [Ke].

(2) Obviously the above result implies that smoothness in the sense of Definition IV.1 is equivalent to the usual definition in the case of Banach spaces (i.e. C_b^∞).

On a manifold M, which we will always assume to be Hausdorff, that is smoothly modelled on a l.c. space E Definition IV.1 can of course be written in terms of the iterated tangent bundles of M. Let $T^{(0)}M = M, T^{(1)}M = TM, T^{(2)}M = T(TM)$ etc. and consider $d^{(1)}f : TM \to F$, given by $d^{(1)}f(x, h) = df(x)(h)$ upon locally identifying TM with $U \times E$, where U is an open set in E. Thus $f : M \to F$ is C^∞ if $d^{(n)}f$ exists and is continuous for all $n \geq 0$. (Observe that we have more "variables" in this approach, but the symmetries of the higher differentials imply that all information is encoded in the maps $d^n f : U \times E^n \to F$.)

Let G be a smooth manifold which is also a group and has the property that the multiplication and the association of the inverse are smooth. By definition the Lie algebra $\mathfrak{g}$ is the vector space $\mathfrak{g} = T_e G$ with the commutator given by identifying $\mathfrak{g}$ with the space of left invariant vector fields on G. Let us state explicitly that a vector field on a manifold M is defined as a smooth section of the tangent bundle and gives rise to an $\mathbb{R}$-linear derivation of the algebra of smooth functions $C^\infty(U, \mathbb{R})$ on each open subset U of M. Furthermore the commutator of two vector fields (as derivations) is again a vector field on M and the commutator of two left invariant vector fields is again left invariant (compare [Mil2] or [N1]). We call G a *Lie group with exponential map* if there exists a smooth "exponential map" $\exp : \mathfrak{g} \to G$ such that for X in $\mathfrak{g}$ the map $t \mapsto \alpha(t) = \exp(tX)$ fullfils

$$\alpha(0) = e \quad \text{and} \quad \frac{d}{dt}\Big|_\tau \alpha(t) = (l_{\alpha(\tau)})_* X \in T_{\alpha(\tau)}G$$

(where $l_g : G \to G$ is the left-multiplication by g), i.e., α is the one-parameter subgroup of G with $\dot\alpha(0) = X$. (Let us observe that the existence of an exponential map is not included in the definition of a *Lie group* in the two aforementioned references.)

In order to put a manifold structure on $TM' = \bigcup_{x \in M}(T_x M)'$, where $(T_x M)'$ is the space of continuous linear functionals on $T_x M$, we have to define a topology on E', the space of continuous linear functionals on E, the local model of M.

Let us first define the space of *alternating k-forms on E* by

$$\Lambda^k E'_{\mathcal{B}} = \{T \in L_{\mathcal{B}}^k(E) \mid T \text{ is skew-symmetric in the } k \text{ variables}\}.$$

This space is closed in $L_{\mathcal{B}}^k(E)$ and thus is a complete l.c. space if, e.g., E is a Fréchet space and $\mathcal{B}$ contains all compact sets. Note that this definition does not make use of the (rather delicate) theory of tensor products of l.c. spaces.

Definition IV.3 Let M be a manifold modelled on a l.c. space E and $\mathcal{B}$ a bornology on E. The $\mathcal{B}$-*cotangent bundle* of M is defined as $TM_{\mathcal{B}}' = \cup_{x \in M}(T_x M)_{\mathcal{B}}'$ and the k-fold exterior product of the $\mathcal{B}$-cotangent bundle as $\Lambda^k TM_{\mathcal{B}}' = \cup_{x \in M}\Lambda^k (T_x M)_{\mathcal{B}}'$.

Remarks.

(1) Obviously $\Lambda^k TM_{\mathcal{B}}'$ is a vector bundle in the category of l.c. spaces if $\mathcal{B}$ is chosen such that all continuous linear maps $T : E \to E$ map $\mathcal{B}$ into $\mathcal{B}$. The local model of $\Lambda^k TM_{\mathcal{B}}'$ is then $E \times \Lambda^k E_{\mathcal{B}}'$.

(2) If M is a Fréchet manifold the local model of $\Lambda^k TM_{\mathcal{B}}'$ is complete if $\mathcal{B}$ contains all compact sets of E.

(3) If M is a Banach manifold then $\Lambda^k TM_b'$ is the usual k-fold exterior power of the cotangent bundle of a Banach manifold.

Definition IV.4 Let M be a manifold modelled on a l.c. space E and $\mathcal{B}$ a bornology on E. A *smooth differential k-form of type $\mathcal{B}$* is a C^∞-section of the bundle $\Lambda^k TM_{\mathcal{B}}'$.

Let us compare this definition to the one given in [Beg].

Definition IV.5 Let M be a manifold modelled on a l.c. space E. A set-theoretic section ω of the bundle $\Lambda^k TM' = \cup_{x \in M}\Lambda^k (T_x M)'$ (without any topology on the fibre, the space of continuous k-linear alternating forms) is called a *smooth differential k-form in the weak sense* iff for all coordinate charts $\varphi : V \to U$ (V open in M, U open in E) the induced maps

$$\tilde{\omega}_U : U \times E^k \to \mathbb{R}, \quad \tilde{\omega}_U(x, h^{(1)}, \ldots, h^{(k)}) = \omega(\varphi^{-1}(x))(\varphi_*^{-1}h^{(1)}, \ldots, \varphi_*^{-1}h^{(k)})$$

are C^∞.

Proposition IV.6 *Let M be a Fréchet manifold modelled on E and $\mathcal{B}$ a bornology containing all compact sets and fulfilling that $T(\mathcal{B}) \subset \mathcal{B}$ for all continuous linear endomorphisms T of E. Then a section ω of $\Lambda^k TM' \to M$ is a smooth differential k-form of type $\mathcal{B}$ iff ω is a smooth differential k-form in the weak sense.*

Remark. The most important bornologies to which the proposition can be applied are of course the families of all bounded respectively of all compact subsets of E.

Proof of Proposition IV.6. Since both smoothness conditions are local we can assume that $M = U$, an open set in E and that ω is a map from U to $\Lambda^k E' = \{T \in L^k(E) | T \text{ is alternating}\}$.

Let us first assume that ω is C^∞ if $\Lambda^k E'$ is topologized as $\Lambda^k E_{\mathcal{B}}'$. In order to show that $\tilde{\omega} : U \times E^k \to \mathbb{R}$ is continuous we consider sequences $x_n \to x_0$ in U and $h_n^{(j)} \to h_0^{(j)}$ in E. The sets $K^{(j)} = \{h_n^{(j)} | n \geq 0\}$ being compact, there is a compact set C in E such that $K^{(j)} \subset C$ for $j = 1, \ldots, k$. We then have with $h_n = (h_n^{(1)}, \ldots, h_n^{(k)})$:

$$|\tilde{\omega}(x_n, h_n) - \tilde{\omega}(x_0, h_0)|$$
$$\leq |\tilde{\omega}(x_n, h_n) - \tilde{\omega}(x_0, h_n)| + |\tilde{\omega}(x_0, h_n) - \tilde{\omega}(x_0, h_0)|$$
$$\leq \sup\{|(\omega(x_n) - \omega(x_0))(h_n)| : n \geq 1\} + |\omega(x_0)(h_n - h_0)|$$
$$\leq \sup\{|(\omega(x_n) - \omega(x_0))(v)| : v \in C^k\} + |\omega(x_0)(h_n - h_0)|.$$

The second term in the last right hand side goes to zero since $\omega(x_0)$ is a continuous k-form on E and the first term tends to zero becaue the map ω is a fortiori continuous with respect to the topology of uniform convergence on products of compact sets.

Obviously $\tilde{\omega}$ is linear in the "second variables" $h^{(1)}, \ldots, h^{(k)}$. Furthermore the partial derivative $d_1\tilde{\omega}$ with respect to the first variable equals $\widetilde{d\omega}$ associated to the continuous map $d\omega : U \times E \to \Lambda^k E'_B$ and thus $d_1\tilde{\omega}$ is continuous by an argument as above for $\tilde{\omega}$. We can now apply Corollary 3.4.4 of [Ha] to conclude that $\tilde{\omega}$ is C^1. Induction shows that $\tilde{\omega}$ is C^∞.

Assume now that $\tilde{\omega} : U \times E^k \to \mathbb{R}$ is smooth and without loss of generality that U is convex. We will first show that $\omega : U \to \Lambda^k E'_b$ is continuous. Let $x_n \to x_0$ in U and set $w_n = x_n - x_0$. Let furthermore B be a bounded set in E and $h^{(1)}, \ldots, h^{(k)}$ be in B. We have

$$(\omega(x_n) - \omega(x_0))(h^{(1)}, \ldots, h^{(k)})$$
$$= (\omega(x_0 + w_n) - \omega(x_0))(h^{(1)}, \ldots, h^{(k)})$$
$$= \tilde{\omega}(x_0 + w_n, h^{(1)}, \ldots, h^{(k)}) - \tilde{\omega}(x_0, h^{(1)}, \ldots, h^{(k)})$$
$$= \int_0^1 d_1\tilde{\omega}(x_0 + tw_n, h^{(1)}, \ldots, h^{(k)}; w_n)dt,$$

where $d_1\tilde{\omega}(x, h^{(1)}, \ldots, h^{(k)}; w)$ is the partial derivative of $\tilde{\omega}$ with respect to the first variable in direction of w and the last equality follows from Thm. 3.2.2 in [Ha].

Since $d_1\tilde{\omega}$ is continuous there are convex open neighborhoods U' of x_0 in U, and V and W of 0 in E such that $d_1\tilde{\omega}(U' \times V^k \times W) \subset \{s \in \mathbb{R} : |s| < 1\}$. Since B is bounded there is $r \geq 1$ such that $B \subset r \cdot V$. Setting $W_\epsilon = \left(\frac{\epsilon}{r^k}\right) \cdot W$ for $\epsilon > 0$, we find

$$(d_1\tilde{\omega})(U' \times B^k \times W_\epsilon) \subset (d_1\tilde{\omega})\left(U' \times (r \cdot V)^k \times \left(\frac{\epsilon}{r^k}\right) \cdot W\right)$$
$$\subset \epsilon \cdot \{(d_1\tilde{\omega})(U' \times V^k \times W)\} \subset \{|s| < \epsilon\},$$

implying that

$$|(\omega(x_n) - \omega(x_0))(h^{(1)}, \ldots, h^{(k)}|$$
$$\leq \int_0^1 |d_1\tilde{\omega}(x_0 + tw_n, h^{(1)}, \ldots, h^{(k)}; w_n)|dt < \int_0^1 \epsilon \, dt = \epsilon,$$

if $x_0 + t w_n$ is in U' and w_n in W_ϵ. Since $w_n \to 0$ and $t \in [0, 1]$ these two conditions are fulfilled for $n \geq n_0(\epsilon)$ with a $n_0(\epsilon) \in \mathbb{N}$ sufficiently big. We conclude that $\omega(x_n) \to \omega(x_0)$ in $\Lambda^k E_b'$.

The derivative $d\omega : U \times E \to \Lambda^k E_b'$ is – as a set-theoretic map – given by

$$(d\omega(x)(v))(h^{(1)}, \dots, h^{(k)}) = (d_1 \tilde{\omega})(x, h^{(1)}, \dots, h^{(k)}; v).$$

Let us assume that $x_n \to x_0$ in U and $v_n \to v_0$ in E, and $h^{(j)}$ are in a bounded set B in E. We have with $h = (h^{(1)}, \dots, h^{(k)})$:

$$\begin{aligned}
&(d\omega(x_n)(v_n) - d\omega(x_0)(v_0))(h) \\
&= (d_1 \tilde{\omega})(x_n, h; v_n - v_0) + ((d_1 \tilde{\omega})(x_n, h; v_0) - (d_1 \tilde{\omega})(x_0, h; v_0)).
\end{aligned}$$

The second summand in the last right hand side tends to zero by using the differentiability of $d_1 \tilde{\omega}$ in the first variable analogously as in the proof of the continuity of ω above. The first summand can be directly estimated by the continuity of $d_1 \tilde{\omega}$ as in the last part of the above argument proving continuity of ω (the "ϵ-r trick"). Thus $d\omega : U \times E \to \Lambda^k E_b'$ is continuous, i.e. ω is C^1.

Induction shows now that $\omega : U \to \Lambda^k E_b'$ is C^∞ and thus a fortiori $\omega : U \to \Lambda^k E_B'$ as well. $\qquad\square$

Motivated by the preceding proposition and in order to include the standard definitions in the case of Banach manifolds we will from now on call smooth differential k-forms of type $\mathcal{B} = \{\text{bounded subsets of } E\}$ simply *smooth k-forms*.

Let us develop now the basic differential operations on a Fréchet manifold. For a smooth vector field ξ and a smooth $(k+1)$-form ω we define the contraction $i_\xi \omega = \xi \lrcorner \omega$ point-wise as in finite dimensions. Obviously $(i_\xi \omega)(x)$ is in $\Lambda^k (T_x M)'$ for all x in M. The only subtle point in proving that $i_\xi \omega$ is a smooth k-form is caused by the fact that the pairing $E \times E_\mathcal{B}' \to \mathbb{R}$ on a locally convex space E is never jointly continuous unless the topology of E is given by a single norm and $E_\mathcal{B}'$ is the strong dual of E. This problem can nevertheless be circumvented by an "hypocontinuity argument" (at least in the Fréchet case):

Lemma IV.7 *Let M be a Fréchet manifold, ξ a smooth vector field and ω a smooth $(k+1)$-form on M. Then $i_\xi \omega$ is a smooth k-form on M.*

Proof. Let us for simplicity give here only the proof for $k = 0$.

We have to show that the function $f : M \to \mathbb{R}$ defined by $f(x) = \omega(x)(\xi(x))$ is smooth. Let $\langle , \rangle$ denote here the pairing between a Fréchet space and its dual. Then we have

$$f(x) = \langle \xi(x), \omega(x) \rangle.$$

Since smoothness is a local condition we may assume that $M = U$ is open in E and $\xi : U \to E$ respectively $\omega : U \to E_c'$ are smooth maps. Since E is a Fréchet space continuity of f is equivalent to sequential continuity. Assume thus that $x_n \to x_0$

in $U \subset E$:

$$|f(x_0) - f(x_n)| = |\langle \xi(x_0), \omega(x_0) \rangle - \langle \xi(x_n), \omega(x_n) \rangle|$$
$$= |\langle \xi(x_0), \omega(x_0) - \omega(x_n) \rangle + \langle \xi(x_0) - \xi(x_n), \omega(x_n) \rangle|$$
$$\leq |\langle \xi(x_0), \omega(x_0) - \omega(x_n) \rangle| + |\langle \xi(x_0) - \xi(x_n), \omega(x_n) \rangle|.$$

The first term on the last right hand side tends to zero, since ω is continuous and $\langle,\rangle$ is separately continuous. The convergence to zero of the second term follows by a standard "hypocontinuity argument": let K be the compact set $\{\omega_n = \omega(x_n) | n \geq 0\}$ in E'_c. By the Banach-Steinhaus theorem K is equicontinuous, i.e. for all $\varepsilon > 0$ there exists an open neighborhood $V = V(\varepsilon, K)$ of 0 in E such that $|\omega(v)| < \varepsilon$ for all ω in K and all v in V. Since $\xi_n = \xi(x_0) - \xi(x_n)$ converges to zero in E there is a $N = N(\varepsilon)$ such that $\xi_n \in V$ for $n \geq N$. It follows that

$$|\langle \xi(x_0) - \xi(x_n), \omega(x_n) \rangle| = |\langle \xi_n, \omega_n \rangle| = |\omega_n(\xi_n)| < \varepsilon$$

for $n \geq N$. Thus $f(x_0) - f(x_n) \to 0$.

Observing that $df : U \times E \to \mathbb{R}$ is given by

$$df(x)(h) = \langle \xi(x), d\omega(x)(h) \rangle + \langle d\xi(x)(h), \omega(x) \rangle$$

a similar argument shows that df is continuous. By induction it follows that f is smooth. $\qquad \square$

As in finite dimensions we know that the commutator of two vector fields is again a vector field (compare [N1], Lemma I.12). Thus, given a smooth k-form ω we can define a $(k+1)$-linear map on vector fields by

$$(d_{\mathrm{dR}}\omega)(\xi_0, \dots, \xi_k) = \sum_{0 \leq i \leq k} (-1)^i \xi_i(\omega(\xi_0, \dots, \hat{\xi}_i, \dots, \xi_k))$$
$$+ \sum_{0 \leq i < j \leq k} (-1)^{i+j} \omega([\xi_i, \xi_j], \xi_0, \dots, \hat{\xi}_i, \dots, \hat{\xi}_j, \dots, \xi_k),$$

where a hat "$\wedge$" means omission.

Lemma IV.8 *Let ω be a smooth k-form on a Fréchet manifold M. Then $d_{\mathrm{dR}}\omega$, the "exterior" or "de Rham-derivative" of ω, is a smooth $(k+1)$-form.*

Proof. Since the assertion is of a local nature we can assume that $M = U$ is open in a Fréchet space E. The associated map $\tilde{\omega} : U \times E^k \to \mathbb{R}$ is smooth by Proposition IV.6 and the exterior derivative reads as

$$\widetilde{d_{\mathrm{dR}}\omega} : U \times E^{k+1} \to \mathbb{R},$$
$$\widetilde{d_{\mathrm{dR}}\omega}(x, v_0, \dots, v_k) = \sum_{0 \leq i \leq k} (-1)^i (d_1 \tilde{\omega})(x, v_0, \dots, \hat{v}_i, \dots, v_k; v_i).$$

Since this expression defines a smooth $(k+1)$-form in the weak sense we conclude by Proposition IV.6 that $d_{\mathrm{dR}}\omega$ is a smooth k-form. $\qquad \square$

Remark. If there is no danger of confusion we will denote the exterior derivative simply by d.

We **define** now the Lie derivative $\mathcal{L}_\xi$ associated to a smooth vector field on smooth differential forms by the homotopy formula of Cartan:

$$\mathcal{L}_\xi = d_{\mathrm{dR}} \circ i_\xi + i_\xi \circ d_{\mathrm{dR}}. \tag{4.2}$$

If ξ has a local flow (which is not even on a Fréchet space always true) this definition coincides of course with the usual one.

2. Symplectic manifolds, group actions and the co-moment map

In this subsection we will first recall the definitions of a symplectic manifold and a symplectic group action as well as that of a co-momentum map and why it gives rise to a Lie algebra cohomology two-cocycle. As a side remark we will explain a beautiful idea of Reznikov that implies that a finite-dimensional Lie group acting almost-effectively on a compact symplectic manifold is already of the type center plus compact semi-simple.

Slightly abusing the language, we will call a Lie group possessing an exponential map (compare the last subsection and [Mil2] or [N1]) simply a "Lie group" in this subsection.

Definition IV.9

(i) A *weakly symplectic manifold* is a manifold with a closed smooth two-form Ω such that the linear continuous map

$$\Omega_x^\# : T_x M \to (T_x M)_b'$$
$$v_x \mapsto \Omega_x(v_x, \cdot) = i_{v_x} \Omega_x \tag{4.3}$$

is injective for all x in M. We say that (M, Ω) is *strongly symplectic* iff $\Omega_x^\#$ is a homeomorphism for all x.

(ii) A vector field ξ_f on a weakly symplectic manifold (M, Ω) is called the *symplectic gradient* of a smooth function f on M if

$$df = -\Omega(\xi_f, \cdot) = \Omega^\#(-\xi_f) \tag{4.4}$$

(Observe that df is here the one-form $d_{\mathrm{dR}} f$ defined by exterior derivation of the function f.) It is obvious that the symplectic gradient is unique if it exists but it may well happen that df is not in the image of $\Omega^\#$.

(iii) If f and g are smooth functions on a weakly symplectic manifold such that ξ_f and ξ_g exist we can define the *Poisson bracket* of f and g by

$$\{f, g\} = \Omega(\xi_f, \xi_g) = \xi_f(g) = -\xi_g(f). \tag{4.5}$$

it is easily verified that the Poisson bracket is $\mathbb{R}$-bilinear, anti-symmetric and fullfills the Jacobi identiy, if all involved symplectic gradients exist. Thus in the case of a strongly symplectic manifold we have a Lie algebra homomorphism

$$C^\infty(M, \mathbb{R}) \xrightarrow{\alpha} \Gamma_{C^\infty}(M, TM), \alpha(f) = \xi_f \tag{4.6}$$

whose kernel consists of the locally constant functions on M.

Remarks.

(1) Since the strong dual E'_b of a locally convex space E is metrizable iff E is normable, a "non-Banach" Fréchet manifold allows only weakly symplectic forms. On the other hand a real Hilbert space has always a constant symplectic form. Since a paracompact manifold modelled on a separable Hilbert space is diffeomorphic to an open subset of its model space by [EE] it carries always a strongly symplectic form.

(2) Let us observe that in the case of weakly symplectic manifolds there might exist linear subspaces in $C^\infty(M, \mathbb{R})$ where the Poisson bracket is defined and yields elements of the given subspace. This will especially happen in the case of group actions.

Given a left-action of a Lie group G on a manifold M, we have an associated anti-Lie homomorphism

$$\tau : \mathfrak{g} \to \Gamma_{C^\infty}(M, TM), \tau(X)_p = \tau(X)(p) = \frac{d}{dt}\Big|_0 (\exp(tX) \cdot p), \qquad (4.7)$$

for all X in $\mathfrak{g}$ and p in M.

If (M, Ω) is weakly symplectic and $\Theta : G \times M \to M$ a left-action, we say that the action is *symplectic* if $\Theta_g^*\Omega = \Omega$ for all g in G ($\Theta_g(p) := \Theta(g, p) = g \cdot p$). By the homotopy formula of Cartan (4.2) we then have

$$\mathcal{L}_{\tau(X)}\Omega = d(\Omega(\tau(X), \cdot)) = 0 \quad \forall\, X \in \mathfrak{g}.$$

(Note that $\tau(X)$ has a flow, namely $\Theta_{\exp(tX)\cdot}$.)

Definition IV.10 A symplectic action is called *Hamiltonian* if there exists a $\mathbb{R}$-linear map $\lambda : \mathfrak{g} \to C^\infty(M, \mathbb{R})$ such that $d(\lambda(X)) = \Omega(\tau(X), \cdot)$ for all X in $\mathfrak{g}$ or, otherwise stated, $\alpha \circ \lambda = -\tau$.

Such a map λ is called a *co-moment* for the action of G on M.

Remark. In infinite dimensions it looks rather natural to demand that λ is continuous. We will thoroughly study this condition in the next subsection, but concentrate on the algebraic aspect here.

Since by the very definition of a co-moment α and thus $\{,\}$ are defined on $\lambda(\mathfrak{g})$ one might ask if λ is a Lie algebra homomorphism, or can be corrected to be one. It is easily checked that

$$c(X, Y) := \{\lambda(X), \lambda(Y)\} - \lambda([X, Y]) \qquad (4.8)$$

is – for fixed X and Y in $\mathfrak{g}$ – a locally constant function on M. Furthermore we have

Lemma IV.11 *Let $G \times M \to M$ be a Hamiltonian action of a Lie group G on a connected weakly symplectic manifold (M, Ω) and $\lambda : \mathfrak{g} \to C^\infty(M, \mathbb{R})$ a co-moment. Then λ defines a class $[c]$ in $H^2(\mathfrak{g}, \mathbb{R})$ that depends only on the action.*

Proof. Obviously c, defined by equation (4.8), is a anti-symmetric bilinear form on $\mathfrak{g}$, i.e. an element of $\Lambda^2\mathfrak{g}^*$. A direct calculation using the Jacobi identity in the various Lie algebras involved shows that $\delta c = 0$, where δ is the coboundary operator of Lie algebra cohomology. Thus $[c]$ is an element of the second Lie algebra cohomology group $H^2(\mathfrak{g}, \mathbb{R})$ (with values in the trivial $\mathfrak{g}$-module $\mathbb{R}$). Since M is connected another co-moment $\tilde{\lambda}$ for the same action can differ from λ only by a linear form b on $\mathfrak{g}$ and thus $[\tilde{c}] = [c + \delta b] = [c]$. $\qquad\square$

It is well-known that $H^2(\mathfrak{g}, \mathbb{R})$ classifies the equivalence classes of central Lie algebra extensions of $\mathfrak{g}$ by $\mathbb{R}$. Thus we arrive in our context at:

Lemma IV.12 *Let G, M and $[c]$ as in Lemma IV.11. Then we have:*

(i) *if $[c] = 0$ there exists a co-moment λ that is a Lie algebra homomorphism and this λ is then equivariant with respect to the subgroup of G generated by $\{\exp(X)|X \in \mathfrak{g}\}$.*

(ii) *if $[c] \neq 0$ there exists a central extension $\tilde{\mathfrak{g}}$ of $\mathfrak{g}$ by $\mathbb{R}$ such that $\tilde{\mathfrak{g}}$ allows for a co-moment that is a Lie algebra homomorphism.*

Proof. Given any co-moment λ_0, one has in the first case $c_0 = \delta b$ for a linear form b on $\mathfrak{g}$. It is now easily verified that $\lambda = \lambda_0 - b$ is a Lie algebra homomorphism. The second assertion of (i) follows from a direct calculation. Let us denote the map $G \times M \to M$ that defines the action again by Θ and let Θ_g be the diffeomorphism of M given by $\Theta_g(p) = \Theta(g, p) = g \cdot p$. then we have to show that

$$\lambda(\mathrm{Ad}(\exp(tX)) \cdot Y) = \lambda(Y) \circ \Theta_{\exp(-tX)} \qquad (*)$$

for all X, Y in $\mathfrak{g}$ and t in $\mathbb{R}$. Obviously $(*)$ holds for $t = 0$ and differentiating the left and right hand side with respect to t yields in any point p in M:

$$\frac{d}{dt}\lambda(\mathrm{Ad}(\exp tX) \cdot Y)(p) = \lambda([X, \mathrm{Ad}(\exp tX)Y])(p)$$
$$= \{\lambda(X), \lambda(\mathrm{Ad}(\exp tX) \cdot Y)\}(p) = -\Omega(\tau(\mathrm{Ad}(\exp tX) \cdot Y)_p, \tau(X)_p)$$

respectively

$$\frac{d}{dt}(\lambda(Y) \circ \Theta_{\exp(-tX)})(p)$$
$$= -d(\lambda(Y))(\tau(X)(\Theta_{\exp(-tX)}(p)))$$
$$= \Omega((\alpha \circ \lambda)(Y)(\Theta_{\exp(-tX)}(p)), \tau(X)(\Theta_{\exp(-tX)}(p)))$$
$$= -\Omega(\tau(Y)(\Theta_{\exp(-tX)}(p)), \tau(X)(\Theta_{\exp(-tX)}(p)))$$
$$= -\Omega((\Theta_{\exp(-tX)})_*(\tau(\mathrm{Ad}(\exp tX) \cdot Y)_p), (\Theta_{\exp(-tX)})_*(\tau(\mathrm{Ad}(\exp tX) \cdot X)_p))$$
$$= -\Omega(\tau(\mathrm{Ad}(\exp tX) \cdot Y)_p, \tau(\mathrm{Ad}(\exp tX) \cdot X)_p)$$
$$= -\Omega(\tau(\mathrm{Ad}(\exp tX) \cdot Y)_p, \tau(X)_p).$$

Regarding assertion (ii) let us first define $\tilde{\mathfrak{g}}$ as the vector space $\mathfrak{g} \oplus \mathbb{R}$ and its bracket by $[(X, a), (Y, b)]_{\tilde{\mathfrak{g}}} := ([X, Y], c(X, Y))$. It is easy to check that $\tilde{\mathfrak{g}}$ is a central Lie algebra extension of $\mathfrak{g}$ by $\mathbb{R}$. Denoting the projection $\tilde{\mathfrak{g}} \to \mathfrak{g}$ by π, $\tilde{\mathfrak{g}}$ acts on M via $\tilde{\tau} = \tau \circ \pi$.

We define now a co-moment $\tilde{\lambda} : \tilde{\mathfrak{g}} \to C^\infty(M, \mathbb{R})$ by $\tilde{\lambda}(X, a) = \lambda(X) + a$, where a denotes the constant function with value a in $\mathbb{R}$ on M as well. It follows that $\tilde{\lambda}$ is a Lie algebra homomorphism and that $\alpha \circ \tilde{\lambda} = -\tilde{\tau}$. $\square$

Remark. We will not go into the question when there exists a corresponding extension of G by $\mathbb{R}$ or $U(1)$ and what is its geometric meaning. At least in finite dimensions these aspects are rather well-understood in the framework of geometric prequantization (see the classical references [Ki], [Ko] and [So]).

 There are many criteria when a symplectic action is already Hamiltonian respectively when then the cocycle $[c]$ vanishes. For the basic results in this direction the reader might consult the recent review [Br] or the standard reference [GS]. We will give here only the following elucidating result in finite dimensions. Though at least parts of it seem to be folkloristic in the field the only proof known to the author is based on an idea of Reznikov ([Re]):

Proposition IV.13 *Let (M, Ω) be a compact symplectic manifold without boundary and G a finite-dimensional Lie group acting almost-effectively on M. If the action has a co-moment λ that is a Lie algebra homomorphism then the Lie algebra $\mathfrak{g}$ of G is a direct product of its center and a semisimple Lie algebra whose Killing form is negative definite.*

Corollary IV.14 *The same conclusion as in Proposition IV.13 holds if the action of G is only assumed to be Hamiltonian.*

Proof of the corollary. Let us assume without loss of generality that M is connected. Since G acts in a Hamiltonian fashion there exists a co-moment λ_0. We "correct" it by setting

$$\lambda(X) := \lambda_0(X) - \left(\int_M \lambda_0(X) \Omega^n \right) \Big/ \left(\int_M \Omega^n \right),$$

where $\dim_{\mathbb{R}} M = 2n$. It is well-known and easily checked that λ is a Lie algebra homomorphism and of course it remains a co-moment. $\square$

Proof of Proposition IV.13. As in the proof of the above corollary we set $2n = \dim_{\mathbb{R}} M$ and assume that M is connected.

 We introduce an inner product $\langle \cdot, \cdot \rangle$ on $\mathfrak{g}$ by defining:

$$\langle X, Y \rangle := \int_M \lambda(X) \cdot \lambda(Y) \Omega^n \text{ for all } X, Y \in \mathfrak{g}.$$

Obviously $\langle \cdot, \cdot \rangle$ is symmetric and the injectivity of λ implies that it is positive definite.

The following calculation shows its ad-invariance:

$$\langle [X, Y], Z \rangle + \langle Y, [X, Z] \rangle$$

$$= \int_M (\lambda([X, Y]) \cdot \lambda(Z) + \lambda(Y) \cdot \lambda([X, Z]) \Omega^n$$

$$= \int_M (\{\lambda(X), \lambda(Y)\} \lambda(Z) + \lambda(Y)\{\lambda(X), \lambda(Z)\}) \Omega^n$$

$$= - \int_M (\tau(X)(\lambda(Y) \cdot \lambda(Z))) \Omega^n$$

$$= - \int_M \mathcal{L}_{\tau(X)}(\lambda(Y) \cdot \lambda(Z) \Omega^n), \text{ since } \mathcal{L}_{\tau(X)} \Omega = 0$$

$$= - \int_M d(\lambda(Y) \cdot \lambda(Z) i_{\tau(X)} \Omega^n) = 0 \text{ by Stokes' theorem.}$$

Let now $\mathfrak{z}(\mathfrak{g})$ be the center and $\mathfrak{g}'$ the derived algebra of $\mathfrak{g}$. It follows easily from the existence of the ad-invariant positive definite inner product $\langle \cdot, \cdot \rangle$ on $\mathfrak{g}$ that $\mathfrak{g}$ in the direct sum of these ideals:

$$\mathfrak{g} = \mathfrak{z}(\mathfrak{g}) \oplus \mathfrak{g}'.$$

Taking the derived algebra of both sides of the last equality we find that $\mathfrak{g}'' = \mathfrak{g}'$. Denoting the radical of $\mathfrak{g}'$ by $\mathfrak{r}$, the existence of the ad-invariant positive definite inner product implies the following Lie algebra decomposition

$$\mathfrak{g}' = \mathfrak{r} \oplus \mathfrak{r}^\perp,$$

where $\mathfrak{r}^\perp = \{X \in \mathfrak{g}' | \langle X, Y \rangle = 0 \, \forall \, Y \in \mathfrak{r}\}$ is an ideal of $\mathfrak{g}'$. The equality $\mathfrak{g}'' = \mathfrak{g}'$ forces now $\mathfrak{r}' = \mathfrak{r}$. Since the radical $\mathfrak{r}$ of $\mathfrak{g}'$ is by definition the maximal solvable ideal of $\mathfrak{g}'$, we conclude that $\mathfrak{r} = 0$, i.e. $\mathfrak{g}'$ is semisimple.

Decomposing now $\mathfrak{g}'$ as the sum of its simple ideals:

$$\mathfrak{g}' = \bigoplus_{j=1}^{m} \mathfrak{g}_j,$$

it follows that $\langle \cdot, \cdot \rangle |_{\mathfrak{g}_j \times \mathfrak{g}_j} = \langle \cdot, \cdot \rangle_j$ is an ad-invariant bilinear form on $\mathfrak{g}_j$ (for $j = 1, \ldots, m$). Thus by Schur's lemma there exist λ_j in $\mathbb{R}$ such that the Killing form of $\mathfrak{g}_j$ equals $\lambda_j \langle \cdot, \cdot \rangle_j$. Since $\langle \cdot, \cdot \rangle_j$ is positive definite all $\lambda_j < 0$, implying that the Killing form of $\mathfrak{g}'$ is negative definite. $\qquad \square$

3. Co-momentum and momentum maps (in infinite dimensions)

In this subsection we will first algebraically dualize the notion of a co-momentum to that of a momentum. Using again the differential calculus à la Hamilton (compare [Ha], or [N1] in this volume) we will study the relation between the continuity of the co-momentum and the smoothness of the momentum in detail in the Fréchet case.

Given a manifold M we always have a map $j : M \to C^\infty(M)^*$, the algebraic dual of the space of smooth functions on M, given by associating to a point x in

M the Dirac-measure in x : $j(x) = \delta_x$. Let us assume that M is weakly symplectic and is acted upon by a Lie group G such that there exists a co-momentum λ : $\mathfrak{g} \to C^\infty(M)$ (compare Definition IV.10). We can dualize λ algebraically to a map $\lambda^* : C^\infty(M)^* \to \mathfrak{g}^*$ ($\mathfrak{g}^*$ being the algebraic dual of $\mathfrak{g}$) and get a map $\Phi = \lambda^* \circ j$: $M \to \mathfrak{g}^*$. Denoting the pairing between $\mathfrak{g}$ and $\mathfrak{g}^*$ here and in the sequel by $\langle,\rangle$ we find $\Phi^X := \langle X, \Phi \rangle = \lambda(X)$ for all X in $\mathfrak{g}$. If λ is a Lie algebra homomorphism formula (∗) in the proof of Lemma IV.12 shows that Φ is equivariant with respect to the G-action on M and the coadjoint action on $\mathfrak{g}^*$ for all g in the subgroup of G generated by the image of the exponential map.

Definition IV.15 Let (M, Ω) be a weakly symplectic manifold and $G \times M \to M$ a symplectic action of a Lie group G. A smooth map $\Phi : M \to \mathfrak{g}'$ is called a *moment map* for the G-action if $\lambda : \mathfrak{g} \to C^\infty(M)$ defined by $\lambda(X) = \langle X, \Phi \rangle$ is a co-moment.

Such a map Φ is called an *equivariant moment map* if it is G-equivariant, i.e. if $\Phi \circ \Theta_g = \mathrm{Ad}^*(g) \circ \Phi$ for all g in G.

Remarks.

(1) The symbol $\mathfrak{g}'$ denotes here the vector space of all continuous linear functions on $\mathfrak{g}$ **together** with a topology that turns it into a l.c. space. If $\mathfrak{g}$ is a Banach Lie algebra $\mathfrak{g}'$ might be the strong dual $\mathfrak{g}'_b$ of $\mathfrak{g}$, being again a Banach space. In case $\mathfrak{g}$ is a Fréchet space there are several natural choices notably $\mathfrak{g}'_b$ and $\mathfrak{g}'_c$ (compare the discussion in Subsection IV.1). Let us remark that these two topologies coincide in the important case that $\mathfrak{g}$ is a Montel space.

(2) By definition the existence of a moment map Φ implies the existence of a co-moment λ. If Φ is G-equivariant it follows furthermore that λ is a Lie algebra homomorphism.

In finite dimensions it is easy to see that the mere existence of a co-moment implies that the algebraically associated moment is already a smooth map. The obvious condition to impose in infinite dimensions is that λ should be continuous. We topologize thus $C^\infty(M)$ – as in finite dimensions – by the topology of uniform convergence of the derivatives up to a finite but arbitrary order on compact subsets, i.e. for K_m compact in $T^{(m)}M$ one defines the seminorm

$$p_{K_m}(f) = \sup\{|d^{(m)}f(x)| : x \in K_m\}.$$

This yields a locally convex topology on $C^\infty(M)$ that is complete if M is a Fréchet manifold.

On the dual of $C^\infty(M)$ we have again several topologies notably the topology of uniform convergence on compact subsets of $C^\infty(M)$ which gives rise to the locally convex space $C^\infty(M)'_c$. Completeness of this space is assured if e.g. $C^\infty(M)$ is barrelled as is the case if e.g. the local model E of M is a nuclear Fréchet space (see [Tho]).

The most direct approach to the smoothness of Φ, given a continuous co-moment, is of course via the map j. Here we have

Lemma IV.16 *Let M be a Fréchet manifold and $j : M \to C^\infty(M)'$ be given by $j(x) = \delta_x$, the Dirac measure in the point x. Then j is smooth if we topologize $C^\infty(M)'$ as $C^\infty(M)'_b$.*

Proof. It is obvious that $j(x) = \delta_x$ is continous for each fixed x in M.

Since smoothness is a local property we may consider an open set U in E, the local model of M, instead of M itself. Assuming that U is the image of a coordinate chart on M we can of course restrict all functions on M to U and a bounded set of functions on M yields by restriction a bounded set of functions on U. Since $U \times E^m$ is open in the metric space $E \times E^m$ it is enough to consider sequential continuity here. Let us furthermore assume that U is convex.

In order to show that j is continuous, let $x_n \to x_0$ in U. For a function f in $C^\infty(U)$ we have

$$|(j(x_0) - j(x_n))(f)| = |f(x_0) - f(x_n)| \leq \int_0^1 |df(x_0 + t(x_0 - x_n))(x_0 - x_n)|dt.$$

The sequence $u_n = x_0 - x_n$ tends to zero in the Fréchet space E. An obvious reformulation of statement (b) of Theorem 1.28 in [Rud] shows that there are positive scalars α_n such that $\alpha_n \to 0$ and $v_n = (\frac{1}{\alpha_n}) \cdot u_n \to 0$. Thus the sets $K' = \{u_n | n \geq 1\} \cup \{0\}$ and $K'' = \{v_n | n \geq 1\} \cup \{0\}$ are compact in E. Setting $K_1 = (x_0 + K') \times K'' \subset U \times E$, we can make the following estimate

$$|(j(x_0) - j(x_n))(f)| \leq \int_0^1 |df(x_0 + tu_n)(\alpha_n v_n)|dt$$

$$= \alpha_n \cdot \left(\int_0^1 |df(x_0 + tu_n)(v_n)|dt \right) \leq \alpha_n \cdot p_{K_1}(f),$$

where $p_{K_1}(f) = \sup\{|df(x)(v)| : (x, v) \in K_1\}$. Given a bounded set $B \subset C^\infty(M)$ there is a constant $M_{K_1}(B)$ such that $p_{K_1}(f) \leq M_{K_1}(B)$ for all f in B. Thus $|(j(x_0) - j(x_n))(f)| \leq \alpha_n \cdot M_{K_1}(B)$ for all f in B. Since $\alpha_n \to 0$ it follows that j is continuous, since

$$q_B(j(x_0) - j(x_n)) = \sup\{|(j(x_0) - j(x_n))(f)| : f \in B\}$$

converges to zero for all B bounded in $C^\infty(M)$ and the q_B define the locally convex topology of $C^\infty(M)'_b$.

We proceed to show the existence of the first derivative: let (x, h) be in $U \times E$ and f in $C^\infty(U)$. Then

$$(dj(x)(h))(f) = \lim_{\varepsilon \to 0} \frac{1}{\varepsilon}\{j(x + \varepsilon h)(f) - j(x)(f)\}$$

$$= \lim_{\varepsilon \to 0} \frac{1}{\varepsilon}\{f(x + \varepsilon h) - f(x)\} = df(x)(h).$$

Thus for (x, h) fixed, $dj(x)(h)$ is in $C^\infty(M)'$. Using an argument as above one finds that dj is continuous.

Similar arguments show the existence and continuity of the higher derivatives $d^m j$. We conclude that $j : M \to C^\infty(M)'_b$ is smooth. $\qquad\square$

Corollary IV.17 *Let M be a weakly symplectic Fréchet manifold and $\mathfrak{g}$ be the Lie algebra of a Lie group modelled on a sequentially complete locally convex space. Then the existence of a continuous co-moment $\lambda : \mathfrak{g} \to C^\infty(M)$ implies that there exists a momentum map $\Phi : M \to \mathfrak{g}'_b$ (and thus a fortiori to $\mathfrak{g}'_c$).*

Proof. Since Φ can be factorized as $\lambda'_b \circ j$ with $j : M \to C^\infty(M)'_b$ and λ'_b the (continuous!) transpose of λ going from $C^\infty(M)'_b$ to $\mathfrak{g}'_b$ the corollary follows from the preceding lemma. $\square$

Though many important Fréchet Lie algebras are Montel spaces and thus fulfil $\mathfrak{g}'_c = \mathfrak{g}'_b$, one might nevertheless be interested in the question if one can replace $\mathfrak{g}'_b$ by $\mathfrak{g}'_c$ in Corollary IV.17. This turns out to be true at least in the Fréchet category:

Proposition IV.18 *Let M be a weakly symplectic Fréchet manifold and $\mathfrak{g}$ the Lie algebra of a Fréchet Lie group acting on M. Then the following are equivalent:*

(i) *There exists a continuous co-moment map $\lambda : \mathfrak{g} \to C^\infty(M)$.*

(ii) *There exists a moment map $\Phi : M \to \mathfrak{g}'_c$.*

Proof. Let us first assume that there exists a moment map.

The map $\mathfrak{g} \xrightarrow{i} (\mathfrak{g}'_c)'_c, i(X)(\alpha) = \alpha(X)$, is easily seen to be linear and continuous. (In fact i is an isomorphism in the case of a Fréchet Lie algebra $\mathfrak{g}$, see e.g. [N1]). A compact subset $\tilde{K}_m \subset T^{(m)}M$ is a finite union of compact sets in $T^{(m)}U$, where U is an open coordinate set in M. Thus the topology on $C^\infty(M)$ is defined by the seminorms

$$P_{K_m}(f) = \sup\{|d^m f(x)(h_1, \ldots, h_m)| : (x, h_1, \ldots, h_m) \in K_m\},$$

where K_m is a compact set in $U \times E^m$ and U is an open set in E, the local model of M. Of course we identified here an open coordinate set in M with its image in E.

The proof of the continuity of λ is now "tautological" in view of the continuity of i. Let us consider explicitly only the cases $m = 0$ and $m = 1$. If K_0 is compact in U the image $\Phi(K_0)$ is of course compact in $\mathfrak{g}'_c$. Thus

$$q_{\Phi(K_0)}(\alpha) = \sup\{|\alpha(\Phi(x))| : x \in K_0\}$$

defines a continuous semi-norm on $(\mathfrak{g}'_c)'_c$ and $q_{\Phi(K_0)} \circ i$ a continuous seminorm on $\mathfrak{g}$. It follows that for X in $\mathfrak{g}$

$$p_{K_0}(\lambda(X)) = \sup\{|\lambda(X)(x)| : x \in K_0\} = \sup\{|i(X)(\Phi(x))| : x \in K_0\}$$
$$= (q_{\Phi(K_0)} \circ i)(X),$$

i.e. for all K_0 compact in U there exists a continuous semi-norm, namely $q_{\Phi(K_0)} \circ i$, such that

$$p_{K_0}(\lambda(X)) \le (q_{\Phi(K_0)} \circ i)(X).$$

For a compact K_1 in $U \times E, d\Phi(K_1)$ is compact in $\mathfrak{g}'_c$ and thus $q_{d\Phi(K_1)} \circ i$ is a continuous semi-norm on $\mathfrak{g}$. It follows that

$$p_{K_1}(\lambda(X)) = \sup\{|d(\lambda(X))(x)(h)| : (x, h) \in K_1\}$$
$$= \sup\{|i(X)(d\Phi(x)(h))| : (x, h) \in K_1\}$$
$$= (q_{d\Phi(K_1)} \circ i)(X).$$

Analogous arguments show that $p_{K_m}(\lambda(X)) = (q_{(d^m \Phi)(K_m)} \circ i)(X)$ and thus λ is continuous.

Assuming now that λ is a continuous co-moment it is obvious that $\Phi(x) = \delta_x \circ \lambda$ is a continuous linear functional on $\mathfrak{g}$. Thus the map $\Psi : M \times \mathfrak{g} \to \mathbb{R}, \Psi(x, X) = \lambda(X)(x)$ has the following properties:

(i) $\Psi_X : M \to \mathbb{R}, \Psi_X(x) = \Psi(x, X) = \lambda(X)(x)$ is smooth,

(ii) $\Psi_x : \mathfrak{g} \to \mathbb{R}, \Psi_x(X) = \Psi(x, X) = (\Phi(x))(X)$ is linear and continuous.

Theorem 3.4.3 and Corollary 3.4.4 in [Ha] imply now that it is enough to show that Ψ is jointly continuous to assure that Ψ is smooth.

Let $x_n \to x_0$ in M and $X_n \to X_0$ in $\mathfrak{g}$, then

$$|\Psi(x_0, X_0) - \Psi(x_n, X_n)| = |\lambda(X_0)(x_0) - \lambda(X_n)(x_n)|$$
$$\leq |\lambda(X_0)(x_0) - \lambda(X_0)(x_n)| + |\lambda(X_0)(x_n) - \lambda(X_n)(x_n)|$$
$$= |\lambda(X_0)(x_0) - \lambda(X_0)(x_n)| + |\lambda(X_0 - X_n)(x_n)|.$$

The first term of the last right hand side converges to zero by the continuity of the function $\lambda(X_0)$ on M and the second can be estimated by

$$p_K(\lambda(X_0 - X_n)) = \sup\{|\lambda(X_0 - X_n)(x)| : x \in K\},$$

where $K = \{x_n | n \geq 0\}$ is compact in M. Since λ is continuous and p_K is a continuous seminorm on $C^\infty(M)$ it follows that the second term goes to zero as well.

Using Theorem III.4 in [N1] we know that the map

$$\hat{\Phi} : M \to C^\infty(\mathfrak{g}), x \mapsto \Psi_x = \Phi(x)$$

is smooth, since Ψ is smooth.

Furthermore $\hat{\Phi}(x)$ is in $\mathfrak{g}'$ for all x in M and one easily checks that $C^\infty(\mathfrak{g}) \cap \mathfrak{g}'$ is closed with respect to the topology of $C^\infty(\mathfrak{g})$ defined by uniform convergence on compact subsets of $\mathfrak{g}$. Furthermore $C^\infty(\mathfrak{g}) \cap \mathfrak{g}'$ with the induced topology equals $\mathfrak{g}'_c$. Thus the smoothness of $\hat{\Phi}$ implies that $\Phi : M \to \mathfrak{g}'_c$ is smooth. $\square$

Remark. Let us observe that the proof of Proposition IV.18 is independent of Lemma IV.16 and Corollary IV.17. Using them the conditions (i) and (ii) in Proposition IV.18 are also equivalent to

(iii) There exists a moment map $\Phi : M \to \mathfrak{g}'_b$.

Corollary IV.19 Let M be a weakly symplectic Fréchet manifold and $\mathfrak{g}$ the Lie algebra of a finite-dimensional Lie group acting on M. Then the existence of a co-moment map λ implies that there is a moment map $\Phi : M \to \mathfrak{g}' = \mathfrak{g}^*$.

Proof. Supplying $\mathfrak{g}$ with any Hausdorff topological vector space structure the map λ is continuous and the algebraic dual $\mathfrak{g}^*$ equals the topological dual $\mathfrak{g}'$. The result follows now from the preceding proposition. $\qquad\square$

Going back to Lemma IV.11 and IV.12, it is obvious that the continuity of a co-moment λ implies that the associated cocycle c in $\Lambda^2\mathfrak{g}^*$ is continuous. Furthermore, if c is not equal to δb for a continuous b, we have a topologically non-trivial continuous central Lie algebra extension

$$\{0\} \to \mathbb{R} \to \tilde{\mathfrak{g}} \to \mathfrak{g} \to \{0\},$$

where $\tilde{\mathfrak{g}} = \mathbb{R} \oplus \mathfrak{g}$ as a vector space and $\tilde{\mathfrak{g}}$ is a s.c.l.c. space (respectively a Fréchet space respectively a Banach space) if $\mathfrak{g}$ is a space of the mentioned types.

4. Examples of symplectic actions and (co-)momementum maps

In this subsection we will briefly sketch some examples of symplectic group actions in finite and infinite dimensions.

Example 1: Abelian actions on tori and $\mathbb{R}^n$

The "worst case scenario" in finite dimensions is given by the torus $\widehat{M} = (S^1)^{2n} = M/(2\pi\mathbb{Z})^{2n}$, where $M = \mathbb{R}^{2n}$ with it standard symplectic structure $\Omega = \sum_{j=1}^{n} dx_j \wedge dy_j$, which induces a symplectic structure $\widehat{\Omega}$ on $\widehat{M}$. The group $\widehat{G} = (S^1)^{2n}$ acts by left-multiplication on $\widehat{M}$. This action is symplectic but not Hamiltonian since the one-form $i_{\tau(\xi)}\widehat{\Omega}$ is not closed for ξ in $\widehat{\mathfrak{g}}\setminus\{0\}$.

Going to the universal covering M of $\widehat{M}$ the group $\widehat{G}$ has to be replaced itself by its universal covering $G = \mathbb{R}^{2n}$. Since $H^1(M,\mathbb{R}) = \{0\}$ the one-forms $i_{\tau(\xi)}\Omega$ have potentials $\lambda(\xi)$ and thus the action $G \times M \to M$ is Hamiltonian. Describing $\mathfrak{g}(\cong \widehat{\mathfrak{g}})$ slightly abusively as the $\mathbb{R}$-linear span of $\frac{\partial}{\partial x_j}$ and $\frac{\partial}{\partial y_j}$ $(1 \leq j \leq n)$ we can give a co-moment by the following formula

$$\lambda\left(\frac{\partial}{\partial x_j}\right) = -y_j \text{ and } \lambda\left(\frac{\partial}{\partial y_j}\right) = x_j.$$

It follows from the definition (4.6) that

$$c\left(\frac{\partial}{\partial x_j}, \frac{\partial}{\partial x_k}\right) = 0 = c\left(\frac{\partial}{\partial y_j}, \frac{\partial}{\partial y_k}\right) \quad \text{and}$$

$$c\left(\frac{\partial}{\partial x_j}, \frac{\partial}{\partial y_k}\right) = \left\{\lambda\left(\frac{\partial}{\partial x_j}\right), \lambda\left(\frac{\partial}{\partial y_k}\right)\right\} - \lambda\left(\left[\frac{\partial}{\partial x_j}, \frac{\partial}{\partial y_k}\right]\right)$$

$$= \Omega\left(\frac{\partial}{\partial x_j}, \frac{\partial}{\partial y_k}\right) = \delta_{jk}.$$

Thus the class $[c]$ is non-zero in $H^2(\mathfrak{g}, \mathbb{R}) \cong \Lambda^2(\mathbb{R}^{2n})^*$ and $\tilde{\mathfrak{g}}$ is a Heisenberg Lie algebra.

Example 2: "Linear" actions on projective varieties
There are nevertheless many situations where compact Lie groups act with an equivariant moment map on symplectic manifolds (see e.g. [GS]). In the case of

Kähler manifolds the geometry of the moment map and the associated "symplectic reduction" of moment levels tie in nicely with geometric invariant theory and the theory of complex-analytic group actions (see, e.g., the references [MFK] and [HeHu]). Let us here only remark the following simple formula for the equivariant moment map of the $SU(n+1)$-action on $\mathbb{P}_n(\mathbb{C})$:

$$\Phi : \mathbb{P}_n(\mathbb{C}) \to \mathfrak{su}(n+1)^*,$$

$$\Phi([z])(X) = \frac{1}{2\pi i} \frac{\bar{z}^t \cdot X \cdot z}{\bar{z}^t \cdot z},$$

where $z^t = (z_0, \ldots, z_n)$ is in $\mathbb{C}^{n+1} \backslash \{0\}$, $[z]$ the associated point in $\mathbb{P}_n(\mathbb{C})$ and X in $\mathfrak{su}(n+1)$ is viewed as a traceless antihermitian $(n+1) \times (n+1)$ matrix.

Example 3: Gauge theory on surfaces
In infinite dimensions the best-known example comes from two-dimensional Yang-Mills theory (compare [AB], and [Au] for the case of surfaces with boundary).

Let S be a smooth oriented closed surface, G a connected finite-dimensional Lie group with an Ad-invariant non-degenerate symmetric form B on its Lie algebra $\mathfrak{g}$ and $P \xrightarrow{G} S$ a fixed smooth principal G-bundle over S. The space $\mathcal{A}$ of smooth connections on P is an affine vector space modelled on the linear space $E = \mathcal{E}^1(\operatorname{ad} P)$. Here $\operatorname{ad} P$ denotes the vector bundle $P \times_G \mathfrak{g}$ associated to the adjoint action of G on $\mathfrak{g}$ and the symbol $\mathcal{E}^k(V)$ denotes the section module $\Gamma_{C^\infty}(S, \Lambda^k T^* S \otimes V)$ if V is a vector bundle over S ($k = 0, 1, 2$ in the surface case at hand).

The bilinear form B induces a smooth map

$$B_* : \mathcal{E}^k(\operatorname{ad} P \otimes \operatorname{ad} P) \to \mathcal{E}^k = \Gamma_{C^\infty}(S, \Lambda^k T^* S)$$

by applying the bilinear form B fibrewise. On $\mathcal{A}$ we have the following two-form

$$\Omega_A(\xi, \eta) = \int_S B_*(\xi \wedge \eta),$$

where ξ, η are in $T_A \mathcal{A} = E$ and $\xi \wedge \eta$ is in $\mathcal{E}^2(\operatorname{ad} P \otimes \operatorname{ad} P)$. It is easy to check that Ω is weakly non-degenerate and closed, i.e. Ω is a weakly symplectic form on $\mathcal{A}$.

The smooth sections of the bundle $\operatorname{Ad} P = P \times_G G$ (constructed from the conjugation action of G on itself), constitute the "gauge group" $\mathcal{G}$. This group acts symplectically on $\mathcal{A}$ and has an equivariant moment map (see [AB] or [Au] for more details)

$$\Phi : \mathcal{A} \to (\operatorname{Lie} \mathcal{G})^* = (\mathcal{E}^0(\operatorname{ad} P))^*,$$

$$\Phi(A)(X) = \int_S B_*(X \otimes F(A)),$$

where X is in $\operatorname{Lie} \mathcal{G} = \mathcal{E}^0(\operatorname{ad} P)$ and $F(A)$, the curvature of A, is in $\mathcal{E}^2(\operatorname{ad} P)$. It is not difficult to check that $\Phi(A)$ is a continuous linear form on $\operatorname{Lie} \mathcal{G}$ and that Φ is a smooth map between l.c.s.c. manifolds. In fact there is even a Banach set-up for this situation since the multiplication in $\mathcal{G}$ is defined by "pointwise multiplication in the target".

The main interest of this example stems from the obvious observation that the zero-level of Φ is exactly the set of flat connections on P. Using "symplectic reduction" one obtains moduli spaces $\Phi^{-1}(0)/\mathcal{G}$ that play an important rôle in mathematics and theoretical physics.

Let us remark that one can modify the above construction in order to replace the surface S by a compact Kähler manifold (see [Don]).

Example 4: Deformation theory for complex and Kähler structures
The above "symplectic approach" to moduli spaces in algebraic geometry can also be applied to the construction of the moduli spaces of polarized algebraic manifolds. Without going into the details we remark only that the space of connections $\mathcal{A}$ has to be replaced by the space $\mathcal{C}_0$ of integrable complex structures on a symplectic manifold and the gauge group $\mathcal{G}$ by a suitable subgroup of the group of symplectic diffeomorphism. For a further study of this construction we propose the articles of Fujiki and co-workers (see [Fuj] and references therein).

Example 5: Reparametrization action on loop spaces
To a given finite-dimensional Riemannian manifold M one can associate its loop space $\mathcal{L}M = C^\infty(S^1, M)$. If M has a Riemannian metric g, $\mathcal{L}M$ carries a closed smooth two-form defined by

$$\Omega_\gamma(u, v) = \int_0^1 g_{\gamma(t)}\left(\frac{\nabla u}{dt}(t), v(t)\right) dt,$$

where γ is a loop in M viewed as a map from $[0, 1]$ to M, u and v are in $T_\gamma \mathcal{L}M = \Gamma_{C^\infty}(S^1, \gamma^* TM)$ and ∇ is the Levi-Civita connection on (M, g). The form Ω_γ is not weakly non-degenerate, but the dimension of its kernel is bounded by the dimension of M. The group $G = \text{Diff}^+(S^1)$ acts naturally on $\mathcal{L}M$ by reparametrization of the loops and preserves the "symplectic structure" Ω. Furthermore there exists an equivariant "moment map"

$$\Phi : \mathcal{L}M \to (\text{Lie } G)^* \cong (C^\infty(S^1, \mathbb{R}))^*,$$

$$\Phi(\gamma)(\xi) = \frac{1}{2}\int_0^1 f(t)\|\dot\gamma(t)\|^2 dt,$$

where $\|\dot\gamma(t)\|^2 = g_{\gamma(t)}(\dot\gamma(t), \dot\gamma(t))$ and Lie G being the space of vector fields of S^1 we identified ξ in Lie G with $f(t)\frac{d}{dt}$ for an f in $C^\infty(S^1, \mathbb{R})$. Despite the finite-dimensional kernels of Ω_γ the map Φ enjoys many properties of moment maps in finite dimensions (see [Wu1] for more details). Let us here only note that Φ is a smooth map from the Fréchet manifold $\mathcal{L}M$ to the complete l.c. space $(C^\infty(S^1, \mathbb{R}))'_c$.

5. The U_{res}-moment map on G_{res} and the Schwinger term

Following Grosse and Maderner ([GM]) we will here describe the Schwinger term explicitly as the cocycle associated to a non-equivariant moment map for the U_{res}-action on $G_{\text{res}} = G_{\text{res}}(K, K_+)$ (the separable Hilbert space K being polarized as $K_+ \oplus K_-$ as usual).

We prepare ourselves with two simple observations:

Lemma IV.20 *The map*

$$G_{\mathrm{res}} \xrightarrow{\rho} (p_+ + \mathcal{L}^2(K)), \quad \rho(W) = p_W = p_+ + (p_W - p_+),$$

associating to an element W of the restricted Grassmannian the orthoprojector p_W on the subspace W of K, is real-analytic.

Proof. Let us first observe that the action of U_{res} on the affine Hilbert space $(p_+ + \mathcal{L}^2(K))$ given by $\vartheta(U, p_+ + T) = U(p_+ + T)U^{-1}$ is real-analytic and that ρ is U_{res}-equivariant: $\rho(\Theta_U(W)) = \rho(U(W)) = U p_W U^{-1} = \vartheta(U, \rho(W))$. By the U_{res}-homogeneity of G_{res} it is thus enough to consider ρ near K_+. For $\varepsilon > 0$ sufficiently small the map

$$\{\gamma \in \mathcal{L}^2(K_+, K_-) : \|\gamma\|_2 < \varepsilon\} \to G_{\mathrm{res}},$$

$$\gamma \mapsto \Theta(g(\gamma), K_+) = g(\gamma) \cdot K_+ = \exp \begin{pmatrix} 0 & -\gamma^* \\ \gamma & 0 \end{pmatrix} \cdot K_+$$

yields real-analytic coordinates near K_+.

Thus near K_+ there is a real-analytic map associating to W a $g(W) = g(\gamma(W))$ such that $W = \Theta(g(W), K_+)$. It follows that $p_W = g(W) \cdot p_+ \cdot g(W)^{-1}$ depends real-analytically on W near K_+. Thus the map $W \mapsto \rho(W) = p_W$ is real-analytic.

$\square$

Corollary IV.21 *Identifying $T_{p_W}(p_+ + \mathcal{L}^2(K))$ with $\mathcal{L}^2(K)$ we have*

$$\frac{d}{dt}\Big|_0 P_{\Theta(\exp(tA), W)} = [A, p_W] \quad \text{for all } A \text{ in } \mathfrak{u}_{\mathrm{res}}.$$

Proof. The formula follows directly from the equivariance of the map ρ. $\square$

We can now state and prove the main result of Grosse and Maderner ([GM]):

Proposition IV.22
(i) *For A in $\mathfrak{u}_{\mathrm{res}}$ the function $\lambda(A)$ defined by*

$$\lambda(A)(W) = -i\, tr(p_W p_- A p_W) + i\, tr(p_{W^\perp} p_+ A p_{W^\perp}) \tag{4.9}$$

is real-analytic and $d(\lambda(A)) = \Omega(\tau(A), \cdot)$ on G_{res}.
(ii) *For a fixed W in G_{res} the map $\mathfrak{u}_{\mathrm{res}} \to \mathbb{R}$, $A \mapsto \lambda(A)(W)$ is linear and continuous.*

Proof. Let us first observe that for W in G_{res} and A in $\mathfrak{u}_{\mathrm{res}}$ the following operators on K are Hilbert-Schmidt: $p_W p_-, p_+ p_{W^\perp}, p_+ - p_W, p_- - p_{W^\perp}, A_{+-} = p_+ A p_-, A_{-+} = p_- A p_+, [p_+, A]$ and $[p_W, A]$.

Rewriting the first trace in the right hand side of (4.9) as

$$(p_W p_-)(p_- A p_+) p_W + (p_W p_-) A (p_- p_W)$$

it follows that this trace exists since the product of two Hilbert-Schmidt operators is of trace class. Using a similar description for the second trace one sees that $\lambda(A)(W)$ is well-defined. Furthermore the function $\lambda(A)$ takes real values. Since

p_W and $p_{W\perp} = 1 - p_W$ are real-analytic in W by Lemma IV.20 the function $\lambda(A)$ is real-analytic as well.

Let us use the notations $\varepsilon = p_+ - p_-$ and $\varepsilon_W = p_W - p_{W\perp}$. It follows that $W = \Theta_U(K_+) = U(K_+)$ for U in U_{res} implies $\varepsilon_W = U\varepsilon U^{-1}$. Recall that for A, B in $\mathfrak{u}_{res}$ one has by (3.3), (3.4) and (2.16)

$$\Omega_{K_+}(\tau(A), \tau(B)) = \frac{i}{4}\operatorname{tr}(\varepsilon \cdot [\varepsilon, A] \cdot [\varepsilon, B])$$

and that by general properties of an action

$$\tau(A)_{U(K_+)} = \tau(A)_{\Theta_U(K_+)} = (\Theta_U)_*(\tau(\mathrm{Ad}(U^{-1})A)_{K_+})$$

for U in U_{res} and A in $\mathfrak{u}_{res}$. A direct calculation now yields that

$$\Omega_W(\tau(A)_W, \tau(B)_W) = \frac{i}{4}\operatorname{tr}(\varepsilon_W \cdot [\varepsilon_W, A] \cdot [\varepsilon_W, B]). \qquad (4.10)$$

Since the $\mathfrak{u}_{res}$-fundamental vector fields give the tangent space in any point of the Grassmannian the following calculation will show that $d(\lambda(A)) = \Omega(\tau(A), \cdot)$:

$$d(\lambda(A))(\tau(B)_W) = \frac{d}{dt}\Big|_0 \lambda(A)(\exp(tB) \cdot W)$$

$$= \frac{d}{dt}\Big|_0 \left\{ -i\operatorname{tr}\left(p_{\exp(tB)\cdot W}p_- Ap_{\exp(tB)\cdot W}\right) \right.$$

$$\left. + i\operatorname{tr}\left(p_{(\exp(tB)\cdot W)^\perp}p_+ Ap_{(\exp(tB)\cdot W)^\perp}\right)\right\}$$

$$= -i\operatorname{tr}\left([B, p_W]p_- Ap_W + p_W p_- A[B, p_W]\right)$$

$$\quad + i\operatorname{tr}\left([B, p_{W\perp}]p_+ Ap_{W\perp} + p_{W\perp}p_+ A[B, p_{W\perp}]\right)$$

$$\text{(where we used Corollary IV.21)}$$

$$= i\operatorname{tr}(p_- Ap_W[B, p_{W\perp}]) + i\operatorname{tr}\left([B, p_{W\perp}]p_W p_- A\right)$$

$$\quad - i\operatorname{tr}\left(p_+ Ap_{W\perp}[B, p_W]\right) - i\operatorname{tr}\left([B, p_W]p_{W\perp}p_+ A\right)$$

(where we used $p_W + p_{W\perp} = 1$ and $\operatorname{tr}(\alpha \cdot \beta) = \operatorname{tr}(\beta \cdot \alpha)$ for α, β Hilbert-Schmidt operators).

Evaluating the last right hand side we find now

$$d(\lambda(A))(\tau(B)_W) = i\operatorname{tr}\left[(p_W Bp_{W\perp})(p_{W\perp} Ap_W) - (p_{W\perp} Bp_W)(p_W Ap_{W\perp})\right]$$

$$= \frac{i}{4}\operatorname{tr}(\varepsilon_W \cdot [\varepsilon_W, A] \cdot [\varepsilon_W, B])$$

$$= \Omega_W(\tau(A)_W, \tau(B)_W).$$

Thus the first assertion is proven.

Recalling formula (2.10) for the norm $\| \;\|_{\mathrm{ures}}$ we can make the following estimate:

$$|\operatorname{tr}(pwp_-Apw)|$$
$$\leq |\operatorname{tr}((pwp_-)(p_-Ap_+)pw)| + |\operatorname{tr}((pwp_-)(p_-A)(p_-pw))|$$
$$\leq \|(pwp_-)A_{-+}pw\|_1 + \|(pwp_-)(p_-A)(p_-pw)\|_1$$
$$\leq \|pwp_-\|_2 \cdot \|A_{-+}\|_2 \cdot \|pw\| + (\|pwp_-\|_2)^2 \cdot |p_-A\|$$
$$\leq \left(\max\left\{\frac{1}{2}\|pwp_-\|_2, (\|pwp_-\|_2)^2\right\}\right) \cdot \|A\|_{\mathrm{ures}}.$$

Using a similar estimate for the second trace in the defining formula (4.9) for $\lambda(A)$ it follows that the obviously linear function $A \mapsto \lambda(A)(W)$ is continuous on $\mathfrak{u}_{\mathrm{res}}$.

$\square$

Corollary IV.23 *The map*

$$\Phi : G_{\mathrm{res}} \to (\mathfrak{u}_{\mathrm{res}})'_b, \ \Phi(W)(A) = \lambda(A)(W)$$

is a moment map for the U_{res}-action on G_{res}.

Proof. We first consider the map

$$\Psi : G_{\mathrm{res}} \times \mathfrak{u}_{\mathrm{res}} \to \mathbb{R}, \Psi(W, A) = \lambda(A)(W).$$

By the above proposition $\Psi_A = \lambda(A)$ is a real-analytic function on G_{res} and $\Psi_W = \Psi(W, \cdot)$ is linear and continuous on $\mathfrak{u}_{\mathrm{res}}$. Let us show that Ψ is jointly continuous: let $W_n \to W_0$ in G_{res} and $A_n \to A_0$ in $\mathfrak{u}_{\mathrm{res}}$, then

$$|\Psi(W_0, A_0) - \Psi(W_n, A_n)|$$
$$\leq |\Psi(W_0, A_0) - \Psi(W_0, A_n)| + |\Psi(W_0, A_n) - \Psi(W_n, A_n)|$$
$$= |\Psi_{W_0}(A_0 - A_n)| + |(\Psi_{W_0} - \Psi_{W_n})(A_n)|.$$

The first summand in the last right hand side goes to zero since Ψ_{W_0} is continuous. Observing that $\alpha_n(A) := (\Psi_{W_0} - \Psi_{W_n})(A) = \lambda(A)(W_0) - \lambda(A)(W_n)$ tends to zero for fixed A by the continuity of the function $\lambda(A)$, we have $|\alpha_n(A)| \leq M(A)$ for all $n \geq 0$ with $M(A) \geq 0$ depending only on A in $\mathfrak{u}_{\mathrm{res}}$. The theorem of Banach-Steinhaus implies that the family $\{\alpha_n | n \geq 0\}$ is equicontinuous and thus the compacity of $\{A_n | n \geq 0\}$ in $\mathfrak{u}_{\mathrm{res}}$ implies that $\alpha_n(A_n) \to 0$. Thus Ψ is jointly continuous and 3.4.3 and 3.4.4 in [Ha] imply that Ψ is C^∞. Theorem III.4 in [N1] then shows that

$$\lambda : \mathfrak{u}_{\mathrm{res}} \to C^\infty(G_{\mathrm{res}}), \ \lambda(A) = \Psi_A$$

is a smooth map. Since λ is linear and $d(\lambda(A)) = \Omega(\tau(A), \cdot) = -\Omega(\alpha(\lambda(A)), \cdot)$ it follows that λ is a continuous co-moment. Corollary IV.17 yields now that $\Phi : G_{\mathrm{res}} \to (\mathfrak{u}_{\mathrm{res}})'_b$ is smooth and thus a moment map for the U_{res}-action on G_{res}. $\square$

We conclude now by

Corollary IV.24 *The moment map $\Phi : G^0_{\mathrm{res}} \to (\mathfrak{u}_{\mathrm{res}})'_b$ is not U^0_{res}-equivariant and the associated cocycle is (i-times) the Schwinger term on $\mathfrak{u}_{\mathrm{res}}$.*

Remark. Since we need connectivity of the manifold, and since the group generated by $\{\exp(A)|A \in \mathfrak{u}_{\text{res}}\}$ is U_{res}^0 we restrict ourselves in the corollary to the connected components.

Proof of Corollary IV.24. Let us observe that $\lambda(C)(K_+) = 0$ for all C in $\mathfrak{u}_{\text{res}}$ (compare (4.9)) and recall that $\alpha \circ \lambda = -\tau$ (see Definition IV.10).

Using formula (3.3) and the fact that $c(A, B)$ is constant on G_{res}^0 we have:

$$c(A, B) = \{\lambda(A), \lambda(B)\} - \lambda([A, B])$$
$$= \Omega(\alpha \circ \lambda(A), \alpha \circ \lambda(B)) - \lambda([A, B])$$
$$= \Omega_{K_+}(\tau(A), \tau(B)) - \lambda([A, B])(K_+)$$
$$= (-i)\tilde{s}(A, B) = i \cdot s(A, B). \qquad \square$$

V. The determinant line bundle on the restricted Grassmannian

The subject of this section is a certain holomorphic line bundle DET on G_{res} whose Chern class is given by the Kähler form of G_{res}, i.e. ultimately by the Schwinger term. We construct it by using the "C^*-algebro-geometric" methods of [SpWu1] and show that it is isomorphic to the "determinant line bundle" à la Pressley-Segal-Wilson (see [PrSe] and [SeWi]). We exhibit some of its properties and close with a subsection on holomorphic sections of its dual, DET*, and related topics as the Plücker embedding. There we will notably realize the Fock space of fermionic second quantization as holomorphic sections of DET*.

1. The C^*-algebraic construction of the determinant bundle DET

Let us recall that a finite-dimensional complex vector space K gives rise to its full Grassmannian $G(K) = \bigcup_{m=0}^{\dim K} G_m(K)$, where $G_m(K) = \{W \subset K| W$ is a complex subspace of dimension $m\}$. Over $G(K)$ there is a tautological vector bundle with total space $\mathcal{T} = \{(W, f) \in G(K) \times K | f \in W\}$ and a determinant line bundle $\text{Det} = \{(W, \xi) \in G(K) \times \Lambda K | \xi \in \Lambda^{\dim W} W\}$, where $\Lambda K = \bigoplus_{m=0}^{\dim K} \Lambda^m K$ denotes the full exterior power of K. Obviously the fibre Det_W over a point W is given by $\Lambda^{\dim W} W$.

Assuming now that K is a Hilbert space we can define for each f in K a contraction operator on ΛK by linearly extending the following definition,

$$f \lrcorner (f_1 \wedge \ldots \wedge f_\ell) =$$

$$\sum_{j=1}^{\ell} (-1)^j \langle f, f_j \rangle_K f_1 \wedge \ldots \wedge f_{j-1} \wedge f_{j+1} \wedge \ldots \wedge f_\ell.$$

This allows the following obvious characterization:

Lemma V.1 *Let K be a finite-dimensional Hilbert space and W in $G(K)$. Then the fibre of the determinant bundle over W is given by*

$$\text{Det}_W = \{\xi \in \Lambda K \,|\, w \wedge \xi = 0 \,\forall\, w \in W \text{ and } w^\perp \lrcorner \xi = 0 \,\forall\, w^\perp \in W^\perp\}. \qquad (5.1)$$

In infinite dimensions we consider again a polarized separable Hilbert space $K = K_+ \oplus K_-$ and replace $G(K)$ by $G_{res} = G_{res}(K, K_+)$. In order to ease the comparison to [PrSe] we consider the representation π of $CAR(K)$ on $\mathcal{F} = \mathcal{H}_{p_+}$ associated to the state ω_{p_+}. (In sections I and II we considered $\mathcal{H}_{p_-}$ to confirm with the physical intuition of avoiding "negative-energy particle states"; see also the remarks in the beginning of Section III, where we changed to the above convention.

The GNS-vector ξ_+ in $\mathcal{F} \backslash \{0\}$ is then, up to a scalar factor, characterized by

$$\pi(a^*(f_+))\xi_+ = 0 \quad \forall\, f_+ \in K_+ \quad \text{and}$$
$$\pi(a(f_-))\xi_+ = 0 \quad \forall\, f_- \in K_-.$$

We can now define a subset of the trivial vector bundle $G_{res} \times \mathcal{F}$ by

$$DET = \{(W, \xi) \in G_{res} \times \mathcal{F} \mid \pi(a^*(w))\xi = 0 \ \forall\, w \in W \quad \text{and}$$

$$\pi(a(w^\perp))\xi = 0 \ \forall\, w^\perp \in W^\perp\}. \tag{5.2}$$

Let us observe that $DET_{K_+} = \mathbb{C} \cdot \xi_+$.

Proposition V.2 [SpWu1] *The subset DET of $G_{res} \times \mathcal{F}$ together with its natural projection on G_{res} forms a holomorphic complex line bundle with a holomorphic, fibrewise linear action of $U_{res}^\sim$ covering its transitive action on G_{res}.*

Proof. Let us first recall from Lemma III.1 that the GNS-representation of $CAR(K)$ associated to the state ω_{pw} can be realized on $\mathcal{F}$ in a unique way up to multiplication of a scalar of modulus one. It follows that DET_W is the GNS-line associated to this representation. Furthermore if $W = U \cdot K_+$ for a U in U_{res} and if $\mathbb{U}$ is an implementer of U on $\mathcal{F}$ then $DET_W = \mathbb{U} \cdot DET_{K_+} = \mathbb{U} \cdot (\mathbb{C} \cdot \xi_+)$. We deduce that DET is given as an associated bundle $U_{res}^\sim \times_{\tilde{H}} \mathbb{C}$, where $H = U(K_+) \times U(K_-)$ and $\tilde{H}$ is the preimage of H under the central extension $U_{res}^\sim \to U_{res}$, i.e.

$$\tilde{H} = \left\{(U, \mathbb{V}) \in U_{res} \times U(\mathcal{F}) \,\middle|\, \rho(U) = [\mathbb{V}] \text{ and } U = \begin{pmatrix} U_+ & 0 \\ 0 & U_- \end{pmatrix} \in H\right\},$$

where $\rho : U_{res} \to \mathbb{P}U(\mathcal{F})$ denotes again the projective representation of U_{res} on the Fock space. Since $\rho(H)$ preserves the GNS-line $\mathbb{C} \cdot \xi_+$, we have a character $\chi : \tilde{H} \to U(1)$ given by $\chi(U, \mathbb{V}) \cdot \xi_+ = \mathbb{V}(\xi_+)$. It follows that $\tilde{H} \to H \times U(1), (U, \mathbb{V}) \mapsto (U, \chi(U, \mathbb{V}))$ is an isomorphism and that DET is an associated bundle $U_{res}^\sim \times_{\tilde{H}} \mathbb{C}$.

Notably we have a non-projective action of H on $\mathcal{F}$ by setting $\tilde{\rho}(U) = \chi^{-1}(U, \mathbb{V}) \cdot \mathbb{V}$, where $\mathbb{V}$ is any element $U(\mathcal{F})$ such that $[\mathbb{V}] = \rho(U)$.

Holomorphicity of DET can now either be deduced from the observation that DET is locally given as the simultaneous zero set of $\mathcal{F}$−valued holomorphic functions or by extending the above description to the complexification $GL_{res}^\sim$ of $U_{res}^\sim$. It follows then that $DET \cong GL_{res}^\sim \times_{\tilde{P}} \mathbb{C}$, where $\tilde{P}$ is the preimage of the stabilizer P of K_+ under the central $\mathbb{C}^*$-extension $GL_{res}^\sim \to GL_{res}$. The character χ extends then to a holomorphic character on $\tilde{P}$. $\qquad\square$

Proposition V.3 *The line bundle* $DET \to G_{\text{res}}$ *has a natural* $U^{\sim}_{\text{res}}$-*invariant her-mitian structure and its Chern class is represented by* $\left(\frac{-1}{2\pi}\right)$ *times the Kähler form on* G_{res}.

Proof. Obviously the scalar product of $\mathcal{F}$ induces a $U^{\sim}_{\text{res}}$-invariant hermitian struc-ture on DET by setting $h_W^{\text{DET}}(\xi, \eta) = \langle \xi, \eta \rangle_{\mathcal{F}}$ for ξ, η in $\text{DET}_W \subset \mathcal{F}$.

Let us calculate the Chern class of DET by defining a $U^{\sim}_{\text{res}}$-invariant principal connection on $\text{DET}^\times = \text{DET} \setminus \{\text{zero-section}\}$.

We first recall that the Schwinger term on $\mathcal{F} = \mathcal{H}_{P_+}$ came from the formula

$$[\mathbb{A}, \mathbb{B}] - \mathbb{C} = \tilde{s}(A, B) \cdot \mathbf{1}_{\mathcal{F}},$$

where A, B and $C = [A, B]$ are in $\mathfrak{u}_{\text{res}}$ and $\mathbb{A}, \mathbb{B}$ and $\mathbb{C}$ their respective second quantizations. Identifying $\mathfrak{u}^{\sim}_{\text{res}}$ with $\mathfrak{u}_{\text{res}} \oplus \mathbb{R}i \cdot \zeta$, where $i \cdot \zeta$ corresponds to $i \cdot \mathbf{1}_{\mathcal{F}}$, we can write the bracket on $\mathfrak{u}^{\sim}_{\text{res}}$ as

$$[(A, \alpha), (B, \beta)]_{\mathfrak{u}^{\sim}_{\text{res}}} = ([A, B], \tilde{s}(A, B) \cdot \zeta).$$

The section $\sigma : \mathfrak{u}_{\text{res}} \to \mathfrak{u}^{\sim}_{\text{res}}, \sigma(A) = (A, 0)$ fulfils then

$$[\sigma(A), \sigma(B)]_{\mathfrak{u}^{\sim}_{\text{res}}} - \sigma([A, B]) = \tilde{s}(A, B) \cdot \zeta. \tag{5.3}$$

Let us furthermore denote the right-action of z in $\mathbb{C}^*$ on $\text{DET}^\times$ by R_z :

$$R_z(W, \xi) = (W, z \cdot \mathbf{1}_{\mathcal{F}} \cdot \xi) = (W, z \cdot \xi),$$

and the left-action of $\mathbb{U}$ in $U^{\sim}_{\text{res}}$ on $\text{DET}^\times$ by $L_{\mathbb{U}}$:

$$L_{\mathbb{U}}(W, \xi) = (UW, \mathbb{U}\xi),$$

where U is the image of $\mathbb{U}$ under the projection $U^{\sim}_{\text{res}} \to U_{\text{res}}$.

We define now horizontal spaces for (K_+, ξ) in $\text{DET}^\times_{K_+}$ by

$$\text{Hor}_{(K_+, \xi)} = \left\{ \tau^{\text{DET}^\times}(\sigma(A))(K_+, \xi) | A \in \mathfrak{u}_{\text{res}} \right\},$$

where $\tau^{\text{DET}^\times}(X)$ denotes the fundamental vector field associated to X in $\mathfrak{u}^{\sim}_{\text{res}}$ on $\text{DET}^\times$. Obviously we have $(R_z)_* \text{Hor}_{(K_+, \xi)} = \text{Hor}_{R_z(K_+, \xi)}$ and $(L_{\mathbb{U}})_* \text{Hor}_{(K_+, \xi)} = \text{Hor}_{L_{\mathbb{U}}(K_+, \xi)} = \text{Hor}_{(K_+, \chi(\mathbb{U}) \cdot \xi)}$ for $\mathbb{U}$ in $\tilde{H}$. Thus we can define a $U^{\sim}_{\text{res}}$-invariant connection by setting

$$\text{Hor}_{L_{\mathbb{U}}(K_+, \xi)} = (L_{\mathbb{U}})_* \text{Hor}_{(K_+, \xi)}$$

for all $\mathbb{U}$ in $U^{\sim}_{\text{res}}$ and all ξ in $\mathbb{C}^* \cdot \xi_+ \cong \text{DET}^\times_{K_+}$.

The associated connection one-form Θ is given by $\Theta|_{\text{Hor}} = 0$ and $\Theta(\frac{d}{dt}|_0 R_{\exp(t\zeta)}(W, \xi)) = \zeta$ for all (W, ξ) in $\text{DET}^\times$. Obviously Θ is a (Lie $\mathbb{C}^*$)-valued, $U^{\sim}_{\text{res}}$-invariant one-form on $\text{DET}^\times$. The (Lie $\mathbb{C}^*$)-valued curvature two-form $R = R^\Theta$ on $\text{DET}^\times$ is then given by the formula $R(X, Y) = (d\Theta)(X, Y)$ for X, Y vector fields on $\text{DET}^\times$. The curvature form R is invariant under the right-action of $\mathbb{C}^*$ and vanishes on vertical vector fields. Thus $R = \pi^* \omega$, where π is the bundle projection and ω is a (Lie $\mathbb{C}^*$)-valued two-form on G_{res}. Since R is $U^{\sim}_{\text{res}}$-invariant, ω is U_{res}-invariant and it is enough to calculate ω in K_+. Let thus A, B be in $\mathfrak{u}_{\text{res}}$

and $\tau^{Gres}(A)$ respectively $\tau^{Gres}(B)$ be their fundamental vector fields on G_{res}. Then

$$\omega_{K_+}(\tau^{Gres}(A), \tau^{Gres}(B))$$

$$= R_{(K_+,\xi)}(\tau^{DET^\times}(\sigma(A)), \tau^{DET^\times}(\sigma(B)))$$

$$= (d\Theta)_{(K_+,\xi)}(\tau^{DET^\times}(\sigma(A)), \tau^{DET^\times}(\sigma(B)))$$

$$= -\Theta_{(K_+,\xi)}([\tau^{DET^\times}(\sigma(A)), \tau^{DET^\times}(\sigma(B))]) \quad \text{(since } \tau^{DET^\times}(\sigma(\cdot)) \text{ is horizontal)}$$

$$= \Theta_{(K_+,\xi)}(\tau^{DET^\times}([\sigma(A), \sigma(B)])) \quad \text{(since } \tau \text{ is an anti-homomorphism)}$$

$$= \Theta_{(K_+,\xi)}(\tau^{DET^\times}(\sigma([A,B]) + \tilde{s}(A,B)\cdot\zeta)) \quad \text{(compare equation (5.3))}$$

$$= \tilde{s}(A,B)\cdot\zeta.$$

Recalling that ζ corresponds to the identity in $\operatorname{End}(\mathbb{C}) = \operatorname{Lie}(\mathbb{C}^*)$ we find on the level of forms:

$$(c_1(DET^\times, \Theta))_{K_+}(\tau^{Gres}(A), \tau^{Gres}(B))$$

$$= (\frac{i}{2\pi} \operatorname{trace}\omega)_{K_+}(\tau^{Gres}(A), \tau^{Gres}(B))$$

$$= (-\frac{1}{2\pi})(-i)\tilde{s}(A,B) = (-\frac{1}{2\pi})\hat{\Omega}_{K_+}(A,B) \quad \text{(compare equation (3.3))}.$$

Since $\hat{\Omega}_{K_+}$ descends to the U_{res}-invariant Kähler form Ω of G_{res} in K_+, we finally arrive at

$$c_1(DET^\times, \Theta) = (-\frac{1}{2\pi})\Omega$$

and thus on the level of cohomology classes at

$$[c_1(DET)] = (-\frac{1}{2\pi})[\Omega] \text{ in } H^2(G_{res}, \mathbb{Z}). \qquad \square$$

Corollary V.4 *The only holomorphic section of DET is the zero-section.*

Proof. Assume that a holomorphic section of DET is non-vanishing in a point W of G_{res}. Using the $\widetilde{U_{res}}$-action on DET we can assume that W is K_+. Lemma III.10 and the naturality of Chern classes imply that we have a non-vanishing section of a holomorphic line bundle with negative Chern class on $\mathbb{P}_1(\mathbb{C})$, the pullback of DET under j. This contradiction shows that there are no non-vanishing sections of DET on G_{res}. $\qquad \square$

2. Comparison to other approaches to the determinant bundle

Aiming to compare the above defined line bundle to the line bundle constructed by Pressley, Segal and Wilson (see [PrSe] and [SeWi]) we recall several notions crucial to their approach. For W in G_{res}^0, they define an "admissible basis" to be a bounded linear map $w : K_+ \to K$ such that w is an isomorphism from K_+ to $W = \operatorname{Im}(w)$ and such that $1_+ - p_+ \circ w = K_+ \to K_+$ is of trace class. It is easy to check that any two admissible basis w and w' for a fixed subspace W fulfil $w' = w \circ L$ for an element L of $GL^1(K_+)$. One defines an equivalence relation on the set $\{(w, \lambda)|w$ admissible basis for a W in G^0_{res} and $\lambda \in \mathbb{C}\}$ by $(w, \lambda) \sim (w', \lambda')$ iff it exists L in $\mathrm{GL}^1(K_+)$ such that $w' = w \circ L^{-1}$ and $\lambda' = (\det L) \cdot \lambda$. Furthermore in [PrSe] and [SeWi] it is shown that the set of equivalence classes $\{[W, w, \lambda]\}$ form a holomorphic line bundle $\mathrm{DET}^\wedge$ and that there is a Hilbert space of holomorphic sections $\mathcal{F}^\wedge$ of $(\mathrm{DET}^\wedge)^*$ allowing a "Plücker embedding" of G_{res} into $\mathbb{P}(\mathcal{F}^\wedge)$. Using the identification of $\mathcal{F}^\wedge$ with the Fock space $\mathcal{F} = \mathcal{H}_{P_+}$ given in [SV], comparison of the two Plücker embeddings arising prove in [SpWu1] that DET and $\mathrm{DET}^\wedge$ are isomorphic. Let us thus give here a more direct isomorphism between $\mathrm{DET}^\wedge$ and DET, avoiding the slightly cumbersome identification of $\mathcal{F}^\wedge$ and $\mathcal{F}$. Instead we will rely on an infinite-dimensional "Stiefel manifold" already constructed in [Mic].

We review first the definition of an admissible basis for W in $G^d_{res} = \{W \in G_{res}| \mathrm{ind}(p_+ : W \to K_+) = d\}$. Let $\{e_j|j \in \mathbb{E}\}$ be as usual a Hilbert basis of K and γ in U_{res} be defined by $\gamma(e_j) = e_{j+1}$ as in Section II.3 after Proposition II.22. We consider $K_d = ((e_{-d}, e_{d+1}, \ldots, e_{-1}, e_0, e_1, \ldots))_\mathbb{C} = \gamma^{-d}(K_+)$ in G^d_{res} (obviously $K_0 = K_+$) and define an admissible basis of W in G^d_{res} as a bounded linear map $w : K_d \to K$ such that w is an isomorphism of K_d onto W and such that $1_{K_d} - p_{K_d} \circ w$ is of trace class. Let us denote the set $\{(W, w)|W \in G^d_{res}$ and w an admissible basis of $W\}$ by St^d_{res} and $\bigcup_{d \in \mathbb{Z}} \mathrm{St}^d_{res}$ by $\mathrm{St}_{res} = \mathrm{St}_{res}(K, K_+)$, the "restricted Stiefel manifold".

Obviously we have a free right-action of $\mathrm{GL}^1(K_+)$ on St^d_{res} by

$$L \cdot (W, w) = (W, w \circ \gamma^{-d} \circ L \circ \gamma^d) \tag{5.4}$$

and St^d_{res} is $\mathrm{GL}^1(K_+)$-equivariantly identified with St^0_{res} by

$$(W, w) \mapsto (\gamma^d(W), \gamma^d \circ w \circ \gamma^{-d}). \tag{5.5}$$

Let us thus first concentrate on St^0_{res}. The Banach Lie group $\mathcal{E}$ (compare again Section II.3) acts as follows on St^0_{res} :

$$\mathcal{E} \times \mathrm{St}^0_{res} \to \mathrm{St}^0_{res}, \quad (A, q) \cdot (W, w) = (A(W), A \circ w \circ q^{-1}). \tag{5.6}$$

The action is easily be seen to be transitive and the stabilizer of $(K_+, 1_+)$ is given by

$$\mathcal{P} = \left\{ \left(\begin{pmatrix} a & b \\ 0 & d \end{pmatrix}, a \right) \middle| \begin{pmatrix} a & b \\ 0 & d \end{pmatrix} \in \mathrm{GL}_{res} \right\}.$$

Since $\mathcal{P}$ is a closed complex subgroup of $\mathcal{E}$ St^0_{res} inherits the structure of a complex Banach manifold which is a holomorphic principal $\mathrm{GL}^1(K_+)$-bundle over G^0_{res}. Identifying St^d_{res} with St^0_{res} as in (5.5) we induce the same structures on St^d_{res}.

Definition V.5 The *determinant bundle à la Pressley-Segal-Wilson* is the associated bundle

$$\mathrm{DET}^\wedge = \mathrm{St}_{res} \times_{\mathrm{GL}^1(K_+)} \mathbb{C} = \bigcup_{d \in \mathbb{Z}} (\mathrm{St}^d_{res} \times_{\mathrm{GL}^1(K_+)} \mathbb{C}), \tag{5.7}$$

where the action of $GL^1(K_+)$ is given as follows

$$L \cdot ((W,w),\lambda) = ((W, w \circ \gamma^{-d} \circ L^{-1} \circ \gamma^d), (\det L) \cdot \lambda).$$

Proposition V.6 *The bundle $DET^\wedge$ is holomorphically isomorphic to $\widehat{GL}_{res} \times_{\hat{P}} \mathbb{C}$, where $\hat{P}$ is the preimage of the stabilizer P of K_+ in GL_{res} in the group $\widehat{GL}_{res}$ and $\hat{P}$, being a product $P \times \mathbb{C}^*$, acts by multiplication by the second factor on $\mathbb{C}$.*

Remark. The preimage of $P = \left\{ \begin{pmatrix} a & b \\ 0 & d \end{pmatrix} \in GL_{res} \right\}$ under the projection $\beta :$ $\mathcal{E} \to GL^0_{res}$ is given by $\beta^{-1}(P) = \left\{ \left(\begin{pmatrix} a & b \\ 0 & d \end{pmatrix}, q \right) \Big| a - q \in \mathcal{L}^1 \right\}$. Thus $\hat{P} =$ $\beta^{-1}(P)/\alpha(SL^1(K_+)) = \left\{ \left[\begin{pmatrix} a & b \\ 0 & d \end{pmatrix}, q \right] \Big| a - q \in \mathcal{L}^1 \right\}$ and it is easily checked that $\hat{P} \to P \times \mathbb{C}^*$, $[A,q] \mapsto (A, \det(aq^{-1}))$ is an isomorphism. The action of $\hat{P}$ on $\mathbb{C}$ is then given by the second projection, i.e. the character $\chi^\wedge : P^\wedge \to \mathbb{C}^*$, $\chi^\wedge([A,q]) = \det(aq^{-1})$.

Proof. Let us first consider the restriction of $DET^\wedge$ over G^0_{res} :

$$DET^\wedge_0 = DET^\wedge|_{G^0_{res}} = St^0_{res} \times_{GL^1(K_+)} \mathbb{C}.$$

We define a left-action of $\widehat{GL}^0_{res} = \mathcal{E}/\alpha(SL^1(K_+))$ on it by

$$[A,q] \cdot [W,w,\lambda] = \vartheta_0([A,q])([W,w,\lambda]) = [A(W), Awq^{-1}, \lambda]. \tag{5.8}$$

Observing that for S in $\mathcal{L}^1(K_+)$ and q in $GL(K_+)$ one has $\det(q^{-1}(1+S)q) = \det(1+S)$ it is straightforward to check that

$$[A(W), AwL^{-1}(q(1+T))^{-1}, (\det L) \cdot \lambda] = [A(W), Awq^{-1}, \lambda]$$

for $1+T$ in $SL^1(K_+)$ and L in $GL^1(K_+)$. Thus the action (5.8) is well-defined and one sees easily that $\vartheta_0([A,q])$ is a fibrewise linear map on $DET^\wedge_0$.

Obviously $\widehat{GL}^0_{res}$ acts transitively on the base G^0_{res} and $\hat{P}$, the stabilizer of K_+, acts linearly on $(DET^\wedge)_{K_+} = \{[K_+, 1_+, \lambda] | \lambda \in \mathbb{C}\}$:

$$\left[\begin{pmatrix} a & b \\ 0 & d \end{pmatrix}, q \right] \cdot [K_+, 1_+, \lambda] = [K_+, aq^{-1}, \lambda] = [K_+, 1_+, \det(aq^{-1}) \cdot \lambda].$$

It follows that $DET^\wedge$ has the asserted form of an associated bundle $\widehat{GL}^0_{res} \times_{\hat{P}} \mathbb{C}$ over G^0_{res}.

Let us now define a $\mathbb{Z}$-action on $DET^\wedge$:

$$\vartheta(n)([W,w,\lambda]) = [\gamma^n(W), \gamma^n w \gamma^{-n}, \lambda], \tag{5.9}$$

where $\gamma(e_j) = e_{j+1}$ as in Section II.3. Obviously the action is fibrewise linear and covers the action $W \mapsto \gamma^n(W)$ of $\mathbb{Z}$ on G_{res}. Denoting $DET^\wedge|_{G^d_{res}}$ by $DET^\wedge_d$ we have $\vartheta(n)(DET^\wedge_d) = DET^\wedge_{d-n}$.

We can extend formula (5.8) to an action of $\widehat{\mathrm{GL}}_{\mathrm{res}}^0$ on all of $\mathrm{DET}^\wedge$ by setting

$$\vartheta_d([A,q]) : \mathrm{DET}_d^\wedge \to \mathrm{DET}_d^\wedge, \vartheta_d([A,q])([W,w,\lambda])$$
$$= \vartheta(-d) \circ \vartheta_0\left(\hat{\sigma}^d([A,q])\right) \circ \vartheta(d)([W,w,\lambda]). \qquad (5.10)$$

Recall from Section II.3 before Definition II.23 that $\hat{\sigma}$ is a group automorphism of $\widehat{\mathrm{GL}}_{\mathrm{res}}^0$ covering the automorphism $\sigma(A) = \gamma A \gamma^{-1}$ of $\mathrm{Gl}_{\mathrm{res}}^0$. It is easily checked that (5.10) defines a fibrewise linear action on $\mathrm{DET}^\wedge$ that extends (5.8).

We combine (5.9) and (5.10) into an action of $\widehat{\mathrm{GL}}_{\mathrm{res}} = \widehat{\mathrm{GL}}_{\mathrm{res}}^0 \rtimes \mathbb{Z}$ (compare again Definition II.23):

$$\vartheta(\hat{A},n) : \mathrm{DET}_d^\wedge \to \mathrm{DET}_{d-n}^\wedge, \vartheta(\hat{A},n) = \vartheta_{d-n}(\hat{A}) \circ \vartheta(n), \qquad (5.11)$$

where $\hat{A}$ is in $\widehat{\mathrm{GL}}_{\mathrm{res}}^0$ and n is in $\mathbb{Z}$.

If $(\hat{A},n)$ and $(\hat{A}',n')$ are in $\widehat{\mathrm{GL}}_{\mathrm{res}}$ a straightforward calculation shows that

$$\vartheta(\hat{A},n)(\vartheta(\hat{A}',n')(\xi_d)) = \vartheta(\hat{A}\hat{\sigma}^n(A'),n+n')(\xi_d)$$
$$= \vartheta((\hat{A},n)\cdot(\hat{A}',n'))(\xi_d)$$

for all d in $\mathbb{Z}$ and all ξ_d in $\mathrm{DET}_d^\wedge$, i.e. ϑ defines an action of $\widehat{\mathrm{GL}}_{\mathrm{res}}$ on $\mathrm{DET}^\wedge$.

Thus we can conclude that $\mathrm{DET}^\wedge = \widehat{\mathrm{GL}}_{\mathrm{res}} \times_{\hat{P}} \mathbb{C}$, where $\hat{P}$ acts on $\mathbb{C}$ as described in the remark preceding this proof. $\qquad\square$

Remark. One might be tempted to try to derive the above result by first extending the action (5.6) to

$$\mathcal{E} \times \mathrm{St}_{\mathrm{res}}^d \to \mathrm{St}_{\mathrm{res}}^d, \quad (A,q)\cdot(W,w,\lambda) = (A(W), Aw\gamma^{-d}q^{-1}\gamma^d),$$

yielding a $\widehat{\mathrm{GL}}_{\mathrm{res}}^0$-action on $\mathrm{DET}_d^\wedge$ by $[A,q]\cdot[W,w,\lambda] = [A(W), Aw\gamma^{-d}q^{-1}\gamma^d,\lambda]$. Trying to combine this action with the $\mathbb{Z}$-action defined by (5.9) is nevertheless not possible since $q^{-1}\circ q_\sigma$ is in general not in $\mathrm{SL}^1(K_+)$. (In fact it is easy to construct q in $\mathrm{GL}(K_+)$ such that $q^{-1}\circ q_\sigma - 1_+$ is not even a compact operator.)

We can now identify the two line bundles

Proposition V.7 *The bundles $\mathrm{DET}^\wedge$ and DET are $\widehat{\mathrm{GL}}_{\mathrm{res}}$-equivariantly isomorphic as holomorphic line bundles over G_{res}.*

Proof. We recall that by Propositions V.2 and V.6 $\mathrm{DET} = \widetilde{\mathrm{GL}}_{\mathrm{res}} \times_{\tilde{P}} \mathbb{C} = \widetilde{U}_{\mathrm{res}} \times_{\tilde{H}} \mathbb{C}$ with $\tilde{P}$ acting by the holomorphic extension of the character $\chi : \tilde{H} \to U(1)$ and $\mathrm{DET}^\wedge = \widehat{\mathrm{GL}}_{\mathrm{res}} \times_{\hat{P}} \mathbb{C}$ with $\hat{P}$ acting by the charakter $\hat{\chi}\left(\left[\begin{pmatrix} a & b \\ 0 & d \end{pmatrix}, q\right]\right)$
$= \det(aq^{-1})$.

Let now $\tau : \widehat{\mathrm{GL}}_{\mathrm{res}} \to \widetilde{\mathrm{Gl}}_{\mathrm{res}}$ be an isomorphism of the central $\mathbb{C}^*$-extensions of $\mathrm{GL}_{\mathrm{res}}$ (compare Proposition II.27 and its obvious "complexification"). It follows that $\tau(\hat{P}) = \tilde{P}$. Furthermore the subgroup

$$\hat{P}_0 = \mathrm{Stab}_{[K_+,1_+,1]} = \left\{\left[\begin{pmatrix} a & b \\ 0 & d \end{pmatrix}, a\right] \middle| \begin{pmatrix} a & b \\ 0 & d \end{pmatrix} \in P\right\}$$

of $\widehat{\mathrm{Gl}}_{\mathrm{res}}$ is the image of the homomorphic section

$$\hat{\sigma} : P \to \widehat{\mathrm{GL}}_{\mathrm{res}}, \quad \hat{\sigma}\left(\begin{pmatrix} a & b \\ 0 & d \end{pmatrix}\right) = \left[\begin{pmatrix} a & b \\ 0 & d \end{pmatrix}, a\right]$$

of the projection $\widehat{\mathrm{GL}}_{\mathrm{res}} \to \mathrm{GL}_{\mathrm{res}}$ over P. Since P is contractible it has no non-trivial characters and thus $\hat{\sigma}$ is the unique homomorphic section. Analogously $P_0^{\sim} = \mathrm{Stab}_{\xi_+} \mathrm{GL}_{\mathrm{res}}^{\sim}$ (ξ_+ is any fixed vector of norm 1 in DET_{K_+} here) is the image of the unique homomorphic section $\tilde{\sigma} : P \to \mathrm{GL}_{\mathrm{res}}^{\sim}$. Since τ covers the identity of $\mathrm{GL}_{\mathrm{res}}$ it follows that $\tau(P_0^{\wedge}) = P_0^{\sim}$.

Let us define (for $\hat{g}$ in $\widehat{\mathrm{GL}}_{\mathrm{res}}$ and λ in $\mathbb{C}$) :

$$(\mathrm{DET}^{\wedge}) \xrightarrow{T} \mathrm{DET}, \quad T(\hat{g} \cdot [K_+, 1_+, \lambda]) = \tau(\hat{g}) \cdot (\lambda \cdot \xi_+). \tag{5.12}$$

We observe that T is well-defined and equivariant with respect to the identification $\tau : \widehat{\mathrm{GL}}_{\mathrm{res}} \to \mathrm{GL}_{\mathrm{res}}^{\sim}$, since $\tau(P_0^{\wedge}) = P_0^{\sim}$. Furthermore T is obviously a morphism of holomorphic line bundles over the identity of G_{res} and the homomorphism τ^{-1} immediately yields the existence of its inverse T^{-1}, thus proving the assertion. $\square$

Remarks. Let us conclude this subsection by briefly sketching other approaches to the determinant line bundle.

(1) Using Lemma III.10 and the homogenity of G_{res} under U_{res} one can prove a priori that $\left[\frac{1}{2\pi}\Omega\right]$ is an integral class on G_{res}. Using the Kirillov-Kostant-Souriau theory of geometric quantization (compare e.g. [Ki], [Ko], [So]) one can construct a "quantizing line bundle" L over G_{res} with the action of a $U(1)$-extension of the group U_{res}, since the latter group acts with a (non-equivariant) moment map. Since the associated cocycle is given by the Schwinger term (compare Corollary IV.24) the $U(1)$-extension is then again isomorphic to $U_{\mathrm{res}}^{\sim}$. The line bundle L will then be the dual DET^* of DET and its "polarized", i.e. holomorphic in our situation, sections form "the geometric quantization" of G_{res}. Furthermore this approach ties in nicely with the easily proven fact that G_{res} is a coadjoint orbit of $U_{\mathrm{res}}^{\sim}$.

(2) As found in [PrSe] (pp. 113-114) one can explicitly give the transition functions defining DET using the covering $\{U_{K_S} | S \in \mathcal{S}\}$ given in the proof of Proposition III.5.

(3) The determinant bundle DET is closely related to determinant bundles à la Quillen arising in the context of families of Fredholm operators, e.g. in the family index theorem. See e.g. [PrSe] (p. 116), [Se] or the work of Booss, Wojciechowski and Scott ([BW], [BSW]) for further reading on this relation. (These recommendations are of course highly subjective and by no means exhaustive for the literature in this very active research area.)

(4) The related "restricted flag manifolds" and line bundles over them are studied in [HeHe].

(5) In [N4] the following approach is advocated: given a separable complex Hilbert space F one has for $p \geq 1$ the groups $\mathrm{GL}(F) \cap \{\mathrm{Id}_F + \mathcal{L}^p(F)\}$ and $\mathrm{U}(F) \cap$

$\{\mathrm{Id}_F + \mathcal{L}^p(F)\}$. These Banach Lie groups have well-behaved coadjoint orbits, including notably the restricted Grassmannian and the flag manifolds of the preceding item. Their representation theory is explored by studying – as in finite-dimensional Borel-Weil theory – holomorphic sections of complex line bundles over these coadjoint orbits. According to a private communication of the author the appendix of [N4] can also serve as a starting point to prove part (ii) of Corollary II.19.

3. Holomorphic sections of the dual of DET

Let us recall that (continuous, smooth, holomorphic, ...) sections of the dual E^* of a complex vector bundle $E \xrightarrow{\pi} M$ can be identified with functions on the total space E that restrict to linear functionals on each fibre $E_x = \pi^{-1}(x)$ of E. In order to avoid the technical subtleties of the more general setting of manifolds modelled on arbitrary locally convex spaces we will always assume that M is a Fréchet manifold and that the typical fibre of E is a Banach space in this subsection. The fibres of E^* are then always given by the strong duals of the fibres of E, $(E^*)_x = (E_x)^*$ and the pairing $(E^*)_x \times E_x \to \mathbb{C}$ (or $\mathbb{R}$) is jointly continuous. If there is no danger of confusion we will denote the strong dual E'_b of a l.c. space E – and notably of a Banach space – simply by E^* here.

For later applications we will stay in the holomorphic category, but everything we say will of course – mutatis mutandis – remain true in the smooth or continuous category. (For the theory of holomorphic maps in infinite dimensions the reader may consult e.g. [He] or [N1] in this volume.) We set $\qquad \mathcal{O}_{\mathrm{lin}}(E) =$

$$\{f : E \to \mathbb{C} \mid f \text{ is holomorphic and } f|_{E_x} : E_x \to \mathbb{C} \text{ is } \mathbb{C}\text{-linear for all } x \in M\}$$

and define

$$\Gamma_{\mathcal{O}}(M, E^*) \to \mathcal{O}_{\mathrm{lin}}(E), \ \sigma \mapsto f_\sigma \tag{5.13}$$

by $f_\sigma(\ell) = \sigma(\pi(\ell))(\ell)$ for all ℓ in E.

Lemma V.8 *Let $E \xrightarrow{\pi} M$ be a holomorphic Banach space bundle over a Fréchet manifold M. Then the map $\sigma \mapsto f_\sigma$ defined by (5.13) is a continuous linear isomorphism.*

Proof. The natural topologies on $\Gamma_{\mathcal{O}}(M, E^*)$ and $\mathcal{O}_{\mathrm{lin}}(E)$ being defined by the uniform convergence of all derivatives on compact subsets of M respectively E we can assume that $E = M \times B$ with B a fixed complex Banach space. A section of E^* is then simply given by a map $\varphi : M \to B^*$ and the associated function f_φ on $M \times B$ reads as $f_\varphi(x, v) = \varphi(x)(v)$. The continuity of the pairing $B^* \times B \to \mathbb{C}$ implies now that f_φ is in $\mathcal{O}_{\mathrm{lin}}(E)$. Since f_φ is linear in the second variable derivatives of f_φ are easily expressed in derivatives of φ, and vice versa. Thus it follows directly from the definition of the respective topologies that the map $\varphi \mapsto f_\varphi$ is continuous.

To prove the map (5.13) is a continuous isomorphism, we consider

$$\mathcal{O}_{\mathrm{lin}}(E) \to \Gamma_{\mathcal{O}}(M, E^*), f \mapsto \sigma_f, \tag{5.14}$$

where $\sigma_f(x)(\ell_x) = f(\ell_x)$ for ℓ_x in E_x. Obviously (5.14) is the inverse of (5.13) and continuity of the former map follows similarly as for the latter.   □

Let us now furthermore assume that we have an action $\Theta^M : G \times M \to M$ by holomorphic transformations of a Fréchet Lie group G which is covered by an action $\Theta^E : G \times E \to E$ by holomorphic bundle automorphisms. There is then a natural linear action

$$G \times \Gamma_\mathcal{O}(M, E) \to \Gamma_\mathcal{O}(M, E), g \cdot \sigma = \Theta_g^E \circ \sigma \circ \Theta_{g^{-1}}^M. \tag{5.15}$$

Analogously we have of course an action on sections of E^* since G acts on E^* as well. Furthermore a direct extension of Theorem III.11 in [N1] shows that the map defined by (5.15) is holomorphic, if G is complex and the action is holomorphic.

Lemma V.9 *Let* $E \xrightarrow{\pi} M$ *be as in Lemma V.8 and* G *a complex Fréchet Lie group acting holomorphically on* M *and* E *such that the action on* E *is given by bundle automorphisms covering the action on* M. *Then the continuous isomorphism* $\Gamma_\mathcal{O}(M, E^*) \to \mathcal{O}_{\mathrm{lin}}(E), \sigma \mapsto f_\sigma$ *is* G-*equivariant with respect to the action (5.15) and the action* $g \cdot f = f \circ \Theta_{g^{-1}}^E$ *on* $\mathcal{O}_{\mathrm{lin}}(E)$.

Proof. Obviously the given action on $\mathcal{O}(E)$ preserves $\mathcal{O}_{\mathrm{lin}}(E)$. The assertion follows now by a direct calculation. Let ℓ be in E :

$$f_{g\cdot\sigma}(\ell) = (g \cdot \sigma)(\pi(\ell))(\ell) = \left(\Theta_g^{E^*} \circ \sigma \circ \Theta_{g^{-1}}^M\right)(\pi(\ell))(\ell)$$

$$= \Theta_g^{E^*}\left(\sigma(\pi(\Theta_{g^{-1}}^E(\ell)))\right)(\ell) = \sigma\left(\pi(\Theta_{g^{-1}}^E(\ell))\right)\left(\Theta_{g^{-1}}^E(\ell)\right)$$

$$= f_\sigma\left(\Theta_{g^{-1}}^E(\ell)\right) = (g \cdot f)(\ell).   □$$

Having in mind the determinant bundle we proceed to a more restricted set-up:

Proposition V.10 *Let* $E \xrightarrow{\pi} M$ *and* G *as in Lemma V.9 and assume that there is a holomorphic* G-*equivariant vector bundle homomorphism* $j : E \to M \times F$ *over the identity of* M, *where* F *is a complex Banach space with a holomorphic linear* G-*action. Then the strong dual* F^* *of* F *maps* G-*equivariantly, continuously linear to the holomorphic sections of* E^*.

Proof. Let us first observe that the map

$$F^* \to \mathcal{O}_{\mathrm{lin}}(M \times F), \varphi \mapsto f_\varphi$$

defined by $f_\varphi(x, v) = \varphi(v)$ is G-equivariant and continuous. We define a "restriction map" r by

$$F^* \to \mathcal{O}_{\mathrm{lin}}(E), r(\varphi)(\ell) = f_\varphi(j(\ell)). \tag{5.16}$$

Thus r is just the pullback of the map $\varphi \mapsto f_\varphi$ by j. Equivariance follows by direct verification:

$$(g \cdot r(\varphi))(\ell) = r(\varphi)\left(\Theta_{g^{-1}}^E(\ell)\right)$$

$$= \varphi\left(j(\Theta_{g^{-1}}^E(\ell))\right) = \varphi\left(\Theta_{g^{-1}}^{M \times F}(j(\ell))\right)$$

$$= \left(\Theta_g^{F^*}\varphi\right)(j(\ell)) \quad (\Theta^{F^*} \text{ denotes the linear action on the dual of } F)$$

$$= r\left(\Theta_g^{F^*}\right)(\ell). \qquad \Box$$

Let us apply the above proposition first in finite dimensions:

Proposition V.11 *Let K be a finite-dimensional complex vector space, $M = G_m(K)$ the Grassmannian of m-dimensional complex subspaces of K, E the line bundle* $\text{Det} \xrightarrow{\pi} G_m(K)$, *and let G be $GL(K)$. Then $\Gamma_{\mathcal{O}}(G_m(K), \text{Det}^*)$ is $GL(K)$-equivariantly isomorphic to $(\Lambda^m K)^*$.*

Proof. By its very definition Det is a $GL(K)$-invariant subbundle of $G_m(K) \times \Lambda^m K$. Thus we have a $GL(K)$-equivariant homomorphism $(\Lambda^m K)^* \xrightarrow{r} \mathcal{O}_{\text{lin}}(\text{Det})$. Since r is obviously injective it remains only to show that it is surjective. The last property follows from a standard application of Riemann's extension theorem for holomorphic functions defined on an open set whose complement is analytic and at least of codimension two. (See e.g. [PrSe], pp. 22-23 for the details.) $\qquad \Box$

More interestingly we can use Proposition V.10 in infinite dimensions:

Proposition V.12 *Let M be the restricted Grassmannian G_{res}, $E = \text{DET} \xrightarrow{\pi} G_{\text{res}}$, and $G = \widetilde{GL}_{\text{res}}$. Then we have a canonical continuous linear, $\widetilde{GL}_{\text{res}}$-equivariant injection $\mathcal{F}^* \xrightarrow{r} \Gamma_{\mathcal{O}}(G_{\text{res}}, \text{DET}^*)$.*

Proof. Since DET is a holomorphic $\widetilde{GL}_{\text{res}}$-invariant subbundle of $G_{\text{res}} \times \mathcal{F}$ application of Proposition V.10 shows all assertions but the injectivity.

Lemma II.6 obviously implies that $\mathcal{F}$ is irreducible as a $\widetilde{GL}_{\text{res}}$-representation. Thus $\mathcal{F}^*$ is an irreducible $\widetilde{GL}_{\text{res}}$-representation as well and the non-triviality of r implies that r is injective. $\qquad \Box$

Remarks. The above proposition shows that the (dual of the) multi-particle Hilbert space $\mathcal{F}$, the starting point of our considerations on fermionic second quantization, is realized by holomorphic sections of a line bundle whose Chern class is given by the Schwinger term. It remains only to characterize intrinsically the image $r(\mathcal{F}^*)$ inside $\Gamma_{\mathcal{O}}(G_{\text{res}}, \text{DET}^*)$ (i.e. without relying on our approach of defining DET as a subbundle of $G_{\text{res}} \times \mathcal{F}$) : Using an appropriate "energy operator" (the second quantization of a one-particle Hamiltonian of the type $H e_n = n e_n$ for a ONB $\{e_n\}$ of K) Pressley and Segal show that $\mathcal{F}^*$ is dense inside the (bigger) space $\Gamma_{\mathcal{O}}(G_{\text{res}}, \text{DET}^*)$. A delicate, though natural, construction of a Gaussian measure allows Pickrell ([Pi1], [Pi2]) to show that an element s of $\Gamma_{\mathcal{O}}(G_{\text{res}}, \text{DET}^*)$ is in $r(\mathcal{F}^*)$ if and only if s is square-integrable.

Proposition V.12 yields now an embedding of G_{res} in a projective space.

Proposition V.13

(i) *The subspace $\mathcal{F}^*$ of $\Gamma_O(G_{res}, DET^*)$ is base-point free, i.e. there is no W in G_{res} such that all elements of $\mathcal{F}^*$ vanish in W.*

(ii) *There is a GL_{res}-equivariant holomorphic embedding $\mathcal{P}\ell : G_{res} \to \mathbb{P}(\mathcal{F})$.*

(iii) *The pullback of the tautological line bundle $T \to \mathbb{P}(\mathcal{F})$ with its canonical hermitian structure is isomorphic to DET with its canonical hermitian structure.*

Proof. The first assertion follows from the obvious fact that for each ξ in $\mathcal{F} \backslash \{0\}$ there is a φ in $\mathcal{F}^*$ such that $\varphi(\xi) \neq 0$.

Given a locally convex topological vector space V and a continuous linear map $\alpha : V \to \mathcal{O}_{\lin}(L)$ to the space of holomorphic sections of a line bundle $L \xrightarrow{\pi} M$ we can consider the dual map $\alpha^* : (\mathcal{O}_{\lin}(L))_b' \to V_b'$, where we use the topology of uniform convergence on bounded sets on both space of linear continuous functionals. If M is furthermore a Banach manifold the map $L \to (\mathcal{O}_{\lin}(L))_b', \ell \mapsto \delta_\ell$ is holomorphic by an easy extension of Lemma IV.16. Thus we have a map:

$$\epsilon_L : L \to V_b', \ell \mapsto \alpha^*(\delta_\ell). \tag{5.17}$$

Base-point freeness of $\alpha(V)$ is now equivalent to the condition that the "evaluation map" ϵ_L is non-vanishing on $L^\times = L \backslash \{\text{zero-section},\}$ i.e. $\epsilon_L(\ell_x) \neq 0$ if $\ell_x \neq 0$.

If this condition is fulfilled ϵ_L descends to a holomorphic map

$$\epsilon_M : M \to \mathbb{P}(V_b'), \epsilon_M(x) = [\epsilon_L(\ell_x)], \tag{5.18}$$

where ℓ_x is any non-zero element of the fibre L_x over x, and $\mathbb{P}(V_b')$ is the projective space associated to V_b'.

If G is acting by holomorphic bundle automorphisms on L, covering a holomorphic action on M and $\alpha : V \to \mathcal{O}_{\lin}(L)$ a G-equivariant map with respect to a linear action on V and the natural G-action on $\mathcal{O}_{\lin}(L)$ the whole construction is of course G-equivariant.

Applying this to $\mathcal{F}^* \to \mathcal{O}_{\lin}(DET)$ we get a $\widetilde{GL}_{res}$-equivariant map $\epsilon_{DET} : DET \to \mathcal{F} = (\mathcal{F}^*)_b'$ since $\mathcal{F}^*$ is the Hilbert space dual of $\mathcal{F}$ and its strong dual $(\mathcal{F}^*)_b'$ is of course canonically isomorphic to $\mathcal{F}$. The map ϵ_{DET} descends by base-point freeness to a $\widetilde{GL}_{res}$-equivariant map $\epsilon_{G_{res}} : G_{res} \to \mathbb{P}(\mathcal{F})$. Since $\widetilde{GL}_{res}$ acts on both manifolds via GL_{res} this map is in fact GL_{res}-equivariant. For its evident analogy with the finite-dimensional analogue, the map $\epsilon_{G_{res}}$ is also called the "Plücker map" $\mathcal{P}\ell$ of the restricted Grassmannian.

The map $\alpha : \mathcal{F}^* \to \mathcal{O}_{\lin}(DET)$ coming from the embedding $j : DET \to G_{res} \times \mathcal{F}$ we can identify ϵ_{DET} with $pr_2 \circ j : DET \to \mathcal{F}$, i.e.

$$\epsilon_{DET}((W, \xi))(\varphi) = \varphi(j(W, \xi)) = \varphi(\xi) = i(\xi)(\varphi),$$

where $i : \mathcal{F} \to (\mathcal{F}^*)_b' = \mathcal{F}^{**}$ is the canonical isomorphism of the Hilbert space $\mathcal{F}$ with its bidual. Since $pr_2(DET_W) \cap pr_2(DET_{W'}) \neq \{0\}$ iff $W = W'$ it follows that

ϵ_{DET} is injective on $\mathrm{DET}^{\times} = \mathrm{DET} \setminus \{\text{zero-section}\}$ and equivariance implies that this map as well as $\mathcal{P}\ell$ are holomorphic embeddings.

Let us recall that the "tautological bundle" over a projective vector space is given as

$$\tau : \mathcal{T} = \{([\xi], \eta) \in \mathbb{P}(\mathcal{F}) \times \mathcal{F} | \eta \in \mathbb{C} \cdot \xi\} \to \mathbb{P}(\mathcal{F}), \; \tau([\xi], \eta) = [\xi].$$

It follows immediately that $\mathcal{P}\ell^{*}(\mathcal{T}) = \{(W, ([\xi], \eta)) \in G_{\mathrm{res}} \times \mathcal{T} | \mathcal{P}\ell(W) = \tau([\xi], \eta)\}$ is canonically isomorphic to DET. Since the hermitian structures on both line bundles are induced from the Hilbert structure of $\mathcal{F}$ they are isomorphic as hermitian line bundles as well. □

Remark. In [SV] the C^{*}-algebraic approach to the Fock space is used to derive further properties of the Grassmannian and its determinant bundle. Let us here only quote that they derive the "Plücker equations" characterising the image $\mathcal{P}\ell(G_{\mathrm{res}})$ inside $\mathbb{P}(\mathcal{F})$. It follows notably that $\mathcal{P}\ell(G_{\mathrm{res}})$ is a closed complex submanifold of $\mathbb{P}(\mathcal{F})$.

References

[Ar]   H. Araki, Bogoliubov automorphisms and Fock representations of canonical anticommutation relations, in: *Operator algebras and mathematical physics (Iowa City, 1985)*, 23–141, Contemp. Math. 62, Amer. Math. Soc., Providence, R.I., 1987.

[AB]   M. F. Atiyah and R. Bott, The Yang-Mills equations over Riemann surfaces, *Philos. Trans. Roy. Soc. London*, Ser. A 308 no. 1505 (1983), 523–615.

[Au]   M. Audin, Lectures on gauge theory and integrable systems, in: *Gauge theory and symplectic geometry* (Montreal, 1995), NATO Adv. Sci. Inst. Ser. C Math. Phys. Sci. 488, 1–48 Kluwer, Dordrecht, 1997.

[Ba]   V. Bargmann, On unitary ray representations of continuous groups, *Ann. of Math.* 59 (1954), 1–46.

[Beg]   E.J. Beggs, The de Rham complex on infinite dimensional manifolds, *Quart. J. Math. Oxford* (2), 38 (1987), 131–154.

[Ber]   F.A. Berezin, *The method of second quantization*, Academic Press, Orlando, 1966.

[BW]   B. Booß-Bavnbek and K.P. Wojciechowski, *Elliptic boundary problems for Dirac operators*, Birkhäuser, Boston, 1993.

[BSW]   B. Booß-Bavnbek, S.G. Scott and K.P. Wojciechowski, *Elliptic boundary problems for Dirac operators II – The heat kernel and determinants, Grassmannians, and Dirac operators on manifolds with boundary*, Birkhäuser, Boston, in preparation.

[Bo]   N. Bourbaki, *Éléments de mathématique. Fasc. XXXVII. Groupes et algèbres de Lie. Chapitre III: Groupes de Lie*, Actualités scientifiques et industrielles, No. 1349, Hermann, Paris, 1972.

[BR]   O. Bratteli and D.W. Robinson, *Operator algebras and quantum statistical mechanics. II. Equilibrium states. Models in quantum statistical mechanics,* Springer, Berlin, 1981.

[Br]   R.L. Bryant, An introduction to Lie groups and symplectic geometry, in: *Geometry and quantum field theory* (Park City, 1991), Ed.D. Freed and K. Uhlenbeck, 5–181, IAS/Park City Math. Ser. 1, Amer. Math. Soc., Providence RI, 1995.

[CH]   A.L. Carey and C.A. Hurst, A note on the boson-fermion correspondence and infinite dimensional groups, *Commun. Math. Phys.* 98 (1985), 435–448.

[CHO'B]   A.L. Carey, C.A. Hurst and D.M. O'Brien, Automorphisms of the canonical anticommutation relations and index theory, *J. Funct. Anal.* 48 (1982), 360–393.

[CMM]   A.L. Carey, J. Mickelsson and M.K. Murray, Bundle gerbes applied to quantum field theory, *Reviews in Mathematical Physics* Vol. 12, No. 1 (2000), 65–90.

[CR]   A.L. Carey and S. N. M. Ruijsenaars, On fermion gauge groups, current algebras and Kac-Moody algebras, *Acta Appl. Math.* 10 (1987), no. 1, 1–86.

[Con]   A. Connes, *Noncommutative geometry*, Academic Press, San Diego, 1994.

[Coo]   J.M. Cook, The mathematics of second quantization, *Trans. of the Amer. Math. Soc.* 74 (1953), 222–245.

[DD]   J. Dixmier and A. Douady, Champs continus d'espaces hilbertiens et de C^*-algèbres, *Bull. Soc. Math. France* 91 (1963), 227–284.

[Don]   S. K. Donaldson, Anti self-dual Yang-Mills connections over complex algebraic surfaces and stable vector bundles, *Proc. London Math. Soc.* (3) 50, no. 1 (1985), 1–26.

[Dou]   R.G. Douglas, *Banach algebra techniques in operator theory*, Pure and Applied Mathematics, Vol. 49, Academic Press, New York London, 1972.

[E]   J. Eells, On the geometry of function spaces, in: *Symposium internacional de topología algebraica*, 303–308, Universidad Nacional Autónoma de México y UNESCO, Mexico City, 1958.

[EE]   J. Eells and K. D. Elworthy, Open embeddings of certain Banach manifolds, *Ann. of Math.*, Sec. Ser. 91 (1970), 465–485.

[Fo]   V. Fock, Konfigurationsraum und zweite Quantelung, *Zeitschrift für Physik* 75 (1932), 622–647.

[Fr]   K.O. Friedrichs, *Mathematical aspects of the quantum theory of fields*, New York, Interscience, 1953.

[FG]   J. Fröhlich and K. Gawędzki, Conformal field theory and geometry of strings, in: *Mathematical quantum field theory I: field theory and many-body theory* (CRM proceedings & lecture notes), ed. J. Feldman, R. Froese, L. Rosen, 57–97 , American. Math. Soc., Providence RI, 1994.

[FGR]   J. Fröhlich, O. Grandjean and A. Recknagel, Supersymmetric quantum theory, non-commutative geometry, and gravitation, in: *Symétries quantiques* (Les Houches, 1995), 221–385, North-Holland, Amsterdam, 1998.

[Fuj]   A. Fujiki, Moduli space of polarized algebraic manifolds and Kähler metrics, *Sugaku expositions* 5 (1992), n.2, 173–191.

[Fuk]   D.B. Fuks, *Cohomology of infinite-dimensional Lie algebras*, Consultants Bu-
          reau, New York, 1986.

[Ga]   S.A. Gaal, *Linear analysis and representation theory*, Springer, Berlin, 1973.

[Gl]   H. Glöckner, *Infinite-dimensional complex groups and semigroups: represen-
          tations of cones, tubes and conelike semigroups*, Dissertation TU Darmstadt,
          2000.

[GHL]   S. Gallot, D. Hulin and J. Lafontaine, *Riemannian geometry*, Sec. Ed., Springer,
          Berlin, 1990.

[GM]   H. Grosse, and W. Maderner, On the classical origin of the fermionic Schwinger
          term *Lett. Math. Phys.* 31 (1994), no. 1, 57–64.

[GS]   V. Guillemin and S. Sternberg *Symplectic techniques in physics*, Cambridge
          University Press, Cambridge New York, 1984.

[Ha]   R.S. Hamilton, The inverse function theorem of Nash and Moser, *Bull. of the
          AMS* (N.S.) Vol.7, No.1 (1982), 65–222.

[dlHJ]   P. de la Harpe and V. Jones, *An introduction to C*-algebras. Chapters 1–9*,
          Publication de l'Université de Genève, 1995.

[HeHu]   P. Heinzner and A.T. Huckleberry, Kählerian potentials and convexity proper-
          ties of the moment map, *Invent. Math.* 126 (1996), no. 1, 65–84.

[HeHe]   G.F. Helminck and A. G. Helminck, The structure of Hilbert flag varieties,
          Publ. Res. Inst. Math. Sci. 30 (1994), no. 3, 401–441.

[He]   M. Hervé, *Analyticity in infinite-dimensional spaces*, de Gruyter Studies in
          Mathematics 10, Walter de Gruyter, Berlin New York, 1989.

[I]   D. Iagolnitzer *The S-matrix*, North-Holland, Amsterdam New York, 1978.

[J]   R. Jackiw, Topological investigations of quantized gauge theories. Notes by
          R. Young, in: *Relativity, groups and topology II*, (Les Houches, 1983), Ed.B.S.
          DeWitt and R. Stora, 221–331, North-Holland, Amsterdam New York, 1984.

[Ke]   H.H. Keller, *Differential calculus in locally convex spaces*, Lect. Notes Math.
          417, Springer, Berlin New York, 1974.

[Ki]   A.A. Kirillov, Geometric quantization, in: *Dynamical systems. IV. Symplectic
          geometry and its applications*, Eds. V.I. Arnol'd and S.P. Novikov, 137–172,
          Encycl. Math. Sci. 4, Springer, Berlin, 1990.

[Ko]   B. Kostant, Quantization and unitary representations I: Prequantization, in:
          Lectures in Modern Analysis and Applications III, Lect. Notes Math. 170, 87–
          207, Springer, Berlin, 1970.

[KM]   A. Kriegl and P. Michor, *The convenient setting of global analysis*, Mathemat-
          ical Surveys and Monographs 53, Amer. Math. Soc., Providence RI, 1997.

[Ku]   N.H. Kuiper, The homotopy type of the unitary group of Hilbert space, *Topology*
          3 (1965), 19–30.

[La]   E. Langmann, On anomalies and noncommutative geometry, in: *Low-dimen-
          sional models in statistical physics and quantum field theory* (Schladming, 1995),
          291–297, Lect. Notes Phys. 469, Springer, Berlin, 1996.

[LM1]   E. Langmann and J. Mickelsson, (3 + 1)-dimensional Schwinger terms and non-
          commutative geometry, *Phys. Lett.* B 338 (1994), no. 2–3, 241–248.

[LM2]   E. Langmann and J. Mickelsson, Scattering matrix in external field problems, *J. Math. Phys.* 37 (1996), no. 8, 3933–3953.

[Lu]   L.-E. Lundberg, Quasi-free "second quantization", *Commun. Math. Phys.* 50 (1976), 103–112.

[Mic]   J. Mickelsson, *Current algebras and groups*, Plenum Monographs in Nonlinear Physics, Plenum Press, New York London, 1989.

[MR]   J. Mickelsson and S.G. Rajeev, Current algebras in $d + 1$-dimensions and determinant bundles over infinite-dimensional Grassmannians, *Commun. Math. Phys.* 116 (1988), no. 3, 365–400.

[Mil1] J. Milnor, *Morse theory. Based on lecture notes by M. Spivak and R. Wells*, Annals of Mathematics Studies, No. 51, Princeton University Press, Princeton, N.J., 1963.

[Mil2] J.Milnor, Remarks on infinite-dimensional Lie groups, in: *Relativity, groups and topology II* (Les Houches, 1983), Ed.B.S. DeWitt and R. Stora, 1007–1057, North-Holland, Amsterdam New York, 1984.

[MFK]   D. Mumford and J. Fogarty and F. Kirwan, *Geometric invariant theory*, Third edition, Ergebnisse der Mathematik und ihrer Grenzgebiete 34, Springer, Berlin, 1994.

[Muj]   J. Mujica, *Complex analysis in Banach spaces*, North-Holland, Amsterdam, 1986.

[Mur]   G.J. Murray, *C*-algebras and operator theory*, Academic Press, Boston, 1990.

[N1]   K.H. Neeb, Infinite dimensional groups and their representations, in: *this volume*.

[N2]   K.H. Neeb, Borel-Weil theory for loop groups, in: *this volume*.

[N3]   K.H. Neeb, Central extensions of infinite-dimensional Lie groups, *Preprint 2084*, TU Darmstadt, 2000.

[N4]   K.H. Neeb, Lecture Notes: Infinite-dimensional groups and their representations, *Preprint*, TU Darmstadt, 2000.

[Ov]   V.Y. Ovsienko, Coadjoint representation of Virasoro-type Lie algebras and differential operators on tensor-densities, in: *this volume*.

[Ot]   J. T. Ottesen, *Infinite dimensional groups and algebras in quantum physics*, Lect. Notes in Physics, m 27, Springer, Berlin, 1995.

[Pa]   R. Palais, On the homotopy type of certain groups of operators, *Topology* 3 (1965), 271–279.

[Pe]   M.E. Peskin and D.V. Schroeder, *An introduction to quantum field theory*, Addison-Wesley, Reading, MA, 1995.

[Pi1]   D. Pickrell, Measures on infinite-dimensional Grassmann manifolds, *J. Funct. Anal.* 70, no. 2 (1987), 323–356.

[Pi2]   D. Pickrell, On the support of quasi-invariant measures on infinite-dimensional Grassmann manifolds, *Proc. Amer. Math. Soc.* 100, no. 1 (1987), 111–116.

[PR]   R.J. Plymen and P.L. Robinson, *Spinors in Hilbert space*, Cambridge University Press, Cambridge, 1994.

[PoSt] R.T. Powers and E. Størmer, Free states of the canonical anticommutation relations, *Commun. Math. Phys.* 16 (1970), 1–33.

[PrSe]   A. Pressley and G. Segal, *Loop groups*, Oxford University Press, New York, 1986.

[RS1]   M. Reed and B. Simon, *Methods of modern mathematical physics. I: Functional analysis*, Academic Press, New York, 1972.

[RS2]   M. Reed and B. Simon, *Methods of modern mathematical physics. II: Fourier analysis, self-adjointness*, Academic Press, New York, 1975.

[RS3]   M. Reed and B. Simon, *Methods of modern mathematical physics. III: Scattering theory*, Academic Press, New York, 1979.

[RS4]   M. Reed and B. Simon, *Methods of modern mathematical physics. IV: Analysis of operators*, Academic Press, New York, 1978.

[Re]   A.G. Reznikov, Characteristic classes in symplectic topology. Appendix D by L. Katzarkov, *Selecta Math.* (N.S.) 3 (1997), no. 4, 601–642.

[Rud]   W. Rudin, *Functional analysis*, McGraw-Hill, New York, 1973.

[Rui]   S. N. M. Ruijsenaars, Index theorems and anomalies: a common playground for mathematicians and physicists, *CWI Quarterly* 3 (1990) no. 1, 3–19.

[Sa]   M. Sato, The KP hierarchy and infinite-dimensional Grassmann manifolds, in: *Theta functions – Bowdoin 1987, Part 1* (Brunswick, 1987), 51–66, Proc. Sympos. Pure Math. 49, Part 1, Amer. Math. Soc., Providence, RI, 1989.

[Sc]   J. Schwinger, Field theoretic commutators, *Physical reviews letters*, Vol. 3, n.6 (1959), 296–297.

[Se]   G. Segal, *The definition of conformal field theory*, unpublished manscript.

[SeWi]   G. Segal and G. Wilson, Loop groups and equations of KdV type, *Inst. Hautes Études Sci. Publ. Math.* No. 61 (1985), 5–65.

[Si]   B. Simon, *Trace ideals and their applications*, Cambridge University Press, Cambridge, 1979.

[ShSt]   D. Shale and W.F. Stinespring, Spinor representations of infinite orthogonal groups, *J. Math. Mech.* 14 (1965) 315–322.

[So]   J.-M. Souriau, Quantification géométrique, *Comm. Math. Phys.* 1 (1966), 374–398.

[SV]   M. Spera and G. Valli, Plücker embedding of the Hilbert space Grassmannian and the CAR algebra, *Russ. Journal of Math. Physics* 2 (1994), 383–392.

[SpWu1] M. Spera and T. Wurzbacher, Determinants, Pfaffians, and quasi-free representations of the CAR algebra, *Reviews in Mathematical Physics* Vol. 10, No. 5 (1998), 705–721.

[SpWu2] M. Spera and T. Wurzbacher, The differential geometry of Grassmannian embeddings of based loop groups, *Diff. Geometry and its Appl.* 13 (2000), 43–75.

[Ta]   M.E. Taylor, *Partial differential equations. II. Qualitative studies of linear equations*, Applied Math. Sciences 116, Springer, New York, 1996.

[Tha]   B. Thaller, *The Dirac equation*, Texts and Monographs in Physics, Springer, Berlin, 1992.

[Tho]   E. G. F. Thomas, Vector fields as derivations on nuclear manifolds, *Math. Nachr.* 176 (1995), 277–286.

[To]   V. Toledano Laredo, Integrating unitary representations of infinite-dimensional Lie groups, *J. Funct. Anal.* 161 (1999), no. 2, 478–508.

[Tr]   F. Trèves, *Topological vector spaces, distributions and kernels*, Academic Press, New York London, 1967.

[Wa]   A. Wassermann, Operator algebras and conformal field theory. III. Fusion of positive energy representations of LSU(N) using bounded operators, *Invent. Math.* 133 (1998), no. 3, 467–538.

[Wu1]   T. Wurzbacher, Symplectic geometry of the loop space of a Riemannian manifold, *J. Geom. Phys.* 16 (1995), no. 4, 345–384.

[Wu2]   T. Wurzbacher, An "elementary" proof of the homotopy equivalence between the restricted general linear group and the space of Fredholm operators, *Preprint*, IRMA Strasbourg, in preparation.

Mathematical Subject Classification (2000)
Primary: 58Bxx.
Secondary: 14M15, 22E65, 53C55, 53D05, 81R10, 81T05, 81T70.

Institut de Recherche Mathématique Avancée
Université Louis Pasteur et C.N.R.S.
7, rue René-Descartes
F-67084 Strasbourg
France
E-mail address: `wurzbach@math.u-strasbg.fr`